Meat Quality Analysis

Meat Quality Analysis
Advanced Evaluation Methods, Techniques, and Technologies

Edited by

Ashim Kumar Biswas
Division of Post-Harvest Technology, ICAR-Central Avian Research Institute, Izatnagar, Bareilly, Uttar Pradesh, India

Prabhat Kumar Mandal
Department of Livestock Products Technology, Rajiv Gandhi Institute of Veterinary Education and Research, Kurumbapet, Puducherry, India

Academic Press is an imprint of Elsevier
125 London Wall, London EC2Y 5AS, United Kingdom
525 B Street, Suite 1650, San Diego, CA 92101, United States
50 Hampshire Street, 5th Floor, Cambridge, MA 02139, United States
The Boulevard, Langford Lane, Kidlington, Oxford OX5 1GB, United Kingdom

Copyright © 2020 Elsevier Inc. All rights reserved.

No part of this publication may be reproduced or transmitted in any form or by any means, electronic or mechanical, including photocopying, recording, or any information storage and retrieval system, without permission in writing from the publisher. Details on how to seek permission, further information about the Publisher's permissions policies and our arrangements with organizations such as the Copyright Clearance Center and the Copyright Licensing Agency, can be found at our website: www.elsevier.com/permissions.

This book and the individual contributions contained in it are protected under copyright by the Publisher (other than as may be noted herein).

Notices
Knowledge and best practice in this field are constantly changing. As new research and experience broaden our understanding, changes in research methods, professional practices, or medical treatment may become necessary.

Practitioners and researchers must always rely on their own experience and knowledge in evaluating and using any information, methods, compounds, or experiments described herein. In using such information or methods they should be mindful of their own safety and the safety of others, including parties for whom they have a professional responsibility.

To the fullest extent of the law, neither the Publisher nor the authors, contributors, or editors, assume any liability for any injury and/or damage to persons or property as a matter of products liability, negligence or otherwise, or from any use or operation of any methods, products, instructions, or ideas contained in the material herein.

British Library Cataloguing-in-Publication Data
A catalogue record for this book is available from the British Library

Library of Congress Cataloging-in-Publication Data
A catalog record for this book is available from the Library of Congress

ISBN: 978-0-12-819233-7

For Information on all Academic Press publications
visit our website at https://www.elsevier.com/books-and-journals

Publisher: Charlotte Cockle
Acquisition Editor: Patricia Osborn
Editorial Project Manager: Redding Morse
Production Project Manager: Vijayaraj Purushothaman
Cover Designer: Mark Rogers

Typeset by MPS Limited, Chennai, India

Contents

List of Contributors xix

Section 1 Current perspectives of meat quality evaluation 1

Chapter 1: Current perspectives of meat quality evaluation: techniques, technologies, and challenges 3
A.K. Biswas and P.K. Mandal

1.1 Introduction 3
1.2 Techniques and technologies 5
    1.2.1 Nutritional composition 5
    1.2.2 Physical and structural quality 6
    1.2.3 Meat traceability and authentication 7
    1.2.4 Food preservatives and additives 9
    1.2.5 Freshness and pathogen identification 9
    1.2.6 Chemical contaminants 10
    1.2.7 Sensory quality 12
1.3 Challenges 13
References 15

Section 2 Advances in carcass quality evaluation and nutritional composition of meat 19

Chapter 2: Methods for nutritional quality analysis of meat 21
Nira Manik Soren and Ashim Kumar Biswas

2.1 Introduction 21
2.2 Meat categorization 22
2.3 Basic nutritional composition of meat 23

2.4 Methodology for assessing the nutritional quality 25
    2.4.1 Determination of moisture 25
    2.4.2 Determination of protein 28
    2.4.3 Determination of amino acids 29
    2.4.4 Determination of lipids 30
    2.4.5 Determination of fatty acids 31
    2.4.6 Determination of ash content and minerals 31
    2.4.7 Determination of vitamins 32
2.5 Conclusion 33
References 34

Chapter 3: Nondestructive methods for carcass and meat quality evaluation 37

K. Narsaiah, A.K. Biswas and P.K. Mandal

3.1 Introduction 37
3.2 Emerging nondestructive methods 38
    3.2.1 Optical methods 38
    3.2.2 Near-infrared spectroscopy 40
    3.2.3 Nuclear magnetic resonance and magnetic resonance imaging 41
    3.2.4 Electronic nose (e-nose) 42
    3.2.5 X-ray and computed tomography 43
    3.2.6 Hyperspectral imaging 44
    3.2.7 Electrical properties 44
    3.2.8 Acoustic methods 44
3.3 Future research needs 46
3.4 Conclusion 47
References 47
Further reading 49

Section 3 Postmortem ageing and meat quality evaluation 51

Chapter 4: Detection techniques of meat tenderness: state of the art 53

Rui Liu and Wangang Zhang

4.1 Introduction 53
4.2 The detection methods 54
    4.2.1 Sensory analysis 54
    4.2.2 Shear force measurements 55
    4.2.3 Texture profile analysis 57
    4.2.4 Star probe measurement 57

	4.2.5	Spectroscopic methods 58
4.3	Conclusion 61
References 62

Chapter 5: Biochemical changes of postmortem meat during the aging process and strategies to improve the meat quality 67

R. Ramanathan, G.G. Mafi, L. Yoder, M. Perry, M. Pfeiffer, D.L. VanOverbeke and N.B. Maheswarappa

5.1	Introduction 67
	5.1.1 Conversion of muscle to meat 68
5.2	Effects of aging on tenderness 68
	5.2.1 Mechanism of postmortem aging 68
	5.2.2 Techniques to improve tenderness 69
	5.2.3 Different types of aging 72
	5.2.4 Techniques to quantify tenderness in plant 74
	5.2.5 Techniques to measure tenderness 75
5.3	Effects of aging on color 76
	5.3.1 Biochemical basis of lower color stability in aged meat 77
	5.3.2 Practical approaches to improve color stability of aged steaks 77
5.4	Effects of aging on buffalo meat quality 77
5.5	Application of proteomics and metabolomics to study meat quality 78
5.6	Conclusion 79
References 79
Further reading 80

Chapter 6: Recent developments in postmortem aging and evaluation methods 81

Kiran Mohan, Rituparna Banerjee and Naveea Basappa Maheswarappa

6.1	Introduction 82
6.2	Aging 82
6.3	Postmortem aging and meat quality 83
	6.3.1 Tenderness 84
	6.3.2 Flavor 84
	6.3.3 Color 85
	6.3.4 Water-holding capacity 85
6.4	Evaluation/assessment of postmortem aging 85
6.5	Advanced methods for evaluation of postmortem proteolysis 89
	6.5.1 Proteoglycans/glycosaminoglycan quantification 89
	6.5.2 Determination of cross-links and decorin 89

　　　　6.5.3　Cathepsin activity 89
　　　　6.5.4　Calpain extraction and casein zymography 90
　　　　6.5.5　Immunological protein quantification by dot-blot 90
　　　　6.5.6　Nuclear magnetic resonance transverse relaxation (T2) measurements 90
　　　　6.5.7　Fluoroscence polarization 91
　　6.6　Biomarkers of postmortem aging 91
　　　　6.6.1　Proteomic markers in the conversion of muscle to meat 91
　　　　6.6.2　Genotyping for marker gene 93
　　　　6.6.3　Metabolomics 93
　　6.7　Conclusion 94
　　References 94
　　Further reading 99

Chapter 7: Calpain-assisted postmortem aging of meat and its detection methods ... 101

A.K. Biswas, S. Tandon and P.K. Mandal

　　7.1　Introduction 101
　　7.2　Structure and functions of calpain and calpastatin 102
　　7.3　Pathways of calpain activity in muscle tissues 103
　　7.4　Postmortem proteolysis of skeletal muscle by calpain 104
　　7.5　Purification of calpain 1, calpain 2, and calpastatin 106
　　　　7.5.1　Hydrophobic interaction chromatography 106
　　　　7.5.2　Ion-exchange chromatography 107
　　7.6　Detection and quantification 108
　　　　7.6.1　Casein zymography 108
　　　　7.6.2　Biochemical method 108
　　　　7.6.3　Fluorometric method 109
　　　　7.6.4　Bioluminescent assay 109
　　　　7.6.5　Proteomic-based analysis 110
　　　　7.6.6　Identification of biomarkers 110
　　7.7　Scope of future work 111
　　References 111

Section 4　Molecular basis of meat colour development and detection 115

Chapter 8: Molecular basis of meat color 117

R.A. Mancini and R. Ramanathan

　　8.1　Introduction 117
　　8.2　Meat chemistry 118

	8.2.1	Red color development	119
	8.2.2	Discoloration	120
	8.2.3	Color deviations and approaches to improve color	121
	8.2.4	Myoglobin and lipid oxidation	122

8.3 Cooked color 122
8.4 Analysis of meat color 123
    8.4.1 Instrumental color analysis (use of handheld devices) 123
    8.4.2 Spectrophotometric techniques to study meat color 124
    8.4.3 Mitochondrial functional analysis 125
    8.4.4 Visual color analysis 128
    8.4.5 Cooked color measurements 128
8.5 Conclusion 129
References 129

Section 5 Meat authenticity and traceability 131

Chapter 9: Molecular techniques for speciation of meat 133

P.S. Girish and Nagappa S. Karabasanavar

9.1 Introduction 133
9.2 Identification of the origin of meat species by DNA hybridization 135
9.3 PCR-based techniques for species identification of meat 137
    9.3.1 Random amplified polymorphic DNA—PCR (RAPD fingerprinting) 138
    9.3.2 Forensically informative nucleotide sequencing 138
    9.3.3 Species-specific PCR 140
    9.3.4 Multiplex PCR 141
    9.3.5 PCR-restriction fragment length polymorphism 141
    9.3.6 DNA microarrays 142
    9.3.7 Other PCR-based methods 143
9.4 Meat species identification using loop-mediated isothermal amplification 144
9.5 Quantitative meat speciation 145
    9.5.1 Quantitative PCR 145
9.6 Scope for future research 147
References 147

Chapter 10: Meat traceability and certification in meat supply chain 153

P.S. Girish and S.B. Barbuddhe

10.1 Introduction 153

10.2 Benefits of livestock traceability system ... 154
10.3 Methods for identification of animals ... 155
    10.3.1 Visual tagging ... 156
    10.3.2 Bar-coded tags ... 156
    10.3.3 Radio frequency identification devices ... 156
    10.3.4 Quick response code-based tags ... 156
10.4 Scenario of livestock traceability around the world ... 157
    10.4.1 Livestock traceability system in the European Union ... 157
    10.4.2 Livestock traceability system in Australia ... 159
    10.4.3 Livestock traceability system in Japan ... 160
    10.4.4 Comparison of traceability regulatory systems in different countries ... 160
10.5 Molecular meat traceability ... 162
    10.5.1 Sample requirements for DNA-based meat traceability ... 162
    10.5.2 Molecular meat traceability using microsatellite genotyping ... 162
    10.5.3 Molecular meat traceability using single nucleotide polymorphism genotyping ... 166
10.6 Future scope of work ... 168
References ... 168

Section 6 Chemical residues in meat and their detection techniques ... 171

Chapter 11: Residues of harmful chemicals and their detection techniques ... 173

Milagro Reig and Fidel Toldrá

11.1 Introduction ... 173
11.2 Environmental contaminants ... 174
11.3 Polycyclic aromatic hydrocarbons ... 175
11.4 Veterinary drug residues ... 176
11.5 *N*-Nitrosamines ... 178
11.6 Oxidation of lipid-derived compounds ... 179
11.7 Oxidation of protein-derived compounds ... 180
References ... 181
Further reading ... 183

Section 7 Food preservatives/additives in meat and their detection.......185

Chapter 12: Use of food preservatives and additives in meat and their detection techniques 187

Meera Surendran Nair, Divek V.T. Nair, Anup Kollanoor Johny and Kumar Venkitanarayanan

12.1 Introduction 188
12.2 Food additives 189
    12.2.1 Antioxidants 190
    12.2.2 Binders 191
    12.2.3 Emulsifiers 191
    12.2.4 Antimicrobials 192
    12.2.5 Curing agents and cure accelerators 192
    12.2.6 Flavoring agents 193
    12.2.7 Coloring agents 193
12.3 Preservatives 194
    12.3.1 Chemical preservatives 194
    12.3.2 Natural preservatives 196
12.4 Federal oversight 199
12.5 Health concerns and safety assessment 200
12.6 Analytical techniques 201
    12.6.1 Detection of sulfites 202
    12.6.2 Detection of nitrites and nitrates 203
    12.6.3 Detection of sorbic acid 203
    12.6.4 Detection of nisin and organic acids 204
    12.6.5 Detection of color additives 204
    12.6.6 Detection of synthetic phenolic antioxidants 204
12.7 Conclusion and future directions 205
References 205

Section 8 Detection and prevention of lipid oxidation products 215

Chapter 13: Analysis of lipids and lipid oxidation products 217

Trinidad Pérez-Palacios and Mario Estévez

13.1 Introduction 217
13.2 Advances in the analysis of meat lipids 218
    13.2.1 Optimization of classic methods for total lipid quantification 218

        13.2.2 Optimization of classic methods for lipid composition 220
        13.2.3 Advanced methodologies for lipid analysis: nondestructive methods 225
   13.3 Advances in the detection of lipid oxidation products 226
        13.3.1 Optimization of classic methods for lipid oxidation assessment 226
        13.3.2 Advanced methodologies for lipid oxidation assessment 230
   13.4 Future perspectives: lipidomics and oxidomics 232
   References 232
   Further reading 239

Chapter 14: Plant antioxidants, extraction strategies, and their application in meat 241

Zabdiel Alvarado-Martinez, Arpita Aditya and Debabrata Biswas

   14.1 Introduction 242
   14.2 Common sources of plant antioxidants 243
   14.3 Common extraction methodologies/strategies of various antioxidants from plant sources 243
        14.3.1 Extraction strategy of plant polyphenolic compounds 246
        14.3.2 Purification and fractionation 249
   14.4 Application of antioxidants in animal products during the postharvest stage 251
        14.4.1 Improve the flavor and the shelf life 251
        14.4.2 Reduce microbiological contamination 251
        14.4.3 Improve the nutritional values 252
   14.5 Mechanisms of action of antioxidants 252
   14.6 Enzymatic antioxidants 253
        14.6.1 Superoxide dismutase 253
        14.6.2 Catalase 253
        14.6.3 Glutathione peroxidase 253
   14.7 Natural nonenzymatic antioxidants 254
        14.7.1 Vitamin E 254
        14.7.2 Vitamin C 254
        14.7.3 Vitamin A 255
        14.7.4 Flavonoids 256
        14.7.5 Phenolic acids 257
   14.8 Conclusion 257
   References 258

Section 9 Strategies for elimination and detection of foodborne pathogens 265

Chapter 15: Strategies for elimination of foodborne pathogens, their influensive detection techniques and drawbacks 267

Sandeep Ghatak

- 15.1 Introduction 268
- 15.2 Physical methods of elimination of foodborne pathogens 268
  - 15.2.1 Preslaughter washing 268
  - 15.2.2 Removal of hair 269
  - 15.2.3 Spot trimming of carcasses 269
  - 15.2.4 Vacuum-steam/water application 269
  - 15.2.5 Carcass washing 270
- 15.3 Chemical processes for elimination of microbial pathogens 270
  - 15.3.1 Acidic compounds 270
  - 15.3.2 Chlorine and related chemicals 271
  - 15.3.3 Ozone 271
  - 15.3.4 Other chemical agents 271
- 15.4 Elimination of microbial pathogens by ultraviolet light 272
- 15.5 Irradiation of meat for eliminating microbial hazards 272
- 15.6 Application of low temperature 273
- 15.7 High-pressure processing for elimination of pathogens 273
- 15.8 Other emerging approaches for elimination of microbial pathogens 273
  - 15.8.1 Nonthermal plasma (cold plasma) 274
  - 15.8.2 Dense phase carbon dioxide 274
  - 15.8.3 Electrolyzed oxidizing water 274
  - 15.8.4 Microwave and radio frequency 274
  - 15.8.5 Infrared heating 275
  - 15.8.6 Biocontrol with bacteriophage 275
- 15.9 Detection of microbial pathogens 275
  - 15.9.1 Conventional culture-based techniques 276
  - 15.9.2 Immunological techniques 277
  - 15.9.3 Nucleic acid-based techniques 278
  - 15.9.4 Matrix-assisted laser desorption ionization-time of flight mass spectrometry 280
  - 15.9.5 Hyperspectral imaging and analysis 280
  - 15.9.6 Nanotechnology-based approaches 281
  - 15.9.7 Other assays 281
- References 281

Chapter 16: Modern techniques for rapid detection of meatborne pathogens 287

Prabhat Kumar Mandal and Ashim Kumar Biswas

- 16.1 Introduction 288
- 16.2 Need for rapid detection method 288
  - 16.2.1 Trends in rapid detection of meatborne pathogens 289
- 16.3 Biosensors-based detection techniques 290
  - 16.3.1 Bioluminescence sensors 291
  - 16.3.2 Fiber optic biosensor 291
  - 16.3.3 Surface plasmon resonance biosensor 293
  - 16.3.4 Electrical impedance biosensor 294
  - 16.3.5 Impedance-based biochip sensor 294
  - 16.3.6 Piezoelectric biosensors 295
  - 16.3.7 Cell-based sensor 295
  - 16.3.8 Fourier transform infrared spectroscopy 295
  - 16.3.9 Flow cytometry 296
  - 16.3.10 Solid phase cytometry 296
- 16.4 Nucleic acid-based assays 297
  - 16.4.1 DNA hybridization 297
  - 16.4.2 Polymerase chain reaction 298
  - 16.4.3 DNA microarrays (gene chip technology) 298
  - 16.4.4 Loop-mediated isothermal amplification 300
- 16.5 Requirements for rapid detection methods 300
- 16.6 Limitations of rapid detection methods 301
- References 301

Section 10 Modern biological concept of meat deterioration and its detection 305

Chapter 17: Spoilage bacteria and meat quality 307

Abraham Joseph Pellissery, Poonam Gopika Vinayamohan, Mary Anne Roshni Amalaradjou and Kumar Venkitanarayanan

- 17.1 Introduction 308
- 17.2 Causes of meat spoilage 308
- 17.3 Microbiome of spoiled meat 309
  - 17.3.1 Microflora of fresh meat 309
  - 17.3.2 Spoilage microflora associated with aerobically packaged meat 310
  - 17.3.3 Spoilage microflora in vacuum-packaged meat 312
  - 17.3.4 Spoilage microflora in modified atmosphere packaged meat 313

17.4 Chemistry of meat spoilage 314
    17.4.1 Nonmicrobial/biochemical spoilage of meat 314
    17.4.2 Microbiological spoilage of meat 314
17.5 Characteristics of spoiled meat 317
    17.5.1 Discoloration 317
    17.5.2 Off-odors and off-flavors 318
    17.5.3 Gas production 318
    17.5.4 Filaments and slime formation 318
17.6 Factors affecting microbial meat spoilage 319
    17.6.1 Intrinsic factors 320
    17.6.2 Extrinsic factors 321
    17.6.3 Implicit factors 322
17.7 Indicators of microbial spoilage 322
17.8 Detection of microbial spoilage in meat 323
    17.8.1 Enumeration 323
    17.8.2 Detection of bacterial metabolites 324
    17.8.3 Molecular methods 326
17.9 Conclusion 328
References 328
Further reading 334

Chapter 18: Modern concept and detection of spoilage in meat and meat products 335

V.J. Ajaykumar and Prabhat Kumar Mandal

18.1 Introduction 335
18.2 Microbial spoilage of meat 336
18.3 Microbial metabolites 337
18.4 Detection of spoilage bacteria 338
    18.4.1 Enumeration methods 338
    18.4.2 Detection methods 339
    18.4.3 Detection by molecular methods 339
18.5 Detection of microbial metabolites 340
18.6 Modern trends in the spoilage detection of meat 340
    18.6.1 Odor sensors and electronic nose technology 340
    18.6.2 Conducting organic polymers 341
    18.6.3 Metal oxide semiconductor 341
    18.6.4 Spectroscopy in advanced forms 342
    18.6.5 Use of lasers in detecting food quality 344
    18.6.6 Smartphone-based food diagnostic technologies 345
    18.6.7 Chemical and biological sensors for food-quality monitoring 345

18.7 Conclusion 346
References 346
Further reading 349

Section 11 Proteomic and genomic tools in meat quality evaluation 351

Chapter 19: Application of proteomic tools in meat quality evaluation 353
M.N. Nair and C. Zhai

19.1 Introduction 353
19.2 Proteomics 354
    19.2.1 Gel-based approaches 355
    19.2.2 Gel-free approaches 357
19.3 Proteomic approaches to meat quality 358
    19.3.1 Meat color 359
    19.3.2 Tenderness 361
    19.3.3 Water-holding capacity 363
19.4 Conclusion 365
References 365

Chapter 20: Application of genomics tools in meat quality evaluation 369
T.K. Bhattacharya

20.1 Introduction 369
20.2 Genomic tools for meat quality assessment 370
    20.2.1 Isolation of nucleic acid (DNA and RNA) 370
    20.2.2 Quantification and purity measurement of DNA 372
    20.2.3 Polymerase chain reaction 372
    20.2.4 Real-time PCR 374
    20.2.5 Microsatellite analysis 375
    20.2.6 Agarose gel electrophoresis 384
    20.2.7 Polyacrylamide gel electrophoresis 385
    20.2.8 SDS-polyacrylamide gel electrophoresis 386
    20.2.9 Western blot 386
    20.2.10 Two-dimensional gel electrophoresis 387
20.3 Some important requisites 388
    20.3.1 Sterilization and disinfection of laboratory wares 388
    20.3.2 Sterilization of glasswares 388
    20.3.3 Dry heat sterilization 388
    20.3.4 Moist heat sterilization 388
    20.3.5 Disinfection and sterilization 388

20.4 Conclusion 389
References 389

Section 12 Sensory evaluation techniques 391

Chapter 21: Innovation in sensory assessment of meat and meat products 393
Sonia Ventanas, Alberto González-Mohino, Mario Estévez and Leila Carvalho

21.1 Introduction 393
21.2 Quick-fast descriptive sensory techniques 394
      21.2.1 Napping 396
      21.2.2 Flash profile 397
      21.2.3 Check-all-that-apply 400
      21.2.4 Rate-all-that-apply 401
21.3 Dynamic sensory techniques 402
      21.3.1 Time—intensity 402
      21.3.2 Temporal dominance of sensation 404
      21.3.3 Temporal check-all-that-apply 406
21.4 Emotions in sensory meat science 407
      21.4.1 Emotions 407
      21.4.2 Emotions elicited by food: methodological approaches 408
      21.4.3 Examining extrinsic factors that influence the emotions elicited by foods 409
      21.4.4 Emotional responses to meat and meat products 410
References 412
Further reading 418

Index 419

List of Contributors

Arpita Aditya   Department of Animal and Avian Sciences, University of Maryland, College Park, MD, United States

V.J. Ajaykumar   Department of Veterinary Public Health, Rajiv Gandhi Institute of Veterinary Education and Research, Puducherry, India

Zabdiel Alvarado-Martinez   Department of Biology-Molecular and Cellular Biology, University of Maryland, College Park, MD, United States

Mary Anne Roshni Amalaradjou   Department of Animal Science, University of Connecticut, Storrs, CT, United States

Rituparna Banerjee   ICAR - National Research Centre on Meat, Chengicherla, Hyderabad, India

S.B. Barbuddhe   ICAR - National Research Centre on Meat, Hyderabad, India

T.K. Bhattacharya   ICAR-Directorate of Poultry Research, Hyderabad, India

Ashim Kumar Biswas   Division of Post-Harvest Technology, ICAR-Central Avian Research Institute, Izatnagar, Bareilly, India

Debabrata Biswas   Center for Food Safety and Security Systems, Department of Animal and Avian Sciences, University of Maryland, College Park, MD, United States

Leila Carvalho   Postgraduate program in Food Science and Technology, Department of Food Engineering, Federal University of Paraiba, João Pessoa, Brazil

Mario Estévez   IProCar Research Institute, University of Extremadura, Caceres, Spain

Sandeep Ghatak   ICAR Research Complex for North Eastern Hill Region, Umiam, India

P.S. Girish   ICAR - National Research Centre on Meat, Hyderabad, India

Alberto González-Mohino   IProCar Research Institute, University of Extremadura, Caceres, Spain

Nagappa S. Karabasanavar   Department of Veterinary Public Health & Epidemiology, Veterinary College, Hassan, India; Karnataka Veterinary, Animal & Fisheries Sciences University, Bidar, India

Anup Kollanoor Johny   Department of Animal Science, University of Minnesota, Saint Paul, MN, United States

Rui Liu   College of Food Science and Engineering, Yangzhou University, Yangzhou, China

G.G. Mafi   Department of Animal and Food Sciences, Oklahoma State University, Stillwater, OK, United States

Naveena Basappa Maheswarappa   ICAR - National Research Centre on Meat, Chengicherla, Hyderabad, India

R.A. Mancini   Department of Animal Science, University of Connecticut, Storrs, CT, United States

Prabhat Kumar Mandal Department of Livestock Products Technology, Rajiv Gandhi Institute of Veterinary Education and Research, Puducherry, India

Kiran Mohan Department of Livestock Products Technology, Veterinary College, KVAFSU, Bidar, India

Divek V.T. Nair Department of Animal Science, University of Minnesota, Saint Paul, MN, United States

M.N. Nair Department of Animal Sciences, Colorado State University, Fort Collins, CO, United States

K. Narsaiah Central Institute of Post-harvest Engineering and Technology, Ludhiana, India

Abraham Joseph Pellissery Department of Animal Science, University of Connecticut, Storrs, CT, United States

Trinidad Pérez-Palacios IProCar Research Institute, University of Extremadura, Caceres, Spain

M. Perry Department of Animal and Food Sciences, Oklahoma State University, Stillwater, OK, United States

M. Pfeiffer Department of Animal and Food Sciences, Oklahoma State University, Stillwater, OK, United States

R. Ramanathan Department of Animal and Food Sciences, Oklahoma State University, Stillwater, OK, United States

Milagro Reig Instituto de Ingeniería de Alimentos para el Desarrollo, Universitat Politècnica de València, Ciudad Politécnica de la Innovación, Valencia, Spain

Nira Manik Soren Animal Nutrition Division, ICAR-National Institute of Animal Nutrition and Physiology, Bangalore, India

Meera Surendran Nair Department of Veterinary Population Medicine, University of Minnesota, Saint Paul, MN, United States

S. Tandon Division of Post-Harvest Technology, ICAR-Central Avian Research Institute, Izatnagar, Bareilly, India

Fidel Toldrá Instituto de Agroquímica y Tecnología de Alimentos (CSIC), Valencia, Spain

D.L. VanOverbeke Department of Animal and Food Sciences, Oklahoma State University, Stillwater, OK, United States

Kumar Venkitanarayanan Department of Animal Science, University of Connecticut, Storrs, CT, United States

Sonia Ventanas IProCar Research Institute, University of Extremadura, Caceres, Spain

Poonam Gopika Vinayamohan Department of Animal Science, University of Connecticut, Storrs, CT, United States

L. Yoder Department of Animal and Food Sciences, Oklahoma State University, Stillwater, OK, United States

C. Zhai Department of Animal Sciences, Colorado State University, Fort Collins, CO, United States

Wangang Zhang Key Laboratory of Meat Processing and Quality Control, Ministry of Education, Jiangsu Collaborative Innovation Center of Meat Production and Processing, Quality and Safety Control, College of Food Science and Technology, Nanjing Agricultural University, Nanjing, China

SECTION 1

Current perspectives of meat quality evaluation

CHAPTER 1

Current perspectives of meat quality evaluation: techniques, technologies, and challenges

Ashim Kumar Biswas[1] and Prabhat Kumar Mandal[2]

[1]Division of Post-Harvest Technology, ICAR-Central Avian Research Institute, Izatnagar, Bareilly, India [2]Department of Livestock Products Technology, Rajiv Gandhi Institute of Veterinary Education and Research, Puducherry, India

Chapter Outline
1.1 Introduction 3
1.2 Techniques and technologies 5
    1.2.1 Nutritional composition 5
    1.2.2 Physical and structural quality 6
    1.2.3 Meat traceability and authentication 7
    1.2.4 Food preservatives and additives 9
    1.2.5 Freshness and pathogen identification 9
    1.2.6 Chemical contaminants 10
    1.2.7 Sensory quality 12
1.3 Challenges 13
References 15

1.1 Introduction

The meat we eat is an integral component of the human diet. It contains essential nutrients which help to maintain normal physiological functions, improve immunity, and prevent certain diseases including malnutrition. When such foods contain unauthorized chemicals or do not meet the required standards for nutritional composition, sensory properties, food preservatives, microbial pathogens, residues of pesticides, veterinary drugs, and/or heavy metals, potential health risks for consumers may arise. The consumers' requirements are healthful muscle foods that not only taste good but are also safe, fresh, natural, and contain fewer chemical additives, such as preservatives. However, all these conditions can be organized into quality criteria indexed to physical, chemical, and microbial properties that can be objectively quantified through robust and sensitive techniques. The traditional approach of meat analysis using mechanical and chemical methods was robust, but all of the methods suffered from overcomplexity. Although the experienced inspectors carrying out meat quality evaluations ensure that the occurrence of misclassifications is rare, this manual method is subjective, time-consuming, and thus not suitable for online monitoring (Xiong et al., 2017). So to overcome this deficiency of the traditional methods and to satisfy the regulatory requirements, modern technologies are to be applied in meat quality evaluation since they are robust, sensitive, selective, qualitative, and quantitative. These techniques may use cost-intensive sophisticated instruments at cutting edge labs, but often they can be fabricated locally using only simple probes derived from those labs. Meat processors outfitted with these kinds of probes could gain tighter process control for production feasibility, quality sorting, and automation and thus provide consumers with certified products bearing quality seals and trust marks (Damez and Clerjon, 2012).

Thus in meat science the meat quality analysis actually integrates systematically all branches of science including mathematical science in order to explain the modest characteristics of meat. Meat quality analysis therefore exploits the basic principles and applications of various advanced techniques since meat science is no longer simply an academic discipline. The impact of this knowledge has also penetrated to the molecular level for better evaluation of meat quality using genome analysis, proteomics tools, sensor-based techniques, DNA microarray techniques, and loop-mediated isothermal amplifications (LAMP). Further, the spectacular technological advances and the rapid expansion of scientific knowledge have revolutionized our understanding of biological processes for the production of safe meat products. Thus many in the scientific community have the belief that the meat industry will become a prominent industry in the coming era.

There is a diverse array of meat quality analysis for which analytical chemistry plays a crucial role, including the identification of meat enzymes or protein markers for optimizing meat maturation; assessing structural integrity including morphology, physical, biophysical characteristics; sensory properties including meat color development; the detection of adulterants and product tempering; meat authenticity; the characterization of the chemical composition of meat; the impact of production and processing practices on the generation or inactivation of toxic chemicals; the compliance with food and trade laws to ensure safety and traceability; thermal properties; and microbiological impedance. All of these analyses have critical roles in assuring product safety, quality, and palatability. Further, in recent years, it has been intimidated "one health program" since animal foods are directly linked with the human health. Thus meat is considered today not only a source of essential nutrients but also an affordable way to prevent future diseases.

To tackle these problems a numbers of opportunities have been sought in various ways that are quite impressive, for example, a tailor-made meat product given to a particular group of people to promote health and well-being based on their genome sequences (Herrero et al., 2012). Another well-known technique, proteomics, is the study of the proteome, the protein complement of the genome that is expressed, and modified following expression, by the entire genome in the lifetime of a cell. Today proteomics is a scientific discipline that promises to bridge the gap between our understanding of the genome sequence and cellular behavior. It can be viewed as more of a biological assay or tool for determining gene function. Thus the application of "omics" technologies such as genomics, proteomics, foodomics, nutrigenomics, and metagenomics may solve certain problems that were untouchable until a few years ago. But success for the application of all these new advanced technologies is reliant on the depth of knowledge in the relevant area and the requirement to use this knowledge in a rational manner.

Thus meat quality evaluation is considered as an important area to assure the quality and safety of finished products. This chapter will discuss the emerging techniques and technologies employed in meat quality analysis together with the current difficulties and future challenges.

1.2 Techniques and technologies

1.2.1 Nutritional composition

The importance of nutritional levelling of fresh and finished products is high for informing and guiding the consumers about the quality of products since many literature reports have indicated the negative influence of nutrition on health. It has been implicated that high fat intake aggravates the risk of coronary heart diseases, cancers, atherosclerosis, etc., while high glucose and sodium (salt) intakes give rise to the risks of diabetes and hypertension, respectively. On the other hand the intake of high amounts of fiber, prebiotics, antioxidants, certain vitamins, and minerals ameliorate the negative effects. X-ray imaging, particularly dual X-ray energy imaging (DXA), offers useful capabilities for successful evaluation of meat quality in terms of fat, bone, and lean meat content and has been in use for 30 years in the meat industry, though now it seems to be a slow measurement technique (Mercier et al., 2006). Nuclear magnetic resonance (NMR) and magnetic resonance imaging (MRI) techniques are now also in use (Clerjon and Bonny, 2011; Xiong et al., 2017) and NMR microimaging on meat samples has the capability to quantitatively characterize the fat, in addition to its inherent function for checking meat quality. The MRI method uses the principle of diffusion-weighted imaging of muscle for the determination of apparent diffusion coefficients of myofibers and lipids. Ultrasound works on the principle of analyzing the acoustic parameters of waves propagating in a medium.

Most of the applications of ultrasound imaging so far have been concentrated on predicting the body composition of live animals, including intramuscular fat percentage, lean content, and fat tissue thickness (Xiong et al., 2017). Ultrasound has the capability to assess and characterize the muscle samples based on acoustic wave propagation through meat. For example, if acoustic waves are propagating through fat and collagen, they produce different waves and this helps in discriminating different muscle types on the basis of fat and collagen content (Morlein et al., 2005). Similarly the near infrared spectroscopy (NIRS) technique has been successfully deployed for rapid nondestructive determination of fatty acid composition in dry-cured sausages (Fernandez-Cabanas et al., 2011). Recently several image processing techniques (MRI, fluorescence imaging, hyperspectral imaging, thermal imaging) were developed and also applied for the determination of muscular tissue (Veberg et al., 2006; Burfoot et al., 2011; Adedeji et al., 2011; Yang et al., 2010). All of these image processing techniques were applied for the determination of the chemical composition of meat from all species with a great degree of variation in success. However, thermal image processing techniques were focused on temperature differences over a large range, which is an indirect method for the determination of the surface fat covering of carcasses, since lower the surface temperature indicated then the lower the fat covering (Costa et al., 2010). Salt distribution analysis in salted meat products by the use of X-ray computed tomography was also a major application (Segtnan et al., 2009), while micro- and macronutrients (protein, fat, amino acid, fatty acid, organic acid, vitamins, minerals) can be detected using DXA imaging, infrared spectroscopy, and nuclear magnetic resonance spectroscopy (Damez and Clerjon, 2012).

1.2.2 Physical and structural quality

In meat processing the determination of physical and structural quality is of the utmost importance for the proper merchandizing of the finished produce. The meat industry around the globe is seeking quick and accurate methods for authenticating quality and keeping trust marks on the packaging to maintain consumers' faith in finished products. Recently Damez and Clerjon (2008) reviewed some fast and robust invasive and noninvasive biophysical techniques for predicting the structural quality of meat. These can be used for the measurement of meat components (collagen content, marbling, water content, fat content, specific proteins detection, salt content, water holding capacity, PSE (pale, soft exudative), DFD (dark, firm, dry), etc.) or their organization (collagen organization, collagen typing, fat organization, myofiber organization, myofiber spacing, myofiber diameter, myofiber density, myofilaments structure changes, Z line degradation, sarcomere length, endomysium structure, etc.), either directly or by calculating them indirectly using the correlations between one or several biophysical measurements and meat components' properties. All these measurements are based on either mechanical, optical, or electrical probing or by using ultrasonic measurements, electromagnetic waves, NMR, near infrared (NIR), and so on. For safety aspects the detection of physical particles like bone fragments, woods, metal, and glasses using ultrasound, X-rays, and image processing techniques was documented (Damez and Clerjon, 2012).

Proteomic tools are now being applied to investigate the proteome changes induced due to compensatory growth in pigs, different preslaughter stressors (Lametsch et al., 2006), postslaughter handling, processing, etc. The lean color is crucial in merchandizing of meat since it may affects due to alterations in the proteome myoglobin (Mb). The dynamics of meat color stability is due to the primary structure of Mb which is mediated via autoxidation, heme retention, structural stability, thermostability, and oxygen affinity. The interactions of pH, temperature, and postmortem time also affect the biochemical dynamics of early Mb discoloration and hence the meat color. Furthermore, variations in the amino acid sequence of Mb influence meat color stability through species-specific interactions with small biomolecules like lactate and aldehydes. For Mb, a heme protein with different redox states, this is extremely critical, because Mb stability and the aforementioned molecular interactions govern meat color/color stability (Sayd et al., 2006).

Similarly the results of proteomic study have revealed that changes of proteins occurred in muscle during postmortem storage. A total of 15 proteins were changed, some increasing and some decreasing in abundance after slaughter. Several of these proteins were identified as fragments of structural proteins such as actin, myosin heavy chain, and troponin T (Hwang et al., 2005). The calpain system is believed to be important for the degradation of myofibrillar proteinsand thereby improves tenderness. The activity of calpain system was elucidated via a proteome study in pork LD muscle for the identification of myofibrillar substrates for μ-calpain (Lametsch et al., 2004). Changes in metabolic protein composition in biopsies from live animals to postmortem samples collected shortly after slaughter in the cattle LD muscle revealed that 24 protein spots were changed (Jia et al., 2006). This reflects the contribution of several factors such as transportation, lairage, stunning, exsanguination, and dehiding on the LD muscle proteome. Identification of the proteins by MALDI-TOF/TOF MS revealed that a wide range of metabolic enzymes and stress proteins increased in abundance after slaughter.

1.2.3 Meat traceability and authentication

Meat authenticity and traceability are important for the labeling and assessment of value, and are therefore necessary to avoid unfair competition in the meat trade and also to assure consumers of safety. In recent years there has been an unprecedented growth in digital literacy, and with this several innovative IT-based tools have been developed and used in the meat processing sector throughout the world. These IT tools assist in evaluating pre- and postslaughter carcass/meat quality and ensure traceability through online computerized data acquisition without the intervention of human errors. Since during meat production animals are routed through different intermediaries, for example, traders, retailers, processors, quality control inspectors, local bodies, it is necessary to bring them under single umbrella so as to judiciously locate the supplies and demands of livestock and meat. The IT-based tools have the potential to provide integrative platforms for the public, corporations, traders, and the government that would enable real-time analysis for helping in decision-making, generating benefits for all stakeholders. It has been reported that IT-based image analysis was used for carcass grading, predicting meat quality, and carcass yield (Tan, 2004). Visoli et al. (2011) developed a spatial decision support system enabling traceability systems that involved automatic recording, several times a day, of an animal at different locations, coupled with system analysis for the decision-making by stakeholders, validation, and hypothesis testing. Furthermore, Mishra et al. (2015) developed a label-free impedimetric immunosensor for the detection of *Escherichia coli* in water. These IT-based analytical tools enable the production feasible, quality shorting and improved product performances in real time and in a most economical and user-friendly manner in the meat industry.

The adulteration or falsification of meat and its possible substitution with poor quality meat or even nonmeat ingredients represent a problem, since unauthorized use of meat or nonmeat ingredients not only give rise to allergic reactions in sensitive individuals, but also damage religious sentiment and quality. This adulteration of meat is misleading the people, causing them to lose their faith in packaged food items. Thus accurate labeling of meat and its products is important in order to safeguard consumers as well as for the sustainability of the meat industry, in addition to meeting the commitments of foreign trade. Several analytical tools have been developed, including physical, anatomical, histological, chemical, biological or serological or immunological, electrophoresis, chromatography, and molecular techniques, and these techniques use a diverse range of equipment with variable degrees of satisfactory results.

Molecular or PCR-based techniques offer greater flexibility and usefulness as compared to others (Vaithiyanathan and Vishnuraj, 2018), since these methods use the amplification of targeted DNA which has conserved gene sequences (Kocher et al., 1989). The digital PCR is gaining wider popularity, as in this method a sample is diluted and partitioned into hundreds to millions of separate reaction chambers, each containing one or no copies of the gene sequence of interest (Baker, 2012). Further development was done to ascertain species speciation with the introduction of droplet digital polymerase chain reaction (Ren et al., 2017). This measures absolute quantities by counting nucleic acid molecules encapsulated in discrete, volumetrically defined water-in-oil droplet partitions. The assay combines water–oil emulsion droplet technology with microfluidics. Later a DNA microarray, popularly known as a low-cost and low-density (LCD) array, was introduced for the simultaneous detection of 32 species of meat samples. This array is based on classical PCR followed by LCD array hybridization. Lateral flow tests, otherwise known as "dipstick" tests, were also developed for on-site in-time verification of meat samples. The applied tests were capable of detecting small quantities of pork, beef, mutton, and poultry in industrial canned meat products (Biswas and Kondaiah, 2014).

1.2.4 Food preservatives and additives

The use of food preservatives is not new; rather the development and application of these compounds has increased with the diversity and progress of meat science. Food preservatives are generally used for the retention of lean color, flavor, odor, texture, etc., and these arrays are interrelated with the factors like temperature, relative humidity, atmospheric oxygen pressure, light, endogenous enzymes, and microbial loads, which actually lead to the spoilage or decay of meat and meat products. While food additives are substances that are not normally consumed as food or not used as a basic food ingredient, they are often used for technological purposes to improve quality. The common additives in meat include antioxidants, binders, emulsifiers, antimicrobials, curing ingredients and cure accelerators, flavoring agents, and coloring agents. However, the consumption of all these ingredients may lead to potential health implications, particularly when present in concentrations greater than the legal thresholds. Thus the use of additives and preservatives in meat requires a stringent food safety policy. So to comply with the requirements many national agencies and international organizations are harmonizing methods to identify and quantify various food additives and preservatives in fresh meat and processed products. However, often the chemical analysis of high concentrations of fats, oils, lipids, proteins, carbohydrates, polysaccharides, salts, surfactants, pigments, emulsions, and many other constituents present in meat is challenging. Thus to overcome the issues several pretreatment methods like homogenization, dilution, centrifugation, distillation, simple solvent extraction, supercritical fluid extraction, stir-bar sorptive extraction, pressurized-fluid extraction, microwave-assisted extraction, and Soxhlet extraction have been developed to remove the maximum possible amounts of interfering compounds (Martins et al., 2018). Although spectrometric and colorimetric methods are widely used for the detection and quantification of many additives and preservatives, the recent strategies are the use of chromatography techniques, mass spectrometry (MS) methods, electrophoresis, electronic spin resonance, and flow injection methods in addition to traditional enzymatic and immunoassays (Iammarino et al., 2017; Martins et al., 2018).

1.2.5 Freshness and pathogen identification

The safety of meat and meat products is an important aspect of modern food trade. Since meat foods are rich in nutrients, they support the growth of many bacterial populations leading to decay or spoilage. The meat spoilage pattern is an important phenomenon directly linked with the freshness indicator, although many other arrays are also responsible for meat spoilage. In the modern era the spoilage of muscle foods can be detected by sensor-based techniques. The sensor-based techniques are usually low-cost on-site in-time detection tools that can comprehensively be applied for the detection of meat freshness. Recently the Institute of Instrumental Analysis at Karlsruhe, Germany, has developed micronose microarray (gas sensor microarray) for the effective evaluation of meat freshness (Musatov et al., 2010), while for the discrimination of frozen−thawed meat, NMR spectroscopy methods have been developed (Ballin and Lametsch, 2008). Another innovation is a noninvasive portable method that works with the principle of Raman and fluorescence spectroscopy detection. This new device is capable of monitoring soluble protein content, the concentration of biogenic amines, and even microbial load in meat (Jordan et al., 2009). The Raman sensor works at an excitation wavelength of 671 nm for the detection of microbial spoilage on the meat's surface even through the packaging foil (Schmidt et al., 2010). The synchronous front-face fluorescence spectroscopy is capable of determining microbial load alongside chemometric methods (Damez and Clerjon, 2012).

The accurate identification of foodborne pathogens has been a long-pending issue but continuous efforts have been sustained to develop assay methods as alternatives to conventional detection methods. In recent years with the highly publicized antimicrobial resistance (AMR) of several foodborne pathogens, rapid assay methods like chemical, biochemical, biophysical, molecular biological, serological, immunological, nucleic acid, or even biosensor-based techniques were tried for the early detection and characterization of isolates of potential health risks (Bhunia et al., 2007; Bhunia, 2008; Biswas et al., 2008; Mandal et al., 2011). However, all these rapid methods are recommended and performed for the initial screening of samples rather than for confirmation.

A new generation of mobile sensing approaches offers significant advantages over traditional platforms in terms of test speed, control, low cost, ease of operation, and data management, and requires minimal equipment and user involvement. Further, the universal presence of mobile phones makes it suitable for on-site testing (Mandal et al., 2018). Recent developments in the field of smartphone-based food diagnostic technologies were reviewed by Ren et al. (2017). These devices typically comprise multiple components, such as detectors, sample processors, disposable chips, batteries, and software, which are integrated with a commercial smartphone. One of the most important aspects of developing these systems is the integration of these components onto a compact and lightweight platform that requires minimal power. To date researchers have demonstrated several promising approaches employing various sensing techniques and device configurations.

1.2.6 Chemical contaminants

The safety of meat food products due to the inadvertent presence of chemical residues of pesticides, veterinary drugs, heavy metals, mycotoxins and their metabolites, metals, plastics, wooden parts, shards of glass, etc., is the key issue of the new pattern of the food trade. These chemical contaminants in meat can be potential sources of several mysterious diseases unheard of in the past. The catalogue of such diseases and disorders includes cancer, epilepsy, liver and kidney dysfunction, somatic cell growth, depression, and neuritis. For this the regulatory authorities around the globe are harmonizing their responses to these contaminants through appropriate statutory measures and setting up standards using risk–benefit relationships (Biswas et al., 2010). This balancing takes into account "the economic, social, and environmental cost as well as potential benefits of the use of any chemicals in relation to its efficacy, inherent toxicity to mammals, wild-life, and plants." The immunoassay techniques (ELISA) are one of the recently developed rapid screening methods for the analysis of residues of some chemical contaminants in foods based on the interactions between antibodies and antigens. Antibodies are highly specific and vary structurally depending on compounds, but can show considerable cross-reactivity for structural analogues, as they recognize only specific chemical groups—the epitope. The only disadvantage of the immunochemical approach is the extensive effort and time required to elicit the antibodies in the vertebrate host.

Lateral flow assay, also known as solid-state immunochromatography (ICA), has been in great demand. Several ICA methods were developed (Suri et al., 2009; Guo et al., 2009; Wang et al., 2005) and all of them had successful applications in residue analysis. For the detection of AFT B_1 in foods an ICA was proposed by Xiulan et al. (2006). The assay used gold-labeled polyclonal antibodies as a detector reagent for AFT B_1, and AFT B_1-BSA as a capture reagent. The enzymatic sensors are the most extended biosensors used for the determination of chemical compounds. However, this biosensing technique, although sensitive enough, is not selective, and therefore it cannot be used for the quantification of either an individual or a class of chemical contaminants. For the detection of mycotoxins a great number of specific sensors have been developed for food and environmental control. Molecular imprinting is another technique that has potential applications in mycotoxin analysis (Shephard et al., 2005). Similarly a microarray technique that is essentially a reverse dot-blotting technique is a new technique for the identification of DNA fragments, such as PCR products, and the differentiation of high numbers of microorganisms and toxins, especially mycotoxins (Nicolaisen et al., 2005). Luminex's xMAP technology is another new invention comprised of existing technologies—flow cytometry, microspheres, lasers, digital signal processing, and traditional chemistry—and now being applied for the determination of mycotoxins in foods. The ranges of applications are considerable throughout the drug-discovery and diagnostics fields, as well as in basic research (Vignali, 2000). However, there are very limited applications of these techniques for the identification of chemical compounds in meat systems.

AMR of foodborne pathogens is widely discussed in today's world. So the harmonization of these compounds is important regarding their presence in the food chain system. Currently six main types of analytical methods are followed—microbial growth inhibition assays, microbial receptor assays, enzymatic colorimetric assays, receptor binding assays, immunoassays, and chromatographic methods. Immunoassays like ELISA enable the simultaneous processing of a large number of samples in a short time, but are difficult to implement for multiresidue screening because of a wide range of sensitivities for the different compounds rather than a generic response for a family of compounds. Biosensors are also used as screening methods (Pellegrini et al., 2004), as are liquid chromatography (LC) techniques equipped with UV, fluorimetric, or electrochemical detectors. But the LC technique coupled to MS is gaining more acceptance for screening purposes as it enables the analysis of a large number of analytes in a single run (Weber et al., 2005; Suarez et al., 2009; Adrian et al., 2009). However, the key point for all screening methods is that the sample treatment should be as simple as possible without compromising the operation of the mass spectrometer. Karsourd et al. (1998) evaluated a number of common bacterial inhibition tests for the screening of antimicrobial residues in tissues.

1.2.7 Sensory quality

With customized meat processing, the meat industry around the world is showing a great deal of interest in the quality evaluation of meat based on dynamic sensory properties as well as the physical and structural properties of different species' meats and their products. Tenderness is probably the most important meat quality that greatly varies according to the type of processing, processing environment, age factors, myofibrillar orientation, connective tissue development, and marbling. Warner-Bratzler shear force or compressive techniques still remain as the benchmark reference methods for the determination of meat toughness, but these are destructive methods and required a long time to get results (Lorenzen et al., 2010). Uses of a number of handful portable apparatuses such as the Armor tenderometer, G2 tenderometer, and tendertec mechanical penetrometer have also been reported (Damez and Clerjon, 2012). However, using modern techniques some instruments can determine tenderness by evaluating bioimpedance, ultrasound, X-ray scanning, and NMR/MRI. Bioimpedance is used for characterizing biological tissue properties (aging-related tenderness), while electrical impedance mainly focuses on measuring the state of muscle fiber and shifts in its anisotropy (Damez and Clerjon, 2008). Since meat and meat products are viscoelastic, acoustic waves generated from ultrasound or even a nondestructive magnetic resonance elastography, optical coherence tomography, or transient elastography could help in determining the localized viscoelastic properties of meat tissues (elasticity).

Image processing techniques like macroscopic imaging consistently characterize complex color, texture, and geometrical properties. X-ray imaging helps in understanding lipid structures, while microscopic imaging techniques are applied for the characterization of the histochemical properties of a muscle. The details of the sensory characteristics of meat are also given in Section 1.2.2. Some image processing techniques also utilize sensors that work like human senses to recognize and discriminate volatile and liquid compounds with high sensitivity and accuracy. These are called arrays-based techniques. The electronic nose (arrays of gas sensors) and electric tongue (arrays of liquid) are examples of such arrays widely used for discriminating smell and taste, respectively. Fluorescence imaging determines the fluorescence in the form of luminescence, by absorbing light or other electromagnetic radiation that can emit fluorescence. This may help in indirect measurement of slime development that is mainly due to bacterial spoilage. A laser-induced fluorescence imaging system for the discrimination of feces-contaminated poultry carcasses is well documented (Chao et al., 2010). Hyperspectral imaging integrates the merits of computer vision and conventional spectroscopy and is capable of determining wholesomeness, contamination, and moisture, in addition to simultaneously obtaining abundant spatial and spectral information (Kandpal et al., 2013). Thermal imaging is another rapid, noninvasive, and accurate technique for the evaluation of sensory quality. This can catch moving targets in real time and create a visual picture, showing temperature differences over a large range (Ibarra et al., 2000). There is extensive literature on the sensory quality evaluation of meat and meat products utilizing modern techniques that are integrated with many other meat quality parameters, so only a few of them can be mentioned here.

1.3 Challenges

The development and application of technologies for meat quality evaluation has grown in parallel with the consumer concern for safety. But researchers are often facing greater challenges due to the complexity of trade-related issues that are arising around the globe (Garcia-Canas et al., 2012). The first and principal goal in meat quality evaluation has traditionally been, and also still now is, to ensure safety. The modern quality evaluation of meat encompasses many recent techniques. These were made possible due to constant efforts; even then, there are a large number of issues that need to be improved substantially. It has been found that many modern image processing techniques are used only for specific applications. Although these techniques are robust, sensitive, and selective enough, they are cumbersome and not cost-effective. Thus the problems arising with the widening of their application through the integration of multiple imaging techniques need to be solved. For instance, many nondestructive methods, like conventional fluorescence imaging, X-ray imaging, and NIRS, have been developed but the integration of these with other nondestructive methods, in particular microscopic imaging, hyperspectral imaging Raman spectroscopy, and electronic nose, could achieve better performance in meat quality evaluation while carcasses are still on the processing line. Furthermore, nearly all the image processing techniques are based on huge spectral data, so the development and exploitation of innovative data-processing algorithms is highly relevant and need to be taken care of in order to reduce the time required for image data processing.

Another area of concern is the identification of pathogens. In the last few decades, reliable culture-based techniques have been developed for several microbiological communities. These are known as "golden tests" for declaring microbiological standards, but they are time-consuming, cumbersome, and require several days to a week to obtain the final result. Thus the introduction of molecular (DNA) or array-based techniques allows for the development of automated and more sensitive, selective, and accurate methods for initial screening of microbial pathogens in real time (Garcia-Canas et al., 2012; Biswas et al., 2008). Fortunately many analytical methods have been developed and some are under use for the initial screening of meat samples, but these are also facing great challenges to achieve detection limits with respect to the standard or official guidelines outlined for pathogens.

Currently miniaturization techniques are gaining wider popularity and acceptability for the effective control and monitoring of food and environmental contaminants, since they have greater flexibility, automation, and multiplexing capabilities than traditional assay methods. It has been shown that varieties of array-based techniques have potential in detecting pathogens, food allergens and adulterants, toxins, antibiotics, and environmental contaminants (Raz and Haasnoot, 2011), but their potential application in meat food systems is yet to be explored. Another important consideration for the application of modern techniques for the evaluation of meat is for the eventual or unlawful presentation of varieties of chemical toxicants, including their residues, due to their increasing production as well as productivity adopting new agricultural practices. So it is logical to develop more robust, sensitive, and fast analytical tools that may fit present-day needs, while keeping in mind the rapid industrialization and its resultant impact on effluents, disposal of nanomaterials, pharmaceutical residues, antibiotics, and other veterinary drugs. Some research communities are also encouraging the development of green analytical techniques in order to reduce the burden of the disposal problems of solvents, reagents, chemical preservatives, and many other hazardous materials that have potential health risks and negative impacts on the environment. Armenta et al. (2008) reported that the use of green analytical techniques, in particularly sample pretreatment or processing, could greatly contribute to reaching the goals of this new green era.

Proteomic approaches will help to identify useful low abundance proteins, which can be studied further to understand their beneficial properties. Global genetics—based analyses provide information regarding which genes an organism contains or which genes are expressed under specific conditions; however, examining the posttranslational protein output of an organism allows one to query the ultimate outcome of the organism's genetic and regulatory activities. Techniques that fall within the category of either proteomics or protein arrays are used for global analysis of cellular protein output under different conditions and are potentially relevant to food and food processing environments. The integration of both genetic- and protein-based approaches provides a global analysis of treatment or environment-related changes at the molecular level, thus presenting a more comprehensive view of cellular activities. The development of high-throughput analytical techniques will make it possible to use multiomics approaches to understand complete biological systems, a field known as systems (integrated) biology.

Although omics technologies are becoming standard research tools that offer tremendous opportunities, there are also significant challenges. There is a need to properly manage the large quantity of complex raw data generated by these technologies in a manner in which it can be adequately analyzed, scrutinized, and compared for the benefit of the scientific community. There are various omics standardization activities underway, which are critical for the integration and interpretation of data from different data sources. Lastly, there is a need to bridge the gap between the knowledge of the genome, proteome, and metabolome, and results obtained in relevant systems by studying the behavior of pathogens in foods and in the animal host, not only in model systems under laboratory conditions. The knowledge garnered from omics-based research in the coming years will play an important role in understanding how pathogens survive food safety barriers and interact with host species. Each new advance in our understanding will potentially give rise to improved and novel strategies for the detection, identification, and control of foodborne pathogens, as well as for the diagnosis and control of infections.

References

Adedeji, A.A., Liu, L., Ngadi, M.O., 2011. Microstructural evaluation of deep-fat fried chicken nugget batter coating using confocal laser scanning microscopy. J. Food Eng. 102 (1), 49–57.

Adrian, J., Pasche, S., Pinacho, D.G., Font, H., et al., 2009. Wavelength-interrogated optical biosensor for multi-analyte screening of sulfonamide, fluoroquinolone, β-lactam and tetracycline antibiotics in milk. Trends Anal. Chem. 28, 769–777.

Armenta, S., Garrigues, S., de la Guardia, M., 2008. Green analytical chemistry. Trends Anal. Chem. 27, 497–511.

Baker, M., 2012. Digital PCR hits its stride. Nat. Methods. 9, 541–544.

Ballin, N.Z., Lametsch, R., 2008. Analytical methods for authentication of fresh vs. thawed meat—a review. Meat Sci. 80 (2), 151–158.

Bhunia, A.K., 2008. Biosensors and bio-based methods for the separation and detection of foodborne pathogens. Adv. Food Nutr. Res. 54, 1–44.

Biswas, A.K., Kondaiah, N., 2014. Flatulent substitution or falsification of meat and its detection, Meat Science and Technology, first ed. Jaya Publishing House, Delhi, pp. 74–87.

Bhunia, A.K., Banada, P., Banerjee, P., et al., 2007. Light scattering, fiber optic- and cell-based sensors for sensitive detection of foodborne pathogens. J. Rapid Method Autom. Microbiol. 15, 121–145.

Biswas, A.K., Kondaiah, N., Anjaneyulu, A.S.R., et al., 2008. Microbial profiles of frozen trimmings and silver sides prepared at Indian buffalo meatpacking plants. Meat Sci. 80, 418–422.

Biswas, A.K., Kondaiah, N., Anjaneyulu, A.S.R., Mandal, P.K., 2010. Food safety concerns of pesticides, veterinary drug residues and mycotoxins in meat and meat products. Asian J. Anim. Sci. 4 (2), 46–55.

Burfoot, D., Tinker, D., Thorn, R., Howell, M., 2011. Use of fluorescence imaging as a hygiene indicator for beef and lamb carcasses in UK slaughterhouses. Biosyst. Eng. 109 (3), 175–185.

Chao, K.L., Yang, C.C., Kim, M.S., 2010. Spectral line-scan imaging system for high-speed non-destructive wholesomeness inspection of broilers. Trends Food Sci. Technol. 21 (3), 129–137.

Clerjon, S., Bonny, J.M., 2011. Diffusion-weighted NMR micro-imaging of lipids: application to food products. In: Renou, J.P., Belton, P.S., Webb, G.A. (Eds.), Magnetic Resonance in Food Science. An Exciting Future. The Royal Society of Chemistry, Cambridge, pp. 182–189.

Costa, L.N., Stelletta, C., Cannizzo, C., et al., 2010. The use of thermography on the slaughter-line for the assessment of pork and raw ham quality. Ital. J. Anim. Sci. 6 (1S), 704–706.

Damez, J.-L., Clerjon, S., 2008. Meat quality assessment using biophysical methods related to meat structure. Meat Sci. 80, 132–149.

Damez, J.-L., Clerjon, S., 2012. Recent advances in meat quality assessment, Handbook of Meat and Meat Processing, second ed. CRC Press, pp. 161–175.

Fernandez-Cabanas, V.M., Polvillo, O., Rodriguez-Acuna, R., Botella, B., Horcada, A., 2011. Rapid determination of the fatty acid profile in pork dry-cured sausages by NIR spectroscopy. Food Chem. 124 (1), 373–378.

Garcia-Canas, V., Simo, C., Herrero, M., Ibanez, E., Cifuentes, A., 2012. Present and future challenges in food analysis: foodomics. Anal. Chem. 84, 10150–10159.

Guo, Y.R., Liu, S.Y., Gui, W.J., Zhu, G.N., 2009. Gold immunochromatographic assay for simultaneous detection of carbofuran and triazophos in water samples. Anal. Biochem. 389, 32–39.

Herrero, M., Simó, C., García-Cañas, V., Ibáñez, E., Cifuentes, A., 2012. Foodomics: MS-based strategies in modern food science and nutrition. Mass Spectrom. Rev 31, 49–69.

Hwang, I.H., Park, B.Y., Kim, J.H., Cho, S.H., Lee, J.M., 2005. Assessment of post-mortem proteolysis by gel-based proteome analysis and its relationship to meat quality traits in pig *Longissimus*. Meat Sci. 69, 79–91.

Iammarino, M., Marino, R., Albenzio, M., 2017. How meaty? Detection and quantification of adulterants, foreign proteins and food additives in meat products. Int. J. Food Sci. Technol. 52 (4), 851–863.

Ibarra, J.G., Tao, Y., Xin, H., 2000. Combined IR imaging-neural network method for the estimation of internal temperature in cooked chicken meat. Opt. Eng. 39 (11), 3032–3038.

Jia, X., Hollung, K., Therkildsen, M., Hildrum, K.I., Bendixen, E., 2006. Proteome analysis of early post-mortem changes in two bovine muscle types: *M. longissimus dorsi* and *M. semitendinosis*. Proteomics 6, 936–944.

Jordan, G., Thomasius, R., Schroder, H., et al., 2009. Noninvasive mobile monitoring of meat quality. J. Verbrauch Lebensm. 4 (1), 7–14.

Kandpal, L.M., Lee, H., Kim, M.S., Mo, C., Cho, B.K., 2013. Hyperspectral reflectance imaging technique for visualization of moisture distribution in cooked chicken breast. Sensors 13 (10), 13289–13300.

Karsourd, G.O., Boison, J.O., Nouws, J.F., MacNeil, J.D., 1998. Bacterial inhibition tests used to screen for antibacterial veterinary drug residues slaughtered animals. J. AOAC-Int. 81, 21–24.

Kocher, T.D., Thomas, W.K., Meyer, A.M., et al., 1989. Dynamics of mitochondrial DNA evolution in animals: amplification and sequencing with conserved primers. Proc. Natl. Acad. Sci. 86, 6196–6200.

Lametsch, R., Roepstorff, P., Molle, H.S., Bendixen, E., 2004. Identification of myofibrillar substrates for l-calpain. Meat Sci. 68, 515–521.

Lametsch, R., Kristensen, L., Larsen, M.R., et al., 2006. Changes in the muscle proteome after compensatory growth in pigs. J. Anim. Sci. 84, 918–924.

Lorenzen, C.L., Calkins, C.R., Green, M.D., Miller, R.K., Morgan, J.B., Wasser, B.E., 2010. Efficacy of performing Warner-Bratzler and slice shear force on the same beef steak following rapid cooking. Meat Sci. 85 (4), 792–794.

Mandal, P.K., Biswas, A.K., Choi, K., Pal, U.K., 2011. Methods for rapid detection of foodborne pathogens: an overview. Am. J. Food Technol. 6 (2), 87–102.

Mandal, P.K., Pal, U.K., Kasthuri, S., 2018. Advances in rapid detection of microbial spoilage of muscle foods. In: Compendium of 8th Conference of Indian Meat science Association and International Symposium on "Technological Innovations in Muscle Food Processing for Nutritional Security, Quality and Safety" held at Kolkata from 22–24 November, 2018, pp. 263–266.

Martins, F.C.O., Sentanin, M.A., de Souza, D., 2018. Analytical methods in food additives determination: compounds with functional applications. Food Chem. 272, 732–750.

Mercier, J., Pomar, C., Marcoux, M., Goulet, F., et al., 2006. The use of dual-energy X-ray absorptiometry to estimate the dissected composition of lamb carcasses. Meat Sci. 73 (2), 249–257.

Mishra, G.K., Bacher, G., Roy, U., Bhand, S., 2015. A label free impedemetric immunosensor for detection of *Escherichia coli* in water. Sci. Lett. J. 4, 76.

Morlein, D., Rosner, F., Brand, S., et al., 2005. Non-destructive estimation of the intramuscular fat content of the longissimus muscle of pigs by means of spectral analysis of ultrasound echo signals. Meat Sci. 69 (2), 187–199.

Musatov, V.Y., Sysoev, V.V., Sommer, M., Kiselev, I., 2010. Assessment of meat freshness with metal oxide sensor microarray electronic nose: a practical approach. Sensors Actuators B: Chem. 144 (1), 99–103.

Nicolaisen, M., Justesen, A.F., Thrane, U., et al., 2005. An oligonucleotide microarray for the identification and differentiation of trichothecene producing and non-producing Fusarium species occurring on cereal grain. J. Microbiol. Methods 62, 57–69.

Pellegrini, G.E., Carpico, G., Coni, E., 2004. Electrochemical sensor for the detection and presumptive identification of quinolone and tetracycline residues in milk. Anal. Chim. Acta 520, 13–18.

Raz, S.R., Haasnoot, W., 2011. TrAC. Multiplex bioanalytical methods for food and environmental monitoring. Trends Anal. Chem. 30, 1526–1537.

Ren, J., Deng, T., Huang, W., Chen, Y., Ge, Y., 2017. A digital PCR method for identifying and quantifying adulteration of meat species in raw and processed food. PLoS One 12 (3), e0173567.

Sayd, T., Morzel, M., Chambon, C., et al., 2006. Proteome analysis of the sarcoplasmic fraction of pig semimembranosus muscle: implications on meat color development. J. Agric. Food Chem. 54 (7), 2732–2737.

Schmidt, H., Sowoidnich, K., Kronfeldt, H.D., 2010. A Prototype hand-held Raman sensor for the *in situ* characterization of meat quality. Appl. Spectrosc. 64 (8), 888–894.

Segtnan, V.H., Høy, M., Sørheim, O., et al., 2009. Noncontact salt and fat distributional analysis in salted and smoked salmon fillets using X-ray computed tomography and NIR interactance imaging. J. Agric. Food Chem. 57 (5), 1705–1710.

Shephard, G.S., van der Westhuizen, L., Gatyeni, P.M., et al., 2005. Do fumonisin mycotoxins occur in wheat? J. Agric. Food. Chem. 53, 9293–9296.

Suarez, G., Jin, Y.H., Auerswald, J., et al., 2009. Lab-on-a-chip for multiplexed biosensing of residual antibiotics in milk. Lab Chip 9, 1625–1630.

Suri, C.R., Boro, R., Nangia, Y., et al., 2009. Immunoanalytical techniques for analyzing pesticides in the environment. Trends Anal. Chem. 28, 29–39.

Tan, J.L., 2004. Meat quality evaluation by computer vision. J. Food Eng. 61 (1), 27–35.

Vaithiyanathan, S., Vishnuraj, M.R., 2018. Advances in PCR platforms for meat species identification with special reference to DNA macro array and droplet digital (dd) PCR techniques. In: Compendium of 8th Conference of Indian Meat science Association and International Symposium on "Technological Innovations in Muscle Food Processing for Nutritional Security, Quality and Safety" held at Kolkata from 22–24 November, 2018, pp. 187–190.

Veberg, A., Sørheim, O., Moan, J., et al., 2006. Measurement of lipid oxidation and porphyrins in high oxygen-modified atmosphere and vacuum-packed minced turkey and pork meat by fluorescence spectra and images. Meat Sci. 73 (3), 511–520.

Vignali, D.A., 2000. Multiplexed practile-based flow cytometric assays. J. Immunol. Methods 243, 243–255.

Visoli, M., Bimonte, S., Ternes, S., et al., 2011. Towards spatial decision support system for animals traceability. In: do Prado, H.A., Alfredo, J.B.L., Filho, H.C. (Eds.), *Computational Methods for Agricultural Research: Advances and Applications*. IGI Global, Publisher, pp. 389–411.

Wang, S., Zhang, C., Wang, J., Zhang, Y., 2005. Development of colloidal gold-based flow-through and lateral-flow immunoassays for the rapid detection of the insecticide carbaryl. Anal. Chim. Acta 546, 161–166.

Weber, C.C., Link, N., Fux, C., et al., 2005. Broad spectrum protein biosensors for class specific detection of antibiotics. Biotechnol. Bioeng. 89, 9–17.

Xiong, Z., Sun, D.-W., Pu, H., Gao, W., Dai, Q., 2017. Applications of emerging imaging techniques for meat quality and safety detection and evaluation: a review. Crit. Rev. Food Sci. Nutr. 57 (4), 755–768.

Xiulan, S., Xiaolian, Z., Jian, T., et al., 2006. Development of immunochromatographic assay for detection of aflatoxin B_1 in food. Food Control. 17, 256–262.

Yang, C.C., Chao, K., Kim, M.S., et al., 2010. Machine vision system for on-line wholesomeness inspection of poultry carcasses. Poult. Sci. 89 (6), 1252–1264.

SECTION 2

Advances in carcass quality evaluation and nutritional composition of meat

CHAPTER 2

Methods for nutritional quality analysis of meat

Nira Manik Soren[1] and Ashim Kumar Biswas[2]

[1]Animal Nutrition Division, ICAR-National Institute of Animal Nutrition and Physiology, Bangalore, India [2]Division of Post-Harvest Technology, ICAR-Central Avian Research Institute, Izatnagar, Bareilly, India

Chapter Outline

2.1 Introduction 21
2.2 Meat categorization 22
2.3 Basic nutritional composition of meat 23
2.4 Methodology for assessing the nutritional quality 25
    2.4.1 Determination of moisture 25
    2.4.2 Determination of protein 28
    2.4.3 Determination of amino acids 29
    2.4.4 Determination of lipids 30
    2.4.5 Determination of fatty acids 31
    2.4.6 Determination of ash content and minerals 31
    2.4.7 Determination of vitamins 32
2.5 Conclusion 33
References 34

2.1 Introduction

Meat is consumed in most parts of the world and it is regarded as a food with high nutritive value. It is a good source of protein, especially essential amino acids, fatty acids, minerals, and vitamins. It also contains a wide range of endogenous antioxidants like α-tocopherol, histidine peptides, antioxidant enzymes, such as glutathione peroxidase, superoxide dismutase, and catalase (Chan and Decker, 1994), and other bioactive substances, such as carnosine, carnitine, choline, taurine, ubiquinone, and creatine (Vongsawasdi and Noomhorm, 2014). Meat is low in carbohydrates and does not contain dietary fiber (Verma and Banerjee, 2010). The chemical components of meat can vary depending upon animal species, breed, age, sex, feed, body weight, and other factors (Beserra et al., 2004). On the other hand, the quality of meat is dependent on its chemical components, namely moisture, protein, fat, and ash (Tariq et al., 2013). In recent years there has been increased awareness from consumers concerning the quality of nutrients, especially the fatty acids content, in the meat.

Ruminants' products such as meat and milk are rich in saturated fatty acids because of the inherent biohydrogenation process undertaken by the microbes present in the rumen (Lourenço et al., 2010). The consumption of dairy products and ruminant meats is often associated with an increased incidence of coronary heart disease (CHD) in humans (Menotti et al., 1999). CHD is one such emerging disease which has affected a significant part of the human population in recent times. As per the World Health Organization's estimate, CHD is the single largest cause of death in the developed countries and also one of the leading causes of disease burden in developing countries (WHO, 2002). Therefore researchers are constantly trying to increase the unsaturated fatty acids content, that is, omega-3-fatty acids, conjugated fatty acids, etc., in ruminant products, either by dietary manipulation or by manipulating the rumen microbes.

Thus nutritional quality analysis of meat not only provides us with the information on the meat quality in terms of amino acids, fatty acids, minerals, vitamins, etc., but also indicates the likely health implications by consuming such foods and their products. In this chapter deep insights into the various aspects of nutritional quality analysis of meat that are generally followed globally are discussed from the point of view of meat consumption.

2.2 Meat categorization

Most meat consumed by humans falls into two categories: red meat and white meat. Meat is defined as skeletal muscle and its associated tissues (including nerves, connective tissues, blood vessels, skin, fat, and bones) and edible offal derived from mammals, avian, and aquatic species deemed as safe and suitable for human consumption (AMSA, 2016). The aquatic species, for example, fish, that are intended for human consumption are also included under meat. The function of muscle in the animal body determines the type of meat it becomes after the slaughter of the animal for food.

In general muscles which are regularly in use for certain activities, such as walking, running, etc., give rise to an increase in the number of red fibers (red muscle) due to the action of greater blood circulation and the presence of myoglobin (Mb), the main oxygen-carrying protein in the muscle, while white meat is derived from muscles used in short and sharp bursts. Myoglobin, the red pigments, are responsible for imparting red color to the meat (Mancini, 2009). The color of meat has been reported to be regulated by several conditions, such as exercise, raising environment, slaughtering processes, and storage conditions (Olsson and Pickova, 2005). In general the myoglobin content varies with age within a species. The myoglobin content of beef cattle at different ages is summarized in Table 2.1. As

Table 2.1: Variation in myoglobin content with age within a species (cattle).

Age class	Myoglobin content (mg/g)
Veal	2
Calf	4
Young beef	8
Old beef	18

Table 2.2: Myoglobin content of meat in different livestock species.

Species	Color	Myoglobin content (mg/g)
Pig (pork)	Pink	2
Sheep (lamb)	Light red	6
Beef	Cheery red	8

the age of the animal increases the concentration of myoglobin also increases. However, across species its content differs considerably (Table 2.2).

The most common type of meat, red meat, has a much stronger flavor than white meat. Red meat also has high levels of zinc and iron, vitamins, such as riboflavin, niacin, thiamin, vitamins B_6 and B_{12}, and amino acids (Williams, 2007). Dark meat (red meat) can be relatively tough due to the narrower muscle fibers, so longer moist methods of cooking are often required to tenderize the meat. Mutton, chevon, beef, carabeef (buffalo meat), pork, etc., are the most common examples of red meat (Astruc, 2014).

White meat on the other hand is meat which is sallow in color both before and after cooking. A common example of white meat is that of poultry coming from the breast, as contrasted with dark meat from the legs. White meat is also obtained from rabbits, pigs (pork), and from the flesh of milk-fed young ruminants (veal and lamb) and pork. White meat in general is leaner and has broader preference due to health concern. Within poultry, there are two types of meats—white and dark. The different colors are based on the different locations and uses of the muscles. White meat can be found within the breast muscle of a chicken or turkey. Dark muscles are used to develop endurance, or long-term use, and contain more myoglobin than white muscles (Muhlisin et al., 2016), allowing the muscle to use oxygen more efficiently for aerobic respiration. White meat contains large amounts of protein.

2.3 Basic nutritional composition of meat

Meat is mostly the muscle tissue of an animal. The muscle of most animals contains 75% water, 20% protein (amino acids), and 5% fat, carbohydrates, and a variety of vitamins and

Table 2.3: Nutritional composition of meat of different livestock species.

Meat	Nutritional composition (per 100 g)				Energy (kJ/100 g)	References
	Water	Protein	Fat	Ash		
Beef (lean) (g)	75.0	22.3	1.8	1.2	485	Heinz and Hautzinger (2007)
Beef carcass (g)	54.7	16.5	28.0	0.8	1351	
Pork (lean) (g)	75.1	22.8	1.2	1.0	469	
Pork carcass (g)	41.1	11.2	47.0	0.6	1975	
Veal (lean) (g)	76.4	21.3	0.8	1.2	410	
Chicken (g)	75.0	22.8	0.9	1.2	439	
Mutton carcass (g)	72.2	21.6	2.5	2.6	–	Soren et al. (2008)
Chevon carcass (g)	75.6	20.3	3.68	4.09	–	Soren et al. (2014) (Unpublished data)
Buffalo carcass (g)	76.3	20.4	1.37	0.98	724	Naveena et al. (2011a,b)

minerals (Listrat et al., 2016). The chemical composition of meat of different livestock species is shown in Table 2.3.

Moisture is the major component and plays an important role in the sensory aspects of the meat. The moisture influences the quality parameters of meat such as the tenderness, juiciness, and processing quality of the meat (Warner, 2017). From an economic point of view moisture contributes to the weight of the meat, if its content in the meat is less then it will affect the weight of meat. The moisture content ranges from 41% to 76% in the meat of different livestock species. Moisture is the only component of meat that is significantly volatile at temperatures above 100°C, thus the moisture content of the meat can be quantified by drying in a hot-air oven. Not all meat has the same water-retaining capacity, in general beef and buffalo meat have the greatest capacity, followed by chevon and pork, with poultry and mutton having the least.

Meat contains about 20% protein of which 12% is structural proteins—actin and myosin (myofibrillar)—6% is the soluble sarcoplasmic proteins found in the muscle juice, and 2% is the connective tissues—collagen and elastin, encasing the structural protein (Bender, 1992). Collagen differs from most other proteins in containing the amino acids hydroxylysine and hydroxyproline but no cysteine or tryptophan. Elastin, also present in connective tissue, has less hydroxylysine and hydroxyproline. Thus the protein content of meat rich in connective tissue is lower than that of connective tissue-free meat. The content of connective tissue in these cuts makes them tough and lowers their economic and eating quality values.

The lipid content of meat varies according to the animal species, age of the animal, and part of the carcass (Irshad et al., 2012). The lipid content and lipid composition of the meat is also influenced by animal feeding. The fatty acid composition of meat can be modified by dietary manipulation in monogastric animals, that is, pig and poultry (Bolte et al.,

2002; Smet et al., 2004). Total fat content of meats varies from around 0.8 to 48 g/100 g. Meat lipids comprise mostly monounsaturated fatty acids (MUFAs) and saturated fatty acids. The commonly found fatty acids in meat are oleic (C18:1), palmitic (C16:0), and stearic (C18:0) acids (Abbas et al., 2009). Poultry and pork contain somewhat more unsaturated fatty acids than beef and mutton, and also a notable amount of polyunsaturated fatty acids (PUFAs). Linoleic acid (C18:2) is the predominant PUFA, followed by α-linolenic acid. *Trans*-fatty acids comprise about 1%–2% of total fatty acids across all types of meat; in ruminant meats they represent 2%–4%. Conjugated linoleic acid, a group of polyunsaturated fatty acids that appear in dairy products and are thought to have beneficial effects on health, are also found at low mg-levels in meats, especially in beef and lamb (Belury, 2002).

In addition to moisture, protein, and fat, meat contains a wide variety of minerals such as iron, zinc, and copper. Ash is the inorganic residue remaining after the water and organic matter (protein, fat, and carbohydrates) have been removed by incineration at high temperature (500°C–600°C) in the presence of oxidizing agents. This provides a measure of the total amount of minerals present in the meat. The proportion may vary considerably in different species. Feeding high levels of minerals in the feed does not necessarily increase the level of minerals in the meat. The ash content of fresh meat rarely exceeds 5% (Table 2.3), although some processed meat products can have ash content as high as 12%.

2.4 Methodology for assessing the nutritional quality

Meat is made up of moisture, protein, fat, minerals, and contains a small quantity of carbohydrates (glycogen). The chemical composition of meat on an average comprises approximately 72% water, 21% protein, 5% fat, and 1% ash (potassium, phosphorus, sodium, chloride, magnesium, calcium, and iron). The moisture content of meat is highly variable and is inversely related to its fat content. The fat content is higher in the entire carcass than in lean carcass cuts, while in the processed meat products its content is generally higher because in processed products higher amounts of fats are used. The most important component of meat from the nutritional point of view is protein. The value of meat is essentially associated with its protein content. In the animal body, approximately 65% of the proteins are skeletal muscle protein, about 30% are connective tissue proteins (collagen, elastin), and the remaining 5% are blood proteins and keratin in hairs and nails. The biological value of animal protein is higher than vegetable protein because the assortment of amino acids in meat is almost similar to that of the human body.

2.4.1 Determination of moisture

Moisture content influences the taste, texture, weight, appearance, and shelf life of meat. Excessive moisture in meat increases the probability of microbial growth leading to the spoilage of meat while too little moisture could affect the consistency of the end product (Dave and Ghaly, 2011). Water is also an inexpensive ingredient for adding to the weight of the final product. Hence to acquire the best analytical value moisture is of great economic significance. For these reasons food analysts often balance delicately the moisture and total solids content of meat to ensure consistent product quality, safety, and profitability.

There are a number of methods for the determination of the moisture content of meat. A summary of different methods for moisture determination are listed below.

2.4.1.1 Gravimetric methods

Oven drying is the most common method for determining the moisture content of meat. The moisture is evaluated by drying the meat samples in an oven at $100°C-105°C$ until the sample reaches a constant weight by release of moisture. The meat sample is cooled in the desiccator before reweighing. Moisture content is calculated by the difference in wet and dry weight. In this process, measuring accuracy and the resolution of the balance are particularly important. Careful consideration must also be given to maintain identical conditions, where temperature and duration are vital for generating precise and reproducible results.

Freeze-drying or lyophilization is also one of the methods of removing moisture from foods containing a high level of moisture. The principle involved in freeze-drying is sublimation, where water passes directly from the solid state (ice) to the vapor state without passing through the liquid state under reduced pressure and temperature conditions.

Moisture can also be determined by employing microwave drying. In this method the moisture is evaporated from the sample by using microwave energy (600 W) for 10 minutes. The loss in weight is determined by electronic balance readings before and after drying and is converted to moisture percentage.

Moisture in a meat sample can also be determined more rapidly using a moisture analyzer that is based on the thermogravimetric method. Here the meat samples are heated rapidly by the absorption of infrared energy. The most important advantage is the rapid measurement time of this method. Results can be obtained within a short span of $2-10$ minutes. Samples are heated quickly and evenly, and the obtained measurements show good repeatability. Handling is also straightforward and the risk of error is reduced. However, all the thermogravimetric methods, including the moisture analyzer, carry the risk of decomposing constituents or the loss of volatile components during heating. This results in a further decrease in weight, which is not explained by the release of water.

2.4.1.2 Chemical analysis

The Karl Fischer titration is a widely used analytical method for quantifying water content in a variety of products. It is based on the reduction of iodine by sulfur dioxide in an aqueous medium. The main reaction of the method is

$$2H_2O + SO_2 + I_2 = H_2SO_4 + 2HI$$

Karl Fischer modified the reaction for the determination of water in a nonaqueous system containing an excess of sulfur dioxide (Fischer, 1935). He used methanol as the solvent, and pyridine as the buffering agent. The alcohol reacts with sulfur dioxide (SO_2) and base to form an intermediate alkylsulfite salt, which is then oxidized by iodine to an alkylsulfate salt. This oxidation reaction consumes water. Classic Karl Fisher reagents contained pyridine (toxic carcinogen) as the base. The Karl Fischer method has become a standard method for the determination of the moisture of liquid and solids due to its selectivity, high precision, and rapidity (Pyper, 1985). It is especially useful for the determination of moisture in foods for which heating methods give erratic results (Pomeranz and Meloan, 1994; Bradley, 1998). The moisture content is calculated from the amount of titrant consumed, and is often expressed in milligrams which can be converted into percentage moisture using the initial sample mass.

For this method of moisture determination, the apparatus consists of a glass burette (automatic filling type), a titration vessel with an agitation device (pressurized with dry inert gas to exclude air), and an electrometric apparatus and galvanometer to record the end point of the titration. The reagents consist of methanol and Karl Fischer reagent which comprises iodine in methanol and sulfur dioxide in pyridine (these reagents are mixed before use). The sample is first weighed into a predried 50 mL round-bottomed flask, then 40 mL methanol is added and the flask is placed on a heater and then connected to a reflux condenser. The content is boiled gently under reflux for 15 minutes. After that heating is stopped and the condenser is allowed to drain for 15 minutes. Then the flask is removed from the heater and stoppered. A volume of 10 mL of the aliquot is pipetted into the titration vessel and titrated with the Karl Fischer reagent until the end point is reached. A blank flask is also run following the above procedure.

2.4.1.3 Spectroscopic analysis

These are indirect methods for moisture determination in meat samples. Spectroscopic methods utilizes the interaction of electromagnetic radiation with materials to obtain information about their composition, for example, X-rays, UV-visible, NMR, microwaves, and infrared. Common spectroscopic methods, include refractometry, infrared absorption spectroscopy, and near-infrared reflectance spectroscopy.

Refractometry is an optical method measuring the refractive index (RI) of a solution, which can be used for determining its moisture content. The moisture content can be rapidly determined by measuring the RI of a solution or semiliquid mixture using a calibration curve (Pomeranz and Meloan, 1994; Torkler, 1990). The sample for which moisture has to be determined is homogenized with an anhydrous solvent and RI of the solution is measured by using a refractrometer. Then a calibration curve is plotted by measuring the RI of solutions containing the same solvent with a known amount of added water. The moisture content in the sample is determined from the calibration curve.

In contrast to the conventional methods which are used to determine the physical and chemical composition of meat, near-infrared spectroscopy (NIRS) is a sensitive, expedient, simple, safe, and nondestructive method for the simultaneous determination of several parameters in meat samples (Tao et al., 2013). NIRS technology uses its high-resolving power of reflectance spectra in the near-infrared (NIR) range (800–2500 nm) as an analytical tool for components analysis. The mid-infrared (IR) range (2500–24,000 nm) has high resolution in the absorption spectrum and can absorb IR radiation effectively from many compounds, but the resolution of the reflectance spectrum is poor (Park, 1981). The NIRS is used to determine the basic components of meat like proteins, fat, water (moisture), and dry matter, as well as sensory properties (Alomar et al., 2003; Ripoll et al., 2008). The successful definition of calibration methods depends on the variability of the analyzed samples. If the range of reference values for the definition of calibration models is too narrow, this may have a negative impact on the predictive value of this method (Su et al., 2014).

2.4.2 Determination of protein

Protein is the main component in meat that contains nitrogen, and the nitrogen content of meat is roughly constant. Therefore the protein content of meat is determined on the basis of total nitrogen content, with the Kjeldahl method being almost universally applied to determine nitrogen content. Nitrogen content is then multiplied by a factor to give the protein content. This approach is based on two assumptions: that dietary carbohydrates and fats do not contain nitrogen and nitrogen recovered during digestion is mainly amino-nitrogen from proteins (total organic nitrogen) and that the contribution of inorganic nitrogen (nitrate, nitrite, ammonium) or other organic nitrogen (nucleotides, nucleic acids) is negligible. The average nitrogen content of proteins has been found to be about 16%, which led to the use of the calculation $N \times 6.25$ ($1/0.16 = 6.25$) to convert nitrogen content into protein content. The factor 6.25 is also used to convert total nitrogen in meat to the total protein content of meat (Benedict, 1987). The Kjeldahl method consists of a digestion step where nitrogen is converted into ammonium (NH_4^+) and an analytical step where NH_4^+ is quantified by titrimetry, colorimetry, or by using an ion-specific electrode.

Dumas' (nitrogen combustion) method was introduced in 1831 by Jean-Baptiste Dumas. In this method meat samples are combusted at high temperatures (700°C–1000°C) with a flow of pure oxygen. All carbon in the sample is converted to carbon dioxide during the flash combustion. Nitrogen-containing components produced include N_2 and nitrogen oxides. The nitrogen oxides are reduced to nitrogen in a copper reduction column at a high temperature (600°C). The total nitrogen (including nitrate and nitrite) released is carried by pure helium and quantified by gas chromatography using a thermal conductivity detector (Sweeney and

Rexroad, 1987; Jones, 1991). Ultrahigh purity acetanilide and EDTA (ethylenediamine tetraacetate) are used as the standards for the calibration of the nitrogen analyzer. The nitrogen determined is converted to protein content in the sample using a protein conversion factor. The combustion method is an alternative to the Kjeldahl method and is suitable for all types of foods including meat. AOAC method 992.15 is used for the determination of the nitrogen content of meat samples using Dumas method.

The nitrogen content of the meat samples can also be determined by infrared spectroscopy. It measures the absorption of radiation (near-infrared regions) by molecules in meat samples. Different functional groups present in meat absorb different frequencies of radiation. For proteins and peptides, various mid-infrared bands (6.47 μm) and NIR bands (3300−3500 nm; 2080−2220 nm; 1560−1670 nm) characteristic of the peptide bond can be used to estimate the protein content of a food including meat (AOAC Method 997.06). By irradiating a sample with a wavelength of infrared light specific for the constituent to be measured, it is possible to predict the concentration of that constituent by measuring the energy that is reflected or transmitted by the sample (which is inversely proportional to the energy absorbed) (O'Sullivan et al., 1999). NIRS is applicable to a wide range of food products including meat and dairy products (AOAC, 2007; Luinge et al., 1993; Krishnan et al., 1994). Due care must be taken while interpreting the results, and calibration of the equipment is essential for precise results. The advantage of this method is that the sample can be analyzed rapidly.

2.4.3 Determination of amino acids

The methodology for analysis of amino acids in meat samples involves three distinct stages: extraction, deproteinization, and analysis. The extraction consists of the separation of the free amino acid fraction from the insoluble portion of the matrix, in this case from the muscle. It is usually achieved by homogenization of the ground sample in an appropriate solvent. The extraction solvent can be hot water, 0.01−0.1 N hydrochloric acid solution, or diluted phosphate buffers. Once homogenized, the solution is centrifuged at least at 10,000g under refrigeration to separate the supernatant from the nonextracted materials (pellet) and filtered through glass wool to retain any fat material remaining on the surface of the supernatant.

The deproteinization process can be achieved through different chemical or physical procedures. Chemical methods include the use of concentrated strong acids such as sulfosalicylic, perchloric, trichloroacetic, picric, or phosphotungstic acids, or organic solvents such as methanol, ethanol, or acetonitrile. Under these conditions, proteins are precipitated by denaturation, whereas free amino acids remain in solution. After sample preparation, the amino acids can be analyzed by any of the methods such as direct spectrophotometric or by chromatographic (high-performance liquid chromatography, HPLC or gas−liquid chromatography) methods. For HPLC separation, amino acids are derivatized to allow their separation or to enhance their detection. Common HPLC separation techniques used are cation-exchange HPLC and reverse-phase HPLC. Cation-exchange HPLC is used for the separation of nonderivatized amino acids, which are then derivatized postcolumn (ninhydrin), whereas reverse-phase HPLC is mainly used to separate precolumn derivatized amino acids (Aristoy and Toldra, 2009).

2.4.4 Determination of lipids

The total lipid content of meat is commonly determined by organic solvent extraction methods or by alkaline or acid hydrolysis followed by Mojonnier extraction. For multicomponent food products, acid hydrolysis is often the method of choice. Both acid hydrolysis and alkaline hydrolysis methods can be performed using Mojonnier extraction equipment (Srigley and Mossoba, 2017). The use of acid hydrolysis for fat estimation eliminates some of the matrix effects that may be exhibited by simple solvent extraction methods. The accuracy of direct solvent extraction methods greatly depends on the solubility of the lipids in the solvent used and the ability to separate the lipids from complexes with other macromolecules. The lipid content of a food determined by extraction with one solvent may be quite different from the content determined with another solvent of different polarity. In addition to solvent extraction methods, there are nonsolvent wet extraction methods and several instrumental methods that utilize the physical and chemical properties of lipids in foods for fat content determination (Min and Ellefson, 2010).

There are basically two main methods to evaluate the fat content, a method based on Soxhlet extraction with or without previous acid hydrolysis and petroleum ether and a method based on the Folch method (Folch et al., 1957)—extracting the fat with a mixture of chloroform and methanol. The Soxhlet method for fat determination in meat (AOAC method 960.39) is an example of the semicontinuous method of extraction. In this method the solvent is accumulated in the extraction chamber for 5–10 minutes and completely surrounds the sample and then is siphoned back to the boiling flask. Fat content is measured by the weight loss of the sample or by weight of the fat removed. This method provides a soaking effect of the sample and does not cause channeling. However, this method requires more time than the continuous method.

The chloroform–methanol procedure is another method to extract lipids from meat samples. The Folch extraction (Folch et al., 1957) is applied to small samples, while the Bligh and Dyer extraction (Bligh and Dyer, 1959) is applied to large samples of high moisture content. Both utilize this combination of solvents to recover lipids from different food samples including meat.

In a modified extraction process, water is replaced with 0.88% potassium chloride aqueous solution which creates two phases (Christie, 1982). In both the modified Folch extraction and Bligh and Dyer procedure, meat samples are mixed/homogenized in a chloroform−methanol solution, and the homogenized mixture is filtered into a collection tube. A 0.88% potassium chloride aqueous solution is added to the chloroform−methanol mixture containing the extracted fats. This causes the solution to break into two phases: the aqueous phase (top) and the chloroform phase containing the lipid (bottom). The phases are further separated in a separating funnel or by centrifugation. After evaporation of the chloroform, the fat is determined by weight. The modified methanol−chloroform extraction procedures are rapid, well-suited to low-fat samples, and can be used to generate lipid samples for subsequent fatty acids compositional analysis of meat through gas chromatography.

2.4.5 Determination of fatty acids

The analysis of fatty acids in meat involves three basic steps: (1) lipid extraction; (2) preparation of fatty acid methyl ester (FAME); and (3) gas chromatographic (GC) analysis. The critical step in GC analysis of fatty acids is methylation of the fatty acids to get FAMEs. Many different methylation methods are described in the literature but the most commonly used are those catalyzed by an acid, base, or boron trifluoride and methylation with diazomethane, each of which have advantages and disadvantages and differ in their applicable range. In this chapter a direct method for FAMEs synthesis in muscle tissue is discussed briefly (O'Fallon et al., 2007).

Appropriate quantity of meat sample (1 g fresh or 0.5 g dry or semifrozen meat sample) are cut it into small pieces and 200 mg tissue samples are taken in a 50 mL tube, to which 1.0 mL of internal standard (tridecanoic acid, C13:0), 0.7 mL 10 N KOH, and 5.3 mL methanol are added. The tube containing the sample is incubated in a water bath at 55°C for 1.5 hours with shaking at intervals of 20 minutes for 5 seconds so as to permeate, dissolve, and hydrolyze the samples. After incubation the tubes are cooled under running tap water. This is followed by the addition of 0.58 mL 24 N H_2SO_4 and proper mixing of the sample. The tube containing the sample is again incubated at 55°C in a water bath for 1.5 hours with shaking as described earlier. The above process results in FAME synthesis. After the tube is cooled, 3.0 mL of hexane is added and mixed by vortex. The tube is then centrifuged at 5000 rpm for 5 minutes and the hexane layer containing the FAME is employed for gas chromatographic analysis (O'Fallon et al., 2007).

2.4.6 Determination of ash content and minerals

Ash refers to the inorganic residue remaining after either ignition or complete oxidation of organic matter in a meat sample. Two major types of ashing are used: dry ashing, primarily for proximate composition and for some types of specific mineral analyses; and wet ashing (oxidation), as a preparation for the analysis of certain minerals (Jones, 2001; Enders and Lehmann, 2012). A basic knowledge of the characteristics of various ashing procedures and types of equipment is essential to ensure reliable results. In dry ashing the sample is oxidized in a muffle furnace at 500°C–550°C. Water and volatile substances present in meat samples are vaporized, and organic substances are burned in the presence of oxygen in air to CO_2 and oxides of N_2. Most minerals are converted to oxides, sulfates, phosphates, chlorides, and silicates. Elements such as Fe, Se, Pb, and Hg may partially volatilize with this procedure, so other methods must be used if ashing is a preliminary step for specific elemental analysis.

In wet ashing the organic substances present in meat are oxidized by using acids and oxidizing agents or their combinations. Sometimes the use of a single acid in wet ashing does not give the complete and rapid oxidation of organic material, so a mixture of acids are often used. Combinations of the following acid solutions are used frequently: (1) nitric acid, (2) sulfuric acid–hydrogen peroxide, and (3) perchloric acid. Different combinations of the acids are recommended for different types of samples. In this method of digestion/ashing minerals are solubilized without volatilization. Wet ashing often is preferable to dry ashing as a preparation for specific (macro- and micro-) mineral analysis.

After the digestion of the meat sample by a suitable method, macro- (calcium, phosphorus, magnesium, sodium, and potassium), micro- (iron, zinc, copper, manganese, selenium, and chromium) and potentially toxic trace (cadmium, lead, aluminum, arsenic, and mercury) minerals can determined by different methods such as spectrophotometry, flame atomic absorption spectrometry, flame atomic emission spectrometry, hydride generation atomic absorption spectrometry, graphite-furnace atomic absorption spectrometry, cold vapor atomic absorption spectrometry, inductively coupled plasma emission spectrometry, neutron activation analysis, anodic stripping voltammetry, and inductively coupled plasma mass spectrometry.

2.4.7 Determination of vitamins

A vitamin is defined as an organic compound and a vital nutrient that an organism requires in limited amounts. Not all the vitamins are always synthesized in adequate quantities in the body, thus they must be obtained through the diet. There are two types of vitamins: fat-soluble and water-soluble. Fat-soluble vitamins are stored in fat cells, consequently requiring fat in order to be absorbed. Fat-soluble vitamins are vitamin A, D, E, and K. Water-soluble vitamins on the other hand are not stored in the body and therefore need to be replenished daily.

Meat is a major source of five of the B complex vitamins: thiamin, riboflavin, niacin, vitamin B6, and vitamin B12. Meat is not a good source of folacin but it does contain biotin and pantothenic acid. The B vitamins are found in a wide variety of meats and other foods.

Most meat is a very good source of thiamin, pork, in recommended serving sizes, provides more thiamin than any other food commonly eaten. Liver is the best food source of riboflavin. Meat is also a good source of niacin and tryptophan. Veal, liver, beef, and lamb are high in vitamin B_{12}. Liver is also a good source of the fat-soluble vitamins D and K. However, meat is not an important source of vitamin E and with the exception of liver is not a particularly good source of fat-soluble vitamins.

Sample preparation (extraction and purification) is very important for the determination of vitamins in biological samples, which is a very complex process. For solid samples like meat the grinding homogeneity of the samples is very important. Different extraction methods, such as protein precipitation centrifugation and filtration, ultrasonic-assisted extraction (UAE), liquid–liquid extraction (LLE), dispersive liquid–liquid microextraction (DLLME), solid-phase extraction (SPE), and supercritical fluid extraction (SFE) methods, are employed for the extraction of vitamins from biological samples. Among all the sample preparation methods, reflux, UAE, and SFE are preferred for solid samples, while for liquid samples LLE, SPE, and DLLME are preferred (Zhang et al., 2018). Reflux extraction methods are traditional methods involving the consumption of large amounts of organic solvents and extraction time. High extraction efficiency can be obtained with SFE, but expensive instruments are required in comparison to UAE. Considering the column passing operation, methods like SPE can be complicated. However, multiple samples can be prepared simultaneously by SPE; thus the total time required can be greatly saved. Moreover, for SPE, it can be coupled with liquid chromatography (LC) to achieve online analysis. Different analysis methods are employed for estimating the vitamin content of meat. Common methods such as liquid chromatography (LC coupled with mass spectrometry and multiclass analysis, HPLC coupled with other techniques such as ultraviolet, photodiode array), electrophoretic methods (capillary electrophoresis, micellar electrokinetic chromatography), microbiological assay, biosensors, and spectrometry (fluorescence spectrometry, NIRS) have been widely used for the determination of vitamins in meat samples (Zhang et al., 2018).

2.5 Conclusion

Meat as a food for humans provides essential nutrients like protein, fat, minerals, and vitamins, which are essential for the growth and maintenance of body functions. To provide a quality food for humans, animal products such as meat and meat products need to maintain very high standards in terms of essential nutrients. Proper analytical methods are required to assess the nutritional quality of meat. A sound knowledge of the advances in the analytical methods will enable us to determine accurately the various nutritional parameters for improving the quality of meat and meat products for human consumption.

References

Abbas, K.A., Mohamed, A., Jamilah, B., 2009. Fatty acids in fish and beef and their nutritional values: a review. J. Food Agric. Environ. 7, 37–42.

Alomar, D., Gallo, C., Castañeda, M., Fuchslocher, R., 2003. Chemical and discriminant analysis of bovine meat by near infrared reflectance spectroscopy (NIRS). Meat Sci. 63, 441–450.

AMSA, 2016. AMSA Meat Science Lexicon. American Meat Science Association, Chicago, IL.

AOAC, 2007. Official Methods of Analysis, eighteenth ed. AOAC International, Gaithersburg, MD, 2005; Current through revision 2, 2007 (online).

Aristoy, M.C., Toldra, F., 2009. Essential amino acids. In: Nollet, L.M.L., Toldra, F. (Eds.), Handbook of Processed Meats and Poultry Analysis. CRC Press, Taylor & Francis Group, pp. 215–227.

Astruc, T., 2014. Muscle fiber types and meat quality. In: Dikeman, C.D.M. (Ed.), Encyclopaedia of Meat Sciences, second ed. Elsevier, Oxford, pp. 442–448.

Belury, M.A., 2002. Dietary conjugated linoleic acid in health: physiological effects and mechanisms of action. Annu. Rev. Nutr. 22, 505–531.

Bender, A., 1992. Meat and Meat Products in Human Nutrition in Developing Countries, 53. FAO Food and Nutrition Paper, Food and Agriculture Organization, Rome, pp. 1–91.

Benedict, R.C., 1987. Determination of nitrogen and protein content of meat and meat products. J. Assoc. Off. Anal. Chem. 70, 69–74.

Beserra, F.J., Madruga, M.S., Leite, A.M., Silva, E.M.C., Maia, E.L., 2004. Effect of age at slaughter on chemical composition of meat from Moxotó goats and their crosses. Small Ruminant Res. 55, 177–181.

Bligh, E.G., Dyer, W.J., 1959. A rapid method of total lipid extraction and purification. Can. J. Biochem. Physiol. 37, 911–917.

Bolte, M.R., Hess, B.W., Means, W.J., Moss, G.E., Rule, D.C., 2002. Feeding lambs high-oleate or high-linoleate safflower seeds differentially influences carcass fatty acid composition. J. Anim. Sci. 80, 609–616.

Bradley Jr., R.L., 1998. Moisture and total solid analysis. In: Nielsen, S.S. (Ed.), Food Analysis, second ed Aspen Publishers, Gaithersburg, MD, pp. 119–139.

Chan, K.M., Decker, E., 1994. Endogenous skeletal muscle antioxidants. Crit. Rev. Food Sci. Nutr. 34, 403–426.

Christie, W.W., 1982. Lipid Analysis. Isolation, Separation, Identification, and Structural Analysis of Lipids, second ed. Pergamon, Oxford.

Dave, D., Ghaly, A.E., 2011. Meat spoilage mechanisms and preservation techniques: a critical review. Am. J. Agric. Biol. Sci. 6, 486–510.

Enders, A., Lehmann, J., 2012. Comparison of wet-digestion and dry-ashing methods for total elemental analysis of biochar. Comm. Soil Sci. Plant Anal. 43, 1042–1052.

Fischer, K., 1935. A new method for the analytical determination of the water content of liquids and solids. Angew. Chem. Int. 48, 394–396.

Folch, J., Lees, M., Stanley, G.H.S., 1957. A simple method for the isolation and purification of total lipides from animal tissues. J. Biol. Chem. 226, 497–509.

Heinz, G., Hautzinger, P., 2007. Meat Processing Technology for Small to Medium Scale Producers. RAP Publication 2007/20. FAO, Bangkok.

Irshad, A., Kandeepan, G., Kumar, S., Ashish Kumar, A., Vishnuraj, M.R., Shukla, V., 2012. Factors influencing carcass composition of livestock: a review. J. Anim. Prod. Adv. 3, 177–186.

Jones Jr., J.B., 1991. Kjeldahl Methods for Nitrogen Determination. Micro-Macro Publishing, Inc, Athens, Georgia, USA, p. 79.

Jones, J.B., 2001. Laboratory Guide for Conducting Soil Tests and Plant Analysis. CRC Press, Boca Raton, FL.

Krishnan, P.G., Park, W.J., Kephart, K.D., Reeves, D.L., Yarrow, G.L., 1994. Measurement and protein and oil content of oat cultivars using near-infrared reflectance. Cereal Foods World 39, 105–108.

Listrat, A., Lebret, B., Louveau, I., et al., 2016. How muscle structure and composition influence meat and flesh quality. Sci. World J. 2016, Article ID 3182746.

Lourenço, M., Ramos-Morales, E., Wallace, R., 2010. The role of microbes in rumen lipolysis and biohydrogenation and their manipulation. Animal 4, 1008–1023.

Luinge, H.J., Hop, E., Lutz, E.T.G., van Hemert, J.A., de Jong, E.A.M., 1993. Determination of the fat, protein and lactose content of milk using Fourier transform infrared spectrometry. Anal. Chim. Acta 284, 419–433.

Mancini, R.A., 2009. Meat color. In: Kerry, J.P., Ledward, D. (Eds.), Improving the Sensory and Nutritional Quality of Fresh Meat. Woodhead Publishing, pp. 89–110.

Menotti, A., Kromhout, D., Blackburn, H., et al., 1999. Food intake patterns and 25-year mortality from coronary heart disease: cross-cultural correlations in the seven countries study. The Seven Countries Study Research Group. Eur. J. Epidemiol. 15, 507–515.

Min, D.B., Ellefson, W.C., 2010. Fat analysis. In: Nielsen, S.S. (Ed.), Food Analysis. Springer, Boston, MA, pp. 117–132.

Muhlisin, Utama, D.T., Lee, J.H., Choi, J.H., Lee, S.K., 2016. Antioxidant enzyme activity, iron content and lipid oxidation of raw and cooked meat of Korean native chickens and other poultry. Asian-Australas. J. Anim. Sci. 29, 695–701.

Naveena, B.M., Kiran, M., Sudhakar Reddy, K., et al., 2011a. Effect of ammonium hydroxide on ultrastructure and tenderness of buffalo meat. Meat Sci. 88, 727–732.

Naveena, B.M., Sen, A.R., Muthukumar, M., Babji, Y., Kondaiah, N., 2011b. Effects of salt and ammonium hydroxide on the quality of ground buffalo meat. Meat Sci. 87, 315–320.

Olsson, V., Pickova, J., 2005. The influence of production systems on meat quality, with emphasis on pork. Ambio 34, 338–343.

O'Fallon, J.V., Busboom, J.R., Nelson, M.L., Gaskins, C.T., 2007. A direct method for fatty acid methyl ester synthesis: application to wet meat tissues, oils, and feedstuffs. J. Anim. Sci. 85, 1511–1521.

O'Sullivan, A., O'Connor, B., Kelly, A., McGrath, M.J., 1999. The use of chemical and infrared methods for analysis of milk and dairy products. Int. J. Dairy Technol. 52, 139–148.

Park, Y.W., 1981. A Study on Predicting Carotene, Nitrate, Soluble N and Fibrous Fractions in Forages and Vegetables With Near-Infrared-Reflectance Spectroscopy, and Effects of Processing Methods on Carotene Contents (Ph.D. dissertation). Utah State University, Logan, UT.

Pomeranz, Y., Meloan, C.E., 1994. Determination of moisture, Food Analysis: Theory and Practice, third ed. Chapman and Hall, New York, pp. 575–600.

Pyper, J.W., 1985. The determination of moisture in solids: a selective review. Anal. Chim. Acta 170, 159–175.

Ripoll, G., Albertí, P., Panea, B., Olleta, J.L., Sañudo, C., 2008. Near-infrared reflectance spectroscopy for predicting chemical, instrumental and sensory quality of beef. Meat Sci. 80, 697–702.

Smet, S.D., Raes, K., Demeyer, D., 2004. Meat fatty acid composition as affected by fatness and genetic factors: a review. Anim. Res. 53, 81–98.

Soren, N.M., Sastry, V.R.B., Saha, S.K., Mendiratta, S.K., 2008. Effect of feeding processed karanj (*Pongamia glabra*) cake on carcass characteristics and meat sensory attributes of fattening lambs. Ind. J. Anim. Sci. 78, 858–862.

Srigley, C.T., Mossoba, M.M., 2017. Current Analytical Techniques for Food Lipids. Food and Drug Administration, Papers. 7. <http://digitalcommons.unl.edu/usfda/7>.

Su, H., Sha, K., Zhang, L., Zhang, Q., et al., 2014. Development of near infrared reflectance spectroscopy to predict chemical composition with a wide range of variability in beef. Meat Sci. 98, 110–114.

Sweeney, R.A., Rexroad, P.R., 1987. Comparison of LECO FP-228 'Nitrogen Determinator' with AOAC copper catalyst Kjeldahl method for crude protein. J. AOAC Int. 70, 1028–1030.

Tao, L.L., Yang, X.J., Deng, J.M., Zhang, X., 2013. Application of near infrared reflectance spectroscopy to predict meat chemical compositions: a review. Spectros. Spec. Anal. 33, 3008–3015.

Tariq, M.M., Eyduran, E., Rafeeq, M., et al., 2013. Influence of slaughtering age on chemical composition of mengali sheep meat at Quetta, Pakistan. Pak. J. Zool. 45, 235–239.

Torkler, K.H., 1990. Rapid determination equipment for food quality control. In: Baltes, W. (Ed.), Rapid Methods for Analysis of Food and Food Raw Material. Technomic Publishing, Lancaster, PA, pp. 59–71.

Verma, A.K., Banerjee, 2010. Dietary fiber as functional ingredient in meat products: a novel approach for healthy living – a review. J. Food Sci. Technol. 47, 247–257.

Vongsawasdi, P., Noomhorm, A., 2014. Bioactive compounds in meat and their functions. In: Noomhorm, A., Ahmad, I., Anal, A.K. (Eds.), Functional Foods and Dietary Supplements: Processing Effects and Health Benefits. Wiley-Blackwell Publisher, pp. 113–138.

Warner, R.D., 2017. The eating quality of meat—IV water-holding capacity and juiciness. In: Toldrá, F. (Ed.), Woodhead Publishing Series in Food Science, Technology and Nutrition, eighth ed. Woodhead Publishing, pp. 419–459. , Lawriés Meat Science.

WHO, 2002. The World Health Report 2002. Reducing Risks, Promoting Healthy Life. World Health Organization, Geneva.

Williams, P., 2007. Nutritional composition of red meat. Nutr. Diet. 64 (Suppl. 4), S113–S119.

Zhang, Y., Zhou, W.E., Yan, J.Q., et al., 2018. A review of the extraction and determination methods of thirteen essential vitamins to the human body: an update from 2010. Molecules 23, 1484.

CHAPTER 3

Nondestructive methods for carcass and meat quality evaluation

K. Narsaiah[1], Ashim Kumar Biswas[2] and Prabhat Kumar Mandal[3]

[1]Central Institute of Post-harvest Engineering and Technology, Ludhiana, India [2]Division of Post-Harvest Technology, ICAR-Central Avian Research Institute, Izatnagar, Bareilly, India [3]Department of Livestock Products Technology, Rajiv Gandhi Institute of Veterinary Education and Research, Puducherry, India

Chapter Outline

3.1 Introduction 37
3.2 Emerging nondestructive methods 38
    3.2.1 Optical methods 38
    3.2.2 Near-infrared spectroscopy 40
    3.2.3 Nuclear magnetic resonance and magnetic resonance imaging 41
    3.2.4 Electronic nose (e-nose) 42
    3.2.5 X-ray and computed tomography 43
    3.2.6 Hyperspectral imaging 44
    3.2.7 Electrical properties 44
    3.2.8 Acoustic methods 46
3.3 Future research needs 46
3.4 Conclusion 47
References 47
Further reading 49

3.1 Introduction

Quality control and monitoring are essential parts of any food industry. Consumers' expectations for lower prices and consistent quality necessitate the need for the development of reliable instruments for both assessing quality and pricing. Meat quality is mainly affected by different processes that occur during the growth of the animal and after slaughter. The color, tenderness, juiciness, and flavor of meat are important factors that affect a consumer's evaluation of meat quality and influence their decision relative to making a repeated purchase. The objective of determining the meat quality is to offer to the consumer wholesome, tasty, and safe meat at a reasonable price. Evaluation of meat quality is also critical for the preparation of good quality meat products. The key elements of success for any evaluation technique in the meat industry are the existence of a real need and an assured benefit, a direct relation to the desired quality traits in the end product, reasonable prediction accuracy, realistic cost, rapidity in order to comply with slaughter, cutting, or packing rates, potential of full automation, and noninvasiveness (Monin, 1998).

The current trend in the monitoring of meat quality is to move the measurements of quality from the laboratories to the processing lines. Different techniques and methodologies based on different principles, procedures, and/or instruments are currently available for measuring different meat quality attributes (ElMasry et al., 2012). However, this must not sacrifice the essential benefits of human grading, that is, intuition. To satisfy the increased awareness, sophistication, and greater expectation of consumers, it is necessary to improve automated quality inspection.

A general definition of nondestructive testing is the evaluation performed on any object, for example, meat cuts, without changing or altering that object in any way, in order to determine the absence or presence of conditions that may have an effect on certain characteristics (Hellier, 2003), for example, quality. Nondestructive quality evaluation has been a subject of interest to researchers for many years (Chen and Sun, 1991) and has seen significant growth. It can be considered one of the fastest growing technologies from the standpoint of uniqueness and innovation. The sophistication of nondestructive methods has evolved rapidly with modern technologies (Mix, 2005). The foremost underlying drivers for using nondestructive technologies are automation and improved rapid operations. Certain visual characteristics that are commonly used to describe quality are size, shape, color, texture, and appearance. These quality characteristics are linked to features that are measurable by nondestructive techniques (Becker, 2002).

Nondestructive methods for determining composition and quality include color measurement, computer image processing, visual and NIR spectrometry, hyperspectral imaging, X-ray imaging, ultrasound, nuclear magnetic resonance imaging (NMRI), e-nose, and biosensors. These methods have the advantage of being nondestructive, fast, inexpensive (after development), and are considered suitable for online determination of many parameters simultaneously. Some of the image processing techniques under use are summarized in Table 3.1.

3.2 Emerging nondestructive methods

3.2.1 Optical methods

These methods include computer/machine vision and image analysis and color measurement using colorimeter. The techniques based on computer/machine vision and image analysis

Table 3.1: Emerging image processing techniques for quality monitoring of meat.

Methods	Application	Reference
Ultrasound imaging	Foreign object detection	Cho and Irudayaraj (2003), Pallav et al. (2009)
X-ray imaging	Bone fragment, bone mineral density	Korver et al. (2004)
Magnetic resonance imaging	Moisture and structure changes	Shaarani et al. (2006)
Fluorescence imaging	Study the fat distribution, Contamination detection, Measurement of lipid oxidation,	Adedeji et al. (2011) Cho et al. (2009) Wold and Kvaal (2000)
Hyperspectral imaging	Wholesomeness, moisture, contaminant detection, bone fragment, skin tumor detection, springiness	Chao et al. (2010), Yang et al. (2010), Kandpal et al. (2013)
	Microbial spoilage	Feng and Sun (2013a,b)
Thermal imaging	Monitoring of doneness, measuring skin temperature	Ibarra et al. (2000)

Modified from: Xiong, Z., Sun, D.W., Pu, H., Gao W., Dai, Q., 2017. Applications of emerging imaging techniques for meat quality and safety detection and evaluation: a review. Crit. Rev. Food Sci. Nutr. 57, 755–768 (Xiong et al., 2017).

can be used to assess meat quality, and grade it based on its appearance. A computer vision system generally consists of five basic components: illumination, a camera, an image capture board (frame grabber or digitizer), computer hardware, and software. A well-designed illumination system is a prerequisite for the success of the image analysis by enhancing image contrast. The cameras used in machine vision are usually based on solid-state charged-coupled device camera technology. The digitizer divides the image into a two-dimensional grid of small regions or pixels. Computer hardware and software is used for image processing and image analysis. The rapid technological advancements in digital cameras and smartphones are expected to pave the way for the development of cost-effective technologies for grading meat based on appearance.

The other simple and cost-effective optical method is the use of a colorimeter. Among the properties widely used for analytical evaluation of any material, color is unique in several aspects. While every material can be said to possess a specific property such as mass, no material is actually colored as such. Color is primarily an appearance property attributed to the spectral distribution of light and, in a way, is related to some source of radiant energy (the illuminant), to the object to which the color is ascribed, and to the eye of the observer. Without light or the illuminant, color does not exist. The property of an object that gives it a characteristic color is its light-absorptive capacity. Various constituents of food products can absorb a certain amount of this radiation. Absorption varies with the constituents, wavelength, and path length of the light. Reflection is a complex action involving several physical phenomena. Depending on how light is reflected back after striking an object, reflection may be defined as regular or specular reflection and diffused reflection. The absorptive and reflective characteristics are used to correlate with the internal quality parameters of meat.

The color is quantified as CIE tristimulus values (red, green, and blue). The tristimulus values indicate the amount of red, green, and blue needed to form a specific color. The other color scales include the Munsell system; Hunter *L, a, b*; CIE *L*, a*, b**; and color Atlases and charts. Hunter *L, a, b* and CIE *L*, a*, b** are widely used for assessing the quality of meat.

3.2.2 Near-infrared spectroscopy

The use of near-infrared (NIR) and visual spectroscopy is rapid and often nondestructive for measuring the composition of biological materials. It works on the principle of absorption, reflection, transmission, and/or scattering of light in or through a food material following the Beer–Lambert law. Now various NIR spectrometers are available and are being used commercially. The absorption or reflectance of light in a known range of wavelengths is measured and correlated with various quality parameters of the food material. The configuration of the NIR spectrometer is shown in transmittance mode in Fig. 3.1 and in reflectance mode in Fig. 3.2. This configuration is the same for the majority of spectrometers measuring changes in radiation energy. NIR spectroscopy in conjunction with chemometrics can be used to determine all constituents (proteins, fat, sugars, etc.) of food products simultaneously.

The basis of ultraviolet, visible, and NIR (UV-vis-NIR) spectroscopy is the interaction of radiant energy with molecules of samples. These interactions in terms of absorption, reflection, transmission, and/or scattering of radiation in or through a sample can be quantified according to the Beer–Lambart law. The simplified configuration of spectroscopy is shown in transmission mode in Fig. 3.1 and in reflectance mode in Fig. 3.2.

NIR and visual spectroscopy is one of widely used nondestructive methods for measuring the composition of meat. The absorption or reflectance of radiation by meat in the NIR region is often complex and normally possesses broad overlapping NIR absorption bands. Multivariate data analysis is used to filter the enormous spectral data in order to correlate a certain property with specific relevant spectral data. Further details on the theory of NIR are given by Jha and Garg (2010).

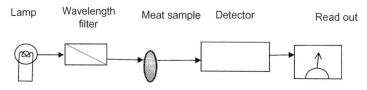

Figure 3.1
Configuration of NIR spectrometer in transmittance mode.

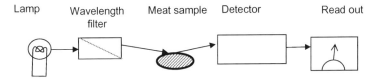

Figure 3.2
Configuration of NIR spectrometer in reflectance mode.

3.2.3 Nuclear magnetic resonance and magnetic resonance imaging

The nuclear magnetic resonance (NMR) technique, often referred to as magnetic resonance imaging (MRI), involves resonant magnetic energy absorption by nuclei placed in an alternating magnetic field. Spin is a fundamental property of matter, like electrical charge or mass. Certain nuclei (spin quantum number $i \neq 0$) have a magnetic moment and align along a strong static magnetic field. MRI thus works on the principle of resonant magnetic energy absorption by nuclei placed in an alternating magnetic field. The amount of energy absorbed by the nuclei is directly proportional to the number of a particular nucleus in the sample such as the protons in water and oil. A detailed theory of NMR is presented elsewhere by Ruan and Chen (2001). Information on experimentation, assembling hardware, conducting laboratory tests, and interpreting the results is also available from Fukushima and Roeder (1981). These authors also provided a detailed theory for a better understanding of what a scientist should seek and what she might expect to find out by using NMR. This method can be used for the noninvasive quality evaluation of many foods.

There are many applications of NMR in agriculture. The simplest among them is the determination of moisture and oil content. But the NMR response many times is not clear and poses problems, especially when constituents other than water are present in the material. Besides the established relationship between the moisture and the output of NMR experiments, various other facts that are helpful in determining the quality of food materials without destroying them are available in the literature. MRI was used for measuring body composition and visualizing distribution in meat and meat products, and in addition it has capabilities for monitoring salt diffusion and water mobility in meat during brine curing. Monitoring the cooking process by the use of magnetic resonance imaging is another important application. It also provides structural information on muscle tissue by utilizing diffusion tensor imaging, which is conducted by measuring diffusion coefficients in at least six directions.

To increase the marketability a longer shelf life is needed and this is achieved by the freezing and secondary processing of the food. During freezing it is natural that ice will form within the food and this may change its characteristics. Ice formation during food freezing can be examined using the NMRI method as the formation of ice has been seen to reduce the spatially located NMR signal. The characteristics of a food can be better controlled as

MRI can serve to assess freezing times and the food structure during the freezing process. The secondary processing changes almost all characteristics of a food, such as physical, thermal, and hygroscopic properties, which in turn, change its key acceptability factors, that is, sensory texture and taste. The sensory texture of cooked foods has been predicted using the NMRI technique. In addition, NMR image intensity, the ratio of the oil and water resonance peaks of the one-dimensional NMR spectrum, and both the spin-lattice relaxation time and spin-spin relaxation time of water in the meat are correlated with water-holding capacity. This important finding has desirable features for high-speed sorting using a surface coil NMR probe that determines the oil/water resonance peak ratio of the signal from one region in intact meat.

An online NMR quality evaluation sensor was designed, constructed, and tested (Kim et al., 2003). The device consists of a superconducting magnet with a 20 mm diameter surface coil and a 150 mm diameter imaging coil coupled to a conveyor system. These spectra were used to measure the oil/water ratio in avocados and this ratio was correlated to percent dry weight. One-dimensional magnetic resonance images of cherries were later used to detect the presence of pits inside.

3.2.4 Electronic nose (e-nose)

As all meat emits characteristic volatiles and aromatic compounds at different maturity stages and during ripening, the aroma of meat is an important factor contributing to purchase decisions by consumers. An electronic nose is used for qualitative and/or quantitative analysis of simple or complex gases, vapors, or odors to determine meat quality objectively.

Let us have a look at the human or biological nose to appreciate the factors leading to the development of the e-nose. Of all the five senses, olfaction uses the largest part of the brain and is an essential part of our daily lives. Indeed the appeal of most flavors is more related to the odor arising from volatiles than to the reaction of the taste buds to dissolved substances. Our olfactory system has evolved not only to enhance taste but also to warn us of dangerous situations. We can easily detect just a few parts per billion of the toxic gas hydrogen sulfide in sewer gas, an ability that can save our life. Olfaction is closely related to the limbic or primitive brain, and odors can elicit basic emotions like love, sadness, or fear.

The electronic nose is an instrument comprising a sampling system, an array of chemical gas sensors with differing selectivity, and a computer with an appropriate pattern-classification algorithm. The entire genus of electronic noses includes those with conductive polymer, polymer composite, quartz crystal microbalance, surface acoustic wave, calorimetric, and other classes of sensors. The term "electronic nose" is used to indicate artificial olfaction. Since many modern electronic noses are constructed with more than one class of sensor in them, these latter instruments are said to employ "heterogeneous" sensor arrays. Many sources of multiparameter chemical data including infrared spectrometers, gas chromatographs, and mass spectrometers have been used to identify odors and therefore have been called E-noses. So even though the above definition is broad, it may not be broad enough to describe this entire field of technology. E-nose can replace some of the existing methods of fresh meat quality evaluation based on expensive and relatively subjective taste panels and slow and invasive chemical tests.

The main steps of odor recognition can be summarized as follows:

- Heating the sample for a certain time generates volatile compounds.
- The gas phase is transferred to a detection device which reacts to the presence of molecules.
- The difference in sensor reactions is revealed using different statistical calculation techniques to classify the odors.
- From this pattern and from previous human input (human training from sensory panels), the system predicts the most likely human response to the new pattern.
- A large number of investigations into the use of an electronic nose for meat maturity and quality are reported in literature.

3.2.5 X-ray and computed tomography

X-rays, because of their high energy, can penetrate through many objects. However there are differences in penetration through different materials due to the differences in the material properties. Photons in an X-ray beam, when passing through a body, are either transmitted, scattered, or absorbed.

X-ray emission or absorption spectra are dependent only on atomic number and not on the physical state of the sample or its chemical composition. Radiography uses the difference in the X-ray absorbing powers of different elements to locate their position in a composite material. Positions where there are elements that strongly absorb the X-ray appear light and positions where there are elements that do not absorb the X-rays appear dark on a film placed behind the sample.

In any type of X-ray imaging there are three basic elements: (1) X-ray converter; (2) imaging medium; and (3) casing for imaging medium. The X-ray converter, for example, phosphor screen, stops X-rays from reaching the imaging medium and produces a visible output proportional to the incident X-ray photons. The imaging medium, for example, photographic medium captures the image while the casing protects the imaging medium from surrounding visible radiation. Historically, X-ray imaging has been done on photographic plates or films by subjectively identifying the feature of interest.

X-ray computed tomography (CT) provides two-dimensional X-ray images of thin slices which are used to reconstruct an image of the entire sample. X-ray CT can be used to image interior regions of meat with varying moisture and, to a limited extent, density states. The images represent maps of X-ray absorption of meat cross-sections.

The changes in internal composition related to salt concentration, lean meat content, and fat for pork (Haseth et al., 2008; Vester-Christensen et al., 2009; Jensen et al., 2011), fat and muscle content for beef (Prieto et al., 2010), and bone fragment and bone mineral density for chicken (Tao and Ibarra, 2000) were determined using X-ray imaging techniques. Similarly, the physiological constituents have been monitored in peaches by CT methods in which X-rays absorbed by the peaches are expressed as CT numbers and used as an index for measuring the changes in the internal quality of the meat (Barcelon et al., 1999). Relationships between the CT number and the physiological contents were determined and it was concluded that X-ray CT imaging could be an effective tool in the evaluation of the internal quality of peach.

3.2.6 Hyperspectral imaging

Hyperspectral imaging has great potential to acquire quick information about the chemical constituents and related physical properties of all kinds of meat. Hyperspectral imaging is an emerging analytical technique which involves adding spectral data to two dimensional spatial image to for a hypercube. This gives spatial distribution of quality parameters of meat sample. Thus it is an improvement over traditional spectroscopy which is point specific. It is extensively explored to grade meat based on both extrinsic (appearance, color, size, intramuscular fat) and intrinsic (maturity and tenderness) properties (Elmasry et al., 2012; Kamruzzaman 2013; Konda Naganathan et al., 2016). An algorithm for grading of pork using hyperspectral imaging is given in Fig. 3.3.

There are certain constraints which are limiting the full exploitation of hyperspectral imaging techniques. The major problem associated with hyperspectral imaging system is that it generates spectral images of substantial amount of data with redundant information. Other analytical drawback is that it needs standardized calibration and model transfer procedures. Also, hyperspectral imaging is not suitable in case of homogenous samples because the value of imaging lies in the ability to resolve spatial heterogeneities in samples. Another constraint is the high initial cost of hyperspectral imaging systems.

3.2.7 Electrical properties

Important electrical properties for the quality evaluation of meat include electrical conductivity, capacitance, impedance, relative permissibility, dielectric loss factor, and loss tangent. Probing the variation of these properties with the application of electromagnetic

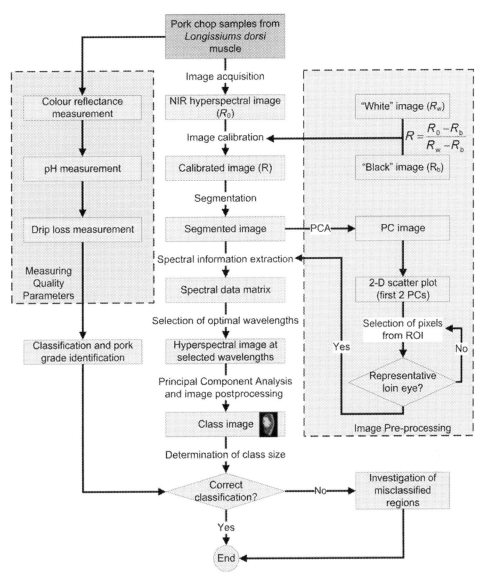

Figure 3.3
The main steps of the processes involved in the nondestructive classification and grading of pork meat samples by hyperspectral imaging. *Adapted from Valous, N.A., Zheng, L., Sun, D.W., Tan, J., 2016. Quality Evaluation of Meat Cuts. doi: 10.1016/B978-0-12-802232-0.00007-4.*

fields gives an indication of meat quality. Grading of water melons was accomplished based on density by estimating volume from electric capacity and using electronic balance for mass determination (Kato, 1997). The electrical resistance of eggplant during storage was determined using a galvanometer to assess freshness (Jha et al., 2004).

A comprehensive review of techniques used for the measurement of electrical properties and their application for quality was presented by Jha et al. (2011). Electrical impedance spectroscopy is also being explored for nondestructive quality evaluation and was adopted by Bauchot et al. (2000) for the assessment of the physiological condition of kiwi fruits.

3.2.8 Acoustic methods

Acoustics deals with the generation and reception of mechanical waves and propagations. As the acoustic phenomena is dependent on the structure, texture, and physical state of the components, the acoustic characteristics can be correlated with maturity, firmness, and other quality parameters of meat.

The frequency range of 20 kHz to 10 MHZ is used for the quality evaluation of foods with ultrasonics. These low intensity ultrasonics should be distinguished from the high intensity ultrasonics that lead to physical/chemical disruption as they travel through the medium. The ultrasonics with higher power content are used for cleaning, homogenization, cell disruption, etc. The square of ultrasonic velocity is inversely proportional to the density and elastic modulus. The ultrasonic velocity and attenuation (of ultrasonic wave) as waves propagate through the medium are correlated to the quality of meat.

The firmness indices expressed as a function of natural acoustic frequency, mass, and density of meat were well correlated with texture and sensory scores of meat. The lipid, protein, moisture, and ash content of Atlantic cod (*Gadus morhua*) fillets were investigated for correlation with acoustic properties. Tenderness of a carcass via live animal ultrasound could tremendously enhance genetic selection and ultimately the uniformity of the beef. But the main drawback is that it is difficult to detect smaller differences and is not applicable in soft tissues.

3.3 Future research needs

Although the noninvasive quality evaluation of meat has made excellent progress, challenges still remain. In order to find reliable quality indicators for evaluating fresh meats, the conventional indicators of color, marbling, and maturity may not be sufficient to predict eating quality such as tenderness. Therefore, many opportunities exist to discover new measurable characteristics that are predictors of quality (Valous et al., 2010). The potential nondestructive methods, which can supplement or replace traditional methods, include CT, MRI, electronic nose, ultrasound, and biosensors (Narsaiah and Jha, 2012). Computer vision technology and hyperspectral imaging are two new promising technologies in this area. Given the complex nature of meat images, one of the most challenging issues is to develop effective image segmentation algorithms (Valous et al., 2016). Segmenting a meat image into regions of interest without human intervention in a reliable and consistent manner is a prerequisite for the success of all subsequent operations, leading to automated meat grading (Jackman et al., 2011).

The best opportunities for improving computer vision solutions lie with hyperspectral imaging, which can provide additional information on meat composition and structure. Spectra could be considered as fingerprints containing valuable information about meat samples (Kamruzzaman et al., 2012). Furthermore, system robustness, real-time capability, sample handling, and standardization are among the issues that remain to be addressed to handle the biological variations in meat products (Valous et al., 2016). In addition, seeking the most sensitive wavebands so that multispectral imaging systems can be built will be the trend in research and development of the technology (Xiong et al., 2014). In essence, multispectral imaging is more suitable for online purposes (Cheng and Sun, 2015). Multispectral imaging allows flexible selection for the number of bands, central wavelengths, and bandwidths. Practically though, detailed internal inspections with imaging devices such as MRI or CT is still slow and expensive for widespread adoption. Further software development for enhancing sensitivity and accuracy would increase the potential for widespread application (Teena et al., 2013).

3.4 Conclusion

This chapter summarized advanced nondestructive methods shown to have great potential for the evaluation of meat quality. Color and computer image processing techniques correlate well with the wholesomeness of meat, marbling of fat, tenderness, diseases and disorders, and the effects of treatments and storage period. The NIR spectroscopy can deliver a wide spectrum of applications ranging from biochemical characteristics, texture profiles, intramuscular fat deposition, etc., while acoustic waves, X-ray imaging, and computerized tomography can be well applied for predicting the body composition of live animals and for online inspection systems. NMRI can be used to find water mobility and distribution, water-holding capacity, intramuscular fat, muscle content of carcass, and postmortem changes in meat. Besides these, the technical merits and demerits of some of the techniques have been also depicted. Finally, it was concluded that the vertical integration of multiple nondestructive methods needs to be evolved while keeping in mind the cost–benefit relationship and the nature of applications.

References

Adedeji, A.A., Liu, L., Ngadi, M.O., 2011. Microstructural evaluation of deep-fat fried chicken nugget batter coating using confocal laser scanning microscopy. J. Food Eng. 102, 49–57.

Barcelon, E.G., Tojo, S., Watanabe, K., 1999. X-ray CT imaging and quality detection of peach at different physiological maturity. Trans. ASAE 42, 435–441.

Bauchot, A.D., Harker, F.R., Arnold, W.M., 2000. The use of electrical impedance spectroscopy to assess the physiological condition of kiwifruit. Postharvest Bio. Technol. 18 (1), 9–18.

Becker, T., 2002. Defining meat quality. In: Kerry, J., Kerry, J., Ledward, D. (Eds.), Meat Processing: Improving Quality. Woodhead Publishing, Cambridge, pp. 3–24.

Chao, K.L., Yang, C.C., Kim, M.S., 2010. Spectral line-scan imaging system for high-speed non-destructive wholesomeness inspection of broilers. Trend. Food Sci. Technol. 21, 129–137.

Chen, P., Sun, Z., 1991. A review of non-destructive methods for quality evaluation and sorting of agricultural products. J. Agric. Eng. Res. 49, 85–98.

Cheng, J.-H., Sun, D.-W., 2015. Recent applications of spectroscopic and hyperspectral imaging techniques with chemometric analysis for rapid inspection of microbial spoilage in muscle foods. Comprehen. Rev. Food Sci. Food Safety 14, 478e490.

Cho, B.K., Irudayaraj, J., 2003. Foreign object and internal disorder detection in food materials using noncontact ultrasound imaging. J. Food Sci. 68, 967–974.

Cho, B., Kim, M.S., Chao, K., Lawrence, K., Park, B., Kim, K., 2009. Detection of fecal residue on poultry carcasses by laser-induced fluorescence imaging. J. Food Sci. 74, E154–E159.

Elmasry, G., Barbin, D.F., Sun, D.-W., Allen, P., 2012. Meat quality evaluation by hyperspectral imaging technique: an overview. Crit. Rev. Food Sci. Nutr. 52, 689–711.

Feng, Y., Sun, D.-W., 2013a. Determination of total viable count (TVC) in chicken breast fillets by near-infrared hyperspectral imaging and spectroscopic transforms. Talanta. 105, 244–249.

Feng, Y.Z., Sun, D.-W., 2013b. Near-infrared hyperspectral imaging in tandem with partial least squares regression and genetic algorithm for non-destructive determination and visualization of pseudomonas loads in chicken fillets. Talanta. 109, 74–83.

Fukushima, E., Roeder, S.B.W., 1981. Experimental Pulse NMR. A Nuts and Bolts Approach. Addison-Wesley Publ. Comp., Inc., Reading, MA.

Haseth, T., Høy, M., Kongsro, J., et al., 2008. Determination of sodium chloride in pork meat by computed tomography at different voltages. J. Food Sci. 73 (7), E333–E339.

Hellier, C.J., 2003. Handbook of Non-destructive Evaluation. McGraw Hill, New York.

Ibarra, J.G., Tao, Y., Xin, H., 2000. Combined IR imaging-neural network method for the estimation of internal temperature in cooked chicken meat. Opt. Eng. 39, 3032–3038.

Jensen, T.H., Bottiger, A., Bech, M., Zanette, I., et al., 2011. X-ray phase-contrast tomography of porcine fat and rind. Meat Sci. 88 (3), 379–383.

Jackman, P., Sun, D.-W., Allen, P., 2011. Recent advances in the use of computer vision technology in the quality assessment of fresh meats. Trends Food Sci. Technol. 22, 185–197.

Jha, S.N., Garg, R., 2010. Non-destructive prediction of quality of intact apple using near infrared spectroscopy. J. Food Sci. Technol. 47 (2), 207–213.

Jha, S.N., Matsuoka, T., Kawano, S., 2004. Changes in electrical resistance of eggplant with gloss, weight and storage period. Biosyst. Eng. 87 (1), 119–123.

Jha, S.N., Narsaiah, K., Basediya, A.L., et al., 2011. Measurement techniques and application of electrical properties for nondestructive quality evaluation of foods—a review. J. Food Sci. Technol. 48 (4), 387–411.

Kamruzzaman, M., 2013. Climatic influences on rainfall and runoff variability in the southeast region of the Murray-Darling Basin. Int. J. Climatol. 33 (2), 291–311.

Kamruzzaman, M., Barbin, D., ElMasry, G., Sun, D.-W., Allen, P., 2012. Potential of hyperspectral imaging and pattern recognition for categorization and authentication of red meat. Innovat. Food Sci. Emerg. Technol. 16, 316–325.

Kandpal, L.M., Lee, H., Kim, M.S., Mo, C., Cho, B.K., 2013. Hyperspectral reflectance imaging technique for visualization of moisture distribution in cooked chicken breast. Sensors 13 (10), 13289–13300.

Kato, K., 1997. Electrical density sorting and estimation of soluble solids content of watermelon. J. Agric. Eng. Res. 67 (2), 161–170.

Kim, K.B., Lee, S.S., Noh, M.S., 2003. On-line measurement of grain moisture content using RF impedance. Trans. ASAE. 46 (3), 861–867.

Konda Naganathan, G., Cluff, K., Samal, A., Calkins, C.R., Jones, D.D., Meyer, G.E., et al., 2016. Three dimensional chemometric analyses of hyperspectral images for beef tenderness forecasting. J. Food Engg. 169, 309–320.

Korver, D., Saunders-Blades, J., Nadeau, K., 2004. Assessing bone mineral density in vivo: quantitative computed tomography. Poul. Sci. 83 (2), 222–229.

Mix, P.E., 2005. Introduction to Non-destructive Testing: A Training Guide, second ed Wiley Interscience, New Jersey.

Monin, H., 1998. Recent methods for predicting quality of whole meat. Meat Sci. 49S, 231–243.

Narsaiah, K., Jha, S.N., 2012. Nondestructive methods for quality evaluation of livestock products. J. Food Sci. Technol. 49, 342–348.

Pallav, P., Hutchins, D.A., Gan, T., 2009. Air-coupled ultrasonic evaluation of food materials. Ultrasonics. 49 (2), 244–253.

Prieto, N., Navajas, E., Richardson, R., et al., 2010. Predicting beef cuts composition, fatty acids and meat quality characteristics by spiral computed tomography. Meat Sci. 86 (3), 770–779.

Ruan, R., Chen, P.L., 2001. Nuclear magnetic resonance techniques. In: Chinachoti, P., Vodovotz, Y. (Eds.), Bread Staling. CRC Press, New York, pp. 113–128.

Shaarani, S.M., Nott, K.P., Hall, L.D., 2006. Combination of NMR and MRI quantitation of moisture and structure changes for convection cooking of fresh chicken meat. Meat Sci. 72, 398–403.

Tao, Y., Ibarra, J., 2000. Thickness-compensated X-ray imaging detection of bone fragments in deboned poultry-model analysis. Trans. ASAE. 43, 453–459.

Teena, M., Manickavasagan, A., Mothershaw, A., El Hadi, S., Jayas, D.S., 2013. Potential of machine vision techniques for detecting fecal and microbial contamination of food products: a review. Food Bioprocess Technol. 6, 162–1624.

Valous, N.A., Drakakis, K., Sun, D.W., 2010. Detecting fractal power-law long-range dependence in pre-sliced cooked pork ham surface intensity patterns using detrended fluctuation analysis. Meat Sci. 86, 289–297.

Valous, N.A., Zheng, L., Sun, D.W., Tan, J., 2016. Quality Evaluation of Meat Cuts. doi: https://doi.org/10.1016/B978-0-12-802232-0.00007-4.

Vester-Christensen, M., Erbou, S.G.H., Hansen, M.F., et al., 2009. Virtual dissection of pig carcasses. Meat Sci. 81, 699–704.

Wold, J.P., Kvaal, K., 2000. Mapping lipid oxidation in chicken meat by multispectral imaging of autofluorescence. Appl. Spectros. 54, 900–909.

Xiong, Z., Sun, D.-W., Zeng, X.-A., Xie, A., 2014. Recent developments of hyperspectral imaging systems and their applications in detecting quality attributes of red meats: a review. J. Food Eng. 132, 1–13.

Xiong, Z., Sun, D.W., Pu, H., Gao, W., Dai, Q., 2017. Applications of emerging imaging techniques for meat quality and safety detection and evaluation: a review. Crit. Rev. Food Sci. Nutr. 57, 755–768.

Yang, C.C., Chao, K., Kim, M.S., Chan, D.E., et al., 2010. Machine vision system for on-line wholesomeness inspection of poultry carcasses. Poult. Sci. 89 (6), 1252–1264.

Further reading

Jha, S.N., Matsuoka, T., Kawano, S., 2001. A simple NIR instruments for liquid type samples. In: Proceedings of the Annual Meeting of the Japanese Society of Agricultural Structures, Paper No. c-20, 146–147.

SECTION 3

Postmortem ageing and meat quality evaluation

CHAPTER 4

Detection techniques of meat tenderness: state of the art

Rui Liu[1] and Wangang Zhang[2]

[1]College of Food Science and Engineering, Yangzhou University, Yangzhou, China [2]Key Laboratory of Meat Processing and Quality Control, Ministry of Education, Jiangsu Collaborative Innovation Center of Meat Production and Processing, Quality and Safety Control, College of Food Science and Technology, Nanjing Agricultural University, Nanjing, China

Chapter Outline
4.1 Introduction 53
4.2 The detection methods 54
    4.2.1 Sensory analysis 54
    4.2.2 Shear force measurements 55
    4.2.3 Texture profile analysis 57
    4.2.4 Star probe measurement 57
    4.2.5 Spectroscopic methods 58
4.3 Conclusion 61
References 62

4.1 Introduction

Meat tenderness is defined as "the ease, perceived by the consumer, along with meat structure disorganization during the masticating" (Lepetit and Culioli, 1994). Tenderness plays a critical role in the contribution of meat organoleptic (sensory) quality. Consumers are willing to pay more for the meat which possesses a guaranteed and consistent tenderness (Feldkamp et al., 2005; Shackelford et al., 2001). The tenderness-based beef classification has been gradually acquired by the sellers to accommodate the consumers' satisfaction for the premium meat (Shackelford et al., 2001). Due to the great importance of tenderness to meat quality attributes, considerable research has been conducted to develop detecting methods for meat tenderness. These include sensory analysis, Warner−Bratzler shear force (WBSF), slice shear force (SSF), texture profile analysis (TPA), star probe, spectroscopic methods [Raman spectroscopy (RS), near-infrared reflectance (NIR) spectroscopy, hyperspectral imaging (HI) system], and others. Thus this chapter describes the basic principles and the application of those detecting methods for meat tenderness.

4.2 The detection methods

4.2.1 Sensory analysis

It is known that consumer sensory analysis provides consistent and valuable consumer preference rankings for meat and meat products (Wheeler et al., 2004). The palatability attributes of meat, including meat tenderness, juiciness, and taste, are originally determined by a trained sensory panel and also a nontrained (consumer) sensory panel. In order to control the variability among studies for the comparable data, the research guidelines for collecting and preparing samples for sensory evaluation for fresh beef, pork, and lamb steaks/chops, roasts, and ground patties have been recommended by the American Meat Science Association (AMSA, 1995). The methods for sensory analysis are primarily categorized into discrimination analysis, descriptive analysis, and consumer evaluation. The accurate and repeatable objective data can be obtained by the trained sensory panel with selection of an appropriate method. The selection and training of candidates for a panel have great significance in sensory evaluation because individuals are different in sensitivity, interest, motivation, and ability to judge differences (Lawless and Heymann, 2010). The general procedures for panelists' selection include recruitment from the surrounding community or company employees, prescreening for their background, and a series of screening activities for sensory evaluation abilities (Belk et al., 2015). The training of the panelists is aimed at improving their familiarity with the test procedures and the abilities to recognize and identify sensory attributes. Standardization, including places of the mouth, cleansing procedures of the palate and mouth, and intervals between samples, should be taught during the training. For meat tenderness, which teeth and how much chewing time are strictly defined for panelists (Belk et al., 2015).

Generally, an 8-ponit scale for meat tenderness evaluation by a trained sensory panel or a nontrained (consumer) sensory panel is used in most researches. Descriptive attributes for muscle fiber tenderness or overall tenderness are 8, extremely tender; 7, very tender; 6, moderately tender; 5, slightly tender; 4, slightly tough; 3, moderately tough; 2, very tough; 1, extremely tough. The trained sensory panel ratings for meat tenderness are always regarded as the reference to evaluate the applicability of other developing methods for meat tenderness judgments (Andrés et al., 2007; Lorenzen et al., 2003; Shackelford, et al., 1999b, 2005) and also to evaluate differences in treatments (Cross et al., 1980; Savell et al., 1977; Tan, 2004; Wheeler et al., 1999, 2000).

4.2.2 Shear force measurements

4.2.2.1 Warner–Bratzler shear force

WBSF is an instrumental method for objectively measuring meat tenderness and was originally devised by Bratzler and Warner (Bratzler, 1949; Warner, 1952). The standard protocol for detecting WBSF was developed by AMSA (1995). Briefly, the shaped meat samples are cooked to the internal temperature of 71°C. After chilling to room temperature, a round core 1.3 cm in diameter is obtained and oriented parallel to the muscle fiber direction. Then the cores are sheared perpendicular to the longitudinal orientation of the muscle fibers by the WBSF machines and this shear force is reported as "kg" or "N" (Belk et al., 2015). However, numerous factors can affect the repeatability of the WBSF measurements. As indicated by Wheeler et al. (1996), the initial meat sample temperature before cooking can significantly impact meat tenderness. Generally, the WBSF decreases as the initial temperature increases. The cooking parameters, including cooking method, rate, and degree of doneness, can also directly affect the cooking loss, juiciness, and meat tenderness. The recommended cooking methods include roasting, broiling, pan broiling, impingement oven, and clam shell, while the other cookery methods are also acceptable if the result is demonstrated to be accurate and repeatable (Belk et al., 2015). Wheeler et al. (1997a,b) suggested an instrument called a belt grill which cooks by passing the sample between two electrically-heated metal plates on teflon-coated conveyor belts. The belt grill largely simulates the cooking method of the consumer and reduces the cooking loss to increase the juiciness ratings (Shackelford et al., 1997, 1999b). In addition, at least six "good" cores should be obtained for WBSF measurement to increase the repeatability. Another factor is crosshead speed when using the electronic testing machines and AMSA (1995) recommends 200–250 mm/min crosshead speeds. It is difficult to compare the WBSF data between institutions (Wheeler et al., 1997a). If the investigations are conducted to detect the differences between treatments within a study, however, it may not be important to have comparative data to other institutions (Wheeler et al., 1997b).

Currently, WBSF seems to be one of the primary instrumental methods to identify meat tenderness and is now the most commonly used method worldwide. The WBSF threshold for consumer acceptability has been increasingly focused over the last three decades. The general research design is to use regression analysis between WBSF values and descriptive tenderness ratings of trained panelists and/or consumers, in order to figure out the separation line between tender and tough meat to build the meat tenderness classification system (Rodas-González et al., 2009). Shackelford et al. (1991) have associated a WBSF value of 4.6 kg with a sensory panel rating of "slighter tender" while the consumer response might be different due to the variable acceptability threshold from consumer to consumer. Miller et al. (2001) reported that consumer perception from tender to tough beef occurred between 4.3 and 4.9 kg of WBSF based on more than 86% consumer acceptability, thus suggesting a meat tenderness of WBSF threshold of 4.3 kg. Huffman et al. (1996) showed that the 4.1 kg of WBSF values for steaks were rated as 6 by the consumer panel and 98% of these were identified as acceptable in tenderness. Furthermore, the steaks were categorized into three groups of <3.0, 3.0–5.7, and >5.7 kg of WBSF value, suggesting that the groups <3.0 and >5.7 kg were 100% acceptable and 100% unacceptable. A similar experimental procedure was also conducted by Destefanis et al. (2008) who suggested the threshold of WBSF value was 47.77 N and further split the range of WBSF into five categories to link with five classes of sensory tenderness. This approach may facilitate better management of the existing variation in tenderness acceptability among consumers.

4.2.2.2 Slice shear force

SSF has been developed as a simplified technique for measuring longissimus shear force for online assessment of meat tenderness by Shackelford et al. (1999b). The general procedures of SSF measurement for meat tenderness are quite similar to the method of WBSF and it seems that SSF is a modified technique of the WBSF method (Wheeler et al., 2005). The modifications can be summarized as the following. The slice is cut immediately after cooking other than cooled down prior to cutting of the cores. There are several cuts to acquire a 1-cm thick and 5-cm long slice. The first cut is made across the width of the longissimus at a point approximately 2 cm from the lateral end of the muscle. The second cut is performed across the longissimus parallel to the first cut at a distance of 5 cm using a sample sizer. The third cut is made using two parallel cuts with 1 cm blades space through the length of the 5-cm long steak portion at a 45-degree angle to the long axis of the longissimus and parallel with the muscle fibers (Wheeler et al., 2005). The shearing of the slice is the same as the method of WBSF, although the blade of SSF is designed to replace the WBSF blade in an automated testing machine. In addition, 500 mm/min is set as the crosshead speed for rapid measurement of the shear force (Belk et al., 2015).

Many studies have been conducted to demonstrate and enhance the accuracy and repeatability of SSF data compared with the WBSF values and trained sensory tenderness ratings. Shackelford et al. (1999a) showed that SSF value was more strongly correlated with trained sensory ratings when SSF detection was conducted immediately after cooking, rather than when steaks were chilled before SSF detection. Moreover, the SSF value showed slightly higher correlation coefficient index with sensory panel tenderness than the WBSF value ($r = -0.82$ vs $r = -0.77$, respectively) and a high repeatability (0.91) over a broad range of tenderness (Shackelford et al., 1999a). That result was further confirmed in the later study which reported a slightly stronger correlation of SSF with trained sensory panel tenderness ratings than that of WBSF ($r = -0.77$ vs $r = -0.66$, respectively) with a 0.90 repeatability (Shackelford et al., 2004). Meanwhile, a stronger correlation was obtained between lateral SSF and average WBSF ($r = 0.64$) suggesting that the SSF and WBSF were the most commonly used and effective measurements for instrumental tenderness

(Derington et al., 2011). However, the deficiency of SSF and WBSF measurements is the difficulty in identifying the differences of meat tenderness among muscles while those differences can be easily detected by the trained sensory panel (King et al., 2009; Rhee et al., 2004; Shackelford et al., 1995). In addition, the SSF value can also be used in the tenderness classification system as evidenced by Shackelford et al. (1999b) who reported that 483 carcasses were categorized into three groups based on SSF value (<23, $23-40$, and >40 kg, $P<.001$) and corresponding trained sensory panel tenderness ratings (7.3 ± 0.04, 6.4 ± 0.06, and 4.4 ± 0.20).

4.2.3 Texture profile analysis

TPA is widely used for texture assessment in a variety of foods including fruits, vegetables, and bakery goods, as well as raw meat and meat products (Chen and Opara, 2013). TPA simulates the mechanical process of mastication, generating the force—deformation curves to obtain the texture attributes data of hardness, cohesiveness, springiness, chewiness, resilience, and adhesiveness (Rosenthal, 2010). Szczesniak (1968) initially correlated textural parameters of TPA to sensory panel tenderness ratings, indicating that TPA could explain 3%—85% variation of meat tenderness. Caine et al. (2003) reported that the TPA values of hardness, cohesiveness, and chewiness were negatively correlated with trained panel sensory tenderness ratings ($r = -0.64$, -0.41, -0.62, respectively). Stepwise regression analysis revealed that hardness and adhesiveness accounted for 47% of variation in sensory tenderness, while WBSF only accounted for 37% of the variation in sensory tenderness (Caine et al., 2003). A similar experiment was also carried out for testing raw meat and cooked meat sensory attributes by DeHuidobro et al. (2005), who reported that the sensory characteristics of hardness were better predicted by TPA than by WBSF. Taken together, TPA seems to be a better predictor of sensory tenderness and explains more variation than the WBSF value.

4.2.4 Star probe measurement

Star probe measurement is also an instrumental indication of texture to describe meat tenderness described in many studies (Anderson et al., 2012a,b, 2014; Lonergan et al., 2007; Malek et al., 2001; Oltrogge and Prusa, 1987; Wang et al., 2012; Wiegand et al., 2002). The sample preparation is much like WBSF measurement, although the core cutting and detection machine has a circular, five-pointed star probe (Lonergan et al., 2007). The value is obtained from the peak load necessary to puncture and compress the sample to 80% of its height. A circle 9 mm in diameter is measured by the star probe attachment with 6 mm between each point. The angle is 48 degrees from the end of each point up to the center of the attachment. The crosshead speed of puncturing is 3.3 mm/s and the maximum force (kg) is recorded for each puncture. It was reported that the star probe values were negatively correlated with sensory tenderness scores and positively correlated with WBSF (Lonergan et al., 2007; Wang et al., 2012). The correlation coefficient (R) of star probe values and the sensory tenderness of pork samples was -0.568, implying that this mechanical measurement correlated moderately with sensory tenderness (Wang et al., 2012). Thus star probe can be an alternative method to WBSF for describing meat tenderness.

4.2.5 Spectroscopic methods

4.2.5.1 Near-infrared reflectance

NIR is an electromagnetic wave between visible light (VIS) and mid-infrared light in the 780–2526 nm wavelength range. The near-infrared spectrum is mainly generated when the molecular vibration transitions from the ground state to the high energy level due to the nonresonance of the molecular vibration. The overtones and combined bands of hydrogen-containing groups such as C–H, O–H, N–H, S–H, and P–H are the main molecular vibration absorption regions in the near-infrared region. Therefore the near-infrared spectrum can be used to predict the concentration of chemical substances containing these bonds. Near-infrared spectroscopy is the use of frequency doubling and frequency combining absorption of hydrogen-containing group vibration. In the near-infrared region, the near-infrared absorption spectrum of the calibration sample is correlated with the concentration of its components or the property data by selecting an appropriate chemometric multivariate calibration method. The position and intensity of the absorption peak in the near-infrared spectrum changes depending upon the composition of the chemical substance and thus it can realize qualitative and quantitative analysis.

By using near-infrared spectroscopy, correlation modeling between the measured WBSF and spectral values can predict meat tenderness. Principal component regression (PCR) analysis was reported between the near-infrared spectrum of beef *Longissimus dorsi* (LD) muscle in the spectral range of 750–1098 nm and meat tenderness, and the predicted correlation coefficients were 0.67 (group 1), 0.72 (group 2), and 0.53 (group combination) (Byrne et al.,1998). Park et al. (2001) assessed the tenderness of beef LD muscle by the NIR spectrum and used principal component regression to analyze the absorption spectrum with a wavelength of 1100–2498 nm. The results showed that if the absorption spectrum was between 1100 and 2498 nm, the measurement coefficient (R^2) of the PCR model for predicting tenderness was 0.692. When the spectrum was between 1100 and 1350 nm, the measurement coefficient (R^2) was 0.612. The precision was much higher in the study of Xiaoyu et al. (2013), who used VIS/NIR reflectance spectroscopy and hand-held probe devices for signal acquisition and spectral preprocessing, and used different data processing methods to establish a beef quality PLSR model in the wavelength range of 400–700 and 700–2000 nm. The results showed that the SNP-treated partial least squares regression (PLSR) model had good performance, and the correlation coefficient of beef tenderness verification set was 0.906. More accurate prediction was conducted by Ripoll et al. (2008) who showed that the correlation coefficient of prediction (R_p^2) of tenderness in the optimal model was 0.981 by using 400–2500 nm near-infrared radiation.

However, some reports have shown the inefficiency of the near-infrared spectrum for the prediction of meat tenderness. Zhang et al. (2015) established a physical property model for near-infrared spectrum prediction by the PLSR model combined with mathematical preprocessing, for example, by orthogonal signal correction and detrended correction. The determination coefficient of the cross-validation of the WBSF prediction model was only 0.433 in yak *Longissimus thoracis* muscle. The determination coefficient $R_{cv}^2 = 0.34$ of WBSF was observed by using visible light and NIR (VIS-NIR, 350–1800 nm) to predict beef meat tenderness (De Marchi et al., 2013). Moreover, the use of NIR for pork meat is not very effective in predicting meat tenderness. The poor determination coefficient of cross-validation (R_{cv}^2) for meat shear force was detected as 0.30 in the study of Balage et al. (2015), 0.20 in the investigation of Geesink et al. (2003), and 0.17 in the report of Chan et al. (2002). The low NIR prediction of WBSF values of samples, especially pork samples, may be due to the limited WBSF reference data and also the characteristics of different animal species.

Compared to the traditional analytical methods, NIR spectroscopy has the advantages of convenience, rapidity, high efficiency, and nondestructivity of the samples. It can predict the chemical properties of the meat and classify meat into quality grades (Prieto et al., 2009). However, many problems emerge due to NIR's own characteristics. For example, NIR is the carrier of sample chemical information and it is difficult to extract the characteristic information related to the analysis target from the spectrum with a low effective information rate. The identification of overlapping spectral information, the filtering of noise signals, the selection of a modeling band, and the design of the modeling algorithm are the key and difficult parts of NIR analysis. Therefore NIR detection technology is still mostly at the laboratory stage and its application in the meat industry for online detection needs further investigation.

4.2.5.2 Hyperspectral imaging technique

HI is a new technique combining spectroscopy with imaging techniques which can simultaneously extract spectral data and spatial information. Such information forms a 3D data cube consisting of one-dimensional spectral information and two-dimensional spatial information (Qin et al., 2013). The HI system contains an illumination, a camera, a spectrograph, a translation stage, and a computer. Due to the differences of chemical composition and physical properties of samples, the light will produce different reflectance and dispersion after it irradiates the sample surface. After that the light is separated into single-wavelength light by beam splitting elements and then projected onto the camera to generate an image (Gowen et al., 2007).

In the last few decades HI has been reported to be able to assess meat tenderness. Naganathan et al. (2008) developed a HI system ($\lambda = 400-1000$ nm) with diffuse lighting. A discriminant model was developed and was based on the characteristics extracted from hyperspectral images by PCA and GLCM. The model was used to predict three beef tenderness classifications with an accuracy of 96.4%, namely tender (SSF $\leq$ 205.80 N), intermediate (205.80 N < SSF < 254.80 N), and tough (SSF $\geq$ 254.80 N). Cluff et al. (2013) used optical scattering HI ($\lambda = 922-1739$ nm) on fresh beef muscle tissue to classify cooked beef tenderness. The linear discriminant model could classify beef tenderness—tough and tender with accuracy of 83.3% and 75.0%, respectively. The result demonstrates HI of optical scattering is a feasible method for tenderness prediction. ElMasry et al. (2012) evaluated the feasibility of a HI system in the NIR region ($\lambda = 900-1700$ nm) to establish a PLSR model which had good performance in tenderness prediction of beef. Similarly, Kamruzzaman et al. (2013) used NIR HI ($\lambda = 900-1700$ nm) to assess the tenderness of lamb, demonstrating that the established model could classify lamb tenderness (WBSF) with reasonable accuracy ($R_{cv} = 0.84$) into tender and tough meat with accuracy of 94.51% and 91.0%. Naganathan et al. (2016) used three-dimensional chemometric analysis combined with discriminant models to analyze hyperspectral images ($\lambda = 400-1100$ nm) in order to predict beef tenderness with a tenderness certification accuracy of 86.7% and an accuracy index value of 66.8%.

These studies have effectively shown that the HI technique has great potential for the prediction of meat tenderness. It is a nondestructive testing technology and can provide spatial and spectral information to establish chemical images. However, HI still has some challenges. Due to the huge amount of information acquired by HI, chemometric methods are needed to extract effective information. The establishment and updating of the model makes it very costly. The huge original image data of HI makes it difficult for the HI system to be widely used in online and real-time application. In addition, HI instruments are relatively expensive compared to conventional methods, thereby increasing the cost of commercial inspection and hindering their adoption (Wang et al., 2018).

4.2.5.3 Raman spectroscopy

Raman spectroscopy (RS) is also a scattering spectrum with specific Raman lines, the magnitude of the displacement and the intensity of the bands. Each functional group and chemical bond of a sample has its own vibrational and rotational energy. The frequency of the line determines the presence of functional groups and chemical bonds and the peak intensity is used for quantitative analysis (Herrero, 2008).

Das and Agrawal (2011) indicated that the increase in the intensity of the two bands of amide I (1669 cm^{-1}) and amide III (1235 cm^{-1}) was related to the increase of the β-sheet content and the ratio of the α-helix and β-fold in the protein of beef. Beattie et al. (2004) reported that partial least squares (PLS) analysis of Raman spectra, shear force, and cooking loss showed strong correlation ($R^2 = 0.65$, RMSEP% of $\mu = 18\%$) based on the data from a set of 52 cooked beef samples. This preliminary interpretation of a regression coefficient curve demonstrated the relationship between the α-helix and β-sheet ratio of the protein and the hydrophobicity of the myofibril environment and the shear strength, tenderness, texture, and overall acceptability of the beef. Fowler et al. (2014a) predicted the shear force of fresh lamb semimembranosus with a Raman probe using the spectra at 1-day postmortem aging. It gave a root mean square error of prediction (RMSEP) of 11.5 N (Null = 13.2) and a squared correlation between observed and cross-validated predicted values of $R^2_{cv} = 0.27$. Further, PLS regression models successively predicted sensory tenderness determined by an untrained consumer panel using 80 fresh lamb Longissimus lumborum (Fowler et al., 2014b) and 45 beef loins (Fowler et al., 2018) with the same Raman probe system. However, Santos et al. (2018) reported a weak correlation ($R^2 = 0.2$) between Raman spectroscopic signatures with sensory tenderness and SSF in pork loins. The Raman spectra range of 1300–2800 cm^{-1} detected in the study of Fowler et al. (2018) showed high predictive performance with all sensory attributes (R^2 value 0.50–0.84, RMSECV value 1.31–9.07). The variations among studies may be due to the different Raman equipment as well as different muscles and animals.

In comparison with the traditional methods for detecting meat tenderness, RS has considerable advantages. The required number of samples is flexible and the instrument operation is simple. Moreover, it presents clear and nonoverlapping spectral peaks with rapid and accurate detection of the multiple indicators (Wang et al., 2018). It is possible to realize the real-time detection. The method of RS for detecting the meat tenderness tends to develop in the direction of low-cost, micro, and hand-held. The technical advantages of RS offer its possibility to be a suitable tool in the meat industry.

4.3 Conclusion

As the meat tenderness contributes significantly to meat quality attributes, consumers and the meat industry desire an easy, accurate, and applicable method to predict meat tenderness. In this chapter we have summarized the commonly used techniques to detect meat tenderness from a variety of investigations. Those techniques include sensory analysis, WBSF, SSF, TPA, star probe, and spectroscopic methods. Advantages and disadvantages emerge among these methods. Sensory analyses provide consistent and valuable consumer preference rankings, although the procedures are complex and time-consuming. Instrumental analyses, including WBSF, SSF, and star probe, showed strong correlation with sensory tenderness, but they are sample-destructive and difficult to facilitate for online detection for screening. The spectroscopic methods can give a simple, rapid, and nondestructive determination to predict meat tenderness, while the current application is limited due to the inconvenience and high cost of the detection equipment. Thus it is necessary for further research to develop a method that can easily detect tenderness of meat with accuracy and repeatability.

References

AMSA, 1995. Research Guidelines for Cookery, Sensory Evaluation and Instrumental Tenderness Measurements of Fresh Meat. National Live Stock and Meat Board, Chicago.

Anderson, M., Lonergan, S., Fedler, C., et al., 2012a. Profile of biochemical traits influencing tenderness of muscles from the beef round. Meat Sci. 91 (3), 247–254.

Anderson, M., Lonergan, S., Huff-Lonergan, E., 2012b. Myosin light chain 1 release from myofibrillar fraction during postmortem aging is a potential indicator of proteolysis and tenderness of beef. Meat Sci. 90 (2), 345–351.

Anderson, M., Lonergan, S., Huff-Lonergan, E., 2014. Differences in phosphorylation of phosphoglucomutase 1 in beef steaks from the longissimus dorsi with high or low star probe values. Meat Sci. 96 (1), 379–384.

Andrés, S., Murray, I., Navajas, E., et al., 2007. Prediction of sensory characteristics of lamb meat samples by near infrared reflectance spectroscopy. Meat Sci. 76 (3), 509–516.

Balage, J.M., e Silva, S. d L., Gomide, C.A., et al., 2015. Predicting pork quality using Vis/NIR spectroscopy. Meat Sci. 108, 37–43.

Beattie, R.J., Bell, S.J., Farmer, L.J., et al., 2004. Preliminary investigation of the application of Raman spectroscopy to the prediction of the sensory quality of beef silverside. Meat Sci. 66 (4), 903–913.

Belk, K., Dikeman, M., Calkins, C., et al., 2015. Research Guidelines for Cookery, Sensory Evaluation, and Instrumental Tenderness Measurements of Meat. AMSA, Champaign, IL, pp. 6–104.

Bratzler, L., 1949. Determining the tenderness of meat by use of the Warner-Bratzler method. In: Proc. Recip. Meat Conf, vol. 2, pp. 117–121.

Byrne, C., Downey, G., Troy, D., Buckley, D., 1998. Non-destructive prediction of selected quality attributes of beef by near-infrared reflectance spectroscopy between 750 and 1098 nm. Meat Sci. 49 (4), 399–409.

Caine, W., Aalhus, J., Best, D., et al., 2003. Relationship of texture profile analysis and Warner-Bratzler shear force with sensory characteristics of beef rib steaks. Meat Sci. 64 (4), 333–339.

Chen, L., Opara, U.L., 2013. Approaches to analysis and modeling texture in fresh and processed foods—a review. J. Food Eng. 119 (3), 497–507.

Chan, D.E., Walker, P.N., Mills, E.W., 2002. Prediction of pork quality characteristics using visible and near–infrared spectroscopy. Trans. ASAE 45 (5), 1519.

Cluff, K., Naganathan, G.K., Subbiah, J., et al., 2013. Optical scattering with hyperspectral imaging to classify longissimus dorsi muscle based on beef tenderness using multivariate modeling. Meat Sci. 95 (1), 42–50.

Cross, H., Berry, B., Wells, L., 1980. Effects of fat level and source on the chemical, sensory and cooking properties of ground beef patties. J. Food Sci. 45 (4), 791–794.

Das, R.S., Agrawal, Y., 2011. Raman spectroscopy: recent advancements, techniques and applications. Vibr. Spectrosc. 57 (2), 163–176.

Destefanis, G., Brugiapaglia, A., Barge, M., Dal Molin, E., 2008. Relationship between beef consumer tenderness perception and Warner–Bratzler shear force. Meat Sci. 78 (3), 153–156.

Derington, A., Brooks, J., Garmyn, A., et al., 2011. Relationships of slice shear force and Warner-Bratzler shear force of beef strip loin steaks as related to the tenderness gradient of the strip loin. Meat Sci. 88 (1), 203–208.

De Huidobro, F.R., Miguel, E., Blázquez, B., Onega, E., 2005. A comparison between two methods (Warner–Bratzler and texture profile analysis) for testing either raw meat or cooked meat. Meat Sci. 69 (3), 527–536.

De Marchi, M., Penasa, M., Cecchinato, A., Bittante, G., 2013. The relevance of different near infrared technologies and sample treatments for predicting meat quality traits in commercial beef cuts. Meat Sci. 93 (2), 329–335.

ElMasry, G., Sun, D.W., Allen, P., 2012. Near-infrared hyperspectral imaging for predicting colour, pH and tenderness of fresh beef. J. Food Eng. 110 (1), 127–140.

Feldkamp, T., Schroeder, T., Lusk, J., 2005. Determining consumer valuation of differentiated beef steak quality attributes. J. Muscle Foods 16 (1), 1–15.

Fowler, S.M., Schmidt, H., van de Ven, R., et al., 2014a. Predicting tenderness of fresh ovine semimembranosus using Raman spectroscopy. Meat Sci. 97 (4), 597–601.

Fowler, S.M., Schmidt, H., van de Ven, R., et al., 2014b. Raman spectroscopy compared against traditional predictors of shear force in lamb m. longissimus lumborum. Meat Sci. 98 (4), 652–656.

Fowler, S.M., Schmidt, H., van de Ven, R., Hopkins, D.L., 2018. Preliminary investigation of the use of Raman spectroscopy to predict meat and eating quality traits of beef loins. Meat Sci. 138, 53–58.

Geesink, G., Schreutelkamp, F., Frankhuizen, R., et al., 2003. Prediction of pork quality attributes from near infrared reflectance spectra. Meat Sci. 65 (1), 661–668.

Gowen, A., O'Donnell, C., Cullen, P., et al., 2007. Hyperspectral imaging—an emerging process analytical tool for food quality and safety control. Trends Food Sci. Technol. 18 (12), 590–598.

Herrero, A.M., 2008. Raman spectroscopy a promising technique for quality assessment of meat and fish: a review. Food Chem. 107 (4), 1642–1651.

Huffman, K., Miller, M., Hoover, L., et al., 1996. Effect of beef tenderness on consumer satisfaction with steaks consumed in the home and restaurant. J. Anim. Sci. 74 (1), 91–97.

Kamruzzaman, M., ElMasry, G., Sun, et al., 2013. Non-destructive assessment of instrumental and sensory tenderness of lamb meat using NIR hyperspectral imaging. Food Chem. 141 (1), 389–396.

King, D., Wheeler, T., Shackelford, S., Koohmaraie, M., 2009. Comparison of palatability characteristics of beef gluteus medius and triceps brachii muscles. J. Anim. Sci. 87 (1), 275–284.

Lawless, H.T., Heymann, H., 2010. Sensory evaluation of food: principles and practices. Springer, New York, pp. 19–50.

Lepetit, J., Culioli, J., 1994. Mechanical properties of meat. Meat Sci. 36 (1–2), 203–237.

Lonergan, S.M., Stalder, K.J., Huff-Lonergan, E., et al., 2007. Influence of lipid content on pork sensory quality within pH classification. J. Anim. Sci. 85 (4), 1074–1079.

Lorenzen, C., Miller, R., Taylor, J., et al., 2003. Beef customer satisfaction: trained sensory panel ratings and Warner-Bratzler shear force values. J. Anim. Sci. 81 (1), 143–149.

Malek, M., Dekkers, J.C., Lee, H.K., et al., 2001. A molecular genome scan analysis to identify chromosomal regions influencing economic traits in the pig. II. Meat and muscle composition. Mamm. Genome 12 (8), 637–645.

Miller, M.F., Carr, M., Ramsey, C., et al., 2001. Consumer thresholds for establishing the value of beef tenderness. J. Anim. Sci. 79 (12), 3062–3068.

Naganathan, G.K., Grimes, L.M., Subbiah, J., et al., 2008. Visible/near-infrared hyperspectral imaging for beef tenderness prediction. Comp. Electro. Agric. 64 (2), 225–233.

Naganathan, G.K., Cluff, K., Samal, A., et al., 2016. Three dimensional chemometric analyses of hyperspectral images for beef tenderness forecasting. J. Food Eng. 169, 309–320.

Oltrogge, M.H., Prusa, K., 1987. Research note: sensory analysis and Instron measurements of variable-power microwave-heated baking hen breasts. Poult. Sci. 66 (9), 1548–1551.

Park, B., Chen, Y.R., Hruschka, W.R., et al., 2001. Principal component regression of near–infrared reflectance spectra for beef tenderness prediction. Trans. ASAE 44 (3), 609.

Prieto, N., Roehe, R., Lavín, P., et al., 2009. Application of near infrared reflectance spectroscopy to predict meat and meat products quality: a review. Meat Sci. 83 (2), 175–186.

Qin, J., Chao, K., Kim, M.S., et al., 2013. Hyperspectral and multispectral imaging for evaluating food safety and quality. J. Food Eng. 118 (2), 157–171.

Rhee, M., Wheeler, T., Shackelford, S., Koohmaraie, M., 2004. Variation in palatability and biochemical traits within and among eleven beef muscles. J. Anim. Sci. 82 (2), 534–550.

Ripoll, G., Albertí, P., Panea, B., et al., 2008. Near-infrared reflectance spectroscopy for predicting chemical, instrumental and sensory quality of beef. Meat Sci. 80 (3), 697–702.

Rodas-González, A., Huerta-Leidenz, N., Jerez-Timaure, N., Miller, M., 2009. Establishing tenderness thresholds of Venezuelan beef steaks using consumer and trained sensory panels. Meat Sci. 83 (2), 218−223.

Rosenthal, A.J., 2010. Texture profile analysis—how important are the parameters? J. Text. Stud. 41 (5), 672−684.

Santos, C., Zhao, J., Dong, X., et al., 2018. Predicting aged pork quality using a portable Raman device. Meat Sci. 145, 79−85.

Savell, J., Smith, G., Dutson, T., et al., 1977. Effect of electrical stimulation on palatability of beef, lamb and goat meat. J. Food Sci. 42 (3), 702−706.

Shackelford, S., Morgan, J., Cross, H., Savell, J., 1991. Identification of threshold levels for Warner-Bratzler shear force in beef top loin steaks. J. Muscle Foods 2 (4), 289−296.

Shackelford, S., Wheeler, T., Koohmaraie, M., 1995. Relationship between shear force and trained sensory panel tenderness ratings of 10 major muscles from *Bos indicus* and *Bos taurus* cattle. J. Anim. Sci. 73 (11), 3333−3340.

Shackelford, S., Wheeler, T., Koohmaraie, M., 1997. Tenderness classification of beef: I. Evaluation of beef longissimus shear force at 1 or 2 days postmortem as a predictor of aged beef tenderness. J. Anim. Sci. 75 (9), 2417−2422.

Shackelford, S., Wheeler, T., Koohmaraie, M., 1999a. Evaluation of slice shear force as an objective method of assessing beef longissimus tenderness. J. Anim. Sci. 77 (10), 2693−2699.

Shackelford, S., Wheeler, T., Koohmaraie, M., 1999b. Tenderness classification of beef: II. Design and analysis of a system to measure beef longissimus shear force under commercial processing conditions. J. Anim. Sci. 77 (6), 1474−1481.

Shackelford, S., Wheeler, T., Meade, M., et al., 2001. Consumer impressions of Tender Select beef. J. Anim. Sci. 79 (10), 2605−2614.

Shackelford, S., Wheeler, T., Koohmaraie, M., 2005. On-line classification of US Select beef carcasses for longissimus tenderness using visible and near-infrared reflectance spectroscopy. Meat Sci. 69 (3), 409−415.

Shackelford, S., Wheeler, T., Koohmaraie, M., 2004. Use of belt grill cookery and slice shear force for assessment of pork longissimus tenderness. J. Anim. Sci. 82 (1), 238−241.

Szczesniak, A., 1968. Correlations between objective and sensory texture measurements. Food Technol. 22, 981−986.

Tan, J., 2004. Meat quality evaluation by computer vision. J. Food Eng. 61 (1), 27−35.

Wang, Q., Lonergan, S.M., Yu, C., 2012. Rapid determination of pork sensory quality using Raman spectroscopy. Meat Sci. 91 (3), 232−239.

Wang, W., Peng, Y., Sun, H., et al., 2018. Spectral detection techniques for non-destructively monitoring the quality, safety, and classification of fresh red meat. Food Anal. Methods 1−24.

Warner, K., 1952. Adventures in testing meat for tenderness. In: Proc. Recip. Meat Conf, vol. 5, pp. 156−160.

Wheeler, T., Shackelford, S., Koohmaraie, M., 1996. Sampling, cooking, and coring effects on Warner-Bratzler shear force values in beef. J. Anim. Sci. 74 (7), 1553−1562.

Wheeler, T., Shackelford, S., Johnson, L., et al., 1997a. A comparison of Warner-Bratzler shear force assessment within and among institutions. J. Anim. Sci. 75 (9), 2423−2432.

Wheeler, T., Shackelford, S., Koohmaraie, M., 1997b. Standardizing collection and interpretation of Warner-Bratzler shear force and sensory tenderness data. In: Proc. Recip. Meat Conf, vol. 50, pp. 68−77, Citeseer.

Wheeler, T., Shackelford, S., Koohmaraie, M., 1999. Trained sensory panel and consumer evaluation of the effects of gamma irradiation on palatability of vacuum-packaged frozen ground beef patties. J. Anim. Sci. 77 (12), 3219−3224.

Wheeler, T., Shackelford, S., Koohmaraie, M., 2000. Variation in proteolysis, sarcomere length, collagen content, and tenderness among major pork muscles. J. Anim. Sci. 78 (4), 958−965.

Wheeler, T., Shackelford, S., Koohmaraie, M., 2004. The accuracy and repeatability of untrained laboratory consumer panelists in detecting differences in beef longissimus tenderness. J. Anim. Sci. 82 (2), 557−562.

Wheeler, T.L., Shackelford, S.D., Koohmaraie, M., 2005. Shear force procedures for meat tenderness measurement. Meat Center, NE: United States Department of Agriculture (USDA). https://www.ars.usda.gov/ARSUserFiles/30400510/protocols/ShearForceProcedures.pdf.

Wiegand, B., Sparks, J., Beitz, D., et al., 2002. Short-term feeding of vitamin D_3 improves color but does not change tenderness of pork-loin chops. J. Anim. Sci. 80 (8), 2116–2121.

Xiaoyu, T., Yang, X., Yankun, P., et al., 2013. Rapid detection model of beef quality based on spectroscopy. Trans. Chin. Soc. Agric. Machinery 44 (10), 171–175.

Zhang, L., Sun, B., Xie, P., et al., 2015. Using near infrared spectroscopy to predict the physical traits of *Bos grunniens* meat. LWT-Food Sci. Technol. 64 (2), 602–608.

CHAPTER 5

Biochemical changes of postmortem meat during the aging process and strategies to improve the meat quality

R. Ramanathan[1], G.G. Mafi[1], L. Yoder[1], M. Perry[1], M. Pfeiffer[1], D.L. VanOverbeke[1] and Naveena Basappa Maheswarappa[2]

[1]Department of Animal and Food Sciences, Oklahoma State University, Stillwater, OK, United States
[2]ICAR-National Research Centre on Meat, Chengicherla, Hyderabad, India

Chapter Outline

5.1 Introduction 67
    5.1.1 Conversion of muscle to meat 68
5.2 Effects of aging on tenderness 68
    5.2.1 Mechanism of postmortem aging 68
    5.2.2 Techniques to improve tenderness 69
    5.2.3 Different types of aging 72
    5.2.4 Techniques to quantify tenderness in plant 74
    5.2.5 Techniques to measure tenderness 75
5.3 Effects of aging on color 76
    5.3.1 Biochemical basis of lower color stability in aged meat 77
    5.3.2 Practical approaches to improve color stability of aged steaks 77
5.4 Effects of aging on buffalo meat quality 77
5.5 Application of proteomics and metabolomics to study meat quality 78
5.6 Conclusion 79
References 79
Further reading 80

5.1 Introduction

Nearly 15% of all US retail meat is discounted in price due to surface discoloration. Thus, failure to optimize muscle color life results in US $1 billion of lost revenue every year. Although meat color is an important quality attribute that influences purchasing decisions at the point of sale, repeat purchases are often dependent on tenderness. It has been estimated that variability in tenderness decreases the value of beef cattle by approximately $7.64 per animal and results in $217 million of unattained revenue for the US beef industry. Studies have shown that consumers are willing to pay a premium price for guaranteed tender beef. Hence among the beef palatability attributes, tenderness is considered as the most important economic trait. Therefore better understanding of the postmortem changes within muscles will help to achieve economic benefits associated with shelf life improvements.

Both pre- and postharvest factors can influence tenderness and meat color. Although aging can improve eating quality, an extended time period can be detrimental to meat quality. According to the National Beef Tenderness Survey, the approximate aging period for strip loins under retail conditions ranged from 2 to 102 days. Enzymes and metabolites involved in muscle metabolism remain active postmortem; hence elucidating the cellular mechanisms involved in meat quality changes are important to understand the deviations and to develop strategies to improve meat quality. The overall goal of this chapter is to provide an overview of factors affecting meat tenderness and color.

5.1.1 Conversion of muscle to meat

Immediately after exsanguination, the metabolism changes from aerobic to anaerobic. The tissue tries to maintain homeostasis for energy production. However, the lack of blood flow leads to the conversion of muscle to meat. When the animal is alive adenosine triphosphate (ATP) is the main energy source for cells. Both aerobic and anaerobic metabolism can produce ATP; however, the aerobic pathway is efficient in producing 35ATP molecules compared with 3ATP during anaerobic metabolism. Phosphocreatine, glycolysis, and mitochondrial respiration can provide energy for cellular activities. Phosphocreatine is plentiful in muscle, and it is the first source that resynthesizes ATP in muscle cells when ATP levels begin to drop.

At lower oxygen levels ATP is consumed rapidly and tissue is no longer able to maintain homeostasis. Glycolysis is the breakdown of glucose, and changes in the rate of glycolysis can affect meat quality. Under these conditions ATP is being consumed more rapidly than it can be produced and lactate accumulates in the tissue, along with NAD+ and hydrogen ions. As glycolysis continues, hydrogen ions will accumulate and the pH of the tissue declines. The decrease in muscle pH is one of the most significant postmortem changes in muscle.

5.2 Effects of aging on tenderness

5.2.1 Mechanism of postmortem aging

Postmortem aging is an essential step in improving tenderness. Proteolysis results in the breakage of myofibrillar proteins. Peptides and amino acids can serve as water-soluble flavoring precursors for taste and flavor characteristics of meat (Khan et al., 2016).

Myofibrillar proteins are broken down by weakening and fragmentation of the Z-disk and the degradation of cytoskeletal proteins such as desmin and titin. These changes result in the fragmentation of myofibrils, which is closely related to an increase in meat tenderness (Koohmaraie, 1994).

Endogenous proteinases can improve meat tenderness (Koohmaraie, 1994). Both calpains and cathepsins can influence postmortem proteolysis. Calpains degrade muscle proteins by breaking down muscle structural proteins such as C-protein, M-protein, and cytoskeletal proteins. Calpains are more effective when pH is neutral. There are two forms of calpains: μ- and m-calpain. Each form requires a different amount of calcium to be active. For example, μ-calpain requires 1–30 μmol of calcium, whereas m-calpain needs 100–750 μmol. Calcium initiates calpain activity, whereas calpastatin acts as an inhibitor. After rigor mortis cathepsins, another group of enzymes, are also available to help improve tenderness. Cathepsins are present in lysosomes, and are found to be effective in long-term aging at acidic pH levels (pH 5.4–5.6). Recently, studies have reported the roles of apoptotic and antiapoptotic heat shock proteins in tenderness.

5.2.2 Techniques to improve tenderness

5.2.2.1 Electrical stimulation

Electrical tenderization has been used commercially since the 1970s in New Zealand, and is now widely used in Australia, New Zealand, and the United States. Electrical stimulation increases the speed of postmortem glycolysis by increasing the production of lactic acid, thus reducing the risk of cold shortening by a rapid drop in pH. Cold shortening occurs when calcium is retained and concentrated in the sarcoplasm, which induces muscle contraction resulting in toughness. Electrical stimulation also accelerates tenderization by physically disrupting muscle fibers, increasing the rate of proteolysis, disrupting membranes of lysosomes, and releasing cathepsins. In addition, the reduction of collagen cross-linking may also contribute to tenderness.

The meat industry uses two ranges of voltage. In high-voltage electric stimulation the voltage ranges from 300 to 1000 V and in low-voltage electrical stimulation the voltage varies between 50 and 120 V. A high voltage results in a greater rate of pH decline which is more consistent with less variability. Low voltage stimulation has been shown to be equally effective, but requires a longer stimulation time. High voltage stimulation is quicker than low voltage (10 vs 40 seconds); however, the safety requirement is more significant and is more expensive than low voltage. A few studies have shown that electrical stimulation can impact other proteins that contribute to color stability and water-holding capacity; however, other studies have shown that electric stimulation improves beef color (Bhat et al., 2018; Bolumar et al., 2013).

5.2.2.2 Tenderstretch

Typically in commercial beef plants, beef carcasses are hung from the Achilles tendon, which pulls the hind leg backward. This position is different than the normal position of the animal when standing upright. Tenderstretching is an alternative way to hang carcasses from the pelvic or hipbone, which prevents muscle rigor contraction and increases sarcomere length. Hostetler et al. (1972) introduced the tenderstretch method. After studying five different suspension types, these authors concluded that tenderstretch was the most efficient in improving sarcomere lengths in muscles of the loin and round. Several versions of tenderstretching have been tested by placements of hooks in the carcass such as in vertebra or weight-bearing areas on the neck. All versions improved tenderness of multiple major muscles compared to the traditional Achilles tendon suspension. Several muscles have been shown to have an increase in tenderness including semimembranosus, semitendinosus, longissimus, and gluteus medius, but not psoas major. The tenderness of muscles that contain high amounts of connective tissue seems to be unaffected by tenderstretch. This technique is usually completed in 45–90 minutes post exsanguination and prior to the onset of rigor. Conversely, this technique has also been shown to increase the risk of cold shortening if the carcass is chilled rapidly. Tenderstretching has been shown to increase tenderness of muscles in beef, reduce drip, and cooking loss. Although this method is easy to implement, only a few processors have adopted as it requires more space, labor costs, and has limited impact on many muscles (Bhat et al., 2018; Sørheim and Hildrum, 2002).

5.2.2.3 Tendercut

Another way of increasing tenderness was introduced in the early 1990s by utilizing the carcass weight to stretch muscles of the loin and round. Tendercut requires two cuts to be made on the carcass through bone, connective tissue, adipose tissue and some minor muscles, and the carcass can be suspended traditionally by the Achilles tendon. These cuts are to be made shortly after slaughter, and prior to the onset of rigor mortis. The first cut is made between the 12th and 13th thoracic vertebra, like when splitting the carcass into fore- and hindquarters; however, it is continued to completely sever the multifidus dorsi and finishes prior to the longissimus. The second cut is made between the sirloin and round and completely severs the ischium of the pelvis and the junction between the fourth and fifth sacral vertebra and connective tissue. Significant gaps should appear between the loin and sirloin–round junction to ensure sufficient stretching of the carcass. Tendercut has shown to improve tenderness by increased sarcomere length and reduced Warner–Bratzler shear force (WBSF) in loin and round muscles including vastuslateralis, rectus femoris, and vastus medialis; in addition to the longissimus and gluteus medius when combined with electrical stimulation. Tendercut requires more hands-on labor, as well as requiring higher railing systems due to lengthier carcasses than traditional beef carcasses. These are most likely the reasons that few commercial systems have adopted this system; however, tendercut has been validated by several studies in the United States, Canada, and Norway.

5.2.2.4 Blade tenderization

Blade tenderization can be used on postrigor meat, and is one of the most effective and efficient ways to improve the tenderness of meat. Typically, blade tenderization is used on raw products prior to packaging at the processing plant; however, there are machines available in the market that can be used at home. Blade tenderization has a larger impact than increased time or temperature on tenderness (King et al., 2009). Blade tenderization can happen with a bank of blades or needles that penetrate the surface of the meat, causing weakening of the protein structures. Blade tenderization is carried out in conjunction with tumbling or massaging. Meat is typically passed through a machine 1–4 times depending on the desired amount of tenderization and the cut of meat. Blade tenderization has an advantage compared to other tenderization processes, as the process is fairly instant without the use of high temperatures, holding time, or the addition of nonmeat ingredients. Jeremiah et al. (1999) reported that blade tenderization significantly improved the initial and overall tenderness of several cuts. Only slight differences in tenderization, measured by WBSF readings, have been found with different blade sizes. Larger blade sizes are suggested due to the ability to slightly improve tenderness while also reducing the operational problems that can occur with smaller needles and blades, such as bending. While blade tenderization is the most practical tenderization method, it can affect the texture and appearance of the meat and can cause color changes where the blade penetrates the meat. Additionally, there is a high potential of microbial contamination if the machinery and equipment are not properly cleaned (Bekhit et al., 2014a,b; Mandigo and Olson, 1982).

5.2.2.5 Plant enzymes

Currently there are five enzymes that have been classified as Generally Recognized As Safe (GRAS) by the United States Department of Agriculture (USDA) Food Safety Inspection Service (FSIS). Most of these enzymes have an optimal activity temperature range from 50°C to 70°C, but each also has a specific range. Enzymes are activated by warming during the cooking process, but can become denatured if the temperature is high. Papain, which is an enzyme that is extracted from papaya fruit, has been studied since the 1940s for its tenderizing effects and is the most common enzyme used in the meat industry. Papain has an active pH range of 4.0–9.0, but performs optimally in a pH of 4.0–6.0. The active temperature range of papain varies between 50°C and 80°C. Papain is efficient in degrading both myofibrillar and collagen proteins. Bromelain (from pineapple) and ficin (from Figs) are other common plant enzymes used in meat systems. These enzymes degrade collagen efficiently when compared to papain, but do not degrade myofibrillar proteins. When using these enzymes, it is suggested to inject them into the product rather than dipping or drenching. The metabololic products from bacillus subtilis can degrade collagen.

The enzyme isolated from *Aspergillus oryzae* can degrade myofibrillar proteins effectively (Calkins and Sullivan, 2012; Bekhit et al., 2014a,b; Aberle et al., 2012).

5.2.3 Different types of aging

5.2.3.1 Dry aging

Dry aging refers to when a beef carcass or primal cuts are hung in a refrigerated room for an extended period of time in a controlled environment where temperature, relative humidity, and airflow are all monitored and controlled. This process not only tenderizes meat, but also develops a distinct flavor that is more nutty and beefy. High-quality carcasses are preferred for dry aging since it is an extensive process and more marbling allows for a more consistent flavor and juiciness. Therefore, the typical quality grades used are USDA Prime and the upper two-thirds of USDA Choice, High Choice with moderate marbling, and average Choice with modest marbling (Dashdorj et al., 2016).

5.2.3.1.1 Tenderness

Dry aging promotes the growth of beneficial mold on the surface of the meat like *Thamnidium*. It has the appearance of a pale gray color and will form patches, called whiskers, on the fatty parts of the carcass or cut. *Thamnidium*'s enzymes are able to break down muscle and connective tissues by penetrating into the meat and emitting proteases and producing collagenolytic enzymes. Proper handling practices, aging time, temperature, relative humidity, and airflow are required to grow beneficial mold and limit other microbial contamination.

5.2.3.1.2 Time

The amount of time beef is dry aged depends upon the desired flavor profile. Most beef carcasses or primal cuts are dry aged for at least 21 days. Usually the carcass will be aged for 21 days and then fabricated into the four main primal cuts (round, loin, rib, and chuck). The round and chuck will be broken down and will not be aged further because most of the meat becomes roasts or trim. The loin and rib could possibly be further aged for 7–28 days and then cut into steaks.

5.2.3.1.3 Temperature

The desired temperature for dry aging beef is between 0°C and 4°C. Even though a higher temperature would increase the enzymatic processes, it can stimulate rapid bacterial growth that could produce off-odors. Therefore dry aging is completed at the lowest possible temperature that will not result in frozen meat.

5.2.3.1.4 Relative humidity

If the humidity is too high, it will result in the growth of spoilage bacteria, which would produce an off-flavor. If the humidity is too low, bacteria growth will be limited but evaporative weight loss will increase causing the meat to dry out quickly, resulting in a decrease in juiciness. The preferred range of relative humidity is 61%–85% and should be recorded and checked daily to maintain control.

5.2.3.1.5 Airflow

It is crucial to have sufficient airflow in the refrigerated room to have adequate air circulation. The meat will not be able to release the required amount of moisture for the drying process without sufficient air circulation. However, if there is too much circulation, the meat could dry out too quickly, causing an increase in trimming losses. The recommended range of airflow is 0.5–2 m/s (1.6–6.6 ft/s) with a velocity of 0.2–1.6 m/s. It is important that airflow and velocity stay consistent through the entire dry aging process.

5.2.3.1.6 Dry aging bag

A new bag technology with a high water vapor transmission rate has led to a reduction in shrinkage for dry aged meat. Using these bags to dry age will still produce the desired flavor and tenderness that results from dry aging, while also increasing yield, decreasing microbial contamination, and trim loss. TUBLIN developed this technology and the bags are sold in the United States under the name of UMAi Dry.

5.2.3.1.7 Cost

Even though dry aging improves tenderness and delivers a certain flavor profile, dry aged products are costly. The majority of the beef sold in grocery stores today is wet aged. Dry aging causes shrinkage and moisture loss in steaks. After 3 weeks of dry aging about 10% of weight is lost for strip loins, that loss increases to about 15% after 30 days, about 23% after 50 days, and about 35% after 120 days. While the percentage of shrinkage will depend on the type of steak, it is clear that the more days of dry aging, the more moisture will be lost. Dry aging also results in a decrease in fabrication yields due to the removal of the crust layer, dried and discolored lean and fat on the outside of the subprimals. Dry aged beef is approximately 25% costlier than traditional wet aged beef. Dry aged beef will most likely be offered in high-end restaurants, gourmet steak companies, upscale grocery stores, and butcher shops. Dry aging is a niche market due to the extensive process, increase in cost, and unique flavor. The flavor and tenderness that comes from dry aging keep the value-added beef in demand.

5.2.3.2 Wet aging

Wet aging takes place within a vacuum package at refrigerated temperatures, and is the most common aging method. Wet aging allows the meat to increase tenderness, without sacrificing as much weight and trim loss compared with dry aging. Wet aging also results in a shorter processing time and lower cost for the consumer (Smith et al., 2008). However, wet aging can produce a sour, metallic, and bloody flavor in beef when compared to dry aging. Irrespective of the aging types, an increase in tenderness can result in greater consumer satisfaction.

5.2.4 Techniques to quantify tenderness in plant

Since over 60% of US beef carcasses are graded USDA low Choice or Select, it is important to further measure tenderness to be able to sort quality more accurately (Boykin et al., 2017). Video image analysis (VIA) systems and colorimeters are two methods used to predict tenderness in plants with online techniques (Woerner and Belk, 2008). Both of these techniques have the ability to help branded beef programs to accurately guarantee tenderness.

5.2.4.1 Video image analysis

VIA systems are able to produce data about beef carcasses through images captured of the longissimus muscle. This technology has the ability to interpret color characteristics of the beef longissimus muscle and predict palatability; therefore this technology helps to sort beef carcasses with minimal differences in marbling. BeefCam is a VIA system that was developed as a portable prototype by the researchers at Colorado State University who collaborated with Hunter Associates Laboratory to analyze lean and fat color on beef carcasses in a packing plant using the L^*, a^*, and b^* color scale. BeefCam is used to sort and certify carcasses as tender. However, this technology isn't completely accurate since not all carcasses that are tender are certified by BeefCam and it doesn't identify tough steaks with 100% accuracy (Woerner and Belk, 2008).

5.2.4.2 Colorimeter

A colorimeter measures the exposed longissimus muscle of beef carcasses by evaluating fat and lean color. There are two methods for measuring color: tristimulus and spectral reflectance. The tristimulus method uses Commission Internationale de l'Eclairage (CIE) L^*, a^*, and b^* values (Aberle et al., 2012). A spectral reflectance method that has been successful in measuring tenderness online is near-infrared reflectance (NIR) spectroscopy. The NIR instrument uses a radiation-powered beam of light, which is pointed at the meat and then reflected back to the spectrometer. This results in a range of light wavelengths that create the NIR spectrum (Reis and Rosenvold, 2014). NIR is not able to quantify tenderness for individual carcasses or beef cuts but is an effective tool to sort tough and tender carcasses with 89% accuracy of segregating extremely tough carcasses (Woerner and Belk, 2008).

5.2.5 Techniques to measure tenderness

5.2.5.1 Warner–Bratzler shear force

The most common procedure to measure tenderness is the WBSF test. This test determines tenderness by using a specific blade that cuts through a sample of meat. The meat sample is cooked and chilled overnight to $2°C-5°C$. Once chilled the sample is cored with the muscle fibers running longitudinally; therefore the blade will shear perpendicular to the muscle fibers. The blade used for WBSF must be a V-notch blade. Lower shear force values indicate more tender beef (Novaković and Tomašević, 2017).

5.2.5.2 Sliced shear force

Sliced shear force (SSF) is a very similar procedure to WBSF. The samples are cooked to an internal temperature of $70°C$ and then chilled for 24 hours at $4°C$ (Shackelford et al., 1999). The steaks are then cut into smaller samples, 5 cm long and 1 cm thick, at a 45-degree angle with a knife that has two parallel blades 1 cm apart. The blade used for SSF is flat with a blunt-end. Higher values signify tougher meat (Rust et al., 2008). SSF is more accurate and faster to complete than WBSF (Woerner and Belk, 2008).

5.2.5.3 Sensory panel

Sensory panels are another way to measure tenderness; although, they are not as common as WBSF or SSF due to the variability in panel members. A sensory panel can be classified as either trained or consumer. Trained panels are used to determine treatment differences in the product being tasted. The objectives for training panelists are to inform the panel member on the test procedure, advance the individual's ability to distinguish sensory characteristics, and improve the panel member's memory resulting in consistent sensory judgments. For trained panels there are three different types of rating scales most commonly used: graphic or line scales, verbal scales, and numerical scales. Consumer panels are used to test if the product will be accepted by the consumer. When selecting a consumer panel it is important to take into consideration the target population, demographics, the number of markets to test, and product usage patterns. Consumer panels will use a ballot with either hedonic scales or Just About Right and intensity scales (Wheeler et al., 2015). Trained panels must have at least 6 members and a consumer panel requires at least 50, but 100 members are preferred. A trained panel may have less because the members are highly selected and trained. Using a sensory panel is effective when available because it defines tenderness from people's perceptions. However, it is slower, more expensive, and requires increased amount of the sample to carry out tests (Purchas, 2014).

5.3 Effects of aging on color

Meat color is an important quality attribute that primarily determines consumers' intent to buy meat. Myoglobin is the primary sarcoplasmic protein responsible for meat color. Depending on the redox state of iron and the ligand attached, myoglobin can exist in three different forms, namely, oxymyoglobin (bright red color), deoxymyoglobin (purple/dark color), and metmyoglobin (brown or causes discoloration). Various studies have shown that 7−14 days of aging time is optimum for eating quality. Extended aging periods can decrease color stability (Fig. 5.1). Oxygen consumption, metmyoglobin reducing activity, and lipid oxidation are the important biochemical processes that influence meat color (Ramanathan and Mancini, 2018).

Immediately after animal harvest, meat will have a darker color. This is primarily due to greater oxygen consumption and less available oxygen for myoglobin. When meat is aged for 7−14 days, beef will develop a brighter red color. A greater postmortem time can limit mitochondrial activity and the number of substrates available for oxygen consumption.

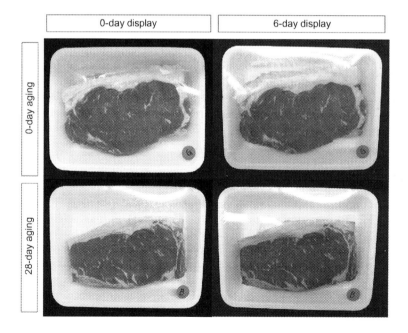

Figure 5.1

Effects of aging and display time on discoloration of steaks. Source: *Photo courtesy: Rachel Mitacek, Oklahoma State University.*

Hence, a shorter aging time can benefit the meat industry by providing a more appealing bright red color.

5.3.1 Biochemical basis of lower color stability in aged meat

According to the National Beef Quality Audit, the approximate time from the processor to retailer can vary from 3 to 300 days. Increased storage time is detrimental to meat color stability, if these steaks are displayed under retail conditions for sale. Greater aging time can increase oxidative changes, depletion of substrates required to generate NADH, and activity of enzymes related to metmyoglobin reducing activity and oxygen consumption. In addition, aging can decrease the activity of mitochondria and increase degeneration. Hence, steaks that have been aged and displayed will have lower color stability.

5.3.2 Practical approaches to improve color stability of aged steaks

Consumers in different countries have various perceptions of color. In the United Kingdom consumers can tolerate intact steak packaged in a vacuum, whereas in the United States consumers associate a bright red color with freshness and wholesomeness. Steaks packaged in vacuum and aged for an extended period of time can be acceptable. However, if vacuum-packaged steaks are repackaged in PVC and displayed under retail conditions it can be detrimental. For example, steaks aged for 21 days and packaged in PVC discolored within 48 hours compared to steaks aged for 7 or 14 days. Modified atmospheric packaging can be used to improve the color stability of aged meat. Studies have shown that packaging steaks in carbon monoxide-modified atmospheric conditions can increase shelf life compared to PVC packaging (English et al., 2016).

5.4 Effects of aging on buffalo meat quality

The Food and Agricultural Organization has named buffalo as an important asset that is "undervalued." Meat produced from buffaloes has gained increased popularity in several southeastern and middle-eastern Asian countries and Africa because of its reduced fat, reduced cholesterol, and other healthier attributes. Composition, physicochemical, nutritional, and functional properties, and sensory attributes of buffalo meat are comparable with beef.

The major attractive features of buffalo meat are red color, reduced fat and cholesterol with poor marbling, low connective tissue, and desirable texture. The myoglobin content of fresh buffalo meat varied from 2.7 to 9.4 mg/g depending on the type of the muscle and animal age. The mean myoglobin content of young buffalo meat (2.36 mg/g) was lower than old buffalo meat (3.59 mg/g). Different authors have reported the redness scores (a^* value) ranging from 12.0 to 20.0 for fresh and frozen buffalo meat of different age groups.

Bloom time had a significant effect on color characteristics of buffalo meat; a^* value increased from 6.47 to 10.01 at 45 minutes. The meat pigment concentration of spent male buffalo meat was significantly higher than young males, which was attributed to a greater content of heme pigment and myoglobin. The heme pigment concentration in meat samples of bulls was 3.59–3.99 mg/g (Maltin et al., 1998).

Muscles from young buffaloes of 1–2 years showed less collagen content (0.91–1.71 g/100 g) compared with old buffaloes of 12 years of age ((1.16–2.23 g/100 g) (Syed Ziauddin et al., 1995)). Collagen solubility of 45.5% was observed in spent buffalo meat chunks. Muscle fiber diameter ranging from 35.32 mm (Anjaneyulu et al., 1985), 60.76 mm (Naveena et al., 2004), and 41.72 mm (Naveena et al., 2011) was reported for fresh buffalo meat. Myofibrillar fragmentation index was reported to be 87.5 in 6-year-old male Murrah buffaloes (Kulkarni et al., 1993). Researchers have reported WBSF values in the range of 22.95–44.08 N (Irurueta et al., 2008) and 20.0–45.0 N (Naveena et al., 2004) for different muscles of buffaloes. Researchers have attempted to improve the tenderness of meat produced from old/spent buffaloes using plant proteases (Naveena et al., 2004) and chemicals (Naveena et al., 2011). Aging also plays a significant role in tenderization. Shear force values have reduced from 41.79 to 26.09 N and 54.28 to 32.06 N in young and old buffalo meat respectively after 6 days of aging. Treatment with ammonium hydroxide (0.1%, 0.5%, and 1.0%) reduced the shear force of buffalo meat chunks by 21.42%, 28.51%, and 42.57%, respectively, in comparison to the control (Naveena et al., 2011).

5.5 Application of proteomics and metabolomics to study meat quality

Both tenderness and meat color are influenced by various interrelated parameters. Hence the use of traditional muscle biology techniques can be challenging to elucidate the mechanistic basis for meat quality changes. Both tenderness and color are influenced by various enzymes. Proteomics techniques allow the quantification and identification of proteome in a biological system. Various studies have utilized proteomics techniques to identify a biomarker for tenderness and also to understand muscle specific differences in color stability (Joseph et al., 2015). Metabolomics is a systematic analysis of small molecules such as amino acids, tricarboxylic intermediates, and fatty acids in a biological system (Fiehn, 2002; Julian, 2004). It is a relatively new omic technique, and the methodology includes metabolite separation, detection, quantification, data analysis, and interpretation. Various studies have shown that the quantification of metabolites provides a real-time snapshot of reactions in a biological system. Metabolites can be identified using a targeted or nontargeted approach. Recent research in meat color analysis reported that aging time could decrease the levels of metabolites that can regenerate NADH (Abraham et al., 2017).

5.6 Conclusion

Aging is a standard method to improve the tenderness, juiciness, and flavor of meat. Consumers are willing to pay premiums for meat products that are guaranteed for their quality. However, various factors can influence aging processes and therefore palatability and color stability. The use of throughput techniques has helped to better understand the biochemical changes involved in aging processes. Standardizing the conditions that can affect tenderness and color are critical to maximize the economic benefits associated with meat quality traits.

References

Aberle, E.D., Forrest, J.C., Gerrard, D.E., Mills, E.W., 2012. Principles of Meat Science, fifth ed. Kendall Hunt.

Abraham, A., Dillwith, J.W., Mafi, G.G., VanOverbeke, D.L., Ramanathan, R., 2017. Metabolite profile differences between beef longissimus and psoas muscles during display. Meat Muscle Biol. 1, 18–27.

Anjaneyulu, A.S.R., Sengar, S.S., Lakshmanan, V., Joshi, B.C., 1985. Meat quality of male buffalo calves maintained in different levels of protein. Buff. Bull. 4, 44–47.

Bekhit, A., Carne, A., Ha, M., Franks, P., 2014a. Physical interventions to manipulate texture and tenderness of fresh meat: a review. Int. J. Food Prop. 17 (2), 433–453.

Bekhit, A., Hopkins, D.L., Geesink, G., Bekhit, A., Franks, P., 2014b. Exogenous proteases for meat tenderization. Crit. Rev. Food Sci. Nutr. 54 (8), 1012–1031.

Bhat, Z.F., Morton James, D., Mason Susan, L., Bekhit, A., 2018. Applied and emerging methods for meat tenderization: a comparative perspective. Compr. Rev. Food Sci. Food Saf. 17 (4), 841–859.

Bolumar, T., Enneking, M., Toepfl, S., Heinz, V., 2013. New developments in shockwave technology intended for meat tenderization: opportunities and challenges. A review. Meat Sci. 95 (4), 931–939.

Boykin, C.A., Eastwood, L.C., Harris, M.K., Hale, D.S., Kerth, C.R., Griffin, D.B., et al., 2017. National beef quality audit–2016: in-plant survey of carcass characteristics related to quality, quantity, and value of fed steers and heifers. J. Anim Sci. 95 (7), 2993–3002.

Calkins, C., Sullivan, G.S., 2012. Adding Enzymes to Improve Beef Tenderness, 6. <https://www.beefresearch.org/CMDocs/BeefResearch/PE_Fact_Sheets/Adding_Enzymes_to_Improve_Beef_Tenderness.pdf>.

Dashdorj, D., Tripathi, V.K., Cho, S., Kim, Y., Hwang, I., 2016. Dry aging of beef; review. J. Anim. Sci. Technol. 58, 20.

Elton, D., Aberle, J.C.F., David, E., Gerrard, Edward, W., Mill, 2012. Principles of Meat Science, fifth ed. Kendall Hunt, Dubuque, IA.

English, A.R., Mafi, G.G., Vanoverbeke, D.L., Ramanathan, R., 2016. Effects of extended aging and modified atmospheric packaging on beef top loin steak color. J. Anim. Sci. 94 (4), 1727–1737.

Fiehn, O., 2002. Metabolomics-the link between genotype and phenotype. Plant Mol. Biol. 48, 155–171.

Hostetler, R.L., Link, B.A., Landmann, W.A., Fitzhugh JR., H.A., 1972. Effect of carcass suspension on sarcomere length and shear force of somemajor bovine muscles. J. Food Sci. 37, 132–135.

Jeremiah, L.E., Gibson, L.L., Cunningham, B., 1999. The influence of mechanical tenderization on the palatability of certain bovine muscles. Food Res. Int. 32 (8), 585–591.

Joseph, P., Nair, M.N., Suman, S.P., 2015. Application of proteomics to characterize and improve color and oxidative stability of muscle foods. Food Res. Int. 76 (P4), 938–945.

Julian, L.G., 2004. Metabolic profile of different Genom. Can we hear the phenotype? Philos. Trans.: Biol. Sci. 359, 857–871.

Khan, M.I., Jung, S., Nam, K.C., Jo, C., 2016. Postmortem aging of beef with a special reference to the dry aging. Korean J. Food Sci. Anim. Resour. 36 (2), 159–169.

King, D.A., Wheeler, T.L., Shackelford, S.D., Pfeiffer, K.D., Nickelson, R., Koohmaraie, M., 2009. Effect of blade tenderization, aging time, and aging temperature on tenderness of beef longissimus lumborum and gluteus medius. J. Anim. Sci. 87 (9), 2952–2960.

Koohmaraie, M., 1994. Muscle proteinases and meat aging. Meat Sci. 36 (1–2), 93–104.

Kulkarni, V.V., Kowale, B.N., Kesava Rao, V., Murthy, T.R.K., 1993. Storage stability and sensory quality of washed ground buffalo meat and meat patties during refrigerated storage. J. Food Sci. Technol. 30, 169–171.

Maltin, C.A., Sinclair, K.D., Warriss, P.D., et al., 1998. The effect of age at slaughter, genotype and finishing system on the biochemical properties, muscle fibre type characteristics and eating quality of bull beef from suckled calves. Anim. Sci. 66, 341–348.

Mandigo, R.W., Olson, D.G., 1982. Effect of blade size for mechanically tenderizing beef rounds. J. Food Sci. 47 (6), 2095–2096.

Naveena, B.M., Mendiratta, S.K., Anjaneyulu, A.S.R., 2004. Tenderization of buffalo meat using plant protease from *Cucumis trigonus ruxb* (Kachri) and *Zingiber officinale roscoe* (Ginger rhizome). Meat Sci. 68, 363–369.

Naveena, B.M., Kiran, M., Sudhakar Reddy, K., Ramakrishna, C., Vaithiyanathan, S., Devatkal, S.K., 2011. Effect of ammonium hydroxide on ultrastructure and tenderness of buffalo meat. Meat Sci. 88, 727–732.

Novaković, S., Tomašević, I., 2017. A comparison between Warner-Bratzler shear force measurement and texture profile analysis of meat and meat products: a review. IOP Conf. Series: Earth Environ. Sci 85 (1), 012063.

Purchas, R.W., 2014. Tenderness Measurement Encyclopedia of Meat Sciences, Elsevier, pp. 452–459.

Ramanathan, R., Mancini, R.A., 2018. Role of mitochondria in beef color: a review. Meat Muscle Biol. 2 (1), 309.

Reis, M.M., Rosenvold, K., 2014. Prediction of meat attributes from intact muscle using Near-Infrared Spectroscopy. Encyclopedia of Meat Sciences, Elsevier, 70–77.

Rust, S., Price, D., Subbiah, J., Kranzler, G., et al., 2008. Predicting beef tenderness using near-infrared spectroscopy. J. Anim. Sci. 86 (1), 211–219.

Shackelford, S.D., Wheeler, T.L., Koohmaraie, M., 1999. Evaluation of slice shear force as an objective method of assessing beef longissimus tenderness: review. J. Anim. Sci. 77 (10), 2693–2699.

Smith, R.D., Nicholson, K.L., Nicholson, J.D., et al., 2008. Dry versus wet aging of beef: retail cutting yields and consumer palatability evaluations of steaks from US Choice and US Select short loins. Meat Sci. 79 (4), 631–639.

Syed Ziauddin, K., Rao, D.N., Amla, B.L., 1995. Effect of lactic acid, ginger extract and sodium chloride on electrophoretic pattern of buffalo muscle proteins. J. Food Sci. Technol. 32, 224–226.

Sørheim, O., Hildrum, K.I., 2002. Muscle stretching techniques for improving meat tenderness. Trends Food Sci. Technol. 13 (4), 127–135.

Wheeler, T.L., Papadopoulos, L.S., Miller, R.K., 2015. Research Guidelines for Cookery, Sensory Evaluation, and Instrumental Tenderness Measuremnts of Meat, second ed. American Meat Science Association.

Woerner, D.R., Belk, K.E., 2008. In: C. S. University (Ed.), The History of Instrument Assessment of Beef. National Cattlemen's Beef Association.

Further reading

Devine, C.E., 2014. Conversion of Muscle to Meat | Aging. In: Dikeman, M., Devine, C. (Eds.), Encyclopedia of Meat Sciences, Second Edition Academic Press, Oxford, pp. 329–338.

Dikeman, M., Devine, C., 2014. Encyclopedia of Meat Sciences, second ed. Elsevier, pp. 329–424.

CHAPTER 6

Recent developments in postmortem aging and evaluation methods

Kiran Mohan[1], Rituparna Banerjee[2] and Naveena Basappa Maheswarappa[2]
[1]Department of Livestock Products Technology, Veterinary College, KVAFSU, Bidar, India
[2]ICAR - National Research Centre on Meat, Chengicherla, Hyderabad, India

Chapter Outline
6.1 Introduction 82
6.2 Aging 82
6.3 Postmortem aging and meat quality 83
    6.3.1 Tenderness 84
    6.3.2 Flavor 84
    6.3.3 Color 85
    6.3.4 Water-holding capacity 85
6.4 Evaluation/assessment of postmortem aging 85
6.5 Advanced methods for evaluation of postmortem proteolysis 89
    6.5.1 Proteoglycans/glycosaminoglycan quantification 89
    6.5.2 Determination of cross-links and decorin 89
    6.5.3 Cathepsin activity 89
    6.5.4 Calpain extraction and casein zymography 90
    6.5.5 Immunological protein quantification by dot-blot 90
    6.5.6 Nuclear magnetic resonance transverse relaxation (T2) measurements 90
    6.5.7 Fluoroscence polarization 91
6.6 Biomarkers of postmortem aging 91
    6.6.1 Proteomic markers in the conversion of muscle to meat 91
    6.6.2 Genotyping for marker gene 93
    6.6.3 Metabolomics 93
6.7 Conclusion 94
References 94
Further reading 99

6.1 Introduction

Meat is particularly appreciated for nutritional values as well as its sensory attributes. Postharvest meat processing practices, particularly aging, play a pivotal role in the development of meat palatability. Aging is a value-adding process and has been extensively practiced by the meat industry for years—either in the form of traditional hanging of carcasses or by packaging primal or subprimal cuts in vacuum bags for a certain period in a cold storage. The postmortem aging of meat is a very important process having a significant effect on its microstructure and quality traits, especially texture, tenderness, and water-holding capacity (WHC). In this chapter we will discuss a brief overview of postmortem aging, the various types and conditions of aging practices, and its impact on meat quality attributes. Traditional and advanced methods of assessment of meat aging will also be elaborated.

6.2 Aging

In the absence of microbial spoilage, the holding of unprocessed meat above its freezing point is called aging or conditioning and it has been associated with an increase in tenderness and flavor. The dominant change during immediate postmortem is glycolysis and simultaneously the other degradation changes commence due to bacterial spoilage.

Aging/conditioning is the term applied to the natural process of tenderization when meat is stored or aged postrigor. Aging periods may vary from 2 to 4 weeks for beef, 6 to 10 days for pork, and 12 to 24 hours for chicken; consequently, toughness is reduced to 50%–60% of the initial value (Takahashi, 1996). Tenderization could be attributable to two types of process: changes in the connective tissue components of the meat, and weakening of the myofibrils. Nishimura et al. (1998) have suggested that aging takes place in two phases. The rapid first phase is caused by changes in the myofibrillar component followed by a slower second phase caused by structural weakening of the intramuscular connective tissue. However, of these, the changes in the myofibrillar component are generally thought to be the more important and, in fact, only very small changes can be seen in the major connective tissue components such as collagen.

In general there are two types of aging methods: wet aging and dry aging. Wet aging (vacuum aging) is the most predominant and successful postmortem aging method with proper vacuum packs and temperature control—primal or subprimal cuts are vacuum-packed in a sealed barrier package and stored at refrigeration temperature (Dikeman et al., 2013; Smith et al., 2008a, b). Advantages associated with this method are significant reduction in product weight loss, extended shelf life without compromising palatability traits, minimal space requirements, and consequently more convenience during storage and transport and less operation facility cost (Kim et al., 2018). However, the development of some negative flavor characteristics, such as bloody, serumy, metallic, and sour are reported with wet aging (Savell, 2008).

Dry aging is a traditional process of aging typically practiced by local meat processors or small meat purveyors (Savell, 2008) where carcasses or cuts are aged under a controlled environment (e.g., temperature, humidity, and air flow) without protective packaging (Kim et al., 2016; Savell, 2008). Dry aging enhances the palatability attributes of meat, especially a unique "dry aged" flavor, but traditional dry aging is an expensive process mainly due to the higher shrinkage and moisture loss and the intensive time or labor cost associated with it (Savell, 2008). A modified form of dry aging also known as stepwise dry/wet aging was reported by Kim et al. (2017). Beef loins are first dry aged for 10 days and then subjected to a wet aging method for 7 days and no significant differences in eating quality attributes were found by the consumer panel between dry aging and stepwise aged beef loin.

An integrated aging system known as "dry aging in a bag" has been reported, in which both the attributes of wet and dry aging are combined. Beef subprimals dry aged in this water vapor-permeable special bag had similar sensory traits as traditionally dry aged cuts, while having a higher yield due to lower weight loss during aging and lower microbial contamination (Dikeman et al., 2013).

6.3 Postmortem aging and meat quality

After rigor mortis, the aging mechanism becomes fully activated across all fibers, however some amount of aging might have been completed in those fibers that have already reached rigor, that is, ATP production ceases. Several factors such as the pH fall and the high temperature can dramatically change the commencement, extent, and subsequent speed of aging.

Tenderization of meat is a result of the postmortem fragmentation of the muscle structure and associated proteins, mediated by endogenous proteases throughout aging (Kim et al., 2014). Widely studied multienzymatic proteolytic systems for meat tenderization include (1) the *cathepsins*, (2) the *calpains*, calcium-activated sarcoplasmic factor; (3) the *proteasomes*; and (4) the *caspases*, a family of cysteine aspartate–specific proteases that mediate apoptosis.

Calpains are calcium-activated proteases with an optimum activity at neutral pH. In skeletal muscle, the calpain system consists of at least three proteases, μ-calpain (microcalpain, activated by low concentration of calcium ions; 50–100 μM), m-calpain (macrocalpain, activated by high concentrations of calcium ions; 1–2 mM), and skeletal muscle-specific calpain, p94, or calpain 3, and an inhibitor of μ- and m-calpain, calpastatin. Once activated, the calpains hydrolyze the myofibrillar and cytoskeletal proteins, such as titin, nebulin, filamin, desmin, and troponin-T, resulting in meat tenderization during aging (Huff-Lonergan et al., 1996). Calpains can cause disruption in the myofibrils and release key proteins such as α-actinin and myosin light chains (Anderson et al., 2012) from the myofibrillar structure. Unlike μ- and m-calpain, calpain 3 cannot be easily extracted from skeletal muscle due to its association with the myofibrillar protein, titin (Sorimachi et al., 1995). High calpastatin activity reduces the extent of proteolysis in muscles. Cathepsins occur in lysosomes in the sarcoplasm. They are released postmortem and have maximum activity in mildly acid conditions. They are known to degrade troponin-T, some collagen cross-links, and mucopolysaccharides of the connective tissue ground substance.

After exsanguination of animals, muscle cells undergo the process of cell death, or apoptosis, which is considered the very first phase in the muscle to meat conversion process (Ouali et al., 2013; Sentandreu et al., 2002). Caspases, a family of cysteine−aspartate proteases and the main enzymes responsible for apoptosis, are becoming of more interest for their role in meat tenderization (Lomiwes et al., 2014; Ouali et al., 2013).

6.3.1 Tenderness

Tenderness is considered to be the most important eating quality of meat (Miller et al., 2001). The rate and extent of the aging response and subsequent meat tenderization are dependent upon various factors such as species, animal age, diet, breed, individual muscle, marbling content, and/or aging condition (Kiran et al., 2015, 2016). Proteolysis is the predominant factor influencing the impact of postmortem aging on meat tenderization (Koohmaraie and Geesink, 2006). Denatured proteins are particularly liable to be attacked by proteolytic enzymes. Though extensive proteolysis of the collagen and elastin fibers of connective tissues might appear to be the most likely changes causing increased tenderness, the proteins of connective tissue do not normally change this way during aging in skeletal muscle but undergo changes due to enzymatic activity. The proteolytic enzymes called cathepsins from the lysosomes and calpains from the sarcoplasm were also reported to break down a minimal amount of collagen in muscles (Zamora et al., 2005). Since neither the proteins of connective tissue nor the myofibrils are subjected to extensive proteolysis during aging, the fall in total soluble protein nitrogen and increase in soluble nitrogen of the total proteins indicate the proteolysis of the sarcoplasmic proteins.

6.3.2 Flavor

In view of the development of flavor which accompanies aging due to liberation of nucleotide compounds, "inosinic acid" gathers importance. The ATP, ADP, and AMPs would break down into inosinic acid, inorganic phosphate, and ammonia. Other flavor related compounds include Maillard reaction-related sugar fragments, such as glucose, volatile compounds, such as *n*-aldehydes (e.g., pentanal and hexanal), and ketones, which also include lipid oxidation-related products (Dashdorj et al., 2015). The breakdown of proteins and fats during aging also contributes to flavor by producing H_2S, ammonia, acetaldehyde, acetone, and diacetyl, but at the same time prolonged aging is associated with the loss of flavor or development of undesirable flavors like bitterness and sourness (Spanier et al., 1997). The greatest reason for dry aging beef is to enhance its flavor. Campbell et al. (2001) extensively studied the effect of dry aging on beef flavor. Higher beefy and brown/roasted flavor intensities have been reported in dry aged steaks compared to unaged or wet aged steaks (Corbin et al., 2015).

6.3.3 Color

Several intrinsic factors may be associated with the development of color and oxidative stability of muscle during the aging process, such as the accumulation of prooxidants (heme or nonheme iron) and/or the depletion of endogenous reducing compounds or antioxidants (Kim et al., 2012). The aging of meat affects the color composition of beef, as aged beef has a brighter and slightly red color due to the enzymatic changes that result from the breakdown of certain proteins (Jayasooriya et al., 2007). Vitale et al. (2014) observed lower color stability in *longissimus thoracis* (LT) muscles of Friesian beef after 14–21 days of wet aging in comparison to 0–8-day aged samples and attributed this difference to the myoglobin and lipid oxidation. The difference in the lipid oxidation and associated discoloration of specific muscle may be due to differences in the development of metabolites such as NAD/NADH, acyl carnitines, nucleotides, nucleosides, and glucuronides in specific muscles with the aging process (Ma et al., 2017).

6.3.4 Water-holding capacity

The distribution and mobility of water may have a profound influence on essential meat quality attributes (Trout, 1988). During aging, the water content of muscle, location, and mobility will change due to numerous factors involved in the conversion of muscle to meat (Honikel, 2004). About 85% of muscle water is located in the protein-dense myofibrillar protein network, and the remaining 15% is extramyofibrillar space (Lawrie, 1998). The key biochemical processes during rigor mortis will influence the conversion of muscle to meat (Honikel, 2004). Proteins play a decisive role in immobilizing water in meat. Initial proteolysis improves WHC, but extended proteolysis reduces the WHC (Xiong, 2004). Several studies have reported a correlation between desmin degradation with WHC during postmortem storage (Melody et al., 2004). Improvement in WHC during aging has been attributed to the destruction of the meat structure and the creation of a "sponge effect" (Farouk et al., 2012). The prerigor myosin denaturation and postrigor cytoskeletal degradation are suggested to be responsible for drip from meat (Rosenvold et al., 2008).

6.4 Evaluation/assessment of postmortem aging

The summary of evaluation methods used to assess postmortem aging has been elaborated in Table 6.1 indicating salient findings from different studies with various aging methods.

Table 6.1: Summary of postmortem aging changes in different meats.

Species	Muscle types	Aging time	Temperatures	Aging evaluation methods	Salient findings	References
Cattle	Psoas major Semitendinosus	Wet aging 12 days	4°C	Light and transmission electron microscopy	• Increased intermyofibrillar space • Continuous degradation of sarcomere structure • Increased sarcomere length	Kolczak et al. (2003)
Chicken	Pectoralis superfacials	Wet aging 1 day		Indirect immunofluoroscent microscopy	• Weakening of rigor linkages by translocated paratropomyosin • Paratropomyosin is a key factor in meat tenderization during postrigor aging. Calcium theory of meat tenderization	Takahashi et al. (1995)
Pork	Pectoralis superfacials	Wet aging 7 days		Sarcomere length		
Beef	Longisimus thoracis	Wet aging 10 days				
Beef	Vastus intermedius	Wet aging 31 days	0°C–2°C	Immunoblotting SDS-polyacrylamide gel electrophoresis (SDS-PAGE)	• Gradual degradation of Troponin-T and the 34 kDa component during aging • Disappearance of Troponin-T at 24 days postmortem, • Appearance of 32 kDa component	Negishi et al. (1996)
Beef	M. semitendinosus	Wet aging 9 days	10°C	Assay of lysosomal enzymes and calpains	Calcium chloride accelerated aging Synergistic nonenzymic tenderization by the addition of high concentrations of salts during aging	Alarcon-Rojo and Dransfield (1995)
Beef	M. semitendinosus	Wet aging 28 days	4°C	Scanning electron microscopy	• The total amount of proteoglycons decreased with time postmortem • Separation of collagen fibrils and fibers from the endomysium and the perimysium results in the partial tenderization of beef during postmortem aging	Nishimura et al. (1996)
Chicken	Semitendinosus	Wet aging-12 hours	4°C	Scanning electron microscopy	Structural weakening of endomysium and perimysium during postmortem aging Disintegration of the intramuscular connective tissue	Liu et al. (1995)
Beef	M. semitendinosus	Wet aging 12 days	4°C	Total and soluble collagen Scanning electron microscopy Texture Profile Analysis	Increase in the quantity of soluble collagen and decrease in the value of textural parameters with aging	Palka (2003)
Beef	M. longissimus thoracis et lumborum	Wet aging 6 days		Phospholipids SDS-PAGE	Phospholipids were liberated from the sarcoplasmic reticulum during aging of meat and calcium ions leak into the sarcoplasm through channels formed by phospholipid liberation	Ji and Takahashi (2006)

Beef	M. longissimus lumborum	Stepwise dry/wet aging	4°C	Shear force Water-holding capacity	Stepwise dry/wet aging could potentially provide beneficial impacts to local/small meat processors	Kim et al. (2017)
Beef	Longissimus thoracis et lumborum	Dry aging 8 days	5.1°C	Sensory evaluation Microbial evaluation	• The dry aged steaks had more umami and butter fried meat taste • Aging improved the sensory traits	Li et al. (2014)
Beef	M. longissimus lumborum	7 days	4°C	Shear force and cook loss Microbial analysis Consumer sensory evaluation and survey	Dry aging could improve the eating quality attributes of grass-fed beef loins with low marbling without inducing any negative impacts on microbial properties	Berger et al. (2018)
Beef	Longissimus dorsi	7 days	4°C	Polymorphism CAPN1 316 and CAPN1 4751 markers	CAPN1 316 showed association with shear force and $L*$ and CAPN1 4751 with $a*$ and $b*$ during aging	Mazzucco et al. (2010)
Beef	M. longissimus dorsi (LD)	28 days	−1.5°C	SDS-PAGE	Fast degradation of titin, nebulin and filamin is a key factor in the improvement of meat tenderness	Wu et al. (2014)
Lamb	M. longissimus lumborum	14 days 8 days 8 days	−1.5°C 3°C 7°C	Western-blot Shear force	The application of elevated aging temperatures could shorten required aging periods prior to freezing	Choe et al. (2016)
Beef	Longissimus lumborum Semimembranosus	42 days 84 days		Calpain extraction and Casein zymography	• Calpain 1 is responsible for early postmortem tenderization • Calpain 2 is responsible for additional tenderization during extended aging	Colle and Doumit (2017)
Beef	M. longissimus dorsi	14 days	−1.5°C	Sarcomere length Western blot	Prerigor stretching will not contribute to tenderness improvement	Pen et al. (2012)
Beef	Longissimus lumborum	21 days	2°C	Myofibril fragmentation index Sarcomere length Electron microscopy	Aging improves the meat quality of cold shortened beef	Li et al. (2012)
Beef	M. infraspinatus	20 days	3°C	Collagen profile	Connective tissue aging might be finished after 10 days	Modzelewska-Kapituła et al. (2015)
Pork	M. longissimus thoracis et lumborum M. biceps femoris	7 days		Sensory and objective measurements	Aging for 7 days did not improve eating quality compared to 2 days postslaughter	Channon et al. (2018)

(Continued)

Table 6.1: (Continued)

Species	Muscle types	Aging time	Temperatures	Aging evaluation methods	Salient findings	References
Lamb	*Biceps femoris*	14 days	2°C	Shear force Collagen content, Desmin degradation Sarcomere length	All the variation in shear force can be explained by varied parameters	Starkey et al. (2017)
Beef	*Longissimus dorsi*	28 days	−1°C	SDS-PAGE and Immunoblot Cathepsin B activity	The development of meat tenderness is likely to be compartmentalized by ultimate pH	Lomiwes et al. (2014)
Beef	*Longissimus dorsi*	21 days	2°C	Two-dimensional gel electrophoresis In-gel digestion and protein identification by MALDI-TOF mass spectrometry	Tenderness and proteolytic changes during aging are related to animal's breed	Marino et al. (2013)
Pork	Loin	4 weeks 12 weeks	2°C–4°C −1°C	Cathepsin B + L activity SDS-page Free amino acids analysis	Superchilling aged pork displays a slower rate of overall proteolytic activity compared to pork aged at chilling conditions	Pomponio et al. (2018)
Beef	*M. longissimus lumborum*	3 weeks	1°C–3°C	Metabolite analysis by NMR spectroscopy and spectral processing	Metabolite analysis showed that flavor precursors, were more abundant in the dry aged beef compared to the wet-aged beef	Kim et al. (2016)
Emu	Thigh Muscle	9 days 15 days	4°C	SDS-PAGE 2-DE Myofibrillar Fragmentation Index	The rate and extent of aging vary considerably between meat stored under aerobically packed conditions Emu meat cubes require 6 and 9 days of postmortem aging under aerobically packed conditions	Naveena et al. (2015)
Buffalo	*M. longissimus lumborum*	6 days	4°C	2-DE MALDI-TOF/TOF MS Transmission electron microscopy Scanning electron microscopy	Complement C1q subcomponent subunit B, uroplakin-1b, aspartate aminotransferase, myosin-IIIa, glycogen phosphorylase, cytosolic carboxypeptidase 3 and phosphatidylinositol transfer protein β isoform may be useful as biomarkers for meat quality in buffalo meat.	Kiran et al. (2016)

6.5 Advanced methods for evaluation of postmortem proteolysis

6.5.1 Proteoglycans/glycosaminoglycan quantification

Collagen fibrils and fibers of intramuscular connective tissue are embedded in ground substances, proteoglycans (PGs) and glycoproteins. Electron microscopic observations of bovine *M. semitendinosus* revealed regularly arranged PGs in the basement membrane and PGs associated with collagen fibrils in the perimysium during postmortem aging (Nishimura et al., 1996). The total amount of PGs decreased with postmortem time suggesting that PGs are degraded during postmortem aging of beef (Liu et al., 1995). During the postmortem period β-glucuronidase and other enzymes are released from lysosomes, and are known to attack PGs and increase the tenderness of beef (Moeller et al., 1976).

6.5.2 Determination of cross-links and decorin

Unreducible (mature) cross-links between the collagen molecules in macromolecular fibrils provide connective tissue with the required physicochemical properties and biomechanical stability (Lepetit, 2008). The known mature cross-links are hydroxylysyl pyridinoline (HP) and lysyl pyridinoline (LP). However, the exact changes that occur in IMCT and the contribution of cross-links to the tenderness during the process of maturity still require further research. Decorin is a major type of PG in striated muscle (Eggen et al., 1994). The interaction of PGs with both collagen and noncollagen materials has been proposed to play an important role in tissue function, architecture, and morphogenesis (Nishimura et al., 1996).

6.5.3 Cathepsin activity

Cathepsins are endopeptidases located inside the lysosomes of the living muscle cell, and when active they catalyze the hydrolysis of internal peptide bonds in the protein (Agarwal, 1990). Cathepsins are divided into cysteine (cathepsins B, H, L, and X), aspartic (cathepsins D and E) and serine (cathepsins G) peptidase families (Sentandreu et al., 2002). Cathepsins are known to hydrolyze the myofibrillar proteins, including troponin-T, I, and C, nebulin, titin, and tropomyosin; during the postmortem conditioning period and positively correlate with tenderness in beef (O'Halloran et al., 1997). The contribution of cathepsins to meat tenderization is opposed by many researchers due to small-scale actin and myosin degradation in the postmortem conditioning (Koohmaraie et al., 1991), lysosomal location of cathepsins (Hopkins and Taylor, 2002), and poor association between cathepsins' activities and the variation in tenderness in meat samples (Whipple et al., 1990).

Residual activity of Cathepsins B and L can be measured fluorometrically. The activities of cathepsin B and B-/L-like enzymes can be measured against the synthetic fluorogenic substrates *N*-carbobenzoxy-arginine-arginine-7-amido-4-methylcoumarin and *N*-carbobenzoxyphenylalanyl-arginine-7-amido-4-methylcoumarin, respectively. The collagenase-like enzymes can be measured against a synthetic fluorogenic substrate, N-succinylglycine-proline-leucine-glycine-proline-7-amido-4-methylcoumarin (Christensen et al., 2011).

6.5.4 Calpain extraction and casein zymography

The contribution of proteolytic calpain activity to meat tenderization is widely accepted in meat science (Koohmaraie and Geesink, 2006; Sentandreu et al., 2002). The calpains are expressed in isoforms μ-calpain, m-calpain, and calpain 3. The μ-calpain is the most important proteolytic enzyme involved in postmortem tenderization (Koohmaraie, 1996).

The purification of calpains and calpastatin was performed using chromatographic techniques (Koohmaraie, 1990) with drawbacks like long elution time, relatively poor calpain activity recovery, etc. The casein zymography method is based on the principle that casein molecules present in the zymogram gel are hydrolyzed by the calpains in the presence of Ca^{2+} ions in the solution and autolysis of casein is used as an indicator of calpain activation. The identification of calpains and calpastatin in turkey meat samples using casein zymography method with appropriate extraction buffer and anion-exchange chromatography technique was suitably standardized (Biswas et al., 2016).

6.5.5 Immunological protein quantification by dot-blot

An antibody was considered specific against the studied protein when only one band at the expected molecular weight was detected by Western blot. The abundance of protein biomarkers belonging to different biological pathways can be determined according to the standard protocol (Guillemin et al., 2009). The conditions for use and specificity of primary antibodies against proteins in bovine muscle may be checked by Western blot (Gagaoua et al., 2015).

6.5.6 Nuclear magnetic resonance transverse relaxation (T2) measurements

The low field nuclear magnetic resonance (LF-NMR) transverse relaxometry (T2) reflects myowater distribution and mobility in meat which directly affect WHC and drip (Pearce et al., 2011). Postmortem pH decline is very crucial for the development of WHC of meat influencing intramyofibrillar and intermyofibrillar water (NMR T_{21}, T_{22}, and their populations) (Bertram et al., 2003). The positive relationship between T_{21} and pH has been shown (Li et al., 2012) suggesting muscle with a lower pH could exhibit more lateral or longitudinal shortening with a higher relaxation time (T_{21}) (Pearce et al., 2011).

6.5.7 Fluoroscence polarization

Among the techniques based on light interactions with biological tissues, fluorescence polarization offers a selective means of characterizing the organization of biological tissues. This rapid and noncontact optical method can investigate the fluorescence polarization of muscle tissues in order to obtain structural information, and specifically the structural modifications caused by meat aging (Clerjon et al., 2011). The fluorescence spectra of meat are modified by several parameters, particularly aging (Dufour et al., 2003). As meat ages, the chemical environment of tryptophan, a major fluorescent amino acid of meat, becomes modified, affecting its fluorescence spectra.

6.6 Biomarkers of postmortem aging

The search for biomarkers to predict the biological mechanisms behind the conversion of muscle to meat and the associated characteristics of meat has been studied extensively (Ouali et al., 2013; Picard et al., 2010). Advanced "omic" technologies like genomics (genes), transcriptomics (gene-transcripts), proteomics (proteomes), and metabolomics (metabolites) are used to identify biomarkers which correlate with meat quality traits (Gagaoua et al., 2015).

6.6.1 Proteomic markers in the conversion of muscle to meat

Proteomics may be defined as the science that studies the entire subset of proteins expressed in a certain cell, tissue, body fluid, organ, or organism (Wilkins et al., 1996). The conversion of muscle to meat implies complex biochemical mechanisms that influence the final meat texture and quality. Postmortem proteolytic degradation of muscle occurs during the meat aging process, resulting in the production of protein fragments. The mechanisms controlling meat quality development are governed by the complex interplay of many cellular processes and conditions (Bendixen, 2005). In the postgenomic era proteomic tools have been utilized to understand the biochemical mechanisms influencing the conversion of muscle to meat in beef and pork (Laville et al., 2009). The effect of aging on sensorial, textural, and proteolytic changes is well documented (Jia et al., 2006; Bjarnadottìr et al., 2012; Kiran et al., 2016). The protein biomarkers identified so far have been sorted and grouped according to their common biological functions. All of them refer to a series of biological pathways including glycolytic (Bouley et al., 2004; Laville et al., 2009) and oxidative energy production (Zapata et al., 2012), cell detoxification (Zapata et al., 2009; Polati et al., 2012), protease inhibitors (Gagaoua et al., 2012), and production of heat shock proteins (Flower et al., 2005; Bjarnadottìr et al., 2012). Some unusual biomarkers, that is, annexins, galectins, and peroxiredoxins, were also reported (Zapata et al., 2009; Bjarnadottìr et al., 2012; Polati et al., 2012). The protein markers of meat aging associated with tenderness development are summarized in Fig. 6.1.

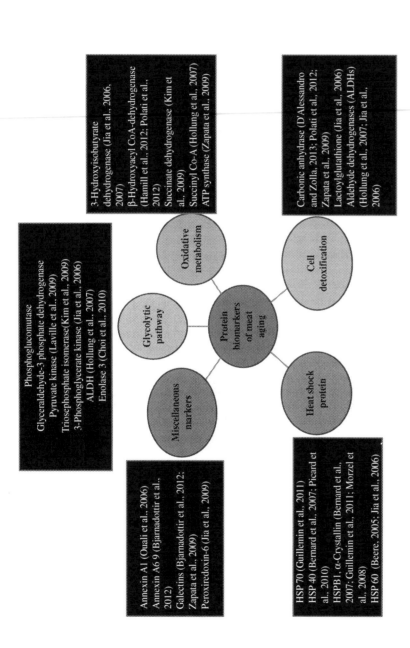

Figure 6.1

Overview of major protein biomarkers associated with important events during conversion of muscle to meat. Beere (2005); Bernard et al. (2007); Bertram et al. (2007); Choi et al. (2010); D'Alessandro, Zolla (2013); Guillemin et al. (2011); Hamill et al. (2012); Hollung et al. (2007); Jia et al. (2007); Jia et al. (2009); Kim et al. (2009); Morzel et al. (2008).

6.6.2 Genotyping for marker gene

μ-Calpain is the most important proteolytic enzyme involved in postmortem tenderization (Koohmaraie, 1996). This enzyme is expressed by the gene known as CAPN1, located on chromosome 29 (Smith et al., 2000). Single nucleotide polymorphisms of this gene, such as CAPN1 316 and CAPN1 4751, have been reported to be associated with meat tenderness (White et al., 2005).

6.6.3 Metabolomics

Metabolomics is the study of biochemical processes involving metabolites in a given biological system at a certain time point and under certain conditions (Fang and Gonzalez, 2014). The "metabolome" indicates all metabolites in a system which may be reactants, intermediates, and end products of metabolism (Dunn et al., 2011). Metabolites are commonly measured using several different analytical techniques, that is, NMR, GC–MS, and LC–MS (Fang and Gonzalez, 2014). Metabolomics has been used to investigate metabolite changes in beef caused by aging (Graham et al., 2012) and metabolite analysis to identify flavor precursors during dry aging (Kim et al., 2016; Ma et al., 2017). The metabolite profiles associated with meat eating quality during aging identified by the metabolomic approach are summarized in Table 6.2.

Table 6.2: Metabolites correlated with meat eating quality traits during conditioning/aging of beef.

Traits	Muscles	Metabolites	Methods	References
Color	Longissimus lumborum, Semimembranosus, Psoas major	Carnitine, free amino acids, nucleotides, nucleosides, glucuronides	HPLC-ESI-MS	Ma et al. (2017)
Flavor	M. longissimus lumborum	Tryptophan, Phenyl alanine, Valine, Tyrosine, Glutamate, IMP, Leucine, Isoleucine	NMR spectroscopy and spectral processing	Kim et al. (2016)
Tenderness		Formic acid, Phe + Trp, Tyr, Tyrm, IMP, α-Glc, α-Man, β-Glc, Ino, Ino + β-Glc, Gln + Glu + α-Glc, αβ-Glc + 2,3-butanodiol, αβ-Glc, Glu + GABA + 3-HBA, Glu, AA + GABA, 2-HBA + BA, α-Ala + Lys + Ile, 3-HBA + EtOH, 2, 3-Butanediol + IBA, Val + ILeu + Leu + 2-HBA	HRMAS-NMR analysis	Castejon et al. (2015)

6.7 Conclusion

The postmortem aging process results in the alteration of the structural, physicochemical, and sensorial characteristics of meat. A number of pre- and postmortem factors and conditions during aging affect the quality characteristics of the aged meat. This chapter provides an overview of different assessment methods to measure aging related changes in meat either directly or indirectly. Among the wide range of assessment methods employed, each method gives different key information related to the postmortem aging of meat. The information provided in this chapter signals toward the potential of using different advanced tools to understand meat quality and optimize the aging/conditioning process for the further improvement of meat quality. Future research in the field of postmortem aging should be focused on optimization of the aging process to shorten the aging period and achieve the desired characteristics of the aged meat.

References

Agarwal, S.K., 1990. Proteases cathepsins-A view. Biochem. Educ. 18, 67–72.

Alarcon-Rojo, A.D., Dransfield, E., 1995. Alteration of post-mortem ageing in beef by the addition of enzyme inhibitors and activators. Meat Sci. 41 (2), 163–178.

Anderson, M.J., Lonergan, S.M., Huff-Lonergan, E., 2012. Myosin light chain 1 release from myofibrillar fraction during postmortem aging is a potential indicator of proteolysis and tenderness of beef. Meat Sci. 90, 345–351.

Beere, H.M., 2005. Death versus survival: functional interaction between the apoptotic and stress-inducible heat shock protein pathways. J. Clin. Invest. 115 (10), 2633–2639.

Bendixen, E., 2005. The use of proteomics in meat science. Meat Sci. 71, 138–149.

Berger, J., Brad Kim, Y.H., Legako, J.F., Martini, S., Lee, J., Ebner, P., et al., 2018. Dry-aging improves meat quality attributes of grass-fed beef loins. Meat Sci. 145, 285–291.

Bernard, C., Cassar-Malek, I., Le Cunff, M., Dubroeucq, H., Renand, G., Hocquette, J.F., 2007. New indicators of beef sensory quality revealed by expression of specific genes. J. Agric. Food Chem. 55 (13), 5229–5237.

Bertram, H.C., Andersen, H.J., Karlsson, A.H., Horn, P., Hedegaard, J., Nogaard, L., et al., 2003. Prediction of technological quality (cooking loss and Napole Yield) of pork based on fresh meat characteristics. Meat Sci. 65, 707–712.

Biswas, A.K., Kripriyalini, L., Tandon, S., Divya, S., Majumder, S., 2016. Simultaneous identification of different domains of calpain from blood and turkey meat samples using casein zymography. Food Anal. Method 9 (10), 2872–2879.

Bjarnadottìr, S.G., Hollung, K., Hoy, M., Bendixen, E., Codrea, M.C., Veiseth-Kent, E., 2012. Changes in protein abundance between tender and tough meat from bovine longissimus thoracis muscle assessed by isobaric Tag for Relative and Absolute Quantitation (iTRAQ) and 2-dimensional gel electrophoresis analysis. J. Anim. Sci. 90, 2035–2043.

Bouley, J., Chambon, C., Picard, B., 2004. Mapping of bovine skeletal muscle proteins using two-dimensional gel electrophoresis and mass spectrometry. Proteomics 4 (6), 1811–1824.

Campbell, R.E., Hunt, M.C., Levis, P., Chambers, E., 2001. Dry-aging effects on palatability of beef longissimus muscle. J. Food Sci. 66, 196–199.

Castejon, D., García-Segura, J.M., Escudero, R., Herrera, A., Cambero, M.I., 2015. Metabolomics of meat exudate: its potential to evaluate beef meat conservation and aging. Anal. Chim. Acta 901, 1–11.

Channon, H.A., D'Souza, D.N., Dunshea, F.R., 2018. Validating post-slaughter interventions to produce consistently high quality pork cuts from female and immunocastrated male pigs. Meat Sci. 142, 14−22.

Choe, J.H., Stuart, A., Kim, Y.H.B., 2016. Effect of different aging temperatures prior to freezing on meat quality attributes of frozen/thawed lamb loins. Meat Sci. 116, 158−164.

Choi, Y.M., Lee, S.H., Choe, J.H., Rhee, M.S., Lee, S.K., Joo, S.T., et al., 2010. Protein solubility is related to myosin isoforms, muscle fiber types, meat quality traits, and postmortem protein changes in porcine longissimus dorsi muscle. Livestock Sci. 127 (2), 183−191.

Christensen, L., Ertbjerg, P., Aaslyng, M.D., Christensen, M., 2011. Effect of prolonged heat treatment from 48°C to 63°C on toughness, cooking loss and color of pork. Meat Sci. 88, 280−285.

Clerjon, S., Peyrin, F., Lepetit, J., 2011. Frontal UV−visible fluorescence polarization measurement for bovine meat ageing assessment. Meat Sci. 88, 28−35.

Colle, M.J., Doumit, M.E., 2017. Effect of extended aging on calpain-1 and -2 activity in beef longissimus lumborum and semimembranosus muscles. Meat Sci. 131, 142−145.

Corbin, C.H., O'Quinn, T.G., Garmyn, A.J., Legako, J.F., Hunt, M.R., Dinh, T.T.N., et al., 2015. Sensory evaluation of tender beef strip loin steaks of varying marbling levels and quality treatments. Meat Sci. 100, 24−31.

D'Alessandro, A., Zolla, L., 2013. Meat science: from proteomics to integrated omics towards system biology. J. Proteomics 78, 558−577.

Dashdorj, D., Amna, T., Hwang, I., 2015. Influence of specific taste-active components on meat flavor as affected by intrinsic and extrinsic factors: an overview. Eur. Food Res. Technol. 241, 157−171.

Dikeman, M.E., Obuz, E., Gök, V., Akkaya, L., Stroda, S., 2013. Effects of dry, vacuum, and special bag aging; USDA quality grade; and end-point temperature on yields and eating quality of beef longissimus lumborum steaks. Meat Sci. 94, 228−233.

Dufour, E., Frencia, J., Elhousseynou, K., 2003. Development of a rapid method based on front-face fluorescence spectroscopy for the monitoring of fish freshness. Food Res. Int. 36 (5), 415−423.

Dunn, W.B., Broadhurst, D.I., Atherton, H.J., Goodacre, R., Griffin, J.L., 2011. Systems level studies of mammalian metabolomes: the roles of mass spectrometry and nuclear magnetic resonance spectroscopy. Chem. Soc. Rev. 40 (1), 387−426.

Eggen, K.H., Malmstrøm, A., Kolset, S.O., 1994. Decorin and a large dermatan sulfate proteoglycan in bovine striated muscle. Biochim. Biophys. Acta 1204 (2), 287−297.

Fang, Z.Z., Gonzalez, F.J., 2014. LC−MS-based metabolomics: an update. Arch. Toxicol. 88 (8), 1491−1502.

Farouk, M.M., Mustafa, N.Md, Wu, G., Krsinic, G., 2012. The "sponge effect" hypothesis: an alternative explanation of the improvement in the water holding capacity of meat with ageing. Meat Sci. 90, 670−677.

Flower, T.R., Chesnokova, L.S., Froelich, C.A., Dixon, C., Witt, S.N., 2005. Heat shock prevents alpha-synuclein-induced apoptosis in a yeast model of Parkinson's disease. J. Mol. Biol. 351 (5), 1081−1100.

Gagaoua, M., Boudida, Y., Becila, S., Picard, B., Boudjellal, A., Sentandreu, M.A., et al., 2012. New caspases' inhibitors belonging to the serpin superfamily: a novel key control point of apoptosis in mammalian tissues. Adv. Biosci. Biotechnol. 3, 740−750.

Gagaoua, M., Terlouw, E.M., Micol, D., Boudjellal, A., Hocquette, J.F., Picard, B., 2015. Understanding early post-mortem biochemical processes underlying meat color and pH decline in the Longissimus thoracis muscle of young blond d'Aquitaine bulls using protein biomarkers. J. Agri. Food Chem. 63 (30), 6799−6809.

Graham, S.F., Farrell, D., Kennedy, T., Gordon, A., Farmer, L., Elliott, C., et al., 2012. Comparing GC−MS, HPLC and H-1 NMR analysis of beef longissimus dorsi tissue extracts to determine the effect of suspension technique and ageing. Food Chem. 134 (3), 1633−1639.

Guillemin, C., Maleszewska, M., Guais, A., Maës, J., Rouyez, M.C., Yacia, A., et al., 2009. Chromatin modifications in hematopoietic multipotent and committed progenitors are independent of gene subnuclear positioning relative to repressive compartments. Stem Cells 27 (1), 108−115.

Guillemin, N., Bonnet, M., Jurie, C., Picard, B., 2011. Functional analysis of beef tenderness. J. Proteomics 75 (2), 352−365.

Hamill, R.M., McBryan, J., McGee, C., Mullen, A.M., Sweeney, T., Talbot, A., et al., 2012. Functional analysis of muscle gene expression profiles associated with tenderness and intramuscular fat content in pork. Meat Sci. 92 (4), 440–450.

Hollung, K., Veiseth, E., Jia, X., Faergestad, E.M., Hildrum, K.I., 2007. Application of proteomics to understand the molecular mechanisms behind meat quality. Meat Sci. 77 (1), 97–104.

Honikel, K.O., 2004. Conversion of muscle to meat. In: Jensen, W.K., Devine, C.E., Dikeman, M. (Eds.), Encyclopedia of Meat Sciences. Elsevier Academic Press.

Hopkins, D.L., Taylor, R.G., 2002. Post-mortem muscle proteolysis and meat tenderization. In: tePas, M., Everts, M., Haagsman, H. (Eds.), Muscle Development of Livestock Animals. Cambridge, MA, pp. 363–389.

Huff-Lonergan, E., Mitsuhashi, T., Beekman, D.D., Parrish, F.C., Olson, D.G., Robson, R.M., 1996. Proteolysis of specific muscle structural proteins by mu-calpain at low pH and temperature is similar to degradation in post-mortem bovine muscle. J. Anim. Sci 74, 993–1008.

Jayasooriya, S.D., Torley, P.J., D'Arcy, B.R., Bhandari, B.R., 2007. Effect of high power ultrasound and ageing on the physical properties of bovine *Semitendinosus* and Longissimus muscles. Meat Sci. 75, 628–639.

Ji, J.R., Takahashi, K., 2006. Changes in concentration of sarcoplasmic free calcium during post-mortem ageing of meat. Meat Sci. 73, 395–403.

Jia, X., Hollung, K., Therkildsen, M., Hildrum, K.I., Bendixen, E., 2006. Proteome analysis of early post-mortem changes in two bovine muscle types: *M. longissimus dorsi* and *M. semitendinosis*. Proteomics 6 (3), 936–944.

Jia, X., Ekman, M., Grove, H., Faergestad, E.M., Aass, L., Hildrum, K.I., et al., 2007. Proteome changes in bovine longissimus thoracis muscle during the early postmortem storage period. J. Proteome Res. 6 (7), 2720–2731.

Jia, X., Veiseth-Kent, E., Grove, H., Kuziora, P., Aass, L., Hildrum, K.I., et al., 2009. Peroxiredoxin-6-A potential protein marker for meat tenderness in bovine longissimus thoracis muscle. J. Anim. Sci. 87 (7), 2391–2399.

Kim, G.D., Jeong, J.Y., Moon, S.H., Hwang, Y.H., Joo, S.T., 2009. Influences of carcass weight on histochemical characteristics and meat quality of crossbred (Korean native black pig × Landrace) pigs. In: Proceedings of 55th International Congress of Meat Science and Technology, vol. PS1.05a, Denmark: Copenhagen.

Kim, Y.H.B., Stuart, A., Black, C., Rosenvold, K., 2012. Effect of lamb age and retail packaging types on the quality of long-term chilled lamb loins. Meat Sci. 90, 962–966.

Kim, Y.H.B., Warner, R.D., Rosenvold, K., 2014. Influence of high pre-rigor temperature and fast pH fall on muscle proteins and meat quality: a review. Anim. Prod. Sci 54 (4), 375–395.

Kim, Y.H.B., Kemp, R., Samuelsson, L.M., 2016. Effects of dry-aging on meat quality attributes and metabolite profiles of beef loins. Meat Sci. 111, 168–176.

Kim, Y.H.B., Meyers, B., Kim, H., Liceaga, A.M., Lemenager, R.P., 2017. Effects of stepwise dry/wet-aging and freezing on meat quality of beef loins. Meat Sci. 123, 57–63.

Kim, S.Y., Yong, H.I., Nam, K.C., Jung, S., Yim, D.G., Jo, C., 2018. Application of high temperature (14°C) ageing of beef *M. semimembranosus* with low-dose electron beam and X-ray irradiation. Meat Sci. 136, 85–92.

Kiran, M., Naveena, B.M., Sudhakar Reddy, K., Shashikumar, M., Ravinder Reddy, V., Kulkarni, V.V., et al., 2015. Muscle-specific variation in buffalo (*Bubalus bubalis*) meat texture: biochemical, ultrastructural and proteome characterization. J. Texture Stud. 46 (4), 254–261.

Kiran, M., Naveena, B.M., Reddy, K.S., Shashikumar, M., Reddy, V.R., Kulkarni, V.V., et al., 2016. Understanding tenderness variability and ageing changes in buffalo meat: biochemical, ultrastructural and proteome characterization. Animal 10 (6), 1007–1015.

Kolczak, T., Pospiech, E., Palka, K., Lacki, J., 2003. Changes in structure of psoas major and minor and semitendinosus muscles of calves, heifers and cows during post-mortem ageing. Meat Sci. 64, 77–83.

Koohmaraie, M., 1990. Inhibition of post-mortem tenderization in ovine carcasses through infusion of zinc. J. Animal Sci 68, 1476.

Koohmaraie, M., 1996. Biochemical factors regulating the toughening and tenderization processes of meat. Meat Sci. 43, S193–S201.

Koohmaraie, M., Geesink, G.H., 2006. Contribution of postmortem muscle biochemistry to the delivery of consistent meat quality with particular focus on the calpain system. Meat Sci. 74, 34–43.

Koohmaraie, M., Whipple, G., Kretchmar, D.H., Crouse, J.D., Mersmann, H.J., 1991. Postmortem proteolysis in longissimus muscle from beef, lamb and pork carcasses. J. Anim. Sci. 69, 617–624.

Laville, E., Sayd, T., Morzel, M., Blinet, S., Chambon, C., Lepetit, J., et al., 2009. Proteome changes during meat ageing in tough and tender beef suggest the importance of apoptosis and protein solubility for beef ageing and tenderization. J. Agri. Food Chem. 57, 10755–10764.

Lawrie, R.A., 1998. Meat Science, sixth ed. Woodhead Publishing Limited, Cambridge.

Lepetit, J., 2008. Collagen contribution to meat toughness: theoretical aspects. Meat Sci. 80, 960–967.

Li, K., Zhang, Y., Mao, Y., Cornforth, D., Dong, P., Wang, R., et al., 2012. Effect of very fast chilling and aging time on ultra-structure and meat quality characteristics of Chinese Yellow cattle M. Longissimus lumborum. Meat Sci. 92, 795–804.

Li, X., Babol, J., Bredie, W.L.P., Nielsen, B., Tomankova, J., Lundstrom, K., 2014. A comparative study of beef quality after ageing longissimus muscle using a dry ageing bag, traditional dry ageing or vacuum package ageing. Meat Sci. 97, 433–442.

Liu, A., Nishimura, T., Takahashi, K., 1995. Structural weakening of intramuscular connective tissue during post mortem ageing of chicken semitendinosus muscle. Meat Sci. 39, 135–142.

Lomiwes, D., Farouka, M.M., Wu, G., Young, O.A., 2014. The development of meat tenderness is likely to be compartmentalised by ultimate pH. Meat Sci. 96, 646–651.

Ma, D., Kim, Y.H.B., Cooper, B., Oh, J., Chun, H., Choe, J., et al., 2017. Metabolomics profiling to determine the effect of postmortem aging on color and lipid oxidative stabilities of different bovine muscles. J. Agric. Food Chem. 65 (31), 6708–6716.

Marino, R., Albenzio, M., della Malva, A., Santillo, A., Loizzo, P., Sevi, A., 2013. Proteolytic pattern of myofibrillar protein and meat tenderness as affected by breed and aging time. Meat Sci. 95, 281–287.

Mazzucco, J.P., Melucci, L.M., Villarreal, E.L., Mezzadra, C.A., Soria, L., Corva, P., et al., 2010. Effect of ageing and μ-calpain markers on meat quality from Brangus steers finished on pasture. Meat Sci. 86, 878–882.

Melody, J.L., Lonergan, S.M., Rowe, L.J., Huiatt, T.W., Mayes, M.S., Huff-Lonergan, E., 2004. Early postmortem biochemical factors influence tenderness and water holding capacity of three porcine muscles. J. Anim. Sci. 82 (4), 1195–1205.

Miller, M.F., Carr, M.A., Ramsey, C.B., Crockett, K.L., Hoover, L.C., 2001. Consumer thresholds for establishing the value of beef tenderness. J. Anim. Sci. 79, 3062–3068.

Modzelewska-Kapituła, M., Kwiatkowska, A., Jankowska, B., Dąbrowska, E., 2015. Water holding capacity and collagen profile of bovine m. infraspinatus during postmortem ageing. Meat Sci. 100, 209–216.

Moeller, P.W., Fields, P.A., Dutson, T.R., Landmann, W.A., Carpenter, Z.L., 1976. Effect of high temperature conditioning on subcellular distribution and levels of lysosomal enzymes. J. Food Sci. 41, 216.

Morzel, M., Terlouw, C., Chambon, C., Micol, D., Picard, B., 2008. Muscle proteome and meat eating qualities of Longissimus thoracis of "Blonde d'Aquitaine" young bulls: a central role of HSP27 isoforms. Meat Sci. 78 (3), 297–304.

Naveena, B.M., Muthukumar, M., Kulkarni, V.V., Praveen Kumar, Y., Usha Rani, K., Kiran, M., 2015. Effect of aging on the physicochemical, textural, microbial and proteome changes in emu (Dromaius novaehollandiae) meat under different packaging conditions. J. Food Process. Preserv. 39 (6), 1745–4549.

Negishi, H., Yamamoto, E., Kuwata, T., 1996. The origin of the 30 kDa component appearing during post-mortem ageing of bovine muscle. Meat Sci. 42 (3), 289–303.

Nishimura, T., Hattori, A., Takahashi, K., 1996. Relationship between degradation of proteoglycans and weakening of the intramuscular connective tissue during post-mortem ageing of beef. Meat Sci. 42 (3), 251–260.

Nishimura, T., Liu, A., Hattori, A., Takahashi, K., 1998. Changes in mechanical strength of intramuscular connective tissue during postmortem aging of beef. J. Anim. Sci. 76, 528–532.

O'Halloran, G.R., Troy, D.J., Buckley, D.J., Reville, W.J., 1997. The role of endogenous proteases in the tenderisation of fast glycolysing muscle. Meat Sci. 47, 187–210.

Ouali, A., Gagaoua, M., Boudida, Y., Becila, S., Boudjellal, A., Herrera-Mendez, C.H., et al., 2013. Biomarkers of meat tenderness: present knowledge and perspectives in regards to our current understanding of the mechanisms involved. Meat Sci. 95 (4), 854–870.

Ouali, A., Herrera-Mendez, C.H., Coulis, G., Becila, S., Boudjellal, A., Aubry, L., et al., 2006. Revisiting the conversion of muscle into meat and the underlying mechanisms. Meat Sci. 74 (1), 44–58.

Palka, K., 2003. The influence of post-mortem ageing and roasting on the microstructure, texture and collagen solubility of bovine semitendinosus muscle. Meat Sci. 64, 191–198.

Pearce, K.L., Rosenvold, K., Andersen, H.J., Hopkins, D.L., 2011. Water distribution and mobility in meat during the conversion of muscle to meat and ageing and the impacts on fresh meat quality attributes — a review. Meat Sci. 89, 111–124.

Pen, S., Kim, Y.H.B., Luc, G., Young, O.A., 2012. Effect of pre-rigor stretching on beef tenderness development. Meat Sci. 92, 681–686.

Picard, B., Berri, C., Lefaucheur, L., Molette, C., Sayd, T., Terlouw, C., 2010. Skeletal muscle proteomics in livestock production. Brief. Funct. Genomics 9 (3), 259–278.

Polati, R., Menini, M., Robotti, E., Millioni, R., Marengo, E., Novelli, E., et al., 2012. Proteomic changes involved in tenderization of bovine Longissimus dorsi muscle during prolonged ageing. Food Chem. 135 (3), 2052–2069.

Pomponio, L., Bukh, C., Ruiz-Carrascal, J., 2018. Proteolysis in pork loins during superchilling and regular chilling storage. Meat Sci. 141, 57–65.

Rosenvold, K., North, M., Devine, C.E., Micklander, E., Hansen, P.W., Dobbie, P.M., et al., 2008. The protective effect of electrical stimulation and wrapping on beef tenderness at high pre-rigor temperatures. Meat Sci. 79 (2), 299–306.

Savell, J.W., 2008. Dry-Aging of Beef. Executive Summary. National Cattlemen's Beef Association, Centennial, CO, 16-11.

Sentandreu, M.A., Coulis, G., Ouali, A., 2002. Role of muscle endopeptidases and their inhibitors in meat tenderness. Trends Food Sci. Technol. 13, 400–421.

Smith, T.P.L., Casas, E., Rexroad III, C.E., Kappes, S.M., Keele, J.W., 2000. Bovine CAPN1 maps to a region of BTA29 containing a quantitative trait locus for meat tenderness. J. Anim. Sci 78, 2589–2594.

Smith, G.C., Tatum, J.D., Belk, K.E., 2008a. International perspective: characterization of United States Department of Agriculture and Meat Standards Australia systems for assessing beef quality. Aust. J. Exp. Agric. 48, 1465–1480.

Smith, R.D., Nicholson, K.L., Nicholson, J.D.W., Harris, K.B., Miller, R.K., Griffin, D.B., et al., 2008b. Dry versus wet aging of beef: retail cutting yields and consumer palatability evaluations of steaks from US Choice and US Select short loins. Meat Sci. 79, 631–639.

Sorimachi, H., Kinbara, K., Kimura, S., Takahashi, M., Ishiura, S., Sasagawa, N., et al., 1995. Muscle-specific calpain, p94, responsible for limb girdle muscular dystrophy type 2A, associates with connectin through IS2, a p94-specific sequence. J. Biol. Chem. 270, 31158–31162.

Spanier, A., Flores, M., McMillin, K., Bidner, T., 1997. The effect of post-mortem aging on meat flavor quality in Brangus beef. Correlation of treatments, sensory, instrumental and chemical descriptors. Food Chem. 59, 531–538.

Starkey, C.P., Geesink, G.H., Ven, R., Hopkins, D.L., 2017. The relationship between shear force, compression, collagen characteristics, desmin degradation and sarcomere length in lamb biceps femoris. Meat Sci. 126, 18–21.

Takahashi, K., 1996. Structural weakening of skeletal muscle tissue during post-mortem ageing of meat: the non-enzymatic mechanism of meat tenderization. Meat Sci. 43 (Suppl. 1), 67–80.

Takahashi, K., Hattori, A., Kuroyanagi, H., 1995. Relationship between the translocation of tropomyosin and the restoration of rigor-shortened sarcomeres during post-mortem ageing of meat-a molecular mechanism of meat tenderization. Meat Sci. 40, 413–423.

Trout, G.R., 1988. Techniques for measuring water-binding capacity in muscle foods-a review of methodology. Meat Sci. 23 (4), 235–252.

Vitale, M., Pérez-Juan, M., Lloret, E., Arnau, J., Realini, C.E., 2014. Effect of aging time in vacuum on tenderness, and color and lipid stability of beef from mature cows during display in high oxygen atmosphere package. Meat Sci. 96, 270–277.

Whipple, G., Koohmaraie, M., Dikeman, M.E., Crouse, J.D., Hunt, M.C., Klemm, R.D., 1990. Evaluation of attributes that affect longissimus muscle tenderness in *Bos taurus* and *Bos indicus* cattle. J. Anim. Sci. 68 (9), 2716–2728.

White, S.N., Casas, E., Wheeler, T.L., Shackelford, S.D., Koohmaraie, M., Riley, D.G., et al., 2005. A new single nucleotide polymorphism in CAPN1 extends the current tenderness marker test to include cattle of *Bos indicus, Bos taurus*, and crossbred descent. J. Anim. Sci. 83, 2001–2008.

Wilkins, M.R., Pasquali, C., Appel, R.D., Ou, K., Golaz, O., Sanchez, J.C., et al., 1996. From proteins to proteomes: large scale protein identification by two-dimensional electrophoresis and amino acid analysis. Biotechnology 14, 61–65.

Wu, G., Farouk, M.M., Clerens, S., Rosenvold, K., 2014. Effect of beef ultimate pH and large structural protein changes with aging on meat tenderness. Meat Sci. 98, 637–645.

Xiong, Y.L., 2004. Protein functionality. In: Jensen, W.K., Devine, C.E., Dikeman, M. (Eds.), Encyclopedia of Meat Sciences. Elsevier Academic Press.

Zamora, F., Aubry, L., Sayd, T., Lepetit, J., Lebert, A., Sentandreu, M.A., et al., 2005. Serine peptidase inhibitors, the best predictor of beef ageing amongst a large set of quantitative variables. Meat Sci. 71, 730–742.

Zapata, I., Zerby, H.N., Wick, M., 2009. Functional proteomic analysis predicts beef tenderness and the tenderness differential. J. Agric. Food Chem. 57 (11), 4956–4963.

Zapata, I., Reddish, J.M., Miller, M.A., Lilburn, M.S., Wick, M., 2012. Comparative proteomic characterization of the sarcoplasmic proteins in the pectoralis major and supracoracoideus breast muscles in 2 chicken genotypes. Poultry Sci. 91, 1654–1659.

Further reading

Huff-Lonergan, E., Zhang, W., Lonergan, S.M., 2005. Biochemistry of postmortem muscle- Lessons on mechanisms of meat tenderization. Meat Sci. 86, 184–195.

Nishimura, T., Hattori, A., Takahashi, K., 1995. Structural weakening of intramuscular connective tissue during conditioning of beef. Meat Sci. 39, 127–133.

CHAPTER 7

Calpain-assisted postmortem aging of meat and its detection methods

Ashim Kumar Biswas[1], S. Tandon[1] and Prabhat Kumar Mandal[2]

[1]Division of Post-Harvest Technology, ICAR-Central Avian Research Institute, Izatnagar, Bareilly, India [2]Department of Livestock Products Technology, Rajiv Gandhi Institute of Veterinary Education and Research, Puducherry, India

Chapter Outline

7.1 Introduction 101
7.2 Structure and functions of calpain and calpastatin 102
7.3 Pathways of calpain activity in muscle tissues 103
7.4 Postmortem proteolysis of skeletal muscle by calpain 104
7.5 Purification of calpain 1, calpain 2, and calpastatin 106
    7.5.1 Hydrophobic interaction chromatography 106
    7.5.2 Ion-exchange chromatography 107
7.6 Detection and quantification 108
    7.6.1 Casein zymography 108
    7.6.2 Biochemical method 108
    7.6.3 Fluorometric method 109
    7.6.4 Bioluminescent assay 109
    7.6.5 Proteomic-based analysis 110
    7.6.6 Identification of biomarkers 110
7.7 Scope of future work 111
References 111

7.1 Introduction

Tenderness of meat is the most influential attribute, considerably affecting the eating quality of meat to the consumers (Koohmaraie, 1994; Koohmaraie and Geesink, 2006; Biswas et al., 2016a). The problem of tenderness though mostly was the concern of the beef industry (Nowak, 2011) until the last decade, but with the consumers' increasing awareness of quality, it is now important for the other animal industries too. Consumers are willing to pay premium prices for meat with guaranteed tenderness. Indeed it is now a real challenge to the researchers and also for the meat industry to achieve a product with standardized and guaranteed tenderness. Many studies have revealed that meat tenderization process is a complex mechanism involving several pathways including pre- and postslaughter factors and their interactions. Thus to acquire knowledge about the biology of meat tenderness a deeper understanding is required, and most importantly, gathering such information is required for its effective applications. Some researchers suggested that proteosomes may be responsible for the meat tenderness, while others felt it is caspases (Bernard et al., 2007; Kemp et al., 2010). Whatsoever the facts, it is now clear that the calpain system has a major role in postmortem proteolysis and meat tenderness which is presumably due to the action of calpains, the calcium-dependent endogenous proteases (Koohmaraie, 1996; Neath et al., 2007; Kemp et al., 2010). Therefore, in this chapter a specific importance is dedicated to the calpains.

Calpains (CAPNs) are a large family of 14 intracellular cysteine proteases, expressed in ubiquitous and tissue-specific forms in all mammals and other organisms including birds. Calpains are involved in variety of pathophysiological activities like postmortem proteolysis including apoptosis, necrosis, cell proliferation, cell motility, cell cycle progression, differentiation, membrane fusion, and platelet activation (Moldoveanu et al., 2002; O'Brien et al., 2005). In the skeletal muscle, the calpain family contains three proteases: calpain 1 or CAPN1 (μ-calpain), calpain 2 or CAPN2 (m-calpain), and CAPN3 (p94), and their potential inhibitor calpastatin (Koohmaraie and Geesink, 2006; Moudilou et al., 2010). Details of the calpain family have been reviewed by Nowak (2011). In skeletal muscle, calpain 1 and calpain 2 are mainly located in myofibrils (70%) and cytosol (Xu et al., 2009), respectively. Likewise, calpastatin is also located in these places and all are eventually distributed in the I-band, A-band, and Z- disk while calpain 3 is situated in the sarcomere only.

Thus to better understand the role of the calpain family in postmortem proteolysis of skeletal muscle, this chapter focuses on the biochemistry and applied aspects of calpain 1, calpain 2, and calpastatin. It also deals with the structure and functions of calpains and calpastatin, the pathways of calpain activity, the postmortem proteolysis of skeletal muscle, the purification and detection of calpains, and the future scope of research in this field.

7.2 Structure and functions of calpain and calpastatin

The structure and function of calpain 1 and 2 are nearly same (Fig. 7.1A and B) and both are active in the presence of calcium ions (Ca^{2+}). Calpain 1 and 2 are activated by using micro- and millimolar concentrations of Ca^{2+}, respectively, hence, they are called μ- and m-calpains. Both the calpains have an identical large catalytic subunit (80 kDa) and a small regulatory subunit (28 kDa). Calpastatin shows variable subunits (60–70 kDa), but the 66 kDa subunit has greater inhibitory activity (Biswas et al., 2016a). The 80 kDa catalytic subunit has four structural domains (I, II, III, and IV). Domain I or the NH_2 terminal domain is known as the autolytic domain as it autolyzed upon calpain activation. Domain II

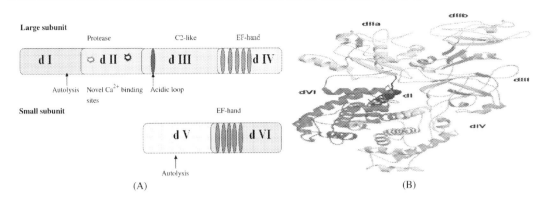

Figure 7.1
Schematic (A) and crystallographic (B) structures of calpain. Source: *The sketch was adopted from Sentandreu, M.A., Coulis, G., Ouali, A., 2002. Role of muscle endopeptidases and their inhibitors in meat tenderness. Trends Food Sci. Technol. 13, 400—421.*

is the cysteine catalytic site which acts as the proteolytic domain and is similar to other cysteine proteases. Domain III is the large subunit which contains the characteristics C2 like domain and is said to be involved in structural changes during the binding of calcium. Domain IV contains five EF hands and is responsible for the "calmodulin"-like binding capacity of calcium. The small subunit has two domains (V and VI). Domain V is 28 kDa, is referred to as the hydrophobic domain, and is reported to bind phospholipids. Domain VI is highly similar to domain IV in the catalytic subunit as it also contains four EF hands homologous to domain IV (Sentandreu et al., 2002; Carragher and Frame, 2002).

Calpastatin contains four inhibitory domains sequences spanning 120—140 amino acid residues (domains I, II, III, and IV) and an N-terminal nonhomologous sequence (domain L). Within the inhibitory domain, in each of the four domains there are three regions A, B, and C that are predicted to interact with calpain. The binding of calcium to calpain causes changes in the calpain molecule enabling it to become active but also allowing calpastatin to interact with the enzyme. The peptide chain helices found in regions A and C, interact with calpain at two separate sites causing the inhibitory domain to wrap around calpain. The region between A and C, region B, then blocks the active site of calpain. The crystallography studies suggest that region B does not directly bind to the active site, thereby preventing it from becoming a substrate for the enzyme (Parr et al., 2001; Goll et al., 2003; Kemp et al., 2010) (Fig. 7.2).

7.3 Pathways of calpain activity in muscle tissues

The calpain activity in muscle is dependent on certain factors such as pH, level of calcium ions, and temperature. Some studies postulated that immediately after slaughter, muscle pH

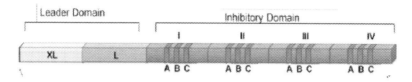

Figure 7.2
Schematic structure of calpastatin. Source: The sketch was adopted from Parr, T., Sensky, P.L., Bardsley, R.G., Buttery, P.J., 2001. Calpastatin expression in cardiac and skeletal muscle and partial gene structure. Arch. Biochem. Biophys. 395, 1–13 and Goll, D.E., Thompson, V.F., Li, H., Wei, W., Cong, J., 2003. The calpain system. Physiol. Rev. 83, 731–801.

remained higher (pH 7.2) (Biswas, 2016) and Ca^{2+} concentration (Dransfield et al., 1992) shows as low as 10^{-7} mol in the sarcoplasm which is not enough to activate the calpain system. On subsequent holding for few hours muscle pH declines to 6.8 with the increase in Ca^{2+} concentration to 10^{-4} mol fueled by calcium pumps or the sarcoplasmic reticulum (SR) that actually help initiate calpain 1 activation. It indicates that a 100- to 1000-fold increase of Ca^{2+} concentration is required for the activation of calpain 1 (μ-calpain). Several in vitro studies suggested that calpain 1 needs at least 0.3 M free Ca^{2+} (in living muscle 0.2 M) to start proteolytic activity, while that for calpain 2 is to be 0.4–0.8 M (Kurebayashi et al., 1993; Goll et al., 2003). Upon activation of the calpain system both the large and small subunits (80 and 28 kDa) undergo hydrolysis, altering their masses to as low 76 and 18 kDa, respectively. Thus the activity of enzymes of the calpain system decreased with the increase of aging time with a concomitant improvement in tenderness (Taylor et al., 1995; Koohmaraie and Geesink, 2006).

The optimum condition of pH for postmortem muscle at 4°C was well-recognized. The activity of calpain 1 in postmortem chicken breast muscle was found until pH 5.9 and the aging process was completed at 48 hours, while in thigh muscle aging was completed at 24 hours postmortem with the resultant pH 6.2. Likewise, in another study it was revealed that breast muscle from chicken species developed postmortem aging as early as 3 hours or even after 24 hours. (Veeramuthu and Sams, 1999). In goat muscle the activity of calpain was noted up to pH 5.92 but decreased gradually from an initial value of 6.92 by 72 hours (Biswas et al., 2016b; Maddock et al., 2005), while in beef its active till pH reached around 5.6 (Huff-Lonergan et al., 1996).

7.4 Postmortem proteolysis of skeletal muscle by calpain

The mechanism of activity of calpain in postmortem proteolysis of skeletal muscle is well understood. But the extent and rate of postmortem proteolysis varies among different muscles and presumably these variations contribute to the large variations in tenderness between muscles. The muscle proteolysis is initiated before and/or after the onset of rigor phase depending on a protease system where meat gradually becomes tender. Several experiments have shown that though calpain 1 and calpain 2 showed same degradation pattern of myofibrillar proteins in knockout mice, calpain 1 rather calpain 2 is responsible for the cleavage of some vital proteins such as tropomyosine, nebulin, titin, troponin T and I, desmin, dystrophin, vinculin, and metavinculin. These are associated with the cytoskeleton maintaining the myofibril's structure, as well as those associated with structures maintaining the myofibril interaction with the sarcolemma (the costameres) (Fig. 7.3) and contribute to the improvement of tenderness, whereas, high levels of calpastatin are related to decreased proteolysis and increased meat toughness (Koohmaraie and Geesink, 2006; Geesink et al., 2006; Kemp et al., 2010). But calpain 3 (p94) did not show postmortem proteolysis of skeletal muscle in knockout mice. It was observed that during the initial phase of proteolysis calpain first degrades the protein of the N_2 line (nebulin), then the Z line (desmin), and finally, frees the α-actinine from the Z line. However, calpain never degrades α-actinine, actin, myosin, or acto-myosin. Thus the degree of cytoskeletal and regulatory proteins degradation indicates the nature of the tenderness of meat. In a study Geesink et al. (2001) reported that the degradation of troponin T leads to the appearance of new fragments of protein with molecular weight of 27–30 kDa, while Hughes et al. (2001) found eight peptides with molecular weight from 14 to 26 kDa but this degradation depends on the species of animals, their breed, and the type of muscle and its activity (Northcutt et al., 1998).

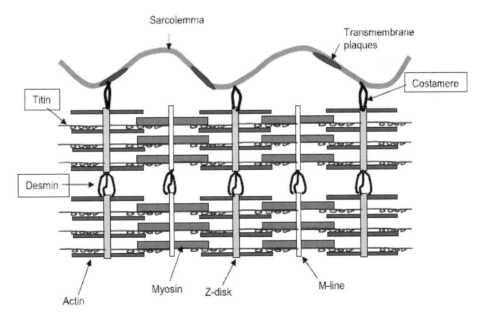

Figure 7.3
Schematic representation of muscle myofibrillar protein. Source: *The sketch adopted from Taylor, R. G., Geesink, G.H., Thompson, V.F., et al., 1995. Is Z-disk degradation responsible for post-mortem tenderization? J. Anim. Sci. 73, 1351–1367.*

In chicken Biswas (2016) reported at least six important modifications. One protein of molecular weight near 110 kDa disappeared at 24 hours postmortem, though it was distinctly visible at 6 hours and fairly visible at 12 and 18 hours. Similarly two proteins of approximately 29 and 30 kDa first appeared immediately after postslaughter, and then the band clearance increased up to 6 hours, but on subsequent aging for 12−24 hours the bands disappeared. The 110 kDa protein could be α-actinin, which is known to be a substrate of calpain 1. The protein bands found near 29 and 30 kDa were results from the partial hydrolysis of troponin T by calpain. This 30 kDa protein is considered to be a good marker of postmortem aging in chicken breast muscle. The bands observed at 43 and 65−70 kDa could be due to the presence of proteins actin and tropomyosine, but additional bands found at 33 and 35.4 kDa might be due to the proteins α-tropomyosine and troponin T, respectively.

For calpastatin activity, Geesink and Koohmaraie (1999) reported meat of callipyge sheep was tough. This could be due to the activity of a high level of calpastatin, a potential inhibitor of calpain 1. This affirmed that calpains of the calpain system are mainly responsible for skeletal muscle degradation, and thereby postmortem tenderization of meat.

7.5 Purification of calpain 1, calpain 2, and calpastatin

The purification and separation of calpains are very difficult tasks since calpains contain identical subunits, while others poorly bind with the column matrix. Several efforts were made for effective separation and purification of calpain 1 and calpain 2 from calpastatin but in most cases hydrophobic interaction and ion-exchange chromatography techniques were applied with a greater degree of selectivity and specificity. Some researchers used single column while others used two column techniques with the variation of column matrices.

7.5.1 Hydrophobic interaction chromatography

Proteins are highly charged molecules, and they are well separated in ion-exchange chromatography using normal phase or reverse phase chromatography techniques. Etherington et al. (1987) first attempted to separate calpastatin from the calpain 1 and calpain 2 by hydrophobic interaction chromatography (phenyl-sepharose), followed by ion-exchange chromatography. But recovery of calpain 1 on phenyl-sepharose columns was poor (Kretchmar et al., 1989) and also a long elution time is required and the number of samples that can be processed in a given time is therefore limited. Later Koohmaraie (1990) reported that purification and separation of all three components, that is, calpain 1, calpain 2, and calpastatin can be achieved in a single step using ion-exchange chromatography. But baseline separation of calpain 1 and calpastatin is very difficult since these enzymes loosely bind with the column matrix, and for this, stepwise elution cannot achieve the desired separation. However, a shallow gradient profile can efficiently be capable of baseline separation of calpastatin from calpain 1 (Koohmaraie, 1990). To further upgrade and for rapid separation, Karlsson (2017) used two column techniques: hydrophobic interaction chromatography and ion-exchange chromatography. This method involves the perfusion of a small amount of tissues using EDTA followed by homogenization, centrifugation, and separation on hydrophobic interaction chromatography using phenyl-sepharose as the column matrix (Karlsson et al., 1985). They reported that calpain 1 can be eluted at 10–25 mM NaCl (pH 7.9) while calpain 2 can be eluted at 50–80 mM NaCl (pH 7.5) using a phenyl-sepharose column but both of these were overlapping with each other and also with other proteins. Since a single column technique seldom purified calpain 1 and calpain 2, these authors again used ion-exchange chromatography (mono Q column) for the separation of calpain molecules (Nilsson and Karlsson, 1990).

7.5.2 Ion-exchange chromatography

Geesink and Koohmaraie (1999) affirmed and compared stepwise and continuous gradient ion-exchange chromatography, in which DEAE-sephacel was used as the column matrix. The continuous gradient was applied from 25 to 400 mM concentration while two- and three-step elutions were conducted at 200 and 400 mM and 100, 200, and 400 mM NaCl concentrations, respectively. According to these authors, the two-step and continuous gradient method yielded similar results over a broad range of activities but calpastatin and calpain 1 could not be separated completely using the three-step chromatography method. Further, the two-step gradient method is a fast and inexpensive method to determine calpain and calpastatin activities in studies designed to quantify the components of the calpain proteolytic system in skeletal muscle. Recently, Biswas et al. (2016a) also purified calpain 1, calpain 2, and calpastatin from blood (RBC) and tissue extracts by centrifugation followed by dialysis (12 kDa MWCO) and a single column separation technique on anion-exchange chromatography (DEAE-sephacel). For elution, calpastatin eluded first, followed by calpain 1 and calpain 2 at 100, 200, and 400 mM concentrations of NaCl, respectively.

In another study Savart et al. (1995) used two-step column separation techniques in which proteins were loaded and separated by anion-exchange chromatography using Q-sepharose column followed by gel filtration chromatography using sephacryl S-300 HR as column matrix. The proteins were eluted with a gradient elution profile of NaCl buffer on a Q-sepharose column equilibrated with equilibrium buffer, and finally, a partially purified fraction containing calpain 1 was loaded on sephacryl S-300 HR for further purification. Purified calpain was separated using 150 mM concentration of NaCl buffer. Calpain activity was assayed by a spectrofluorimeter using fluorescein isothiocyanate-labeled casein as substrate, and was confirmed using immunoblotting techniques.

7.6 Detection and quantification

7.6.1 Casein zymography

The activity of calpain 1 and calpain 2 were effectively identified due to the discovery of zymography techniques. Zymography is a protein chemistry technique used for measuring nondenatured enzyme activity, first coined by Granelli-Piperno and Reich in 1978 and has been subsequently adapted for use with a number of different enzymes (Huessen and Dowdle, 1980; Koohmaraie, 1992; Ono and Sorimachi, 2012). Different types of zymography include (1) substrate-based (gelatine, casein, collagen, heparin, etc); (2) reserved zymography; or (3) in situ zymography. Casein zymography is a type of substrate zymography. In this technique, enzymes within a sample are separated by polyacrylamide gel electrophoresis under denaturing, nonreducing conditions. After electrophoresis, the enzymes are allowed to renature and then consume casein within the polyacrylamide gel that lyses the proteins (Biswas et al., 2016d). The gel is then stained with Coomassie Blue R-250 to identify bands corresponding to the enzymatic degradation of the substrate. The location and size of the clear bands within the gel provide information about which enzyme degraded the substrate and how much activity was present in the sample.

Since casein is enzymatically degraded by calpains, it is considered to be substrate zymography. This technique can be used to identify the calpain 1 and calpain 2 from a crude extract from tissues or cultured cells, even in the presence of endogenous calpastatin (Raser et al., 1995) provided with the some protease inhibitors during sample extraction and preparation, however for confirmation, their separation/purification and Western blotting indeed required (Biswas et al., 2016c). Biswas et al. (2016a) described a method for simultaneous identification of calpain 1 and calpain 2 from skeletal muscle and blood samples from diversified animal and poultry species.

7.6.2 Biochemical method

The biochemical method is designed to quantify the purified or partially purified calpains and calpastatin. It is a simple, low-cost method but time-consuming and sometimes not amenable in the laboratory for the time and speed at which it is required. In this method, two types of assay buffer are used for the determination of calcium-dependent and -independent activity separately. First, the activity of the substrate was measured in the presence of $CaCl_2$ and EDTA in the reaction mixture, whereas for the determination of calcium-independent activity, $CaCl_2$ in the reaction mixture was replaced by EDTA. To determine Ca^{2+}-dependent proteolytic activity, absorbance at A_{278} in the presence of EDTA was subtracted from that of the $CaCl_2$ reactions. Blanks are made using 200 and 400 mM NaCl in an equilibration buffer for calpain 1 and calpain 2, respectively, in an assay buffer containing either $CaCl_2$ or EDTA. Total activity is calculated by multiplying Ca^{2+}-dependent proteolytic activity by the dilution factor. CDP Activity (Units/g) = (A_{278} CaCl$_2$ buffer − A_{278} EDTA buffer) × dilution factor. One unit of calpain activity was defined as the amount of enzyme that catalyzed an increase of 1.0 absorbance unit at A_{278} after 60 minutes at 25°C.

For calpastatin activity, three tubes were used to assay inhibitory activity: (1) calpain 2 pooled fractions, incubated with assay buffer containing CaCl$_2$; (2) calpain 2 fraction plus calpastatin fraction, incubated in assay buffer as described; and (3) calpastatin fraction alone, incubated in assay buffer containing EDTA. Total inhibitor activity is calculated according to the formula: Total inhibitor activity (Units/g) = 1−(2−3) × dilution factors. One unit of calpastatin activity is defined as the amount of calpastatin that inhibits one unit of m-calpain (Koohmaraie, 1990; Salem et al., 2004).

7.6.3 Fluorometric method

Use of the fluorometric method is more advantageous than the biochemical method in terms of time and sensitivity but faces the drawback of being more complex and costlier. Karlsson (2017) reported calcium-dependent activity of partially purified calpain 1 and calpain 2 using synthetic peptide substrate succinyl-Leu-Tyr-7-amino-4-methylcoumarin (Succ-Leu-Tyr-AMC) in the presence of dithiotriotol, dimethyl sulfoxide, calcium ion, and excess EDTA. They also determined calcium-independent activity in the presence EDTA but in the absence of calcium. The absorbance of hydrolysis products was measured using a spectrophotometer with an excitation wavelength of 380 nm and an emission wavelength of 460 nm, while for a microtiter plate calpain assay with casein substrate and fluorescamine dye it measured within 40 minutes and subtracted with zero time value. In a later experiment they used glutamate as the standard for activity. Similarly Biswas et al. (2016c) used fluorogenic calpain substrate *N*-Succinyl-Leu-Tyr-7-amido-4-methylcourmarin for the determination of calpain 1 activity of crude extract but reported unsatisfactory results, but when they tried the biochemical method both calpains (μ and m) and calpastatin were determined with a greater degree of satisfaction.

7.6.4 Bioluminescent assay

Bioluminescent assay also known as Calpain-Glo protease assay was developed by O'Brien et al. (2005) and is to used to measure calpain 1 and calpain 2 activities in the homogenous mixture of the sample matrix. The test is based on the principle that a luminogenic succinyl calpain substrate (Suc-LLVY aminoluciferin) reacts with Ultra-Glo recombinant luciferase in a buffer system specially developed for calpain and produces glow-type luminescent signals which is measured spectrophotometrically. The luminescent signal generated by the homogenous mixture is proportional to the amount of calpain present in the sample.

7.6.5 Proteomic-based analysis

Several independent proteomic analyses were affirmed by some researchers to identify endogenous calpain substrates (Cuerrier et al., 2005; Bozoky et al., 2009; Brule et al., 2010; Caberoy et al., 2011), but their role in the proteolysis of substrates is not clear. In comparison, Kim et al. (2013) reported gel-based techniques using two-dimensional gel electrophoresis that were easy to understand and identify target calpain substrates. They used MN9D dopaminergic neuronal cell line, prepared cell lysates for extraction, and finally separated them by isoelectric focusing. The proteins were immobilized on a gel strip, incubated with or without a recombinant calpain and then separated by molecular weight. On separation of protein by 2D gel electrophoresis, several protein spots were selected for analysis in MALDI-TOF for confirmation and validation in cultured cells and in rat models of neurodegenerative diseases.

A novel bioinformatics-based proteomic approach was also adopted for the prediction of breakdown products of calpain-2 and caspase 3 proteases (El-Assaad et al., 2017). According to these authors, calpain-2 and caspase-3 proteases generate BDPs fragments that are indicative of different neural cell death mechanisms in different injury scenarios. Thus this method can be utilized for the identification of postmortem proteolysis of skeletal muscle too. The method utilizes state-of-the-art sequence matching and the alignment algorithms start by locating consensus sequence occurrences and their variants in any set of protein substrates, generating all the fragments resulting from cleavage.

7.6.6 Identification of biomarkers

Many experiments were conducted to quantify calpastatin as a biomarker of meat tenderness. These include enzyme-linked immunosorbent assay, immunosensors-based techniques (surface plasmon resonance, fluorescence resonance energy transfer, etc), optical fiber biosensors, capillary-based biosensors, etc. (Zor et al., 2009). However, they vary in degree of sensitivity and accuracy. The most recent development elucidated the calpain activity through gene expression (Meyers and Beever, 2008). According to these authors, the calpastatin gene in the skeletal muscle is regulated via some promoters associated with the 5′ exons: 1xa, 1xb, which are in tandem and 1u, which is the 3′ end distal to 1xa and 1xb. Thus it is assumed that the difference between transcriptional activity of the calpastatin gene promoters between species and their different responses to stimuli is probably partly responsible for the variation in calpastatin expression that contributes to variations in meat tenderness (Kemp et al., 2010), however, it requires further study for confirmation.

Recently, gel-based techniques, biochemical tests, genetic tests, and even kit-based tests have been widely used for the detection of calpain. Marker-based tests are more prevalent commercially for the assessment of meat quality (Nowak, 2011).

7.7 Scope of future work

Calpains are probably the most extensively studied endogenous enzyme family of the skeletal muscle system. In most studies it is affirmed that the role of the calpain family in the postmortem aging of meat is mainly dependent on the activity of calpain 1, calpain 2, and their inhibitor calpastatin, while the role of calpain 3 (earlier known as p94) in postmortem aging is still unknown. Although in some studies it is revealed that CAPN3 has unique features that are not found in any of the other 14 calpains, or even in other proteases, that may influence postmortem aging (Pandurangan and Wang, 2012). It is recently reported that CAPN3 is the first and only intracellular enzyme that requires Na^+ for its activation, and is rapidly autolyzed, but has the ability to regain protease function after its autolytic dissociation. Biswas (2016) while studying calpain 1 reported a negligible amount of calpain 1 with postmortem muscle degradation. So it is presumed that some other enzymes like caspases (cysteine proteases) may influence the activity of calpain system, and therefore detailed study of this caspase family in the presence of calpain is important. Skeletal muscle atrophy is another phenomenon associated with the degradation of proteins and also decreased protein synthesis (Jackman and Kandarian, 2004). Skeletal muscle atrophy leads to muscular inactivity, thereby influencing several adverse functional consequences. However, very little is known about the triggers or the molecular signaling events underlying this process. It is assumed that decreases in protein synthesis and increases in protein degradation both have been shown to contribute to muscle protein loss due to disuse, and recent work has delineated elements of both synthetic and proteolytic processes underlying muscle atrophy (Huang and Zhu, 2016). Further, it was reported that phosphorylation of protein kinase K and dephosphorylation of alkaline phosphatase positively influence the activity of calpain 1 (Du et al., 2018). So the ability of calpain 1 in protein degradation in the presence of phosphorylation needs to be explored further. For detection, a novel proteomic approach, sensor-based techniques like immunosensors-based techniques, optical fiber biosensors, capillary-based biosensors, and even gene expression techniques may be the heading in the right direction for the determination of calpain activity on-site in time with a greater degree of sensitivity and accuracy.

References

Bernard, C., Cassar-Malek, I., Le Cunff, M., et al., 2007. New indicators of beef sensory quality revealed by expression of specific genes. J. Agric. Food Chem. 55, 5229–5237.

Biswas, A.K., 2016. Micro-calpain assisted post-mortem ageing of meat and its influence on quality. In: Chatli, M.K., Mehta, N., Wagh, R.V., et al. (Eds.), Proceeding of International Symposium and Seventh Conference of IMSA on New Horizons for Augmenting Meat Production and Processing to Ensure Nutritional Security, Food Safety and Environmental Sustainability and Workshop on Food Quality and Safety, Ludhiana, India, November 10–12, 2016, pp. 299–303.

Biswas, A.K., Tandon, S., Beura, C.K., 2016a. Simple extraction method for determination of different domains of calpain and calpastatin from chicken blood and their role in post-mortem ageing of breast and thigh muscles at 4 ± 1 °C. Food Chem. 200, 315−321.

Biswas, A.K., Tandon, S., Sharma, Divya, 2016b. Identification and characterization of different domains of calpain and their influence on post-mortem ageing of goat meat during holding at 4 ± 1 °C. LWT-Food Sci. Technol. 71, 60−68.

Biswas, A.K., Tandon, S., Giri, A.K., et al., 2016c. Role of calpain system in post-mortem tenderization of meat and its influence on quality-a review. Indian J. Poult. Sci. 51 (2), 123−132.

Biswas, A.K., Kripriyalini, L., Tandon, S., Sharma, D., Majumder, S., 2016d. Simultaneous identification of different domains of calpain from blood and turkey meat samples using casein zymography. Food Anal. Methods 9, 2872−2879.

Bozoky, Z., Alexa, A., Dancsok, J., et al., 2009. Identifying calpain substrates in intact S2 cells of *Drosophila*. Arch. Biochem. Biophys. 481, 219−225.

Brule, C., Dargelos, E., Diallo, R., et al., 2010. Proteomic study of calpain interacting proteins during skeletal muscle aging. Biochimie 92, 1923−1933.

Caberoy, N.B., Alvarado, G., Li, W., 2011. Identification of calpain substrates by ORF phage display. Molecules 16, 1739−1748.

Carragher, N.O., Frame, M.C., 2002. Calpain: a role in cell transformation and migration. Int. J. Biochem. Cell Biol. 34, 1539−1543.

Cuerrier, D., Moldoveanu, T., Davies, P.L., 2005. Determination of peptide substrate specificity for mu-calpain by a peptide library-based approach. The importance of promed side interactions. J. Biol. Chem. 280, 40632−40641.

Dransfield, E., Wakefield, D.K., Parkman, I.D., 1992. Modelling post-mortem tenderisation − I: Texture of electrically stimulated and non-stimulated beef. Meat Sci. 31, 57−73.

Du, M., Li, X., Li, Z., Shen, Q., Wang, Y., Li, G., et al., 2018. Phosphorylation regulated by protein kinase A and alkaline phosphatase play positive roles in μ-calpain activity. Food Chem. 252, 33−39.

El-Assaad, A., Dawy, Z., Nemer, G., Kobeissy, F., 2017. Novel bioinformatics−based approach for proteomic biomarkers prediction of calpain-2 and Caspase-3 protease fragmentation: application to β II-spectrin protein. Sci. Rep. 7, 41039.

Etherington, D.J., Taylor, M.A., Dransfield, E., 1987. Conditioning of meat from different species. Relationship between tenderising and the levels of cathepsin B, cathepsin L, calpain I, calpain II and β-glucuronidase. Meat Sci. 20, 1−18.

Geesink, G.H., Koohmaraie, M., 1999. Post-mortem proteolysis and calpain/calpastatin activity in callipyge and normal lamb *Biceps femoris* during extended post-mortem storage. J. Anim. Sci. 77, 1490−1501.

Geesink, G.H., Taylor, R.G., Bekhit, A.E.D., Bickerstaffe, R., 2001. Evidence against the non-enzymatic calcium theory of tenderization. Meat Sci. 59, 417−422.

Geesink, G.H., Kuchay, S., Chishti, A.H., Koohmaraie, M., 2006. Micro-calpain is essential for post-mortem proteolysis of muscle proteins. J. Anim. Sci. 84, 2834−2840.

Goll, D.E., Thompson, V.F., Li, H., Wei, W., Cong, J., 2003. The calpain system. Physiol. Rev. 83, 731−801.

Huang, J., Zhu, X., 2016. The molecular mechanisms of calpains action on skeletal muscle atrophy. Physiol. Res. 65, 547−560.

Huessen, C., Dowdle, E.B., 1980. Electrophoretic analysis of plasminogen activator in polyacrylamide gels containing sodium dodecylsulfate and co-polymerized substances. Anal. Biochem. 102, 196−202.

Huff-Lonergan, E., Mitsuhashi, T., Beekman, D.D., et al., 1996. Proteolysis of specific muscle structural proteins by μ-calpain at low pH and temperature is similar to degradation in post-mortem bovine muscle. J. Anim. Sci. 74, 993−1008.

Hughes, M.C., Geary, S., Dransfield, E., et al., 2001. Characterization of peptides released from rabbit skeletal muscle troponin-T by μ-calpain under conditions of low temperature and high ionic strength. Meat Sci. 59, 61−69.

Jackman, R.W., Kandarian, S.C., 2004. The molecular basis of skeletal muscle atrophy. Am. J. Physiol. Cell Physiol. 287, 834–843.

Karlsson, J.O., 2017. A simple protocol for separation and assay of μ-calpain, m-calpain, and calpastatin from small tissue samples. In: Elce, J.S. (Ed.), Calpain Methods and Protocols, vol. 144. Humana Press Inc., Totowa, NJ, pp. 17–23. , "Methods in Molecular Biology" Series.

Karlsson, J.O., Gustavsson, S., Hall, C., Nilsson, E., 1985. A simple one step procedure for the separation of calpain I, calpain II and calpastatin. Biochem. J. 231, 201–204.

Kemp, C.M., Sensky, P.L., Bardsley, R.G., et al., 2010. Tenderness-an enzymatic view. Meat Sci. 84, 248–256.

Kim, C., Yun, N., Lee, Y.M., et al., 2013. Gel-based protease proteomics for identifying the novel calpain substrates in dopaminergic neuronal cell. J. Biol. Chem. 51, 36717–36732.

Koohmaraie, M., 1990. Quantification of Ca^{2+} dependent protease activities by hydrophobic and ion-exchange chromatography. J. Anim. Sci 68, 659–665.

Koohmaraie, M., 1992. Ovine skeletal muscle multicatalytic proteinase complex (proteasome): purification, characterization, and comparison of its effects on myofibrils with μ-calpains. J. Anim. Sci. 70, 3697–3708.

Koohmaraie, M., 1994. Muscle proteinases and meat ageing. Meat Sci. 36, 93–104.

Koohmaraie, M., 1996. Biochemical factors regulating the toughening and tenderisation processes of meat. Meat Sci. 43, 193–201.

Koohmaraie, M., Geesink, G.H., 2006. Contribution of post-mortem muscle biochemistry to the delivery of consistent meat quality with particular focus on the calpain system. Meat Sci. 74, 34–43.

Kretchmar, D.H., Hathaway, M.R., Dayton, W.R., 1989. In-vivo effect of a β-adrenergic agonist on activity of calcium-dependent proteinases, their specific inhibitor, and cathepsins B and H in skeletal muscle. Arch. Biochem. Biophys. 275, 228–235.

Kurebayashi, N., Harkins, A.B., Baylor, S.M., 1993. Use of fura red as an intracellular calcium indicator in frog skeletal muscle fibers. Biophys. J. 64, 1934–1960.

Maddock, K.R., Huff-Lonergan, J., Lonergan, E., Steven, M., 2005. The Effect of pH on L-Calpain Activity and Implications in Meat Tenderness. Anim. Ind. Report. AS 651, ASL R1988.

Meyers, S.N., Beever, J.E., 2008. Investigating the genetic basis of pork tenderness: genomic analysis of porcine CAST. Anim. Genet. 39, 531–543.

Moldoveanu, T., Hosfield, C.M., Lim, D., et al., 2002. A Ca^{2+} switch aligns the active site of calpain. Cell 108, 649–660.

Moudilou, E.N., Mouterfi, N., Exbrayat, J.M., Brun, C., 2010. Calpains expression during *Xenopus laevis* development. Tissue Cell 42 (5), 275–281.

Neath, K.E., Del Barrio, A.N., Lapitan, R.M., et al., 2007. Protease activity higher in post-mortem water buffalo meat than Brahman beef. Meat Sci. 77, 389–396.

Nilsson, E., Karlsson, J.O., 1990. Slow anterograde axonal transport of calpain I and II. Neurochem. Int. 17, 487–494.

Northcutt, J.K., Pringle, T.D., Dickens, J.A., et al., 1998. Effects of age and tissue type on the calpain proteolytic system in turkey skeletal muscle. Poult. Sci. 77, 367–372.

Nowak, D., 2011. Enzymes in tenderization of meat – the system of calpains and other systems – a review. Pol. J. Food Nutr. Sci. 61 (4), 231–237.

O'Brien, M., Scurria, M., Rashka, K., Daily, B., Riss, T., 2005. A bioluminescent assay for calpain activity. Promega 91, 6–9.

Ono, Y., Sorimachi, H., 2012. Calpains-an elaborate proteolytic system. Biochim. Biophys. Acta 1824 (1), 224–236.

Pandurangan, M., Wang, I., 2012. The role of calpain in skeletal muscle. Anim. Cells Syst. 16 (6), 431437.

Parr, T., Sensky, P.L., Bardsley, R.G., Buttery, P.J., 2001. Calpastatin expression in cardiac and skeletal muscle and partial gene structure. Arch. Biochem. Biophys. 395, 1–13.

Raser, K.J., Posner, A., Wang, K.K., 1995. Casein zymography: a method to study μ-calpain, m-calpain, and their inhibitory agents. Arch. Biochem. Biophys. 319, 211–216.

Salem, M., Kenney, P.B., Killefer, J., Nath, J., 2004. Isolation and characterization of calpains from rainbow trout muscle and their role in texture development. J. Muscle Foods 15, 245–255.

Savart, M., Verret, C., Dutaud, D., et al., 1995. Isolation and identification of a µ-calpain-protein kinase Cα complex in skeletal muscle. FEBS Lett. 359, 60–64.

Sentandreu, M.A., Coulis, G., Ouali, A., 2002. Role of muscle endopeptidases and their inhibitors in meat tenderness. Trends Food Sci. Technol. 13, 400–421.

Taylor, R.G., Geesink, G.H., Thompson, V.F., et al., 1995. Is Z-disk degradation responsible for post-mortem tenderization?. J. Anim. Sci. 73, 1351–1367.

Veeramuthu, G.I., Sams, A.R., 1999. Post-mortem pH, myofibrillar fragmentation, and calpain activity in *Pectoralis* from electrically stimulated and muscle tensioned broiler carcasses. Poult. Sci. 78, 272–276.

Xu, X.X., Shui, X., Chen, Z.H., et al., 2009. Development and application of a real-time PCR method for pharmacokinetic and biodistribution studies of recombinant adenovirus. Mol. Biotechnol. 43, 130–137.

Zor, K., Ortiz, R., Saatci, E., Bardsley, R., Parr, T., Csoregi, E., et al., 2009. Label free capacitive immunosensor for detecting calpastatin – a meat tenderness biomarker. Bioelectrochem 76, 93–99.

… SECTION 4

Molecular basis of meat colour development and detection

CHAPTER 8

Molecular basis of meat color

R.A. Mancini[1] and R. Ramanathan[2]
[1]Department of Animal Science, University of Connecticut, Storrs, CT, United States
[2]Department of Animal and Food Sciences, Oklahoma State University, Stillwater, OK, United States

Chapter Outline
8.1 Introduction 117
8.2 Meat chemistry 118
    8.2.1 Red color development 119
    8.2.2 Discoloration 120
    8.2.3 Color deviations and approaches to improve color 121
    8.2.4 Myoglobin and lipid oxidation 122
8.3 Cooked color 122
8.4 Analysis of meat color 123
    8.4.1 Instrumental color analysis (use of handheld devices) 123
    8.4.2 Spectrophotometric techniques to study meat color 124
    8.4.3 Mitochondrial functional analysis 125
    8.4.4 Visual color analysis 128
    8.4.5 Cooked color measurements 128
8.5 Conclusion 129
References 129

8.1 Introduction

Meat color has a significant role in consumer purchasing decisions because surface discoloration is typically used as a determinant of product wholesomeness. More specifically, consumers view a bright-red color as indicative of fresh meat and brown surface discoloration as an indicator of spoilage. As a result the overall profit and the amount of annual food and packaging waste are partially dependent on meat color. The molecular basis of color is due to several biochemical changes in meat, particularly those associated with myoglobin and mitochondria. For example, the valence state and ligand bound to iron within myoglobin are important for color determination and can influence color development and stability by forming either purple deoxymyoglobin, red oxymyoglobin and carboxymyoglobin, or brown metmyoglobin. Mitochondria can consume oxygen in postmortem muscle and darken meat by maintaining deoxymyoglobin. As a result, meat with excessive mitochondrial oxygen consumption (OC) will not produce a desirable bright cherry-red color. Mitochondrial OC, specifically enzymes such as cytochrome c and the electron transport chain, also can have a significant role in color stability via metmyoglobin reduction. Modified atmosphere packaging and product-enhancement with lactate have relied on these concepts to promote color stability. Conversely, antemortem stress can cause dark-cutting beef, a major quality issue that results from mitochondria's influence on myoglobin. Cooked color is due to myoglobin denaturation; therefore the presence of dexy-, oxy-, carboxy-, and metmyoglobin in raw meat can influence potential safety and quality issues, such as premature browning and persistent pink.

There are two primary methods available for measuring meat color: visual and instrumental. Protein extraction and reflectance are commonly used instrumental techniques, whereas consumer and trained panels can be used to evaluate visual color. Mitochondrial functional analysis also can be used in meat color research. This includes OC and metmyoglobin reducing activity (MRA) measurements. This book chapter will provide an outline of the factors involved in meat color as well as introduce methodology that can be used to measure meat color.

8.2 Meat chemistry

Meat color is determined by myoglobin, a relatively small sarcoplasmic protein that contains 153 amino acid residues within 8 α-helices. More specifically, iron within a centrally located heme iron can influence color via ligand binding and valence state.

Iron can form six bonds and within myoglobin's heme ring, four of these bonds interact with pyrrole nitrogens of the protoporphyrin and the fifth bond connects the heme group to the proximal histidine (also referred to as H93 or F8). This histidine, in addition to histidine 64, can influence protein structure, functionality, and meat color stability (Mancini and Hunt, 2005). However, it is likely that most important to color determination is the sixth bond or coordination site on iron, which can be vacant as well as bind ligands such as oxygen, carbon monoxide, and nitric oxide. Depending on the ligand, this binding can be reversible and preferential.

Iron within myoglobin also can exist in several valence states. The two most often associated with meat color include ferrous (Fe^{++} or Fe^{+2}) and ferric (Fe^{+++} or Fe^{+3}). Both oxidation and reduction interconversions between these two valence states influence color stability or the maintenance of a desirable red color as well as discoloration or the accumulation of brown surface color (Hunt et al., 2012).

The color on the surface of meat products is due to four common forms of myoglobin that result from the ligand (or lack of) attached to iron and the redox state of iron. These myoglobin forms include deoxymyoglobin, oxymyoglobin, carboxymyoglobin, and metmyoglobin.

8.2.1 Red color development

Deoxymyoglobin has no ligand attached to ferrous iron, resulting in a purplish-red color when meat is deoxygenated or not exposed to oxygen. This includes vacuum packaged meat, modified atmosphere packages without oxygen, and pigments within the interior of freshly-cut meat. The development of a bright cherry-red color occurs when deoxymyoglobin is exposed to oxygen, which produces oxymyoglobin via oxygenation or the addition of diatomic oxygen as the ligand attached to iron. This often is referred to as bloom and there is no effect on iron's valence state as oxymyoglobin maintains a ferrous redox state.

Mitochondria will metabolize oxygen in postmortem muscle. As a result, OC by mitochondria darkens muscle by maintaining deoxymyoglobin via decreased oxygen partial pressure and limited oxygen availability to myoglobin (Tang et al., 2005). This will cause meat to experience a limited bloom, failing to produce the characteristic bright cherry-red color of oxygenated beef. A greater than normal ultimate muscle pH promotes mitochondrial activity and is partially responsible for dark-cutting beef because the formation of oxymyoglobin depends on both the amount of oxygen diffusion beneath the meat surface and OC by mitochondria. In addition to pH, other factors that influence mitochondrial OC also will alter red color development, including time postmortem, temperature, and ingredients. More specifically, increased time postmortem and decreased temperatures will minimize mitochondrial activity and promote red color development, whereas ingredients such as lactate will darken meat and limit red color development by promoting mitochondrial activity.

In addition to preventing myoglobin oxygenation, mitochondrial respiration can convert oxymyoglobin to deoxymyoglobin by decreasing oxygen partial pressure. However, this conversion of oxymyoglobin to deoxymyoglobin is a two-step process that requires both OC and metmyoglobin reduction. More specifically, oxymyoglobin is first oxidized to metmyoglobin and then reduced to deoxymyoglobin, which requires sufficient mitochondrial activity for OC and metmyoglobin reduction.

An intense bright-red color also can be achieved by packaging atmospheres that contain carbon monoxide, which can serve as a ligand and bind to deoxymyoglobin (Cornforth and Hunt, 2008). Myoglobin has a strong affinity for carbon monoxide, resulting in a red color that is extremely stable in packaging that contains carbon monoxide. Conversely, when steaks containing carboxymyoglobin are exposed to the atmosphere and oxygen, they tend to discolor due to interconversion(s) between carboxy- and other myoglobin forms. More specifically, myoglobin likely will release carbon monoxide and subsequently bind oxygen, eventually discoloring due to oxidation.

8.2.2 Discoloration

Brown discoloration results from metmyoglobin formation due to oxidation of either deoxy- or oxymyoglobin (oxidation produces ferric iron). Metmyoglobin formation often initiates within the interior of meat products where oxygen partial pressure is not sufficiently anaerobic enough to cause deoxymyoglobin and not great enough to form oxymyoglobin. This metmyoglobin will move outward from the interior to the surface where consumers use it as an indicator of freshness. Surface discoloration depends on several factors that influence myoglobin redox chemistry, including packaging atmosphere, temperature, ultimate muscle pH, time postmortem, mitochondrial activity, and microbial growth.

Given the detrimental effect of myoglobin oxidation and metmyoglobin accumulation on meat shelf life, promoting ferrous deoxy- and oxymyoglobin is critical to maintain desirable meat color. This can be accomplished, albeit somewhat limited in postmortem muscle, by metmyoglobin reduction or MRA. Although the donation of an electron to metmyoglobin is a process that can occur either with or without enzyme catalysts, metmyoglobin reductase (a NADH: cytochrome b_5 complex) has a significant role in color stability. This mechanism relies upon reducing equivalents, such as NADH, which are often limited in meat. Therefore, the postmortem production of NADH by oxygen scavenging enzymes and reducing enzyme systems can influence beef color stability.

In addition to the ability of mitochondria to consume oxygen and influence red color development, mitochondria also have a significant role in color stability via metmyoglobin reduction. More specifically, mitochondrial enzymes such as cytochrome c and the electron transport chain can promote the transfer of available electrons to metmyoglobin (Arihara et al., 1995). Substrates involved with the electron transport chain and the tricarboxylic acid cycle also can form NADH that can be used to reduce metmyoglobin. Mitochondrial OC will promote metmyoglobin reduction because this process is more efficient in anaerobic conditions.

In addition to myoglobin's role in meat discoloration, the hemoglobin redox state can influence bone marrow discoloration and consumer acceptance. Exposed hemoglobin within the marrow of bone-in beef steaks will interact with oxygen and result in a bright-red color due to the formation of oxyhemoglobin (Mancini et al., 2007). Subsequent oxidation will produce methemoglobin and a gray or black bone marrow color. Fat discoloration, primarily yellow subcutaneous beef fat, is attributed to carotenoids in cattle diets.

8.2.3 Color deviations and approaches to improve color

Deviations from bright-red color lead to consumer rejection. More specifically, lactate-induced darkening and dark-cutters are two scenarios where freshly-cut meat will not have a bright-red color.

8.2.3.1 Lactate-induced darkening

The introduction of centralized packaging has allowed meat purveyors to enhance meat with ingredients such as lactate or ascorbate, and package meat in modified atmospheric conditions to improve shelf life. Lactate is a commonly used antimicrobial; however, the addition of lactate can darken meat color (lower L^* values). Research elucidating the darkening effect indicated that the addition of lactate could increase NADH formation via lactate dehydrogenase activity (Kim et al., 2006). The NADH formed can be used to increase mitochondrial activity, leading to less oxygen available to myoglobin. Therefore lactate-enhanced steaks will have more deoxymyoglobin and be darker in color. The NADH formed can be used for mitochondria-mediated, enzymatic, and nonenzymatic MRA (Ramanathan et al., 2010). The role of mitochondria in lactate-induced darkening was validated by the addition of mitochondrial inhibitor. For example, the addition of rotenone (mitochondrial complex I inhibitor) reversed the darkening effect. In addition to biochemical pathways, lactate can increase the refractive index of sarcoplasm, leading to more absorbance of light and lower L^* values. Further, lactate can hold more water, leading to muscle swelling and decreased reflectance.

8.2.3.2 Dark-cutting beef

Dark-cutting beef or a dark firm dry condition is one of the major beef quality issues worldwide due to antemortem stress. Although the biochemical basis for dark-cutting beef is not known, current research indicates that a decreased glycogen content and defective glycolytic enzymes lead to greater than normal muscle pH. High-pH (>6.0) can affect both biochemical and physical properties of meat. For example, a greater pH is favorable for enzymes involved in OC. Hence, OC will be greater in dark-cutting beef. Further, a greater pH can enhance muscle fibers' ability to hold water, resulting in cell swelling. Therefore, the muscle can absorb more light and reflect less.

8.2.3.3 Packaging techniques to improve the appearance of dark-cutting beef

Various strategies, such as decreasing the meat pH or modified atmospheric packaging, have been utilized to improve the value and appearance of dark-cutting beef. Decreasing pH by enhancing meat with acidic-pH solutions, such as lactic or citric acid, can improve the color of dark-cutting beef. Dark-cutting beef has greater OC than normal-pH beef. Hence, incorporating greater oxygen concentration within the package can meet the oxygen requirements of myoglobin and mitochondria. Dark-cutting steaks enhanced with rosemary and packaged in high-oxygen had similar red color as normal-pH beef (Mitacek et al., 2018). In addition to oxymyoglobin, two other myoglobin forms that can impart bright-red color are carboxy- and nitric oxide myoglobin. Packaging dark-cutting steaks in carboxy- and nitrite-embedded packaging improved the red color of dark-cutting steaks compared to PVC packaging. Similarly, packaging lactate-enhanced steaks in high-oxygen and carbon monoxide packages decreased the darkening effect.

8.2.4 Myoglobin and lipid oxidation

Myoglobin and lipid oxidation are interrelated and can catalyze each other. More specifically, lipid oxidation can form aldehydes that are reactive and can bind to proteins including myoglobin (Suman et al., 2006). This covalent binding to histidine residues can decrease color stability by making myoglobin more prone to oxidation and less likely to be reduced by metmyoglobin reductase. Aldehydes produced by lipid oxidation also can limit color stability by inactivating enzymes involved in the production of NADH necessary for metmyoglobin reduction. Mitochondrial enzymes involved in color stability also are influenced by lipid oxidation. Use of high-throughput mass spectrometry has helped to elucidate the binding of aldehyde to amino acid residues.

8.3 Cooked color

Myoglobin denaturation results in cooked color due to heat-induced unfolding of the protein. The development of cooked color prior to the inactivation of pathogens is a potential food safety concern referred to as premature browning. Conversely, a limited development of cooked color or persistent pink is typically more of a quality issue as overcooking often is required to achieve a desired degree of color doneness. Both of these issues are dependent on the myoglobin form in the raw meat prior to cooking, which influences the pigment's thermal stability and resistance to denaturation (Suman et al., 2010). More specifically, of the four common myoglobin forms in raw meat, metmyoglobin is the least thermally stable, and therefore the most likely to result in premature browning. Conversely, carboxymyoglobin is the most heat stable of the four myoglobin forms and the most likely to result in persistent pink colors. This is evident in beef cooked over smoke, which has a characteristic pink "smoke ring" on the exterior of the meat.

Deoxymyoglobin is more heat stable than oxymyoglobin; nevertheless, the two myoglobin forms are more resistant to heat than metmyoglobin. As a result, packaging will influence cooked color as high-oxygen modified atmosphere packaging tends to increase and vacuum packaging tends to decrease the incidence of premature browning. Packaging that contains carbon monoxide predisposes meat to persistent pinking.

The role of myoglobin form in thermal stability also is partially responsible for the effects of dark-cutting beef on cooked color. This is further complicated by the fact that an increased ultimate meat pH will increase myoglobin's thermal stability. Therefore dark-cutting beef is less likely to prematurely brown when cooked because it has both thermally stable deoxymyoglobin and an increased meat pH.

Nitrite is an ingredient that inhibits the outgrowth of *Clostridium botulinum* spores and results in the characteristic color of cooked meat products that are cured such as hams, hot dogs, bacon, and bologna. Nitrite can be reduced to NO, which serves as a ligand and forms ferrous NO–myoglobin. Heat-induced denaturation will unfold this pigment and produce a pink color that is stable in vacuum packaging. However, this pink color will fade to gray in the presence of oxygen; therefore, anaerobic packaging is critical for the maintenance of cured color.

8.4 Analysis of meat color

A complete review of meat color measurement methodology is in the American Meat Science Association Color Guidelines (Hunt et al., 2012). Nevertheless, there are two overall options available for measuring both raw and cooked meat color. These include visual and instrumental methods and researchers should select the method or methods most appropriate for their particular situation.

8.4.1 Instrumental color analysis (use of handheld devices)

Color and color stability can be assessed using instrumental methodology, including protein extraction and reflectance. Extraction often does not allow for repeated measures, whereas surface reflectance methods are rapid, allow for repeated measures, and possibly most important, are likely a better indicator of consumer perception. Meat is a complex matrix that can absorb, reflect, and scatter light. Hence K/S ratios are used to (1) generate linear data, and (2) account for absorbance and light scattering produced by the muscle's surface and subsurface microstructure; where K is the absorption coefficient and S is the scattering coefficient. Various handheld devices, such as HunterLab or Minolta Colorimeter, can be used to obtain color measurements. When reporting results in a manuscript, care should be taken to include details of the instruments. For example, angle of observer and the type of light can influence color reading.

Reflectance variables often measured include L^* (indicates lightness), a^* (assesses redness), and b^* (evaluates yellowness), hue angle (calculated as $\tan^{-1} b^*/a^*$ to assess discoloration), and saturation index or chroma (calculated as $(a^2 + b^2)^{1/2}$ to measure red color intensity). Spectral reflectance also can quantify myoglobin forms using isosbestic wavelengths including 474, 525, 572, and 610 nm. Oxymyoglobin and metmyoglobin values obtained from spectral data can be used to calculate and indirectly estimate OC and MRA of meat samples.

OC represents the ability of the postmortem muscle to consume oxygen, primarily mitochondrial or oxygen-consuming enzymes. Reflectance measurement utilizes changes in the oxymyoglobin level of bloomed steak (predominant oxymyoglobin form) after vacuum as an indirect indication for OC. A greater OC results in the conversion of oxymyoglobin to deoxymyoglobin. This approach may not be utilized in high-pH beef as the greater OC limits oxymyoglobin formation. Hence, reporting OC as changes in oxymyoglobin in bloomed steak and vacuum packaged steak may not provide realistic values. Therefore reporting initial oxymyoglobin level in bloomed steak can be used as an indicator of OC. For example, a lower oxymyoglobin content represents greater OC. Penetration of oxymyoglobin layer into the interior of meat can also be used an indicator for OC. For example, a greater OC results in minimal penetration of the oxymyoglobin layer into the interior. A digital caliper can be used to measure oxymyoglobin layer. Precautions should be taken to have multiple readings.

Nitrite-induced metmyoglobin reduction is often used as a method to quantify total metmyoglobin reduction (i.e., combination of enzymatic, nonenzymatic, and mitochondria-mediated). This methodology involves the reflectance method to quantify metmyoglobin reduction under anaerobic conditions. Greater changes in metmyoglobin content indicate greater MRA. The quantification of metmyoglobin using changes in metmyoglobin after incubation cannot be applicable to high-pH beef. A greater pH can resist the formation of metmyoglobin content. Hence, resistance to metmyoglobin formation can be used as an indicator for MRA. A lower metmyoglobin represents greater MRA. Especially with larger sample numbers, care must be taken to get uniform conditions for different days of experiments.

8.4.2 Spectrophotometric techniques to study meat color

Bench top UV-Vis spectrophotometers are used to study in vitro myoglobin samples. For example, isolated bovine or equine myoglobins can elucidate the role of lipid oxidation or the glycolytic and tricarboxylic cycle on oxymyoglobin oxidation. Most researchers use the oxymyoglobin form to study discoloration. Hence, precautions should be taken while preparing oxymyoglobin. Hydrosulfite-mediated reduction of metmyoglobin yields deoxymyoglobin, and subsequent oxygenation leads to oxymyoglobin. The residual hydrosulfite is removed by passage through a chromatographic column. The proportion of different myoglobin forms can be quantified using wavelength maxima of each myoglobin form. For example, 503 nm represents the wavelength maxima for metmyoglobin, 557 nm for deoxymyoglobin, and 582 nm for oxymyoglobin. All three forms meet at 525 nm, which is the isosbestic point of three myoglobin forms. The wavelength maxima at 503, 557, and

582 nm in relationship with the isosbestic wavelength (525 nm) are used to quantify deoxy-, oxy-, and metmyoglobin forms.

The case-ready packaging has allowed the use of carbon monoxide gas. Both carboxy- and oxymyoglobin have similar absorbance spectra. Hence the quantification of carbon monoxide oxidation can be challenging when oxymyoglobin coexists in the solution. The Browning index (503/581 nm) is used to quantify carbon monoxide oxidation.

Various researchers have utilized meat extracts to quantify myoglobin forms. The researcher should be aware that the colored meat extracts can be a combination of myoglobin, hemoglobin, and cytochrome c. Hence, precaution should be taken when interpreting the results.

Meat color research also involves various kinetic assays to quantify enzymes related to meat color. For example, one of the characterized enzymatic metmyoglobin systems is NADH-dependent reductase. Hence, to determine the activity of NADH-dependent enzyme, a kinetic mode in the spectrophotometer can be helpful. The reaction mixture includes metmyoglobin, an enzymatic extract from muscle, and cofactors. The reaction is initiated by the addition of NADH, and an increase in reduction as indicated by an increase in absorbance at 582 nm for a specific time. Similarly, to characterize nonenzymatic metmyoglobin reduction, various components, such as metmyoglobin, methylene blue, and EDTA, are added in a reaction mixture. The nonenzymatic metmyoglobin reduction is initiated by the addition of NADH. The main difference between enzymatic and nonenzymatic metmyoglobin reduction is the change in the component that carries electron from a reducing group. For example, in enzymatic metmyoglobin reduction, NADH-dependent enzyme carries electrons from NADH, while in nonenzymatic metmyoglobin reduction, methylene blue carries electrons.

8.4.3 Mitochondrial functional analysis

More direct measurements of OC and MRA can be obtained with isolated mitochondria (Tang et al., 2005). Differential centrifugation is routinely used to isolate intact mitochondria. The isolated mitochondria can be used to determine the effects of oxymyoglobin or metmyoglobin on different glycolytic or tricarboxylic acid substrates. In vitro studies utilizing mitochondria and myoglobin can provide valuable insights into muscle darkening or color stability. Further, isolated mitochondria and myoglobin can be utilized to study the effects of pH, different gas combinations, and lighting conditions on meat color. An integrated spectrophotometer may be used to study the effects of the mitochondria—myoglobin mixture on meat color. An integrated sphere spectrophotometer assembly can be used to measure absorbance of turbid solutions. Hence, selecting a spectrophotometer that can hold an integrated sphere assembly will help to get additional data.

8.4.3.1 Mitochondria isolation

Mitochondria can be isolated from bovine cardiac and *longissimus* muscles depending on the objective of the research. Cardiac tissue will have greater concentration than skeletal muscles. When the study requires a lesser amount of mitochondria, researchers can start with a lesser amount of tissue. The recommended dilution between tissue and buffer should be 1:10. Mince tissue devoid of all visible fat and connective tissue (30 g) can be washed twice with 250 mM sucrose and suspended in 60 mL of mitochondria isolation buffer (250 mM sucrose, 10 mM HEPES, (1 mM EGTA for cardiac tissue, 1 mM EDTA for *longissimus*), and 0.1% BSA, pH 7.2). The suspension is stirred slowly and hydrolyzed with protease (protease/tissue, 0.5 mg/g of tissue) for 20 minutes; the pH is maintained between 7.0 and 7.2. If the research involves any soft tissue, there is no need to add proteolytic enzymes.

After proteolytic digestion, the suspension is diluted to 300 mL with mitochondria isolation buffer and homogenize using a Kontes Duall grinder followed by a Wheaton Potter–Elvehjem grinder. The homogenate is centrifuged ($1200 \times g$ for cardiac tissue and $900 \times g$ for *longissimus*) for 20 minutes with a refrigerated centrifuge, and the resulting supernatant is again centrifuged ($26,000 \times g$ for cardiac tissue and $14,000 \times g$ for *longissimus*) for 15 minutes. Mitochondrial pellets are washed twice and suspended in mitochondria suspension buffer (250 mM sucrose, 10 mM HEPES, pH 7.2). All steps should be performed at $0°C - 4°C$. All the glassware must be washed with deionized water. Mitochondrial protein content can be determined using a bicinchoninic acid protein assay.

8.4.3.2 Oxymyoglobin preparation

Myoglobin is purified via ammonium sulfate precipitation and gel filtration chromatography (Yin et al., 2011). The researchers can either buy lyophilized equine myoglobin or isolate from respective species. Briefly, beef cardiac muscle devoid of fat and connective tissue is homogenized in buffer (10 mM Tris-HCl, 1 mM EDTA, pH 8.0, 4°C) and centrifuged at $5000 \times g$ for 10 minutes. The supernatant is brought to 70% ammonium sulfate saturation and the resulting solution is stirred for 1 hour at 4°C and later centrifuged at $18,000 \times g$ for 20 minutes. The resulting supernatant is saturated with ammonium sulfate (100%) and centrifuged at $20,000 \times g$ for 1 hour. The precipitate is resuspended in homogenization buffer and dialyzed (three volumes) against 10 mM Tris-HCl, 1 mM EDTA, at pH 8.0, 4°C for 24 hours. Myoglobin is separated from hemoglobin using a Sephacryl 200-HR gel filtration column (2.5×100 cm). The elution buffer contain 5 mM Tris-HCl, 1 mM EDTA at pH 8.0, and the flow rate is 60 mL/h.

Isolated myoglobin solution is passed through a PD-10 column pre-calibrated with buffer at pH 5.6, 6.4, or 7.2. Myoglobin is reduced by sodium hydrosulfite-mediated reduction (0.1 mg sodium hydrosulfite to 1 mg myoglobin). Residual hydrosulfite is removed using a PD-10 column and reduced myoglobin is converted to oxymyoglobin by bubbling with oxygen. Myoglobin concentration is confirmed using absorbance at 525 nm ($A_{525\ nm} = 7.6\ mM^{-1}\ cm^{-1}$).

8.4.3.3 Oxygen consumption measurements

Bovine mitochondrial oxygen uptake is measured using a Clark oxygen electrode attached to a Rank Brothers digital model 20 oxygen controller. Reaction components (incubation buffer, mitochondria, and substrates) are added to an 8 mL incubation chamber and stirred with a 10 mm magnetic bar at 600 rpm. The incubation chamber is maintained at 25°C by a water jacket and Lauda RE120 circulating water bath. The researchers can adjust the temperature depending on the objective. A lower temperature is not recommend as the measurement using a Clark electrode may not be accurate.

OC can be recorded over time by suspending mitochondria in the Clark oxygen electrode with the addition of substrates such as succinate, pyruvate-malate, glutamate, or lactate-LDH-NAD at pH 5.6, 6.4, or 7.2 (incubation buffer = 250 mM sucrose, 5 mM KH_2PO_4, 5 mM $MgCl_2$, 0.1 mM EDTA, 0.1% BSA, and 20 mM maleic acid). Control mitochondria samples consist of only mitochondria; no added substrates or antimycin A. To assess the effect of a complex III inhibitor on mitochondrial respiration, mitochondria were preincubated with antimycin A (0.02 mM) for 2 minutes in the Clark oxygen electrode. For other treatments without antimycin A, an equal volume of ethanol is added. Other mitochondrial inhibitors such as rotenone, sodium azide, and cyanide can be added to determine the functionality of mitochondrial complexes.

8.4.3.4 Effects of mitochondria on oxymyoglobin

To study the direct interaction between mitochondria and oxymyoglobin, mitochondria (2 mg/mL) and oxymyoglobin (2.5 mg/mL) are combined in the presence of substrates and inhibitors. Control mitochondria samples consist of mitochondria combined with only oxymyoglobin; no added substrates. The treatments can be added to screw-capped vials, to avoid potential diffusion of air from outside into the mitochondria–oxymyoglobin mixture (2.6 mL capacity, quartz screw-capped cuvette), and incubated at 25°C or the temperature of the particular research objective. Treatments can be repeatedly scanned from 650 to 500 nm with a spectrophotometer with an integrating sphere assembly. The relative proportion of myoglobin contents can be calculated using wavelength maxima at 503, 557, and 582 nm, representative of metmyoglobin, deoxymyoglobin, and oxymyoglobin, respectively.

8.4.3.5 Electron transport-mediated metmyoglobin reduction

The pH of myoglobin solution can be adjusted by passing myoglobin solution through PD-10 columns precalibrated with either citrate/phosphate (pH 5.6) or phosphate buffer (pH 7.4). Metmyoglobin (2.5 mg/mL) reduction was conducted in a glass open top tube at pH 7.4 (120 mM KCl, 5 mM KH_2PO_4, 30 mM K_2HPO_4) or 5.6 (120 mM KCl, 5 mM KH_2PO_4, 30 mM maleic acid) at 25°C. Bovine heart or skeletal mitochondria (3 mg/mL) can be combined with both metmyoglobin and substrates. At specific time points, samples are removed and centrifuged (12,000g) with a centrifuge for 5 minutes. The resulting supernatant can be scanned from 650 to 500 nm with a spectrophotometer. The relative proportions of deoxymyoglobin, oxymyoglobin, and metmyoglobin can be calculated.

8.4.3.6 Enzymatic metmyoglobin reductase activity

The mitochondrial outer membrane contains NADH-dependent reductase activity. In order to determine metmyoglobin reductase activity, mitochondria (3 mg/mL) is preincubated with antimycin (0.01 mM) and rotenone (0.02 mM) for 10 minutes in an incubation buffer at pH 7.4 (120 mM KCl, 5 mM KH_2PO_4, 30 mM K_2HPO_4) or 5.6 (120 mM KCl, 5 mM KH_2PO_4, 30 mM maleic acid) at 25°C in order to inhibit ETC-mediated metmyoglobin reduction. Following incubation, mitochondria are combined with bovine metmyoglobin (0.15 mM), potassium ferrocyanide (3 mM), and EDTA (5 mM) at pH 5.6 and 7.4. The reaction is initiated by the addition of either (1) NADH (0.2 mM) or (2) a combination of lactate (50 mM), LDH (0.2 mM), and NAD (0.2 mM). Absorbance at 580 nm (the wavelength at which the difference in absorbance for oxymyoglobin and metmyoglobin is maximal) is measured for 30 minutes. Metmyoglobin reduction rate is indicated by nanomoles of metmyoglobin reduced per min per mg of mitochondria.

8.4.4 Visual color analysis

Consumer panels can be used to estimate purchasing decisions such as acceptance and satisfaction. However, these panels require many individuals (50 or more) and are not suitable for characterizing meat color, which should be done using a descriptive panel consisting of 4–10 trained panelists. Color scales often used by trained panels include descriptors such as red, pink, gray, tan, and brown, as well as quantification of discoloration assessed as the percentage of the surface that contains metmyoglobin.

8.4.5 Cooked color measurements

Cooked color measurements are important to study premature browning or persistent pinking as discussed in Section 8.3. The methodology for cooked color measurements are similar to raw meat color as discussed in the previous section. Cooked meat samples should be submerged in ice to avoid postcooking temperature rise. This can be critical when researchers are measuring the effects of temperature on cooked color. Various parameters such as internal color, myoglobin denaturation, reducing activity, and reductase activity can be performed using cooked meat samples. Both visual and handheld instruments such as HunterLab or Minolta colorimeter can be used to determine interior cooked color.

Myoglobin denaturation studies include extracting the undenatured pigments using centrifugation and quantification using UV-Vis spectrophotometer.

8.5 Conclusion

Postmortem muscle is biochemically active, and several biomolecules such as mitochondria and myoglobin interact reciprocally. Further, oxidative processes can decrease the functionality of enzymes involved in OC and metmyoglobin reduction. Hence, understanding the molecular basis of myoglobin redox changes can be challenging. A combination of in vitro and in situ models can provide valuable insights into the fundamental changes in meat.

References

Arihara, K., Cassens, R.G., Greaser, M.L., Luchansky, J.B., Mozdziak, P.E., 1995. Localization of metmyoglobin-reducing enzyme (NADH-cytochrome b5reductase) system components in bovine skeletal muscle. Meat Sci. 39 (2), 205–213.

Cornforth, D., Hunt, M., 2008. Low-oxygen packaging of fresh meat with carbon monoxide: meat quality, microbiology, and safety. Am. Meat Sci. Assoc. 2, 1–10.

Hunt, M.C., King, A., Barbut, S., Clause, J., Cornforth, D., Hanson, D., et al., 2012. AMSA Meat Color Measurement Guidelines, vol. 61820. American Meat Science Association, Champaign, IL.

Kim, Y.H., Hunt, M.C., Mancini, R.A., Seyfert, M., Loughin, T.M., Kropf, D.H., et al., 2006. Mechanism for lactate-color stabilization in injection-enhanced beef. J. Agric. Food Chem. 54 (20), 7856–7862.

Mancini, R.A., Hunt, M.C., 2005. Current research in meat color. Meat Sci. 71 (1), 100–121.

Mancini, R.A., Hunt, M.C., Seyfert, M., Kropf, D.H., et al., 2007. Effects of ascorbic and citric acid on beef lumbar vertebrae marrow colour. Meat Sci. 76 (3), 568–573.

Mitacek, R.M., English, A.R., Mafi, G.G., VanOverbeke, D.L., Ramanathan, R., 2018. Modified atmosphere packaging improves surface color of dark-cutting beef. Meat Mus. Biol. 2 (1), 57.

Ramanathan, R., Mancini, R.A., Naveena, B.M., 2010. Effects of lactate on bovine heart mitochondria-mediated metmyoglobin reduction. J. Agric. Food Chem. 58 (9), 5724–5729.

Suman, S.P., Faustman, C., Stamer, S.L., Liebler, D.C., 2006. Redox instability induced by 4-hydroxy-2-nonenal in porcine and bovine myoglobins at pH 5.6 and 4°C. J. Agric. Food Chem. 54 (9), 3402–3408.

Suman, B.S.P., Mancini, R.A., Ramanathan, R., Konda, M.R., 2010. Modified atmosphere packaging influences premature browning in beef Longissimus lumborum steaks. Fleischwirt. Int. 3, 54–55.

Tang, J., Faustman, C., Hoagland, T.A., Mancini, R.A., Seyfert, M., Hunt, M.C., 2005. Postmortem oxygen consumption by mitochondria and its effects on myoglobin form and stability. J. Agric. Food Chem. 53 (4), 1223–1230.

Yin, S., Faustman, C., Tatiyaborworntham, N., Ramanathan, R., Naveena, B.M., Mancini, R.A., et al., 2011. Species-specific myoglobin oxidation. J. Agric. Food Chem. 59 (22), 12198–12203.

SECTION 5

Meat authenticity and traceability

CHAPTER 9

Molecular techniques for speciation of meat

P.S. Girish[1] and Nagappa S. Karabasanavar[2,3]

[1]ICAR - National Research Centre on Meat, Hyderabad, India [2]Department of Veterinary Public Health & Epidemiology, Veterinary College, Hassan, India [3]Karnataka Veterinary, Animal & Fisheries Sciences University, Bidar, India

Chapter Outline
9.1 Introduction 133
9.2 Identification of the origin of meat species by DNA hybridization 135
9.3 PCR-based techniques for species identification of meat 137
    9.3.1 Random amplified polymorphic DNA—PCR (RAPD fingerprinting) 138
    9.3.2 Forensically informative nucleotide sequencing 138
    9.3.3 Species-specific PCR 140
    9.3.4 Multiplex PCR 141
    9.3.5 PCR-restriction fragment length polymorphism 141
    9.3.6 DNA microarrays 142
    9.3.7 Other PCR-based methods 143
9.4 Meat species identification using loop-mediated isothermal amplification 144
9.5 Quantitative meat speciation 145
    9.5.1 Quantitative PCR 145
9.6 Scope for future research 147
References 147

9.1 Introduction

The adulteration of costlier meat with cheaper meat is a fraudulent practice encountered across the globe. Ensuring the authenticity of meat and meat products forms the basis to promote fair trade, to enable consumers to make informed choices, to hasten the implementation of statutory slaughter regulations, to protect public health, and to safeguard religious sentiments. Coordinated meat testing in the European Union performed in the year 2013 indicated widespread adulteration of meat. The uncovered scandal (referred to as "horsegate" in the global media) was associated with the mixing of horse meat into beef; of the 7259 tests carried out by competent authorities in 27 EU countries, horse DNA was detected in 4144 instances. In addition, reports by the member states of another 7951 tests performed for the detection of horse meat by the food business operators (producers, processors, and distributors), showed that horse DNA was detected in 110 samples indicating 1.38% adulteration of meat (Brooks et al., 2017). In India suspected meat samples submitted for species identification based on DNA analysis to the ICAR—National Research Centre on Meat, Hyderabad identified intermixing of species meat (cattle and buffalo), meat from a legally restricted animal (cow), and nonconventional species such as camel and dog (Vaithiyanathan et al., 2018). Therefore in order to deter such fraudulent adulteration practices, authentic species identification methods need to be developed for use by the food analysts and forensic laboratories.

Several techniques have been developed for the purpose of animal species identification, that is, anatomical, histological, and organoleptic methods, that are reliable only in unprocessed raw meats. Chemical methods that mostly rely on lipid or carbohydrate analysis prove unsuitable for mixed or processed meats due to chemical variations in the constituents. Electrophoretic methods such as SDS-PAGE (Bhilegaonkar et al., 1990); isoelectric focusing (King, 1984), or immunological methods, that is, counterimmunoelectrophoresis (Sherikar et al., 1993), peroxidase antiperoxidase technique (Karkare et al.,1989), and ELISA (Martin et al., 1991) have inherent limitations; for instance, ELISA and other immunological tests are cumbersome (isolation and identification of species-specific proteins) or cannot distinguish closely related species (immunoassays). Also conventional protein electrophoresis, isoelectric focusing, or SDS-PAGE fail to differentiate closely related species due to similarities between the species proteins (Koh et al., 1998).

Nucleic acid-based molecular techniques that involve analysis of specific DNA targets (unlike protein-based conventional tests) have presented unprecedented advantages owing to the stability of DNA molecules at higher temperatures, the conserved structure within all tissues of an individual, and the availability of unique species signatures in nucleotide sequences. Genomics has emerged as a versatile discipline in the recent past and encompasses the scientific study of structure, function, and interrelationships of individual genes and the genome in its entirety (Bazer and Spencer, 2005). Genomics is fundamentally strengthened by the developments in the fields of molecular biology, bioinformatics, and biotechnology. Consequently, enormous amounts of information contained in the DNA (from short nucleotide stretches to the whole genome of organism) aid in the application of suitable DNA-based tools for the purpose of meat species identification. The availability of extensive DNA data in the global database (open access) and genome sequencing of animal (chicken, fish, cow, buffalo, etc.) species (Fadiel et al., 2005) aid in designing DNA-based molecular methods for meat species identification. As a result DNA-based techniques have emerged as one of the most preferred choices of animal species identification tools in the last two decades. DNA-based tools could be applied for meat speciation of samples that have been subjected to extreme processing or transformed into a variety of products, including rendered animal products. In addition to species identification, such molecular techniques could also be used for the detection of sex (Ennis and Gallagher, 1994), breed (Bheemashankar et al., 2017), and even individuals using meat or tissue as samples (Negrini et al., 2008).

The versatile applications of molecular techniques is essentially attributed to the DNA molecule (information repository) and the reasons that make DNA the choice molecule for the purpose of species identification include (1) DNA structure is conserved in all tissues of an individual and its composition remains the same in every cell (any animal cell such as muscle, visceral organ, skin, bone, excretions or secretions contains the same information); (2) DNA carries an organism's total genetic information; (3) DNA is a highly stable molecule (compared to protein and RNA) and allows species identification even in heated/processed products; and (4) identification of a species is possible to any level, that is, class, family, genus, species, subspecies, breed/strain, sex, and even individual (Karabasanavar et al., 2012).

9.2 Identification of the origin of meat species by DNA hybridization

One of the initial genetic approaches used for species identification was dot-blot hybridization technique. DNA gets denatured to a single-stranded molecule upon exposure to higher temperatures or alkali treatment. The temperature at which the target DNA gets half (50%) denatured is known as its melting temperature (Tm). If denatured DNA is cooled below Tm (or alkali is neutralized) complementary strands anneal to each other, that is, zipping back, to form double-stranded DNA. The DNA hybridization technique makes use of this property, that is, denaturation and annealing of complementary DNA (Chikuni et al., 1990).

DNA hybridization technique has been used for animal species identification; a species-specific probe is added at a concentration higher than the target for the detection of a species. A hybridization reaction can either be performed in solution (high rate of annealing) or in an immobilized inert solid matrix (low annealing rate) such as nitrocellulose. The hybridization rate is affected by temperature (usually 10°C−20°C lower than Tm), duration, nucleotide composition, and concentration of salt used. Unlabeled DNA used in hybridization is called the "target" and it could be the total DNA isolated from a cell suspension or a specific probe (DNA sequence labeled with a marker that is complementary to the target DNA). Usually radio isotope (^{32}P)- or fluorescent dye-labeled nucleotides are used as probes. DNA hybridization could be used as qualitative or semiquantitative method of animal speciation (Ebbehoj and Thomsen, 1991a) (Fig. 9.1).

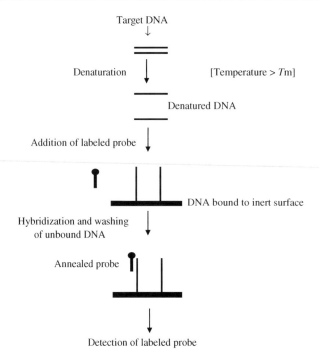

Figure 9.1
Principle of DNA hybridization assay (Karabasanavar et al., 2012).

The earliest blot DNA hybridization used biotin-labeled chromosomal DNA fragments for the detection of chicken, pig, goat, sheep, and beef meat (Chikuni et al., 1990). The method was useful in both fresh and cooked meats. Although chicken and pig probes did not react with other species, the ruminant probes showed cross-reactions with related species. To overcome cross-reactivity, specific genomic DNA probes were designed and unlabeled DNA from cross-hybridizing species was included in the assay (Ebbehoj and Thomsen, 1991b). The suitability of oligonucleotide hybridization was also extended to processed meat products and optimized for performance for samples subjected for repeated freeze–thaw cycles (Buntjer et al., 1999). Generalized DNA hybridization assays are shown in Table 9.1.

The sensitivity of the DNA hybridization technique varies widely depending upon the probe and target and a reasonable sensitivity of about 0.01% could be achieved. The DNA hybridization technique can also be used for the detection of processed meats. Heating meat at 100°C or 120 °C for 30 minutes reduces the signal (time dependent). Nevertheless, tissue type influences the hybridizing signals; a hybridization assay, for instance, could detect admixed sausage luncheon meat and liver pie but not corned beef samples. Broadly, DNA hybridization technique suffers from some limitations including (1) time-consuming; (2) labor-intensive; (3) cross-reactions with closely related species, and (4) interference in

Table 9.1: Application of DNA hybridization for animal species identification.

Sr.	Species	Assay	Sensitivity	Reference
1	Cattle	–	–	Wintero and Thomsen (1990)
2	Pig	P^{32}-labeled porcine probe	0.1% (raw); 0.5% (processed)	Ebbehoj and Thomsen (1991a)
3	Monkey, human, cattle, goat, and sheep	Slot-blot hybridization	0.01%	Ebbehoj and Thomsen (1991b)
4	Rabbit, sheep, pig, cattle, and goat	Species-specific probes	–	Blackett and Keim (1992)
5	Chicken, pig, and cattle	–	–	Buntjer et al. (1999)

processed samples. In view of the complicated procedures and hazardous chemicals involved in DNA hybridization and due to the availability of better and simpler alternative molecular tools (such as polymerase chain reaction, PCR), the application of DNA hybridization for the purpose of species identification is limited.

9.3 PCR-based techniques for species identification of meat

Kary Mullis (Cetus Corporation, USA) invented PCR in the year 1989. It later turned into one of the most versatile techniques in bioscience having a wide range of applications. PCR multiplies specific target DNA sequences (in vitro DNA synthesis) into millions of copies; rapidly, with sensitivity and specificity to make PCR an ideal diagnostic tool for the purpose of animal species identification. PCR makes use of short complimentary oligonucleotide sequences as "primers" and in vitro enzymatic amplification of species-specific target DNA into millions of copies in a span of a few hours, yielding amplified fragments (amplicon/PCR product) that are of diagnostic value in animal species identification, making PCR a state-of-the-art tool. The PCR reaction is performed in a specially engineered machine "thermocycler" (PCR machine) programmed to change the chamber temperature in a predetermined cyclical manner for in vitro DNA amplification that mimics in vivo DNA replication (Karabasanavar et al., 2012).

The discrimination power of PCR is able to detect species even in mixed and degraded samples. During thermal cycling, the heating and cooling processes induce melting of dsDNA, annealing of primers to target DNA, followed by in vitro DNA synthesis by enzymatic replication. As PCR proceeds newly synthesized DNA copies act as templates for upcoming replication leading to a "chain of reactions"; PCR ensures exponential amplification of the target DNA. Due to this unique feature, PCR has revolutionized the modern molecular biology and bioinformatics; the greatest advantage of this technique lies in its scope for modification as per the application requirements. PCR can be coupled or integrated with other techniques to get better end point results. The following applications of PCR have been used for animal species identification.

9.3.1 Random amplified polymorphic DNA—PCR (RAPD fingerprinting)

Random amplified polymorphic DNA (RAPD) is a fingerprinting technique based on arbitrary DNA amplification. A similar technique, that is, AP-PCR (arbitrarily primed PCR) is also based on the same principle; both techniques can be used for animal species identification. Several nonspecific primers are used in RAPD to simultaneously amplify multiple amplicons leading to a definite banding pattern as evidenced after gel electrophoresis; the interpretation of specific banding patterns aids in the identification of animal species. RAPD band patterns vary between species due to the variability of nucleotide sequences. Reproducible band patterns observed after gel electrophoresis are referred as "RAPD fingerprints." Target DNA sequences that yield reproducible fingerprints and help in animal species identification are known as "RAPD markers." Animal species can be identified and discriminated using such unique species-specific fingerprints (markers) generated by the RAPD-PCR (Lee and Chang, 1994). Nevertheless, the basic prerequisite of RAPD-PCR assay is the requirement of purest DNA.

The RAPD fingerprinting used for species identification consists of three basic steps: (1) PCR amplification of target using random primers; (2) gel electrophoresis; and (3) analysis of the resultant fingerprint. Species-specific RAPD fingerprints were used for the detection of a wide variety of samples originating from cattle, buffalo, pig, sheep, horse, mule, donkey, elk, reindeer, goat, kangaroo, and ostrich species and meat products (Martinez and Malmheden Yman, 1998). Min et al. (1996) used RAPD-PCR to identify meat originated from Korean cattle, deer (venison), sheep, and goats species. The RAPD technique has been used for species identification for cattle and buffalo meat (Ganai et al., 2000); different beef breeds (Choy et al., 2001); pork, chicken, duck, turkey, and goose meats (Calvo et al., 2001); and Indian zebu cattle, buffalo, sheep, and goat (Rao Appa et al., 1996). On the contrary, the limited utility of RAPD-PCR for species differentiation has been reported by Koh et al. (1998); wherein generated RAPD fingerprints of red meat species lacked reproducibility owing to intraspecies polymorphisms or PCR cycling conditions. The relative merits and limitations of the RAPD technique are given in Table 9.2.

9.3.2 Forensically informative nucleotide sequencing

PCR amplification of a conserved target (gene) following sequencing is one of the most popular approaches used for meat species identification. Target sequence information is aligned with known sequences stored in the database using computer software or a globally accessible online database, such as the National Center for Biotechnology Information (NCBI). The NCBI's "Basic Local Alignment Search Tool (BLAST)" (http://www.ncbi.nlm.nih.gov/blast/Blast.cgi) is the most popular database used by scientists worldwide. For sequence analysis, "Nucleotide blast" (BlastN) is selected and then the target nucleotide

Table 9.2: Merits and limitations of RAPD fingerprinting for animal speciation.

Merits	Limitations
1. The technique is rapid 2. Many species can be detected simultaneously 3. It does not require previous information about the target species DNA sequence 4. It yields species-specific fingerprint	1. It cannot be used for detecting mixed samples (derived from many species) 2. It lacks quantification ability; level of inclusion or extent of adulteration cannot be detected 3. While detecting several species, computer software is required for interpreting results 4. Intraspecies polymorphisms alter fingerprint and affect decision 5. PCR conditions may affect amplification

sequence is uploaded by selecting "Nucleotide collection" and "highly similar sequences" (Megablast) options in the BLAST page. After submitting the given target sequence, the "BLASTn" tool aligns the sequence in question with all existing sequences in the database. Based on the distance and alignments with query sequence, the nearest species is deduced based on the maximum score; an unknown meat sample DNA nucleotide sequence with the highest score will probably be the species under investigation (Girish et al., 2004).

Since mitochondrial DNA is highly conserved, its sequences are usually targeted in the development of the forensically informative nucleotide sequencing (FINS)-based speciation approach. Of the several mitochondrial targets, cytochrome b is one of the most commonly used molecular targets for the purpose of meat species identification (Chikuni et al., 1994). Forrest and Carneige (1994) amplified and sequenced a region of cytochrome b gene and the sequence obtained was used to identify buffalo, emu, and crocodile meat by phylogenetic comparisons using a computer program. Brodmann et al. (2001) reported PCR-sequencing of target fragments followed by the identification of an unknown game species using a database accessible through the internet. A database search shows a list of sequences in the order of highest percentage of homology. Expected relationships of different vertebrate classes, order, families, and species are also shown. Mitochondrial 12S rRNA gene sequence analysis of a forensic case was sequenced and a 12S rRNA gene sequence was compared to unambiguously prove the origin of skin as bovine and not the tiger (Prakash et al., 2000). Girish et al. (2004) reported the suitability of mitochondrial 12S rRNA sequence analysis for meat animal species identification; the procedure involved the amplification of part of the mitochondrial 12S rRNA gene by PCR, sequencing of 456 bp amplicons, and the analysis of the sequence for a score using a database. This led to unambiguous identification of cattle (*Bos indicus*), buffalo (*Bubalus bubalis*), sheep (*Ovis aries*), goat (*Capra hircus*), and mithun (*Bos frontalis*). The PCR-sequencing was not affected by routine additives or cooking temperatures (72°C, 90°C, 120°C, and 180°C). Further, closely related species such as cattle–buffalo and sheep–goat were also differentiated decisively by the sequence analysis method. FINS-based speciation is dependable as it can withstand legal challenges since sequence data is a clear indicator of the meat species. However, this technique is costly as the sequencing of amplicons requires sophisticated instrumentation, consumables, or outsourcing. Nevertheless, sequencing could be easily accomplished using commercial custom nucleotide sequencing facilities and performed at reasonable prices making this technique the food analyst's choice.

9.3.3 Species-specific PCR

Species-specific-PCR makes use of specific primers that are designed for the definite amplification of the unique target (sequence or gene); the presence of amplification indicates the meat species. This approach requires prior knowledge of nucleotide sequences to be used as a unique target; monospecific primers need to be designed and validated (Meyer et al., 1994). Nuclear as well as mitochondrial genes could be used as unique targets for meat species identification. However, highly conserved regions/genes of mitochondria such as cytochrome b gene, 12S rRNA gene, and D-loop region have been the most widely used targets for species identification using a pair of specific primers so as to amplify species-specific diagnostic amplicons using the end point PCR (Bellis et al., 2003). Of these targets, mitochondrial sequences are preferred owing to their high rate of mutation that allows species identification and differentiation from closely related species. Species identifications using species-specific primers and their target gene are given in the Table 9.3.

Table 9.3: Identification of meat species using species-specific PCR.

Sr.	Species	Target	Sensitivity	Reference
1	Pig	SINE	0.005% in beef; 1% in duck	Calvo et al. (2001)
2	Pig	GH gene	<2%	Wolf and Luthy (2001)
3	Pig	SINE	1%	Calvo et al. (2002)
4	Chicken, turkey	α Cardiac actin	1%	Lockley and Bardsley (2000)
5	Many species	Genomic (β-actin; TP53), mt (cyt-b; D-loop; 28S rRNA)	–	Bellis et al. (2003)
6	Goose, duck	α-actin gene	1%	Rodriguez et al. (2003)
7	Cow, sheep, goat	mt-12S rRNA gene	0.1%	Lopez-Calleja et al. (2004)
8	Partridges	Ribosomal- ITS	1 ng	Tejedor et al. (2006)
9	Beef	1.709 satellite DNA	33.6 fg (raw meat DNA); 0.32 pg (cooked meat DNA)	Zhang-Guoli et al. (1999)
10	Beef	Mitochondrial D loop	0.1%	Karabasanavar et al. (2017)
11	Pork	18S ribosomal gene	2%	Meyer et al. (1994)
12	Sheep and goat	Mitochondrial D loop gene	–	Girish et al. (2016)
13	Sheep	Mitochondrial D loop	0.1%	Nagappa et al. (2011a)
14	Buffalo	Mitochondrial D loop	0.1%	Nagappa et al. (2011b), Girish et al. (2013)

9.3.4 Multiplex PCR

Primers are complementary oligonucleotides that specifically bind to the target sequence (template) as PCR proceeds. A typical "simplex" PCR reaction mixture contains two (one pair consisting of forward and reverse) primers to amplify the discrete target region (specific PCR product). In "duplex" PCR two primer pairs are used; however, more than two primers are used in multiplex PCR for the simultaneous amplification of several targets. Customarily, multiplex PCR applied for species identification makes use of one common primer (either forward or reverse) and several species-specific primers. Different kinds of multiplex PCR assays are in vogue for meat species identification. Multiplex PCR is advantageous to simplex PCR because it can simultaneously detect different meat animals in a single PCR reaction.

The popular multiplex assay developed by Matsunaga et al. (1999) has cytochrome b gene targeted primers; where the forward primer has been designed for a conserved DNA sequence and there are species-specific reverse primers for each of the six species analyzed so as to yield six species-specific PCR products of size 157, 227, 274, 331, 398, and 439 bp specific to goat, chicken, cattle, sheep, pig, and horse, respectively. The multiplex assay was able to detect adulteration of up to 1% of chevon/chicken/beef, 2.2% mutton, and 3.6% horsemeat in the pork. Chevon/chicken/beef were detected at 1% level in mutton/horse base meat and the assay performed well in the detection of meat products too. Likewise, Rodriguez et al. (2003) targeted the mitochondrial 12S rRNA gene and developed multiplex PCR for the identification of goose, mule duck, chicken, turkey, and swine in *foie gras* with a limit of detection at 1% level for each species.

9.3.5 PCR-restriction fragment length polymorphism

PCR-RFLP involves PCR amplification of the targeted gene followed by digestion with specific restriction enzymes (REs). These REs splice the target (PCR product) sequence at specific sites producing fragments that enable the differentiation of species. PCR-RFLP is a convenient, rapid, sensitive, and versatile assay for the differentiation of meat species (Verkaar et al., 2002; Girish et al., 2005). Further, PCR-RFLP has greater potential for application for cooked meat samples and has also proved valuable for the confirmation of results of other speciation tools. Highly conserved regions of mitochondrial DNA, such as cytochrome b gene, 12S rRNA gene, 16S rRNA gene, and hypervariable D-loop regions, are the commonly used targets for species differentiation by PCR-RFLP.

For instance, Meyer et al. (1995) PCR-amplified a 359 bp fragment of cytochrome b gene and spliced it using *Alu*I, *Rsa*I, *Taq*I, and *Hinf*I so as to identify the meats of pig, cattle, wild boar, buffalo, sheep, goat, horse, chicken, and turkey. This method detected pork in heated beef mixtures at up to 1% levels including processed products (marinated, heat-treated, or fermented products). Girish et al. (2004) used a PCR-RFLP technique for the identification of beef, buffalo meat, mutton, and chevon. A 456-bp fragment of mitochondrial 12S rRNA gene was amplified using PCR (universal primers) and digested with *Alu*I, *Hha*I, *Apo*I, and *Bsp*TI REs. The resultant patterns upon electrophoresis of the digested samples identified and differentiated the species. The differentiation of mutton and chevon was accomplished by the use of PCR-RFLP, where satellite I DNA was amplified followed by restriction digestion by *Apa*I enzyme (Chikuni et al., 1994). This enzyme has an RE site only in sheep but not in goat, and therefore sheep and goat species were differentiated in agarose gel electrophoresis. Wolf et al. (1999) used cytochrome b gene-based PCR-RFLP for the identification of game (meat) species, where a 464 bp PCR product was cut using REs for the differentiation of closely related deer species.

Hopwood et al. (1999) amplified a single actin gene locus so as to produce a band only in chicken or turkey but not in other species. Further, chicken was differentiated from that of turkey by RE digestion and assay detected cooked at less than 1% in admixture of meats. The PCR-RFLP technique was used for the differentiation of meat derived from different breeds of cattle and related species such as taurine cattle (*Bos taurus*), zebu (*B. indicus*), banteng (*Bos javanicus*), bison, wisent (*Bison bonasus*), water buffalo (*B. bubalis*), and African buffalo (*Syncerus caffer*) based on the DNA sequence polymorphism (Verkaar et al., 2002). PCR-RFLP is a cost-effective and time-saving technique for the simultaneous detection of several meat species, but the major drawback associated with the technique is linked to intraspecies mutations in the targeted gene that may affect results of the RFLP.

9.3.6 DNA microarrays

The DNA microarray is based on the preferential binding of complementary single-stranded nucleic acid sequences. The technique involves immobilization of a single-stranded DNA (ssDNA) probe onto a transducer surface to attach to its complementary target DNA sequence via hybridization (Rahmati et al., 2016) which will help in identification of meat species. DNA microarrays are essentially arrays of spots that are ordered with probe DNA molecules used for measuring the quantity of target nucleic acid molecules (Javanmard et al., 2010). The DNA hybridization (probe-target) is detected and quantified by labeled targets (fluorophores, silver, or chemiluminescence) leading to the detection of thousands of probes (microarray). In the standard microarray, probes are attached to a solid surface and used for (1) detection of single nucleotide polymorphisms and (2) genotype or sequence mutant genomes. DNA microarrays are considered to be the best for specific detection systems and are fast, reusable, continuous, selective, and sensitive (Sassolas et al., 2008). Ballin et al. (2009) reported a lower sensitivity of DNA microarray hybridization than PCR and also lesser sensitivity to sample cross-contamination.

9.3.7 Other PCR-based methods

9.3.7.1 Direct amplified length polymorphisms

This method in principle is similar to AP-PCR. It is based on genomic fingerprint analysis that occurs virtually in any species (polymorphisms in DNA). The advantage of direct amplified length polymorphism (DALP) is that prior knowledge of the target is not required, there is no need for cloning of the sequence and many species can be differentiated. The limitation of the technique is that sometimes DALP results are not reproducible (Desmais et al., 1998).

9.3.7.2 Single-strand conformation polymorphism

PCR-SSCP (single-strand conformation polymorphism) is a molecular technique applied for species identification of meat based on DNA secondary structure analysis. SSCP is usually coupled with PCR as PCR-SSCP. Based on the differential migration of a single strand upon electrophoreses, the species is identified and differentiated. Strands differing by single base substitutions form different conformations and migrate differently upon nondenaturing electrophoresis such as PAGE (Tejedor et al., 2006).

9.3.7.3 Amplified fragment length polymorphism

Amplified fragment length polymorphism (AFLP) is a wider scan of the genome. Restriction digested genomic DNA is used as a template for PCR reactions, where the primers themselves contain restriction sites. The AFLP technique can even discriminate animal breeds. About 500 primer combinations can yield a selected marker, which is converted into a single nucleotide marker. Usually AFLP is used for high-throughput genotyping. The AFLP has been used for the discrimination of cattle breeds and high-throughput genotyping of the cattle (Sasazaki et al., 2004).

9.3.7.4 Interretrotransposon amplified polymorphisms

Retrotransposons located in the genome are analyzed for speciation. They show genetic differences when amplified using specific primers and also yield reproducible banding patterns. Apart from species authentication, interretrotransposon amplified polymorphism can also be used to indicate the geographical origin of the animal species.

9.3.7.5 DNA invader assays

New generation DNA invader assays detect species without amplification. This is a generic enzymatic technique based on signal amplification through fluorescence resonance energy transfer (FRET).

9.4 Meat species identification using loop-mediated isothermal amplification

Loop-mediated Isothermal Amplification (LAMP) is a novel nucleotide amplification technique developed by Notomi et al. (2000). The assay enables isothermal amplification of the target gene, and thereby obviates the requirement for expensive thermal cyclers. In the LAMP reaction gene amplification proceeds through the repetition of two types of elongation reactions that occur via loop regions (i.e., template self-elongation from the stem loop structure formed at three prime terminals and subsequent binding and elongation of new primers to the loop region).

The LAMP reaction uses pairs of inner and outer primers. Each of the inner primers possesses a sequence complementary to one chain of the amplification region at the three prime terminal and identical to the inner region of the same chain at the five prime terminal. The elongation reactions are sequentially repeated by DNA polymerase-mediated strand-displacement synthesis using the aforementioned stem loop regions as a stage. This method operates on the fundamental principle of the production of a large quantity of DNA amplification products with a mutually complementary sequence and an alternating, repeated structure. LAMP can amplify a few copies of DNA to 10^9 copies in less than an hour under isothermal conditions with great specificity. Visualization of amplification can be done using the naked eye using the dyes (Notomi et al., 2015). The LAMP result can be visualized by the naked eye by turbidity produced by magnesium pyrophosphate, which is produced as a by-product of amplification. Further, amplification can be visualized by adding different dyes like SYBR green, hydroxynapthol blue, and calcein. LAMP is evolving to be an alternative to the PCR-based methods for species identification of meat. LAMP amplifies target DNA at isothermal temperatures and obviates the need for thermal cyclers and postamplification procedures for signal detection (Erwanto et al., 2011).

As in PCR, LAMP assay could amplify target DNA several folds (10^9 copies in a span of an hour) under isothermal conditions without compromising specificity and having the capability of the visual detection of amplified targets using specific dyes (Yang et al., 2014). Consequently, LAMP assay has emerged as an alternative tool to PCR-based techniques for the purpose of testing food safety hazards including the detection of meat adulteration (Abdullahi et al., 2017). LAMP is a field-friendly technique as it does not involve expensive instrumentation and can be undertaken by a semiskilled person. Sensitivity and specificity of LAMP is also high compared to other methods. The technique is quick as it takes only 3 hours for the completion of the assay. The technique is simple as the amplification can be undertaken at constant temperature and no postamplification process is involved (Kumar et al., 2017).

9.5 Quantitative meat speciation

Quantification of the extent of adulteration is required to establish the level of incorporation of inferior meat into the superior one. In particular, it is very important in the cases of adulteration of valued animal products. The detection of the extent of meat adulteration is possible using DNA-based techniques. This approach is based on a number of DNA targets available in the sample matrix (tissue); counting the copy numbers can deduce extent of adulteration.

9.5.1 Quantitative PCR

As the conventional PCR relies on end point detection (i.e., concentration of target), it cannot measure the actual amount of the target DNA (copy number) present in the sample at the beginning of amplification (Tanabe et al., 2007). This limitation of conventional PCR is attributed to nonlinearity in the amplification process. Although it is linear at the beginning, as reactants become limiting in vitro synthesis of DNA tends to take a nonlinear progression. Therefore due to this nonlinearity in product synthesis, the conventional PCR cannot be used for quantitative meat speciation. On the other hand, quantitative PCR has been engineered to measure the initial copy number of target DNA present in the sample. There are two basic techniques in Q-PCR-based quantitative animal speciation (1) quantitative-competitive PCR (QC-PCR) and (2) real-time PCR.

9.5.1.1 Quantitative-competitive PCR

In QC-PCR along with the target DNA an "internal standard" is amplified simultaneously (Wiseman, 2002). During the amplification process, the internal standard (included in the reaction) acts as a "competitor" to the "target DNA." Hence by comparing the products generated, DNA is quantified (Wolf and Luthy, 2001). For example, QC-PCR has been designed for detecting adulteration in meat by targeting a growth hormone gene using a competitor that differs by 20 bp. Usually single copy genes (e.g., glyceraldehyde-3-phosphate dehydrogenase-GAPDH) are targeted for developing QC-PCR assays rather than multicopy targets (Woolfe and Primrose, 2004).

9.5.1.2 Real-time PCR

The advent of "real-time PCR" has revolutionized the field of molecular biology. It is one of the most powerful tools for animal specification; apart from sensitive detection it has the capability of the accurate detection of the extent of adulteration, that is, quantification capability (Higuchi et al., 1993). The real-time PCR has a higher level of specificity, sensitivity, accuracy, and analytical precision than any other animal speciation technique known. Real-time PCR is the most preferred molecular biology tool suitable for quantitative speciation or adulteration detection.

For DNA quantification, a linear (calibration) curve is plotted by taking a known DNA concentration on x-axis and Ct values on y-axis. This yields a linear equation, $y = mx + c$ (where "y" is Ct, "x" is log DNA concentration, "m" is slope, and "c" is a constant). Based on acceptable linearity (coefficient r^2 more than 0.99), the quantification of the unknown DNA is undertaken (Rodriguez et al., 2005). The real-time PCR shows reaction kinetics in real time and is used for the quantification of target DNA. There is no post-PCR sample handling, such as electrophoresis or exposure to hazardous dyes (ethidium bromide), hence the reaction could be completed in less time and thus leading to the screening of a large number of samples. The real-time PCR can be developed for speciation of any meat animal (domestic animals, poultry, fish, etc.). Using different chemistries, the real-time PCR detects adulteration of the sample to the level of femto-, nano-, and picograms (Tables 9.4 and 9.5).

Table 9.4: Application of real-time PCR for animal species identification.

Sr.	Species	Target	Chemistry	LOD	Reference
1	Cattle	GH gene	TaqMan	0.01% (0.02 ng)	Brodmann and Moor (2003)
2	Cattle	16S rRNA	SYBR Green I, Scorpion probes	0.1%	Sawyer et al. (2003)
3	Pig, chicken, ruminants	SINE/LINE	SYBR Green I	0.01–1 pg	Walker et al. (2004)
4	Cattle, pig, sheep, chicken, turkey	cyt-b gene	TaqMan	0.1%–0.5%	Dooley et al. (2004)
6	Aves, equine, canine, feline, rodents	SINE/LINE	SYBR Green I	0.1 ng–0.1 pg	Walker et al. (2003)
7	Bovine, porcine, sheep, chicken, turkey, ostrich	mt tRNA^{Glu}; ND5; Cyt b	TaqMan MGB	0.03–0.8 pg (1%–>5%)	López-Andreo et al. (2005)
8	Pig	12S rRNA	TaqMan	0.5%–5%	Rodriguez et al. (2005)
11	Bovine	cyt-b gene	TaqMan (FAM-labeled)	35 pg	Zhang et al. (2007)
12	Pig, chicken, cattle, sheep, horse	cyt b gene	TaqMan MGB probes	100 fg	Tanabe et al. (2007)

Table 9.5: Relative merits and limitations of real-time PCR.

Merits	Limitations
1. Quantitative analysis 2. Amplification in real time 3. Earliest detection of PCR product 4. High efficiency 5. Multiplex capability 6. Freedom from post-PCR sample handling	1. Equipment (real-time PCR machine) is expensive 2. Chemicals required are expensive 3. More technically demanding as compared to conventional PCR

Different types of real-time PCR chemistries are

1. SYBR Green: a fluorescent dye is used in place of oligonucleotide probes. The SYBR green-based chemistries are used as alternatives to costly TaqMan-based chemistries and the results of uniplex and duplex reactions are comparable to those of probes.
2. TaqMan (MGB/FAM−TAMRA): the specificity of real-time PCR is due to primers, in TaqMan chemistry, MGB probes ensure genuine amplification and reduces the uncertainty associated with the background signals.
3. Scorpion probes: these probes can also be used for specific quantification of the target DNA.

Single as well as multicopy targets can be amplified for species identification and quantification. However, amplicon size has a direct bearing on sensitivity—smaller sized amplicons are better quantifiers (Karabasanavar et al., 2012).

9.6 Scope for future research

A vast array of molecular techniques have been developed for species identification of meat. Most of the techniques require sophisticated equipment, a laboratory, and chemicals for species identification, hence they often cannot be used at field level. LAMP-based techniques that have emerged in the last decade have created the possibility of developing field applications for species identification of meat. But their accuracy and repeatability need to be further improved. To save time, cost, and expenditure on the storage of meat awaiting testing, rapid techniques which can be used as platform tests need to be developed. In addition, more research on quantitative meat species identification need to be undertaken. Combining DNA-based techniques with recent developments in information technology tools could help in improving the applicability of the techniques in the future.

References

Abdullahi, U.F., Igwenagu, E., Aliyu, S., Mu'azu, A., Naim, R., Wan-Taib, W.R., 2017. A rapid and sensitive Loop-mediated isothermal amplification assay fordetection of pork DNA based on porcine tRNA lys and ATPase 8 genes. Int. Food Res. J. 24, 1357−1361.

Ballin, N.Z., Vogensen, F.K., Karlsson, A.H., 2009. Species determination - can we detect and quantify meat adulteration? Meat Sci. 83, 165−174.

Bazer, F.W., Spencer, T.E., 2005. Reproductive biology in the era of genomics biology. Theriogenology 64, 442−456.

Bellis, C., Ashton, K.J., Freney, L., Blair, B., Griffiths, L.R., 2003. A molecular genetic approach for forensic animal species identification. Forensic Sci. Int. 134, 99−108.

Kannur, B.H., Fairoze, Md. N., Girish, P.S., Karabasanavar, N., Rudresh, B.H., 2017. Breed traceability of buffalo meat using microsatellite genotyping technique. J. Food Sci. Technol. 54, 558−563.

Bhilegaonkar, K.N., Sherikar, A.T., Khot, J.B., Karkare, U.D., 1990. Studies on characterization of thermostable antigens of adrenals and muscle tissue of meat animals. J. Sci. Food Agric. 51, 545−553.

Blackett, R.S., Keim, P., 1992. Big game species identification by deoxyribonucleic acid (DNA) probes. J. Forensic Sci. 37, 590−596.

Brodmann, P.D., Moor, D., 2003. Sensitive and semi-quantitative TaqManTM real-time polymerase chain reaction systems for the detection of beef (*Bos illia*) and the detection of the family Mammalia in food and feed. Meat Sci. 65, 599–607.

Brodmann, P., Nichola, A.S., Schaltenbrand, P., Ilg, E., 2001. Identifying unknown game species: experience with nucleotide sequencing of the mitochondrial cytochrome b gene and a subsequent basic local alignment search tool search. Eur. Food Res. Technol. 212, 491–496.

Brooks, S., Elliott, C.T., Spence, M., Walsh, C., Dean, M., 2017. Four years post-horsegate: an update of measures and actions put in place following the horsemeat incident of 2013. npj Sci. Food. 1. Available from: https://doi.org/10.1038/s41538-017-0007-z.

Buntjer, J.B., Lamine, A., Haagsma, N., Lenstra, J.A., 1999. Species identification by oligonucleotide hybridization: the influence of processing of meat products. J. Sci. Food Agric. 79, 53–57.

Calvo, J.H., Zaragoza, P., Osta, R., 2001. Technical note: a quick and more sensitive method to identify pork in processed and unprocessed food by PCR amplification of a new specific DNA fragment. J. Anim. Sci. 79, 2108–2112.

Calvo, J.H., Osta, R., Zaragoza, P., 2002. Quantitative PCR detection of pork in raw and heated ground beef and pate. J. Agric. Food Chem. 50, 5265–5267.

Chikuni, K., Ozutsumi, K., Koishikawa, T., Kato, S., 1990. Species identification of cooked meats by DNA hybridization assay. Meat Sci. 27, 119–128.

Chikuni, K., Tabata, T., Saito, M., Monma, M., 1994. Sequencing of mitochondrial cytochrome b genes for the identification of meat species. Anim. Sci. Technol. 65, 571–579.

Choy, Y.H., Oh, S.J., Kang, J.O., 2001. Application of RAPD methods in meat for beef identification. Asian-Australasian J. Anim. Sci. 14, 1655–1658.

Desmais, E., Lanneluk, I., Langer, J., 1998. Direct amplification of length polymorphisms (DALP), or how to get and characterize new genomic markers in many species. Nucleic Acids Res. 26, 1458–1465.

Dooley, J.J., Pain, K.E., Garrett, S.D., Brown, H.M., 2004. Detection of meat using TaqMan real time PCR assays. Meat Sci. 68, 431–438.

Ebbehoj, K.F., Thomsen, P.D., 1991a. Species differentiation of heated meat products by DNA hybridization. Meat Sci. 30, 221–234.

Ebbehoj, K.F., Thomsen, P.D., 1991b. Differentiation of closely related species by DNA hybridization. Meat Sci. 30, 359–366.

Ennis, S., Gallagher, T.F., 1994. A PCR- based sex-determination assay in cattle based on the bovine amelogenin locus. Anim. Genet. 25, 425–427.

Erwanto, Y., Abidin, M.Z., Rohman, A., Sismindari, M., 2011. PCR-RFLP using *Bse*DI enzyme for pork authentication in sausage and nugget products. J. Anim. Sci. Technol. 34, 14–18.

Fadiel, A., Anidi, I., Eichenbaum, K.D., 2005. Farm animal genomics and informatics: an update. Nucleic Acids Res. 33, 6308–6318.

Forrest, A.R.R., Carnegie, P.R., 1994. Identification of gourmet using FINS (Forensically informative nucleotide sequencing). BioTechniques 17, 24–26.

Ganai, T.A.S., Singh, R.K., Butchiah, G., 2000. DNA amplification fingerprinting of cattle and buffalo genome by RAPD-PCR utilizing arbitrary oligonucleotide primers. Buffalo J. 3, 331–339.

Girish, P.S., Anjaneyulu, A.S.R., Viswas, K.N., Anand, M., Rajkumar, N., Shivakumar, B.M., et al., 2004. Sequence analysis of mitochondrial 12S rRNA gene can identify meat species. Meat Sci. 66, 551–556.

Girish, P.S., Anjaneyulu, A.S.R., Viswas, K.N., Shivakumar, B.M., Anand, M., Patel, M., et al., 2005. Meat species identification by polymerase chain reaction-restriction fragment length polymorphism (PCR-RFLP) of mitochondrial 12S rRNA gene. Meat Sci. 70, 107–112.

Girish, P.S., Haunshi, S., Vaithiyanathan, S., Rajitha, Ramakrishna, C., 2013. A rapid method for authentication of Buffalo (*Bubalus bubalis*) meat by alkaline lysis method of DNA extraction and species specific polymerase chain reaction. J. Food Sci. Technol. 50, 141–146.

Girish, P.S., Vaithiyanathan, S., Karabasanavar, N., Bagale, S., 2016. Authentication of sheep (*Ovis aries*) and goat (*Capra hircus*) meat using species specific polymerase chain reaction. Ind. J. Anim. Sci. 86, 1172−1175.

Higuchi, R., Fockler, C., Dollinger, G., Watson, R., 1993. Kinetic PCR: real time monitoring of DNA amplification reactions. Biotechnol. 11, 1026−1030.

Hopwood, A.J., Fairbrother, K.S., Lockley, A.K., Bardsley, R.G., 1999. An actin gene-related polymerase chain reaction (PCR) test for identification of chicken in meat mixtures. Meat Sci. 53, 227−231.

Javanmard, M., Ronaghi, M., Davis, R.W., 2010. Electrical Detection of Biomarkers Using Bioactivated Microfluidic Channels. Google Patents.

Karabasanavar, N.S., Singh, S.P., Umapathi, V., Kumar, D., Patil, G., et al., 2011a. A highly specific PCR assay for identification of raw and heat treated mutton (*Ovis aries*). Small Rum. Res. 100, 153−158.

Karabasanavar, N.S., Singh, S.P., Umapathi, V., Patil, G., Shebannavar, S.N., et al., 2011b. Authentication of carabeef (water buffalo, *Bubalus bubalis*) using highly specific polymerase chain reaction. Eur. Food Res. Technol 233, 985−989.

Karabasanavar, N.S., Patil, G., Santosh Haunshi, S., Viswas, K.N., 2012. DNA Based Animal Species Identification. Hind Publisher, Hyderabad.

Karabasanavar, N., Girish, P.S., Deepak Kumar, S.P., Singh, 2017. Detection of beef adulteration by mitochondrial D-loop based species-specific polymerase chain reaction. Int. J. Food Prop. 20, 2264−2271.

Karkare, U.D., Sherikar, A.T., Bhilegaonkar, K.N., 1989. Meat speciation by unlabeled antibody peroxidase anti peroxidase (PAP) technique. J. Bombay Vet. Coll. 1, 21−26.

King, N.L., 1984. Species identification of cooked meats by enzyme staining of isoelectric focusing gels. Meat Sci. 11, 59−64.

Koh, M.C., Lim, C.H., Chua, S.B., Chew, S.T., Phang, S.T.W., 1998. Random amplified polymorphic DNA (RAPD) fingerprints for identification of red meat animal species. Meat Sci. 48, 275−285.

Kumar, Y., Bansal, S., Jaiswal, P., 2017. Loop-mediated isothermal amplification (LAMP): a rapid and sensitive tool for quality assessment of meat products. Compr. Rev. Food Sci. Food Saf. 16, 1359−1378.

Lee, J.C., Chang, J.G., 1994. Random amplified polymorphic DNA polymerase chain reaction (RAPD-PCR) fingerprints in forensic species identification. Forensic Sci. Int. 67, 103−107.

Lockley, A.K., Bardsley, R.G., 2000. DNA-based methods for food authentication. Trends Food Sci. Technol. 11, 67−77.

López-Andreo, M., Lugo, L., Garrido-Pertierra, A., Prieto, M.I., Puyet, A., 2005. Identification and quantitation of species in complex DNA mixtures by real-time polymerase chain reaction. Anal. Biochem. 339, 73−82.

Lopez-Calleja, I., Fajardo, M.A., Rodriguez, M.A., Hernandez, P.E., Garcia, T., Martin, R., 2004. Rapid detection of cow's milk in sheeps' and goats' milk by a species-specific polymerase chain reaction technique. J. Dairy Sci. 87, 2839−2845.

Martinez, I., Malmheden Yman, I., 1998. Species identification of meat products by RAPD analysis. Food Res. Int. 31, 459−466.

Martin, R., Urdale, R.J., Jones, S.J., Hernandez, Y.E., Patterson, R.L.S., 1991. Monoclonal antibody sandwich ELISA for the potential detection of chicken meat in mixtures of raw beef and pork. Meat Sci. 30, 23−31.

Matsunaga, T., Chikuni, K., Tanabe, R., Muroya, S., Shibata, K., Yamada, J., et al., 1999. A quick and simple method for the identification of meat species and meat products by PCR assay. Meat Sci. 51, 143−148.

Meyer, R., Candrian, U., Luthy, J., 1994. Detection of pork in heated meat products by the polymerase chain reaction. J. Assoc. Off. Anal. Chem. 77, 617−622.

Meyer, R., Hofelein, C., Lüthy, J., Candrian, U., 1995. Polymerase chain reaction-restriction fragment length polymorphism analysis: a simple method for species identification in food. J. Assoc. Off. Anal. Chem. Int. 78, 1542−1551.

Min, J.S., Min, B.R., Han, J.Y., Lee, M., 1996. The identification of species of meat (Korean cattle, beef, deer meat, sheep meat and goat meat) using random amplified polymorphic DNAs. Korean J. Anim. Sci. 38, 231−238.

Negrini, R., Nicoloso, L., Crepaldi, P., Milanesi, E., Colli, L., Chegdani, F., et al., 2008. Assessing SNP markers for assigning individuals to cattle populations. Anim. Genet. 40, 18–26.

Notomi, T., Okayama, H., Masubuchi, H., Yonekawa, T., Watanabe, K., Amino, N., et al., 2000. Loop-mediated isothermal amplification of DNA. Nucleic Acids Res 28, E63.

Notomi, T., Mori, Y., Tomita, N., Kanda, H., 2015. Loop-mediated isothermal amplification (LAMP): principle, features, and future prospects. J. Microbiol. 53, 1–5.

Prakash, S., Patole, M.S., Ghumatkar, S.V., Nandode, S.K., Shinde, B.M., Shouche Yogesh, S., 2000. Mitochondrial 12S rRNA sequence analysis in wild life forensics. Curr. Sci. 78, 1239–1241.

Rahmati, S., Julkaplia, N.M., Yehyea, W.A., Basirun, W.J., 2016. Identification of meat origin in food products - a review. Food Control 68, 379–390.

Rao Appa, K.B.C., Bhat, K.V., Totey, S.M., 1996. Detection of species-specific genetic markers in farm animals through random amplified polymorphic DNA (RAPD). Genet. Anal.: Biomol. Eng. 13, 135–138.

Rodriguez, M.A., Garcia, T., Gonzalez, I., Asensio, L., Mayoral, B., Lopez-Callejal, et al., 2003. Development of polymerase chain assay for species identification of goose and mule duck in *foie gras* products. Meat Sci. 65, 1257–1263.

Rodriguez, M.A., Gracia, T., Gonzalez, I., Hernandez, P.E., Martin, R., 2005. TaqMan real time PCR for the detection quantification of pork in meat mixtures. Meat Sci. 70, 113–120.

Sasazaki, S., Itoh, K., Arimitsu, S., Imada, T., Takasuga, A., Nagaishi, H., et al., 2004. Development of breed identification markers derived from AFLP in beef cattle. Meat Sci. 67, 275–280.

Sassolas, A., Leca-Bouvier, B.D., Blum, L.J., 2008. DNA biosensors and micro-arrays. Chem. Rev. 108, 109–139.

Sawyer, J., Wood, C., Shanahan, D., Gout, S., McDowell, D., 2003. Real time PCR for quantitative meat species testing. Food Control 14, 579–583.

Sherikar, A.T., Karkare, U.D., Khot, J.B., Jayarao, B.M., Bhilegaonkar, K.N., 1993. Studies on thermostable antigens, production of species-specific antiadrenal sera and comparison of immunological techniques in meat speciation. Meat Sci. 33, 121–136.

Tanabe, S., Hase, M., Yano, T., Sato, M., Fujimura, T., Akiyama, H., 2007. A real-time quantitative PCR detection method for pork, chicken, beef, mutton and horseflesh in foods. Biosci. Biotechnol. Biochem. 71, 3131–3135.

Tejedor, M.T., Monteagudo, L.V., Arruga, M.V., 2006. DNA single strand conformation polymorphisms (SSCPś) studies on Spanish red-legged partridges. Wildl. Biol. Pract 2, 8–12.

Vaithiyanathan, S., Vishnuraj, M.R., Reddy, G.N., Kulkarni, V.V., 2018. Application of DNA technology to check misrepresentation of animal species in illegally sold meat. Biocatal. Agric. Biotechnol. 16, 564–568.

Verkaar, E.L.C., Nijman, I.J., Boutaga, K., Lenstra, J.A., 2002. Differentiation of cattle species in beef by PCR - RFLP of mitochondrial and satellite DNA. Meat Sci. 60, 365–369.

Walker, J.A., Hughes, D.A., Anders, B.A., Shewale, J., Sinha, S.K., Batzera, M.A., 2003. Quantitative intra-short interspersed element PCR for species-specific DNA identification. Anal. Biochem. 316, 259–269.

Walker, J.A., Hughes, D.A., Anders, B.A., Shewale, J., Sinha, S.K., Batzer, M.A., 2004. Quantitative PCR for DNA identification based on genome-specific interspersed repetitive elements. Genomics 83, 518–527.

Wintero, A.K., Thomsen, P.D., 1990. A comparison of DNA-hybridization, immuno-diffusion, counter-current immuno-electrophoresis and iso-electric focusing for detecting the admixture of pork and beef. Meat Sci. 27, 75–85.

Wiseman, G., 2002. State of the art and limitations of quantitative polymerase chain reaction. J. AOAC. Int. 85, 792–796.

Wolf, C., Luthy, J., 2001. Quantitative competitive PCR for quantification of porcine DNA. Meat Sci. 57, 161–168.

Wolf, C., Rentsch, J., Hubner, P., 1999. PCR-RFLP analysis of mitochondrial DNA: a reliable method for species identification. J. Agric. Food Chem. 47, 1350–1355.

Woolfe, M., Primrose, S., 2004. Food forensics: using DNA technology to combat mis description and fraud. Trends Biotechnol. 22, 222–226.

Yang, L., Fu, S., Peng, X., Li, L., Song, T., Li, L., 2014. Identification of pork in meat products using real-time loop-mediated isothermal amplification. Biotechnol. Biotechnol. Equip. 28, 882–888.

Zhang-Guoli, Mingguang, Z., Zhijiang, Z., Hongsheng, O., Qiang, L., 1999. Establishment and application of a polymerase chain reaction for the identification of beef. Meat Sci. 51, 233–236.

Zhang, C.L., Fowler, M.R., Scott, N.W., Graham, L., Slater, A., 2007. A TaqMan real-time PCR system for the identification and quantification of bovine DNA in meats, milks and cheeses. Food Control 18, 1149–1158.

CHAPTER 10

Meat traceability and certification in meat supply chain

P.S. Girish and S.B. Barbuddhe
ICAR - National Research Centre on Meat, Hyderabad, India

Chapter Outline
10.1 Introduction 153
10.2 Benefits of livestock traceability system 154
10.3 Methods for identification of animals 155
    10.3.1 Visual tagging 156
    10.3.2 Bar-coded tags 156
    10.3.3 Radio frequency identification devices 156
    10.3.4 Quick response code-based tags 156
10.4 Scenario of livestock traceability around the world 157
    10.4.1 Livestock traceability system in the European Union 157
    10.4.2 Livestock traceability system in Australia 159
    10.4.3 Livestock traceability system in Japan 160
    10.4.4 Comparison of traceability regulatory systems in different countries 160
10.5 Molecular meat traceability 162
    10.5.1 Sample requirements for DNA-based meat traceability 162
    10.5.2 Molecular meat traceability using microsatellite genotyping 162
    10.5.3 Molecular meat traceability using single nucleotide polymorphism genotyping 166
10.6 Future scope of work 168
References 168

10.1 Introduction

Livestock traceability means the ability to follow an animal or group of animals during all stages of its life (OIE, 2018). Globally, animal identification and traceability are recognized as important components of the management of animal and human health, and food safety (Ted and Glynn, 2012). Traceability is emerging as a major benchmark for meat quality assurance which necessitates that producers, packers, processors, wholesalers, exporters, and retailers assure that livestock and meat are identified and that record keeping guarantees traceability through all or parts of life cycle, and that such information is authentic, visible, and can be verified (Smith et al., 2008). More and more countries are adopting livestock traceability systems with an eye on creating acceptability for their produce in the international market. Traceability requires foolproof animal identification with unique numbering, a system for registration, the identification of premises involved in the raising of the meat animals and their harvesting, a robust database to upload, update, and retrieve the information, and sufficient equipment and personnel to implement and constantly ensure that the above tasks are done in earnest. Developed countries are leading the way in the implementation of livestock traceability systems, which may be due to their financial wherewithal allowing them to employ the required number of trained personnel to keep the system running and also the higher awareness level of stakeholders. Some of the countries which have adopted livestock traceability are Brazil, Australia, the United States, New Zealand, Canada, Argentina, Uruguay, Japan, the European Union, Mexico, and South Korea (Ted and Glynn, 2012). Meat traceability may help in quality assurance, food recall, disease control, protecting public health, implementation of the subsidy programs, and encouraging hygienic production practices. Techniques for verifying traceability are also an important requirement for making the system robust. Molecular techniques like microsatellite genotyping and single nucleotide polymorphism (SNP) can be used to confirm the accuracy that traceability requires with respect to animal of origin. This chapter provides insights into the concept of livestock and meat traceability, its components, the benefits of the system, the status of implementation in different countries, and different molecular techniques available for traceability verification.

10.2 Benefits of livestock traceability system

Livestock traceability has a range of applications and utilities. Registration of the animals and their identification are basic requirements of the traceability system. Generally, animals are identified using laser-printed visible ear tags or radio frequency identification devices (RFIDs) and bar-coded ear tags. This helps in ownership ascertainment, eases the hassles involved in getting transportation clearances, and enables precision farming, as well as scientific feeding and health management. Animal identification systems can also be used for recording performance parameters which in turn help in the selection of the livestock for breeding. The continued practice of stock selection for breeding purposes would improve the overall quality of the species or breed (germplasm) (Herrero et al., 2013). Registration of premises like farms and abattoirs helps in centralized monitoring of activities which enables the achievement of comprehensive quality assurance. The centralized availability of information on animals, farms, and abattoirs will help in the effective implementation of disease control programs. If a disease (or disease-causing agent) is detected during the meat inspection, the traceability system enables the tracking of its farm-of-origin (Johnston,

2005). Tracking the sources of disease-causing agents helps in understanding the epidemiological patterns, in preventing spread, and in planning biosecurity measures to be put in place to prevent a recurrence of the disease. Similarly, traceability records can help in the detection of the source of chemical contaminants in meat. Farm-to-fork traceability helps in achieving a comprehensive quality assurance system which enables the production of quality and safe meat free from physical, chemical, and biological hazards (pathogens, chemical residues, etc.). Traceability-based quality assurance programs help in the documentation of food safety hazards (Thakur and Hurburgh, 2009). Countries which have adopted traceability systems expect that countries which export meat to these countries must also comply with the traceability systems on a par with the domestic regulations (Shackell et al., 2001). Hence meat produced under completely traceable environments can easily find market access in different countries, thereby boosting the export prospects of the commodity.

10.3 Methods for identification of animals

Animal identification is the combination and linking of the identification and registration of an animal individually with a unique identifier, or collectively by its epidemiological unit or group with a unique group identifier (OIE, 2018). Identification of animals with a tamper-proof system is a core requirement for any livestock traceability system. Animals can be identified individually or as a batch depending on the requirements. Retention of the code over the animal throughout its lifetime is one of the challenges of the traceability system. Hence in the European Union two sets of tags are issued per animal which are tagged on both ears. Even if one falls from one ear, the animal can be identified with the other tag until the lost tag is replaced. Any ideal animal identification method must be resistant to varying environmental conditions, as well as being economical, easily applicable, and tamper-proof (Frewer et al., 2005). Based on these requirements several animal identification methods have been followed across the world, that is, branding, tattooing, visual tags, bar code tags, RFID tags, implants, etc. (Musa et al., 2014). Tags must be centrally produced and distributed to the farmers so as to maintain uniformity and also to inscribe numbering pattern in line with the approved national policy or international guidelines, such as International Committee for Animal Identification and recording, Rome (www.icar.org/ICARfacts) (Girish et al., 2017). Where ear tags are used for animal identification, they must be made of flexible material and be tamper-proof, easy to read, reusable, and designed in such a way that they remain attached to the animal without being harmful to it and must carry only nonremovable inscriptions.

Brief details of the different methods commonly used for the identification of animals are given below.

10.3.1 Visual tagging

Visual tags are the simplest tagging method, where an animal's number is printed on a plastic tag and the number is clearly visible. A good quality tag applied skillfully can last for the animal's lifetime. However, poor quality tags often fall or get bleached making the number unreadable. In the European Union animals are mostly identified using visible plastic ear tags that have a laser-printed code. Ear tags are provided in duplicate to farmers for placing on both ears so as to avoid possible confusion that could arise due to dropping of the tags.

10.3.2 Bar-coded tags

A bar code is a machine-readable optical label that contains information regarding the item to which it is attached. The code can be read using a bar code scanner. However, if the bar code is combined with visible numbers, the tag can be read visually also. The possibility of human error can be eliminated using scanners. Nevertheless, scanning becomes difficult when tags get dirty and this requires cleaning of the tag prior to the scan. The bar code method also involves the additional cost of a computer, software, and a scanner for reading the codes. But it is also possible to read bar codes using smartphone applications. The national animal (cattle and buffalo) identification program of India uses visible-cum-bar-coded tags for the identification of animals.

10.3.3 Radio frequency identification devices

The RFID is a convenient noncontact electronic data-reading automation technology (Costa et al., 2013). RFID is not affected by the dirt and has the advantage of long distance reading and high reading accuracy. For animal identification the RFID is one of the ideal options (Liang et al., 2015; Falco et al., 2017) but it requires an RFID reader for reading the identification number. RFID will be especially useful in farms using automated systems for individual animal-based feeding, watering, milking, etc., along with the help of advanced information technology tools and the quick reading of the identification numbers in the tags and online updating of information using hand-held devices.

10.3.4 Quick response code-based tags

The quick response (QR) code is a matrix bar code system (Tarjan et al., 2014). The QR code makes use of four standardized encoding modes (numeric, alphanumeric, byte/binary, and kanji) to efficiently store data. Reading of the QR code does not require sophisticated equipment; smartphone software can easily read such codes. QR code can be read using the smartphone with the support of smartphone apps.

10.4 Scenario of livestock traceability around the world

The need for the implementation of livestock traceability was felt in the last decade of the 20th century after the emergence of bovine spongiform encephalopathy (BSE) and dioxin contamination of food products. To meet this need different countries developed and implemented livestock traceability systems which helped to win back consumers' trust and the confidence of importing countries. It is a defensive quality and safety management tool. Livestock traceability systems are being used in various forms in different countries such as Japan, the European Union, Uruguay, Australia, New Zealand etc. Brief details about the mode of implementation of the livestock traceability systems in a few countries are given below.

10.4.1 Livestock traceability system in the European Union

The European Union (EU) is a conglomeration of 28 member countries that operate and negotiate as a unit. Traceability became a concern for the EU in the 1990s as a result of BSE. Since the discovery of BSE in cattle as the probable cause of the deadly human form, known as new variant Creutzfeldt–Jakob Disease, there was a large-scale crisis in the European cattle sector (Sugiura and Onodera, 2008). This forced the EU to legislate for an mandatory animal traceability system to protect consumers and producers. The EU introduced its Trade Control and Expert System (TRACES) in April 2004. The system provides a central database to track the movement of animals within the EU and from third countries. Regulation EC 178/2002 Article 18 implies that the producer must know enough information (i.e., keep sufficient records) to be able to trace forward one step and trace back one step. Article 11 of Regulation EC 178/2002 adds that all food and feed imported into the EU for placement on the market must be at least equal to the EU standards. This means that to export to the EU, a product must be traceable in the same way that products are traceable in the EU.

At present the EU is implementing one of the best traceability models in the world; a brief detail about the mode of implementation of the livestock traceability system in the EU is given below.

10.4.1.1 Identification of animals

As per the EU regulations every animal needs to be identified by approved ear tags applied to each ear. Ear tags shall be applied at the latest when the calf reaches the age of 6 months or when it is separated from its mother or when it leaves the holding. Birth, death, and movement of animals must be reported to authorities. If any animal is not tagged and if it is not possible to prove the identification of the animal within 2 working days the animal will be destroyed under the supervision of veterinary authorities without the provision of any compensation. If on one holding the number of animals for which the identification and registration requirements are not fully complied with is in excess of 20%, restrictions will be immediately imposed on the movement of all the animals present in the holding. Ear tags include information on the Member State of origin together with information on the individual animal. Keepers shall be authorized to acquire in advance, if they so wish and in compliance with the applicable national provisions, a quantity of ear tags proportionate to their needs for a period not exceeding 1 year. In the case of holdings which keep no more than five animals, the competent authority may not provide in advance more than five pairs of ear tags. In case of replacements due to loss of tags, it must contain, in addition to the information provided for and distinct from it, a mark expressing in Roman numerals the version number of the replacement ear tag.

10.4.1.2 Passport

Every animal after tagging will be issued a passport which contains details of the owner, parental ear tag number, animal ear tag number, signature of the last keeper, name of issuing authority, and the date of issue of the passport. The passport containing the animal and ownership details must be obtained after tagging the animal. The information contained in the passport and the register should be in a form which allows animals to be traced. Whenever an animal is moved, it shall be accompanied by its passport. In the case of the death of an animal, the passport shall be returned by the keeper to the competent authority within 7 days after the death of the animal. Each animal keeper shall complete the passport immediately on arrival and prior to departure of each animal from the holding and ensure that the passport accompanies the animal. In case a calf under 4 weeks of age needs to be moved its navel must be healed. In such a case, Member States may provide for it to be accompanied by a temporary passport containing necessary information approved by the competent authority.

10.4.1.3 Registers to be maintained in holding

The register kept on each holding shall contain at least the following information: date of birth of the animal on the holding; in the case of animals departing from the holding, the name and address of the keeper, with the exception of the transporter, or the identification code of the holding, to whom/which the animal is being transferred, as well as the date of departure; in the case of animals arriving on the holding, the name and address of the keeper, with the exception of the transporter, or the identification code of the holding, from whom/which the animal was transferred, and the date of arrival; and the name and signature of the representative of the competent authority checking the register and the dates on which such checks were carried out.

10.4.1.4 Inspection by authorities

The competent authority of each Member State carries out on-the-spot inspections, which in general are unannounced. Those inspections shall each year cover at least 10% of holdings situated in the territory of each Member State.

10.4.1.5 Beef labeling

A compulsory beef labeling system was introduced and has been obligatory in all Member States since January 1, 2000. Under this compulsory system operators and organizations marketing beef should indicate on the label information about the beef and the point of slaughter of the animal or animals from which that beef was derived. Under this compulsory system operators and organizations marketing beef should, in addition, indicate on the label information concerning origin, in particular where the animal or animals from which the beef was derived were born, fattened, and slaughtered. Information additional to the information concerning where the animal or animals from which the beef was derived were born, fattened, and slaughtered may also need to be provided under the voluntary beef labeling system.

10.4.2 Livestock traceability system in Australia

Australia has a traceability-based quality control program coordinated by Meat and Livestock Australia (MLA) (http://www.mla.com.au/Meat-safety-and-traceability). The National Feedlot Accreditation Scheme (NFAS) of MLA is an industry self-regulatory, quality assurance scheme, initiated by the Australian Lot Feeders' Association and managed by the Feedlot Industry Accreditation Committee. The objective of the NFAS is to develop a quality assurance system for beef feedlots that impacts positively on red meat quality and acceptability. The red meat integrity system working toward meat food safety has three elements: the National Livestock Identification System (NLIS), the Livestock Production Assurance program (LPA), and the LPA National Vendor Declaration (LPA NVD). Australia introduced the NLIS for traceability of livestock in 1999 to track cattle during disease and food safety incidents. It was expanded to sheep and goats in 2009. The NLIS combines three elements to enable the lifetime traceability of animals: (1) an animal identifier (a visual or electronic ear tag known as a device); (2) identification of a physical location by means of a property identification code (PIC); and (3) a web-accessible database to store and correlate movement data and associated details. LPA is an independently audited, on-farm assurance program covering food safety, animal welfare, and biosecurity. It provides evidence of livestock history and on-farm practices when transferring livestock through the value chain. Producers declare this information on LPA NVDs, which are required for any movement of stock to processors, sale yards, or between properties if they have different PICs.

10.4.3 Livestock traceability system in Japan

The first case of BSE was reported in Japan in August 2001. Consequently the consumption of beef reduced by 58% in just 2 months forcing the industry and the government to take serious steps regarding livestock traceability. A project was initiated on an emergency basis and all 4.5 million bovines were ear-tagged with a unique identification number. From January 1, 2003 (date of enforcement of law) it was made mandatory for all cattle owners/keepers to apply ear tags with unique identification codes on to bovines and report all birth, death, and transportation details to the National Livestock Breeding Center. In June 2003 Japan passed legislation requiring traceability from the farm through to retail sale. Under the new law, processors, distributors, and retailers are required to provide traceability information from the slaughterhouse to the retail outlet. The law applied to beef muscle meats and excluded offal, trimmings, ground beef, and processed products. Wholesalers and retailers need to provide traceability information by individual animal or by lot numbers. Penalties for noncompliance range from warnings to fines and making violators' names public. The government provided assistance (low-interest loans and credits) to help companies cover the cost of the computer and labeling technologies required to implement the system (Clemens, 2003). Every meat package will have a tracking number on it. Using the tracking number the consumer can get the following information: date of birth, gender, breed, farm, meat processor, and family tree.

10.4.4 Comparison of traceability regulatory systems in different countries

The importance of livestock traceability systems was realized by various countries and have been adopted by several countries in the last decade. A list of the countries adopting a livestock traceability system, the year of launch, and other details is given in Table 10.1. Control of contagious diseases like BSE and foot and mouth disease (FMD) and the consequent effects on export demand are the major triggering factors for meat traceability in different countries. The United States is a notable exception in this list of countries with comprehensive traceability systems.

Sylvain et al. (2014) have examined the food traceability system in different countries around the world based on 10 different parameters and ranked them based on the extent of implementation. The list of traceability assessment questions used for ranking the food traceability systems of different countries by Sylvain et al. (2014) is as follows:

1. Are there specific regulations/policies on the national level for domestic products? When did these policies come into effect?
2. Are there specific regulations/policies for imported products? What documents are required for import products to address traceability?
3. What is the clarity of the system of authority responsible for traceability regulations?

Table 10.1: Overview of cattle traceability systems in different countries.

Country	System name	Launch date	Mandatory	Motivation
Brazil	ERAS, SISBOV, and GTA	2002	For export animals. Unclear for rest	Control FMD and market access to EU
Australia	NLIS (National Livestock Identification System)	1999	Yes	Market access, food safety, animal disease
United States	None	2012	For animals crossing state lines only	Control disease for animals crossing states
New Zealand	NAIT (National Animal Identification and Tracing)	2006	Yes	Market access and animal health
Canada	CCIA (Canadian Cattle Identification Agency)	2002	Yes	Market access accelerated with BSE
Argentina	Argentina Animal Health Information System—Sistema de Gestion Sanitaria (SGS)	2007	Yes for young animals	Control FMD and market access
Uruguay	Division de Contralor de Semovientes (DICOSE) and National Livestock Information System (SNIG)	2006	Yes	Control FMD and market access
Japan	Cattle Traceability Law	2003	Yes	Response to BSE discovery
European Union	Each Member State has own system name	2000	Yes	Animal health and BSE response
Mexico	National Livestock Individual Animal Identification System (SINIGA)	2003	No	Animal health, census, traceability
South Korea	South Korea beef traceability system	2004	Yes	Consumer food safety assurance and animal health

Source: From Ted, C.S., Glynn, T.T., 2012. International cattle ID and traceability: competitive implications for the US. Food Pol. 37, 31–40.

4. If no specific regulations, are there voluntary practices by industry?
5. What products or commodities are being regulated for traceability?
6. What kinds of identifiers are being used for tracking/registering of imports (e.g., ear tags, bar codes, RFID)?
7. Are GFSI benchmark standards recognized?
8. Are GS1 services (i.e., traceability tools and coding standards) available?
9. Is there an electronic database system used for monitoring imports/export and their traceability? Are these systems accessible by importing countries?
10. What information on packaging labels is available for the consumer to understand traceability?

Detailed scoring of different countries is given in Fig. 10.1. The study ranked the traceability system followed by the European Union as the best in terms of the considered questions. The traceability systems of Australia, Brazil, Canada, Japan, New Zealand, and the United States all scored an "Average" grading.

10.5 Molecular meat traceability

Currently, most traceability systems rely on the integrity of the inventory trails, which although auditable are difficulty to verify unequivocally (Shackell et al., 2005), hence periodic testing of market samples with the preserved reference samples will ensure the appropriateness of the traceability labeling. Molecular meat traceability is a technological solution to address this problem. It enables traceability verification by comparing labeled meat samples with the reference sample preserved during the slaughter process (Negrini et al., 2008). The prerequisite for the molecular meat traceability is preservation of reference samples of each animal or batch of the product until the period of its consumption by the consumers. Molecular meat traceability is based on the variability within the DNA of individuals (Orru et al., 2006; 2009). DNA is a preferred molecule for traceability as it is present in every cell of the animal, remains the same in different tissues, does not change with age, is unalterable, and is stable even after storage and processing to different temperatures. Cost is the major limiting factor for utilizing molecular traceability systems as a routine test in traceability systems.

Identifying and confirming the source of meat to a particular individual when it is packed individually or as a batch is the challenge to be resolved by molecular meat traceability techniques. Microsatellite genotyping and SNP genotyping are the commonly used methods for molecular meat traceability (Oh et al., 2014).

10.5.1 Sample requirements for DNA-based meat traceability

Molecular traceability is used as a tool for the verification of traceability. It verifies the label claim of the meat package and the authenticity of the traceability code. DNA traceability is not merely sample DNA analysis, rather it involves a comparison of the market sample with the live animal sample that is preserved (origin of meat). Hence blood, tissue, or hair follicle samples are collected from the animal prior to or during the slaughter of the animal and preserved. Analysis of the extracted DNA from the reference sample and the corresponding meat sample using DNA markers authenticates and verifies the traceability claims of the meat sample.

10.5.2 Molecular meat traceability using microsatellite genotyping

Microsatellite markers are short DNA fragments that are usually less than 100 bp. They consist of motifs of one to six nucleotides, repeated several times, and have a characteristic

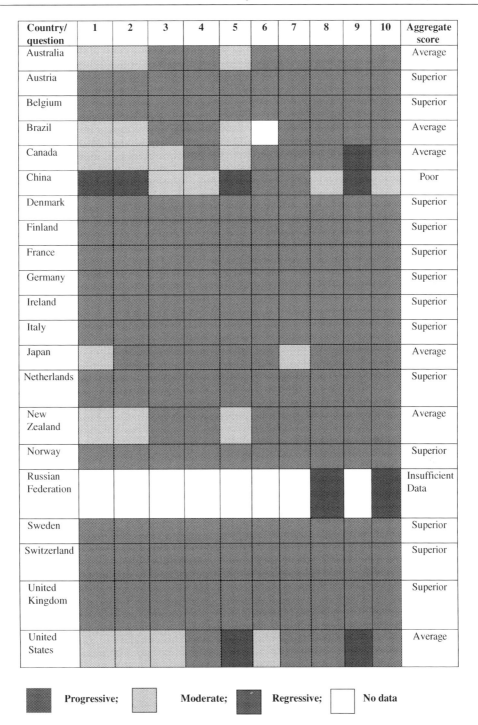

Figure 10.1
Overall world ranking of countries based on comprehensive traceability regulations for domestic and imported products (Sylvain et al., 2014).

mutational behavior. They may be dinucleotide, trinucleotide, or tetranucleotide, and the majority of the repeats are found in noncoding regions. These are also called short tandem repeats, single sequence repeats, or single sequence length polymorphisms. The origin of such polymorphism is most likely due to slippage events occurring during DNA replication. Microsatellites are classified according to the type of repeat sequence as perfect, imperfect, interrupted, or composite.

Microsatellites can be analyzed by PCR amplification of a single tandem repeat locus using primers that anneal at its flanking region (Shackell et al., 2005). The PCR-amplified fragments expressing the size polymorphism are the alleles at the given microsatellite locus. They are phenotypically neutral and developmentally and environmentally stable. The qualities of microsatellites that make them preferred molecular markers include (1) they display a high level of allelic variation that can be analyzed easily, (2) they are codominantly inherited, and (3) they possess versatility of application.

The methodology employed for molecular meat traceability using microsatellite markers is briefly as follows: microsatellite markers required for the assay are usually selected from the panel recommended by ISAG-FAO (International Society for Animal Genetics-Food and Agriculture Organization) advisory group (FAO, 2011). Among the suggested markers, the best suitable markers can be selected based on the polymorphism information content (PIC) of the markers (Orru et al., 2006). Microsatellite markers can be used for both breed traceability, that is, detecting the breed of meat, and individual traceability of meat, that is, identifying the individual from which meat is derived.

For genotyping, genomic DNA has to be extracted from both the meat sample to be traced and the reference samples using the standard protocols. For individual traceability, the genotypic pattern of the test sample and reference sample collected during slaughter has to be compared, while for breed traceability the genotypic pattern of the test sample has to be compared with the already available genotypic pattern of different breeds. It is necessary to develop a repository of microsatellite genotypic data of different breeds to achieve breed traceability. Selected microsatellite markers need to be amplified by polymerase chain reaction employing the extracted DNA. A multiplex PCR can be set for amplification of multiple markers having an annealing temperature in a close range. Otherwise individual PCRs can be undertaken separately for each set of the microsatellite markers. The resultant PCR amplicons need to be analyzed using suitable platforms. Polyacrylamide gel electrophoresis determines the amplicon size and shows polymorphisms between the individuals. Ideally capillary electrophoresis or genetic analysis-based methods should be used for getting accurate results of microsatellite genotyping. Once the size of the amplicon of each marker and individual are known, the results between the test meat and the reference samples can be compared to verify the authenticity of the origin labeling of the meat samples (Table 10.2).

Table 10.2: Reports on utilization of microsatellite genotyping method for meat traceability.

Sl. No.	Species	Level of traceability	Number of microsatellite markers	Brief findings	References
1.	Cattle	Six breeds	16 markers	Could assign six breeds: Japanese Black, Anduo yak, Limousin, Jiaxian Red, Nanyang Yellow, and Luxi Yellow	Jie et al. (2017)
2.	Sheep	One breed	14 markers	Detection of Sambucana sheep	Liliana et al. (2017)
3.	Cattle	Breed	30 markers	Nine Portuguese cattle breeds assigned	Mateus and Russo-Almeida (2015)
4.	Cattle	Breed	21 markers	Four native Italian beef breeds: Chianina, Marchigiana, Romagnola, and Piemontese	Dalvit et al. (2008)
5.	Cattle	Breed	13 markers	Suggested selecting markers based on breed genetic structure for individual assignment	Orru et al. (2006)
6.	Cattle	Individual	19 markers	Identified individual in beef mixtures prepared from less than 10 individuals	Shackell et al. (2005)
7.	Cattle	Breed	22 markers	Test could differentiate meat from Chinese Yellow Cattle from foreign breeds	Rogberg-Muñoz et al. (2014)
8.	Buffalo	Breed	8 markers	Seven Indian buffalo breeds were assigned	Bheemashankar et al. (2017)
9.	Pig	Breed composition	25 markers	Could detect extent of Duroc genome in Iberian pig ham	García et al. (2006)

(Continued)

Table 10.2: (Continued)

Sl. No.	Species	Level of traceability	Number of microsatellite markers	Brief findings	References
10.	Cattle	Breed	20 markers	Four Italian beef breeds could be assigned and differentiated from Holstein Friesian breed beef	Ciampolini et al. (2000)
11.	Cattle	Individual	11 markers	Study concluded that for traceability in consanguineous animals two more markers must be taken to achieve traceability	Baldo et al. (2010)

Microsatellite genotyping methodology can also be used for breed authentication of meat. A method for breed identification of buffalo meat using eight sets of microsatellite markers was developed (Bheemashankar et al., 2017). For individual assignment of meat, microsatellites can be used, provided the genetic structure of the breed is taken into account. It is known that breeds can show differences in frequencies and number of fixed alleles in the same loci (Orru et al., 2006). An evaluation of microsatellite-based technique for the detection of product tracing in ground mixtures revealed DNA microsatellites to be useful for DNA traceability of ground beef mixtures prepared from fewer than 10 individuals, however, the method was inconsistent in cases where a larger number of animals contributed to a mixture (Shackell et al., 2005).

10.5.3 Molecular meat traceability using single nucleotide polymorphism genotyping

SNPs are single base changes in DNA sequences (Kim et al., 2015). SNPs are biallelic and have advantages over microsatellites. The SNPs are highly abundant in the genome (an average of one SNP at every 100–500 base pairs) and several technologies have now been established for SNP genotyping, such as MALDI TOF assay, primer extension, TaqMan, and several microchip techniques, that allow high-throughput automated analysis (Goffaux et al., 2005; Ramos et al., 2011; Dimauro et al., 2013; Heaton et al., 2014).

Basic process involved in the identification and analysis of SNP markers is as follows:

10.5.3.1 Selection of single nucleotide polymorphism markers

Nucleotide sequencing is the method of choice for the identification of SNP markers. Once the candidate gene is identified (targeted resequencing can be done for unrelated individuals), it can be aligned to screen for the presence of SNPs. Extensive data have been generated on SNPs of various livestock and mammalian species (Choi et al., 2016). A public database, *dbSNP*, the most popular database for the SNPs, is hosted by the National Center for Biotechnology (www.ncbi.nlm.nih.gov/snp/).

10.5.3.2 Single nucleotide polymorphism genotyping methods

Various SNP genotyping methods are used for achieving molecular traceability. Direct sequencing of the candidate gene after PCR amplification can assist in the identification of SNPs and also SNP genotyping for individual or breed assignment. However, although sequencing is an expensive and laborious technique, it is the method of choice for the identification of SNPs. Utilizing the variation in the sequence, SNP—restriction fragment length polymorphism (RFLP) can be used if the SNP under question has a restriction site. The target gene is amplified by PCR and digested using a specific restriction enzyme; the resultant restriction profile generated after agarose gel electrophoresis indicates the presence or absence of the SNP in the sample. The SNP—RFLP is more economical as compared to the nucleotide sequencing. A PCR—RFLP-based method targeting SNPs for differentiation of different cattle breeds was developed in Japan (Suekawa et al., 2010). Out of the 30 SNP markers tested, 9 markers were considered to be the best candidate markers in terms of their frequencies. With extensive research undertaken in the area of identification of SNPs, the DNA chips have been developed to look for the DNA sequences that differ by SNPs. In this the DNA sequence of oligos differs only at the last position. To determine which alleles are present, genomic DNA from an individual is isolated, fragmented, tagged with a fluorescent dye, and applied to the chip. The genomic DNA fragments anneal only to those oligos to which they are perfectly complimentary. The computer reads the position of the two fluorescent tags and identifies the individual as a C/T heterozygote. Several commercial SNP chips are available for livestock for the purpose of genotyping and molecular traceability (Karniol et al., 2009). A foolproof method for traceability of beef by combination of ear tagging and SNP genotyping involving a simple animal tagging system, collection of blood sample from cattle using a simple device, and collection of meat samples during slaughter was reported (Zhao et al., 2019). Genetic traceability was achieved by using a panel of 12 SNP loci which could distinguish individuals with a matching probability of 1.70×10^{-5}. The exact individual animal was identified by comparing the SNP genotype bar codes between the meat and blood samples derived from the recording system to further validate authenticity of the recording system. Hanwoo cattle breed could be assigned with 100% accuracy using 90 SNP markers (Cheong et al., 2013). A SNP-based molecular traceability system for detection of four European Geographic Indication protected beef products has been developed (Negrini et al., 2008).

10.6 Future scope of work

Livestock traceability remains a phenomenon of developed countries at present and a major driving force for such initiatives is to get market access and to promote export trade. Research needs to be undertaken to simplify the traceability process, and to bring down the cost of implementation and the efforts involved in updating information at different stages. Researchers must keep track of the latest developments in the area of information technology and must adopt user-friendly techniques for ensuring the flow of information from farm-to-fork. Collective international efforts need to be made to bring about uniformity in livestock and meat traceability in different countries. Reports of adulteration scandals in the EU in spite of the robust traceability system indicate that there are still gaps which need to be fixed to ensure that the system is meeting the intended objectives. Also simple and rapid tests need to be developed for traceability verification to avoid any fraudulent practices.

Modern technological innovations such as scanning and digital technology for product identification, nondestructive testing, and biosensors for quality and safety assessment; and geospatial technology (GIS, GPS) for mobile assets tracking and site-specific operations can be applied to develop and implement an integrated traceability system. The development of appropriate traceability technology for small-scale farmers, particularly in the less developed countries, is challenging and offers research opportunities. Innovations in emerging technologies, such as DNA fingerprinting, nanotechnology, and retinal imaging, and their integration into livestock industries have considerable potential for improving the momentum and accuracy of traceability in knowledge-based agriculture.

References

Baldo, A., Rogberg-Muñoz, A., Prando, A., et al., 2010. Effect of consanguinity on Argentinean Angus beef DNA traceability. Meat Sci. 85, 671–675.

Bheemashankar, H.K., Nadeem Fairoze, M., Girish, P.S., Karabasanavar, N., Rudresh, B.H., 2017. Breed traceability of buffalo meat using microsatellite genotyping technique. J. Food Sci. Technol. 54, 558–563.

Cheong, H.S., Kim, L.H., Namgoong, S., Shin, H.D., 2013. Development of discrimination SNP markers for Hanwoo (Korean native cattle). Meat Sci. 94, 355–359.

Choi, J.S., Jin, S.K., Jeong, Y.H., Jung, Y.C., Jung, J.H., Shim, K.S., et al., 2016. Relationships between single nucleotide polymorphism markers and meat quality traits of duroc breeding stocks in Korea. Asian-Australasian J. Anim. Sci. 29, 1229–1238.

Ciampolini, R., Leveziel, H., Mozzanti, E., Grohs, C., Cianci, D., 2000. Genomic identification of an individual or its tissue. Meat Sci. 54, 35–40.

Costa, C., Antonucci, F., Pallottino, F., Aguzzi, J., Sarri, D., Menesatti, P., 2013. A review on agri-food supply chain traceability by means of RFID technology. Food Bioprocess Technol. 6, 353–366.

Dalvit, C., De Marchi, M., Targhetta, C., Gervaso, M., Cassandro, M., 2008. Genetic traceability of meat using microsatellite markers. Food Res. Int. 41, 301–307.

Dimauro, C., Cellesi, M., Steri, R., Gaspa, G., Sorbolini, S., Stella, A., et al., 2013. Use of the canonical discriminant analysis to select SNP markers for bovine breed assignment and traceability purposes. Anim. Genetics 44, 377–382.

Falco, A., Salmerón, J.F., Loghin, F.C., Lugli, P., Rivadeneyra, A., 2017. Fully printed flexible single-chip RFID tag with light detection capabilities. Sensors 17, 534.

FAO, 2011. Molecular Genetic Characterization of Animal Genetic Resources. FAO Animal Production and Health Guidelines, No. 9, Rome.

Frewer, L.J., Kole, A., Van De Kroon, S.M.A., De Lauwere, C., 2005. Consumer attitudes towards the development of animal-friendly husbandry systems. J. Agric. Environ. Ethics 18, 345–367.

García, D., Martínez, A., Dunner, S., Vega-Pla, J.L., Fernández, C., Delgado, J.V., et al., 2006. Estimation of the genetic admixture composition of Iberian dry-cured ham samples using DNA multilocus genotypes. Meat Sci. 72, 560–566.

Goffaux, F.B., China, L.D., Clinquart, A., Daube, G., 2005. Development of a genetic traceability test in pig based on single nucleotide polymorphism detection. Forensic Sci. Int. 151, 239–247.

Heaton, M.P., Leymaster, K.A., Kalbfleisch, T.S., Kijas, J.W., Clarke, S.M., McEwan, J., et al., 2014. SNPs for parentage testing and traceability in globally diverse breeds of sheep. PLoS One 9, e94851.

Herrero, M., Havlík, P., Valin, H., Notenbaert, A., Rufino, M.C., Thornton, P.K., et al., 2013. Biomass use, production, feed efficiencies, and greenhouse gas emissions from global livestock systems. Proc. Natl. Acad. Sci. U.S.A. 110, 20888–20893.

Jie, Z., Chao, Z., Zhenzhen, X., Xiaoling, J., Shuming, Y., Ailiang, C., 2017. Microsatellite markers for animal identification and meat traceability of six beef cattle breeds in the Chinese market. Food Control 78, 469–475.

Johnston, M., 2005. Meat inspection and chain information at part of the farm to fork approach, Food Safety Assurance and Veterinary Public Health, vol. 3. Wegeningen AcademicPublishers, Netherlands, pp. 257–270.

Karniol, B., Shirak, A., Baruch, E., Singrun, C., Tal, A., Cahana, A., et al., 2009. Development of a 25-plex SNP assay for traceability in cattle. Anim. Gen. 40, 353–356.

Kim, K., Seo, M., Kang, H., Cho, S., Kim, H., Seo, K.S., 2015. Application of LogitBoost classifier for traceability using SNP chip data. PLoS One 10, e0139685.

Liang, W., Cao, J., Fan, Y., Zhu, K., Dai, Q., 2015. Modeling and implementation of cattle/beef supply chain traceability using a distributed RFID - based framework in China. PLoS One 10, e0139558.

Liliana, D.S., Piergiovanni, P., Edoardo, F., Stefano, C., Daniele, B., Emiliano, L., et al., 2017. Lamb meat traceability: the case of Sambucana sheep. Small Rum. Res. 149, 85–90.

Mateus, J.C., Russo-Almeida, P.A., 2015. Traceability of 9 Portuguese cattle breeds with PDO products in the market using microsatellites. Food Control 47, 487–492.

Musa, A., Gunasekaran, A., Yusuf, Y., 2014. Supply chain product visibility: methods, systems and impacts. Expert Sys. Appl. 41, 176–194.

Negrini, R., Nicoloso, L., Crepaldi, P., Milanesi, E., Marino, R., Perini, D., et al., 2008. Traceability of four European protected geographic indication (PGI) beef products using single nucleotide polymorphisms (SNP) and Bayesian statistics. Meat Sci. 80, 1212–1217.

Oh, J.D., Song, K.D., Seo, J.H., Kim, D.K., Kim, S.H., Seo, K.S., et al., 2014. Genetic traceability of black pig meats using microsatellite markers. Asian-Australasian J. Anim. Sci. 27, 926–931.

OIE, 2018. Terrestrial Animal Health Code, General Provisions, vol. I. World Organization for Animal Health (OIE), p. XIII.

Orru, L., Napolitano, F., Catillo, G., Moioli, B., 2006. Meat molecular traceability: how to choose the best set of microsatellites? Meat Sci. 72, 312–317.

Orru, L., Catillo, G., Napolitano, F., De Matteis, G., Scata, M.C., Signorelli, F., et al., 2009. Characterization of a SNPs panel for meat traceability in six cattle breeds. Food Control 20, 856–860.

Girish, P.S., Nagappa, K., Saikia, T., 2017. Farm-to-fork livestock traceability for quality meat production: an overview. J. Meat Sci. 12, 1−10.

Ramos, A., Megens, H., Crooijmans, R., Schook, L., Groenen, M., 2011. Identification of high utility SNPs for population assignment and traceability purposes in the pig using high-throughput sequencing. Anim. Genet. 42, 613−620.

Rogberg-Muñoz, A., Wei, S., Ripoli, M.V., Guo, B.L., Carino, M.H., Castillo, N., et al., 2014. Foreign meat identification by DNA breed assignment for the Chinese market. Meat Sci. 98, 822−827.

Clemens, Roxanne, 2003. Meat traceability in Japan. Iowa Ag. Rev. Online 09, 1−5.

Shackell, G.H., Tate, M.L., Anderson, R.M., 2001. Installing DNA based traceability system in the meat industry. In: Proceedings of the Australasian Association for the Advancement of Breeding and Genetics, vol. 14, pp. 533−536.

Shackell, G.H., Mathias, V.M., Cave, V.M., Dodds, K.G., 2005. Evaluation of microsatellites as a potential tool for product tracing of ground beef mixtures. Meat Sci. 70, 337−345.

Smith, G., Pendell, D., Tatum, J., Belk, K., Sofos, J., 2008. Post-slaughter traceability. Meat Sci. 80, 66−74.

Sugiura, K., Onodera, T., 2008. Cattle traceability system in Japan for bovine spongiform encephalopathy. Vet. Ital. 44, 519−526.

Suekawa, Y., Aihara, H., Araki, M., Hosokawa, D., Mannen, H., Sasazaki, S., 2010. Development of breed identification markers based on a bovine 50K SNP array. Meat Sci. 85, 285−288.

Sylvain, C., Brian, S., Sanaz, H., Sandi, K.N., 2014. Comparison of global food traceability regulations and requirements. Compr. Rev. Food Sci. Food Safety 13, 1104−1123.

Tarjan, L., Šenk, I., Tegeltija, S., Stankovski, S., Ostojic, G., 2014. A readability analysis for QR code application in a traceability system. Comp. Electronics in Agric. 109, 1−11.

Ted, C.S., Glynn, T.T., 2012. International cattle ID and traceability: competitive implications for the US. Food Pol. 37, 31−40.

Thakur, M., Hurburgh, C.R., 2009. Framework for implementing traceability system in the bulk grain supply chain. J. Food Eng. 95, 617−626.

Zhao, J., Xu, Z., You, X., Zhao, Y., He, W., Zhao, L., et al., 2019. Genetic traceability practices in a large-size beef company in China. Food Chem. 277, 222−228.

SECTION 6

Chemical residues in meat and their detection techniques

CHAPTER 11

Residues of harmful chemicals and their detection techniques

Milagro Reig[1] and Fidel Toldrá[2]

[1]Instituto de Ingeniería de Alimentos para el Desarrollo, Universitat Politècnica de València, Ciudad Politécnica de la Innovación, Valencia, Spain [2]Instituto de Agroquímica y Tecnología de Alimentos (CSIC), Valencia, Spain

Chapter Outline
11.1 Introduction 173
11.2 Environmental contaminants 174
11.3 Polycyclic aromatic hydrocarbons 175
11.4 Veterinary drug residues 176
11.5 *N*-Nitrosamines 178
11.6 Oxidation of lipid-derived compounds 179
11.7 Oxidation of protein-derived compounds 180
References 181
Further reading 183

11.1 Introduction

There is a wide range of chemicals that may be present in meat as residues, either during processing or during cooking and household management. Analytical technologies are constantly being developed and improved to allow the correct detection and determination of such compounds, thus contributing to a better meat safety. Some toxic compounds that may be present in meat are the compounds resulting from the oxidation of lipids and proteins, and other substances resulting from the raw materials used in animal production like veterinary drug residues or environmental contaminants. Furthermore, meat products may also contain other harmful compounds like *N*-nitrosamines as a consequence of nitrite use as a preservative, polycyclic aromatic hydrocarbons (PAHs) generated from certain smoking processes, or biogenic amines produced by microorganisms used for fermentation having decarboxylase activity.

Corrective measures have to be taken, especially for prevention. These measures may include the reduction and careful control of the addition of nitrite, the rigorous control of raw materials, the control of mircoorganisms used for meat fermentation, etc. Adequate analytical support is necessary for an effective control of the presence of such compounds in meat and processed meats. There are some reviews that can be found elsewhere of the analytical tools that are commonly used for the detection of such harmful compounds (Toldrá and Reig, 2012).

All these harmful compounds and available detection analysis are briefly described in this chapter.

11.2 Environmental contaminants

A huge number of environmental contaminants, such as dioxins, organophosphorus, and organochlorine compounds including PCBs, mycotoxins, and heavy metals among others, might be present in meats. Each environmental contaminant may exert a different level of toxicity. The assessment of environmental contaminants must be performed because most of them remain in the animal and the resulting products, which can be ingested by consumers even though they can exert great potential toxicity (Heggum, 2004).

Dioxins are a wide group of environmental contaminants responsible for considerable concern within the European Union (EU) and other countries in recent years. But such a group is quite broad in terms of the number and types of substances and their respective toxicity, and this is why a new concept of toxic equivalency factors (TEFs) was introduced. TEF is an approach that sums the toxicity of the different types of dioxins, such as dibenzo-p-dioxins (PCDDs), dibenzofurans (PCDFs), and polychlorinated biphenyls (PCBs). EC Regulation 1881/2006 (EC, 2006) set the maximum levels of the sum of dioxins and dioxin-like PCBs (WHOPCDD/F-PCB-TEQ) to be 4.5 pg/g fat for bovine, 4.0 pg/g fat for poultry, and 1.5 pg/g fat for pork. PCBs (polychlorinated biphenyls) were phased for use even though their presence will be detected in the environment for many years (Moats, 1994).

The way for environmental contaminants to reach the consumer is via the feed given to farm animals. Once the feed is ingested, such contaminants are distributed within the animal, in most cases accumulating in fat, and constituting a direct cause of contamination of the meat (Croubels et al., 2004). In other cases contaminants like mycotoxins may reach the feed due to the growth of molds like species of *Fusarium*, *Aspergillus*, and *Penicillium* that can lead to contamination with mycotoxins. This is due to an inappropriate control of the raw materials. It must be taken into account that some mycotoxins are toxic, for example, aflatoxin B_1 which is a genotoxic and carcinogenic substance.

Methods of detection of mycotoxins include HPLC coupled to ultraviolet, diode array, fluorescence or mass spectrometry detectors, thin-layer chromatography, gas chromatography (GC) coupled to electron capture, and flame ionization or MS detectors (Turner et al., 2015). Previously screenings were carried out with enzyme-linked immunoassay tests (ELISA). Newer approaches include the use of biosensors and optical techniques which are becoming more prevalent (Turner et al., 2015).

The maximum levels of heavy metals in foods is set by Regulations. For instance, the EC Regulation 1881/2006 (EC, 2006) sets the maximum levels in meats as 0.10 mg/kg wet weight (ww) for lead, 0.050 mg/kg ww for cadmium, and 1.0 mg/kg ww for mercury. Classical atomic absorption spectroscopy with graphite furnace or respectively cold vapor, ICP-OES, and ICP-MS techniques are most frequently in use for determining the presence of heavy metals at the lowest detection limits, that is, trace amounts in farm animals or in foods. Toxic (lead, cadmium, and mercury) and essential trace (copper and zinc) metals were measured in muscle, liver, and kidney samples of food-producing animals, and some relationship was found with the heavy metal content in the consumed feed (Hashemi, 2018)

11.3 Polycyclic aromatic hydrocarbons

PAHs have emerged as an important contaminant group in processed food including meat products. The level of PAHs depends on factors like distance from heat source, type of wood or fuel, level of processing, accessibility to oxygen, temperature and time of cooking, and cooking methods (Singh et al., 2016). Different cooking processes and processing techniques applied to meat like roasting, barbecuing, grilling, smoking, heating, drying, baking, etc. contribute towards their formation. The depth of penetration also depends on the conditions of the process.

A traditional route for PAH condensation and adsorption on the surface of the meat product is smoking, where the meat product is exposed to smoke from the combustion of natural woods, and/or aromatic herbs and spices. Whereas an attractive flavor is given to the meat product, the problem of smoke is its content of hazardous compounds like PAHs, phenols, and formaldehyde (Bem, 1995). Formaldehyde has been identified as the cause of cancerous tumors. One of the most well-known PAHs is benzo(a)pyrene which has cancer-inducing and carcinogenic properties. Other relevant PAHs are benz(a)anthracene, benzo(b)fluoranthene, and chrysene (Reig and Toldrá, 2015). The content and proportion of PAH is highly variable because of the many variables involved as mentioned above.

In addition, some smoke phenols could react to form highly toxic nitrosophenols that could further react to form toxic reaction products like nitrophenols, polymeric nitrosic compounds, and other toxic compounds or even catalyze the formation of nitrosamines (Bem, 1995). In any case, there is low concern for consumer health at the average estimated dietary exposures in Europe using the margin of exposure approach, even though high consumers might be close to or less than a margin of exposure of 10,000, which indicates a potential concern for consumer health (EFSA, 2008).

At first, only benzo(a)pyrene was used as marker of PAH but this was not considered of enough relevance, and thus new markers were proposed to be more representative of the PAH content in foods, for example, PAH4 which is the sum of four substances: benzo(a) pyrene, benz(a)anthracene, benzo(b)fluoranthene, and chrysene, and PAH8, which is the sum of eight specific substances: benzo(a)pyrene, crysene, benzo(a)anthracene, benzo(b) fluorantene, benzo(k)fluorantene, benzo(g,h,i)perylene, dibenzo(a,h)anthracene, and inden(1,2,3-cd)pyrene (EFSA, 2008). Both PAH4 and PAH8, in addition to the previous marker benzo(a)pyrene were incorporated into the EU Regulation (EC, 2011). The maximum levels (μg/kg) set in the Regulation for smoked meat and smoked meat products from September 1, 2014 were: 2.0 μg/kg for benzo-pyrene and 12.0 μg/kg for PAH4.

The detection of PAHs is complex. First an extraction is needed either by saponification or ultrasonication, followed by liquid—liquid extraction with solvents like hexane, cleanup using a silica solid-phase extraction cartridge, and then it can be injected into the gas chromatograph (Lee et al., 2019). The analysis is usually performed with either GC coupled to a flame ionization detector or HPLC coupled to ultraviolet or fluorescence detectors (Zachara et al., 2017). Mass spectrometry detectors can also be coupled to both chromatographic methodologies for identification and confirmation purposes. In comparison to LC, however, GC is preferred, as it typically affords greater selectivity, resolution, and sensitivity for separation, identification and quantification of PAHs (Lee et al., 2018). A survey of 600 smoked meat products revealed contents below 0.20 ng/g for benzo-pyrene, 1.5 ng/g for PAH4, and 1.8 ng/g for PAH8 (EFSA, 2008).

11.4 Veterinary drug residues

The veterinary drugs used in food-producing animals may exert adverse health effects in humans if exposed to residues (Baynes et al., 2016). Antimicrobials are still a regulatory concern not only because of their acute adverse effects but also because their use as growth promoters have been linked to antimicrobial resistance. Many of these substances are no longer approved for use in food-producing animals.

The uses of veterinary drugs in animals have different purposes (Dixon, 2001; Delaseleire et al., 2017) like therapeutic agents to control infectious diseases, prophylactic agents to prevent outbreaks of diseases and control parasitic infections, growth-promoting/anabolic agents to improve the feed conversion efficiency and thus increase the lean to fat ratio, and antimicrobial agents to make more nutrients available to the animal. In the last case, the excessive use of antimicrobials has raised concerns because of the development of bacteria resistant to antibiotics (Butaye et al., 2001).

In view of the genotoxic, immunotoxic, carcinogenic, or endocrine effects on humans, the European Union banned the administration of most veterinary drugs to protect consumers' health. In this way, the use of such substances was restricted to limited therapeutic purposes and under the control of a responsible veterinarian (Van Peteguem and Daeselaire, 2004). The major concern with such substances if illegally added to farm animals is that they can remain in the animal and enter the food chain through the meat. For these reasons, these substances must be controlled in farm animals and foods of animal origin (Croubels et al., 2004; Boenke, 2002).

The EC Directive 96/23/EC on measures to monitor certain substances and residues in live animals and animal products regulates the presence of these substances in foods and determine the number of samples to be tested each year in the European Union. The analytical methodology for the monitoring of compliance was given in Decisions 93/256/EEC and 93/257/EEC. The Council Directive 96/23/EC was implemented by the Commission Decision 2002/657/EC. This decision provided rules for the analytical methods to be used in testing of official samples and specific common criteria for the interpretation of analytical results of official control laboratories for such samples. In the United States the Food Safety and Inspection Services (FSIS) establishes the surveillance programmes (Croubels et al., 2004) and the FDA Center for Veterinary Medicine issues the analytical criteria. An Additional Testing Program for the control of residues of these substances in the meats exported to the European Union was set up by the USDA (Croubels et al., 2004).

The analytical methods for drug residue control in edible matrices, at either parts per billion (μg/kg) or even parts per trillion (ng/kg), in combination with the legislative limits on consumer safety require very sensitive analytical methods for the detection, identification, and quantification of the veterinary residues in animal-derived foods (Delaseleire et al., 2017; Hubert et al., 2017; Gajda et al., 2019).

Initial screening controls for the detection of these substances are usually based on screening tests, most of the times ELISA test kits or antibody-based automatic techniques. Specific veterinary drugs may be partly purified with immunochromatography, increasing its sensitivity. The advantages of screening tests are the short time needed for a large number of samples but they can only give qualitative or semiquantitative data that require further confirmation. For suspicious samples, the next step is the confirmatory analysis that can be carried out either by GC or coupled to mass spectrometry or other sophisticated methodologies for accurate characterization and confirmation (Toldrá and Reig, 2006). Further methodologies for the analysis of growth promoters are reported elsewhere (Reig and Toldrá, 2009, 2011).

11.5 N-Nitrosamines

Most processed meats include the use of nitrite as a preservative but also for its contribution to the cured color and flavor. Nitrite is a powerful inhibitor of *Clostridium botulinum* and of the outgrowth of its spores and this is the main purpose and the main reason for its acceptability in countries' regulations. Other technological roles that are highly appreciated by meat industries consist of its contribution to the typical pinky cured color, which is achieved through the formation of nitrosylmyoglobin, as well as its contribution to the development of typical and distinctive cured meat flavor by exerting an antioxidant effect, protecting lipids and proteins from oxidation. The levels of residual nitrite are rapidly decreased during processing with small amounts remaining in the final product if the initial amounts were complying with Regulations.

The controversy about the use of nitrite in meat processing comes from several decades ago, in the 1970s, where there was a serious debate on residual nitrite and the need for it in cured meats. The amount of allowed nitrite was lowered in the regulations in the United States as well as the recommendation to add ascorbate or isoascorbate (erythorbate) to inhibit the formation of nitrosamines formation because they compete with amines for nitrite (Cassens, 1997).

Nevertheless, nitrite and nitric oxide constitute normal human metabolites since nitric oxide can be generated through the enzymatic reaction of arginine (nitric oxide synthase) and then be converted to nitrate and nitrite (Leaf et al., 1989). In the gastrointestinal tract there is also an endogenous generation of *N*-nitrosamines from precursors (Vermeer et al., 1998).

The real nitrosating agent is nitric oxide that is rapidly generated from nitrite. The reaction of nitric oxide with secondary amines produces nitrosamines, some of them being potent carcinogenic compounds. An example is *N*-nitrosodimethylamine that has been proved to be carcinogenic in toxicological assays with numerous animal species. The most important volatile *N*-nitrosamines are *N*-nitrosodimethylamine, *N*-nitrosopirrolidine, *N*-nitrosopiperidine, *N*-nitrosodiethylamine, *N*-nitrosodi-*n*-propylamine, *N*-nitrosomorpholine, and *N*-nitrosoethylmethylamine. But also non-volatile nitroso compounds, which are higher in molecular weight and more polar, are generated, for example, *N*-nitrosoamino acids such as *N*-nitrososarcosine and *N*-nitrosothiazolidine-4-carboxylic acid, hydroxylated *N*-nitrosamines, *N*-nitroso sugar amino acids, and *N*-nitrosamides like *N*-nitrosoureas, *N*-nitrosoguanidines, and *N*-nitrosopeptides (Pegg and Shahidi, 2000).

The formation of *N*-nitrosamines during the processing of meat products depends on the amount of residual nitrite, especially if some heating is applied during processing, other processing conditions like pH, and the presence of any substance that might act as a catalyst or inhibitor (i.e., reducing substances like ascorbic acid) (Hotchkiss and Vecchio, 1985; Walker, 1990). Nitrosodimethylamine and nitrosopiperidined constitute those found at higher amounts, >1 μg/kg. In European fermented sausages, nitrosamines were found at much lower concentrations (Demeyer et al., 2000). Some N-nitrosamines, like N-nitrosodiisobutylamine (NDiBA) and N-nitrosodibenzylamine (NDBzA), may be formed in cured meats packaged in elastic rubber netting due to the reaction of nitrite with the amine additives forming part of the rubber netting (Sen et al., 1987; Fiddler et al., 1998).

The detection of nitrosamines is complex and requires complex analytical instrumentation (Park et al., 2015). Volatile N-nitrosamines can be extracted through aqueous distillation, solid-phase extraction with specific solvents (Raoul et al., 1997), or supercritical fluid extraction (Fiddler and Pensabene, 1996). Once extracted, such nitrosamines are analyzed by GC coupled to thermal energy analyzer that is based on the chemiluminescence generated by the decay of the NO_2 group when it is electronically excited (Sannino and Bolzoni, 2013). Other suitable detectors to be coupled are mass spectrometry detectors. An alternative is the use of liquid chromatography atmospheric pressure chemical ionization mass spectrometry and tandem mass spectrometry (Eerola et al., 1998; Rath and Reyes, 2009). A recent method which is sensitive, low cost, effective, and uses an ecofriendly extraction method consists of extraction with dispersive micro-SPE optimized for the extraction of N-nitrosamines from meat products. The limits of quantitation ranged from 0.03 to 0.36 ng/g in 5 g of meat product (Huang et al., 2013).

A survey made with 101 fermented sausages sampled from the Belgian market and analyzed through GC coupled to a thermal energy analyzer revealed that the total amount of nitrosamines was below 5.5 μg/kg in all samples except one, with N-nitrosopiperidine reported at amounts above 2.5 μg/kg for only 30 of the samples (De Mey et al., 2013). Nitrosamines were also analyzed by GC coupled to a mass selective detector in 386 samples of raw, fried, grilled, smoked, pickled, and canned meat products sampled in the Estonian market. The results revealed that a mean value of 0.85 μg/kg N-nitrosodimethylamine and 4.14 μg/kg N-nitrosopyrrolidine were reported for most of the samples. Other nitrosamines found in nearly one-third of the samples were 0.36 μg/kg N-nitrosodiethylamine, 0.98 μg/kg N-nitrosopiperidine, and 0.37 μg/kg N-nitrosodibutylamine (Yurchenko and Mölder, 2007). Furthermore, 70 products on the Danish market and 20 products on the Belgian market were also sampled and analyzed through HPLC coupled to MS/MS for nitrosamines (Herrmann et al., 2015). The reported amounts for volatile NAs were found to be below 0.8 μg/kg while nonvolatile nitrosamines were reported at higher amounts near 118 μg/kg.

11.6 Oxidation of lipid-derived compounds

The major lipids in meats are triacylglycerols, phospholipids, lipoproteins, and cholesterol, which being phospholipids are more susceptible to oxidation because of their high content of polyunsaturated fatty acids. Lipid oxidation follows a free radical mechanism, which is favored by meat manipulation, processing, and storage, that consists of three steps:

initiation, propagation, and termination. The primary products of oxidation are hydroperoxides and the secondary products can generate toxic compounds (Kanner, 1994). Oxidation by hydrogen peroxide generated in fermented meats by peroxide-forming bacteria constitutes another mechanism for lipid oxidation. Some products from lipid oxidation can be toxic and contribute to aging, cancer, and cardiovascular diseases (Hotchkiss and Parker, 1990).

A strategy to decrease or prevent lipid oxidation in meats is the addition of antioxidants that improve the sensory quality, nutritional value, and safety. Some of the latest and more attractive antioxidants that are being incorporated into meat products to prevent oxidation are phytochemicals (Estévez and Lorenzo, 2019).

The determination of 2-thiobarbituric acid reactive substances is a well-established method for determining lipid oxidation in meat and meat products. This assay, however, is time-consuming and generates undesired chemical waste. Near-infrared reflectance spectroscopy using selected wavelengths is a technique reported to be able to detect lipid oxidation in meat. Selected analytes can be detected at concentrations of parts per million (Ripoll et al., 2018).

Another relevant oxidation with detrimental effects for health is cholesterol oxidation, which can happen through an autooxidative process but is also linked to fatty acid oxidation (Hotchkiss and Parker, 1990). The resulting cholesterol oxides are damaging due to their role in arteriosclerotic plaque but they can be also mutagenic, carcinogenic, and cytotoxic (Bösinger et al., 1993; Guardiola et al., 1996). No cholesterol oxides have been detected in cooked pork sausages (Baggio and Bragagnolo, 2006) and low amounts up to 1.5 μg/g in dry-cured sausages (7-ketocholesterol), while 5,6α-5,6-epoxycholesterol was the main cholesterol oxide found in Italian sausage (Demeyer et al., 2000).

11.7 Oxidation of protein-derived compounds

Sulfur-containing amino acids of proteins, that is, cystine, cysteine, and methionine, are prone to oxidation by peroxide reagents, like hydrogen peroxide (Guyon et al., 2016). So proteins may be oxidized by hydrogen peroxide generated by peroxide-forming bacteria grown in fermented meats. As a result carbonyl groups are generated with thiol oxidation and aromatic hydroxylation (Morzel et al., 2006). Cystine is partly oxidized to cysteic acid and methionine is oxidized to methionine sulfoxide and methionine sulfone (Slump and Schreuder, 1973). Cysteine may be oxidized into sulfinic and cysteic acids (Finley et al., 1981). The oxidation of homocystine can generate homolanthionine sulfoxide (Lipton et al., 1977). Peptides like the reduced glutathione can also be oxidized with cysteine oxidized to the monoxide or dioxide forms.

References

Baggio, S.R., Bragagnolo, N., 2006. The effect of heat treatment on the cholesterol oxides, cholesterol, total lipid and fatty acid contents of processed meat products. Food Chem. 95, 611–619.

Baynes, R.E., Dedonder, K., Kissell, L., Mzyk, D., et al., 2016. Health concerns and management of select veterinary drug residues. Food Chem. Toxicol. 88, 112–122.

Bem, Z., 1995. Desirable and undesirable effects of smoking meat products. Die Fleischerei 3, 3–8.

Boenke, A., 2002. Contribution of European research to anti-microbials and hormones. Anal. Chim. Acta 473, 83–87.

Bösinger, S., Luf, W., Brandl, E., 1993. Oxysterols: their occurrence and biological effects. Int. Dairy J. 3, 1–33.

Butaye, P., Devriese, L.A., Haesebrouck, F., 2001. Differences in antibiotic resistance patterns of *Enterococcus faecalis* and *Enterococcus faecium* strains isolated from farm and pet animals. Antimicrob. Agents Chemother. 45, 1374–1378.

Cassens, R.G., 1997. Composition and safety of cured meats in the USA. Food Chem. 59, 561–566.

Croubels, S., Daeselaire, E., De Baere, S., De Backer, P., Courtheyn, D., 2004. Feed and drug residues. In: Jensen, W., Devine, C., Dikemann, M. (Eds.), Encyclopedia of Meat Sciences. Elsevier, London, pp. 1172–1187.

Delaseleire, E., Van Pamel, E., Van Poucke, C., Croubles, S., 2017. Veterinary drug residues, Chemical Contaminants and Residues in Food, second ed. Woodhead Publishing, UK, pp. 117–153.

Demeyer, D.I., Raemakers, M., Rizzo, A., Holck, A., et al., 2000. Control of bioflavor and safety in fermented sausages: first results of a European project. Food Res. Int. 33, 171–180.

Dixon, S.N., 2001. Veterinary drug residues. In: Watson, D.H. (Ed.), Food Chemical Safety. Volume 1: Contaminants. Woodhead Publishing Ltd, Cambridge, pp. 109–147.

EC, 2006. Commission Regulation (EC) No 1881/2006 of 19 December 2006 Setting Maximum Levels for Certain Contaminants in Foodstuff. O.J. EU. L 364/5.

EC, 2011. Commission Regulation (EU) No 835/2011 of 19 August 2011 Amending Regulation (EC) No 1881/2006 as Regards Maximum Levels for Polycyclic Aromatic Hydrocarbons in Foodstuffs. O.J. EU. L 215/4.

Eerola, S., Otegui, I., Saari, L., Rizzo, A., 1998. Application of liquid chromatography atmospheric pressure chemical ionization mass spectrometry and tandem mass spectrometry to the determination of volatile nitrosamines in dry sausages. Food Addit. Contam. 15, 270–279.

EFSA, 2008. Polycyclic aromatic hydrocarbons in food. Scientific opinion of the panel on contaminants in the food chain. EFSA J. 724, 1–114.

Estévez, M., Lorenzo, J.M., 2019. Impact of antioxidants on oxidized proteins and lipids in processed meat. Encyclopedia of Food Chemistry. Elsevier, UK, pp. 600–608.

Fiddler, W., Pensabene, J.W., 1996. Supercritical fluid extraction of volatile N-nitrosamines in fried bacon and its drippings: method comparison. J. AOAC Int. 79, 895–901.

Fiddler, W., Pensabene, J.W., Gates, R.A., Adam, R., 1998. Nitrosamine formation in processed hams as related to reformulated elastic rubber netting. J. Food Sci. 63, 276–278.

Finley, J.W., Wheeler, E.L., Witt, S.C., 1981. Oxidation of glutathione by hydrogen peroxide and other oxidizing agents. J. Agric. Food Chem. 29, 404–407.

Gajda, A., Nowacka-Kozack, E., Gbylik-Sikorska, M., Posyniak, A., 2019. Multi residues UHPLC-MS/MS analysis of 53 antibacterial compounds in poultry feathers as an analytical tool in food safety assurance. J. Chromatogr. B 1104, 182–189.

Guardiola, F., Codony, R., Addis, P.B., Rafecas, M., Boatella, P., 1996. Biological effects of oxysterols: current status. Food Chem. Toxicol. 34, 193–198.

Guyon, C., Maried, A.M., Lamballerie, A., 2016. Protein and lipid oxidation in meat: a review with emphasis on high-pressure treatments. Trends Food Sci. Technol. 50, 131–143.

Hashemi, M., 2018. Heavy metal concentration bovine tissues (muscle, liver and kidney) and their relationship with heavy metal contents in consumed feed. Ecotoxicol. Environ. Saf. 14, 263–267.

Heggum, C., 2004. Risk analysis and quantitative risk management. In: Jensen, W., Devine, C., Dikemann, M. (Eds.), Encyclopedia of Meat Sciences. Elsevier, London, pp. 1192–1201.

Herrmann, S.S., Duedahl-Olensen, L., Granby, K., 2015. Occurrence of volatile and non-volatile N-nitrosamines in processed meat products and the role of heat treatment. Food Cont. 48, 163–169.

Hotchkiss, J.H., Parker, R.S., 1990. In: Pearson, A.M., Dutson, T.R. (Eds.), Meat and Health. Elsevier AppliedScience, London, pp. 105–134.

Hotchkiss, J.H., Vecchio, A.L., 1985. Nitrosamines in fired-out bacon fat and its use as a cooking oil. Food Technol. 39, 67–73.

Huang, M.C., Chang, H., Fu, S.C., Ding, W.-H., 2013. Determination of volatile N-nitrosamines in meat products by microwave-assisted extraction coupled with dispersive micro solid-phase extraction and gas chromatography – chemical ionisation mass spectrometry. Food Chem. 138, 227–233.

Hubert, C., Roosen, M., Levi, Y., Karolak, S., 2017. Validation of an ultra-high –performance liquid chromatography – tandem spectrometry method to quantify illicit drug and pharmaceutical residues in wastewater using accuracy profile approach. J. Chromatogr. A 1500, 136–144.

Kanner, J., 1994. Oxidative processes in meat and meat products: quality implications. Meat Sci. 36, 169–189.

Leaf, C.D., Wishnok, J.S., Tannenbaum, S.R., 1989. L-arginine is a precursor for nitrate biosynthesis in humans. Biochem. Biophys. Res. Commun. 163, 1032–1037.

Lee, J., Jeong, J.H., Park, S., Lee, K.G., 2018. Monitoring and risk assessment of polycyclic aromatic hydrocarbons (PAH) in processed foods and their raw materials. Food Cont. 92, 286–292.

Lee, Y.-N., Lee, S., Kim, J.-S., Patra, J.K., Shin, H.-S., 2019. Chemical analysis techniques and investigation of polycyclic aromatic hydrocarbons in fruit, vegetables and meat and their products. Food Chem. 277, 156–161.

Lipton, S.H., Bodwell, C.E., Coleman Jr., A.H., 1977. Amino acid analyzer studies of the products of peroxide oxidation of cystine, lanthionine and homocystine. J. Agric. Food Chem. 25, 624–628.

De Mey, E., De Klerck, K., De Maere, H., Dewulf, L., Derdelinckx, G., Peeters, M.C., 2013. The occurrence of N-nitrosamines, residual nitrite and biogenic amines in commercial dry fermented sausages and evaluation of their occasional relation. Meat Sci. 96, 821–828.

Moats, W.A., 1994. Chemical residues in muscle foods. In: Kinsman, D.M., Kotula, A.W., Breidenstein, B.C. (Eds.), Muscle Foods. Meat, Poultry and Seafood Technology. Chapman and Hall, New York, pp. 288–295.

Morzel, M., Gatellier, P., Sayd, T., Renerre, M., Laville, E., 2006. Chemical oxidation decreases proteolytic susceptibility of skeletal muscle myofibrillar proteins. Meat Sci. 73, 536–543.

Park, J.-E., Seo, J.-E., Lee, J.-Y., Kwon, H.J., 2015. Distribution of seven N-nitrosamines in food. Toxicol. Res. 31, 279–288.

Pegg, R.B., Shahidi, F., 2000. Nitrite Curing of Meat. Food & Nutrition Press, Trumbull, CT, pp. 175–208.

Raoul, S., Gremaud, E., Biaudet, H., Turesky, R.J., 1997. Rapid solid-phase extraction method for the detection of volatile nitrosamines in food. J. Agric. Food Chem. 45, 4706–4713.

Rath, S., Reyes, F.G., 2009. Nitrosamines. In: Nollet, L.M.L., Toldrá, F. (Eds.), Handbook of Processed Meats and Poultry Analysis. CRC Press, Boca Raton, FL, pp. 687–705.

Reig, M., Toldrá, F., 2009. Veterinary drug residues. In: Nollet y, L.M.L., Toldrá, F. (Eds.), Handbook of Muscle Foods Analysis. CRC Press, Boca Raton, FL, pp. 837–853.

Reig, M., Toldrá, F., 2011. Growth promoters. In: Nollet, L.M.L., Toldrá, F. (Eds.), Safety Analysis of Foods of Animal Origin. CRC Press, Boca Raton, FL, pp. 227–249.

Reig, M. & Toldrá, F., Chemical origin toxic compounds, In: Toldrá F., Hui YH., Astiasarán I., Sebranek JG. and Talon R., (Eds.), Handbook of fermented meat and poultry, 2nd edition, 2015, Wiley-Blackwell, Chichester, West Sussex, UK, 429–434.

Ripoll, G., Lobón, S., Joy, M., 2018. Use of visible and near infrared reflectance spectra to predict lipid peroxidation of light lamb meat and discriminate dam's feeding systems. Meat Sci. 143, 24–29.

Sannino, A., Bolzoni, L., 2013. GC/CI–MS/MS method for the identification and quantification of volatile N-nitrosamines in meat products. Food Chem. 141, 3925–3930.

Sen, N.P., Baddoo, P.A., Seaman, S.W., 1987. Volatile nitrosamines in cured meats packaged in elastic rubber nettings. J. Agric. Food Chem. 35, 346–350.

Singh, L., Varshney, J.G., Agarwal, T., 2016. Polycyclic aromatic hydrocarbons formation and occurrence in processed food. Food Chem. 199, 768–781.

Slump, P., Schreuder, H.A.W., 1973. Oxidation of methionine and cystine in foods treated with hydrogen peroxide. J. Sci. Food Agric. 24, 657–661.

Toldrá, F., Reig, M., 2006. Methods for rapid detection of chemical and veterinary drug residues in animal foods. Trends Food Sci. Technol. 17, 482–489.

Toldrá, F., Reig, M., 2012. Analytical tools for assessing the chemical safety of meat and poultry. Springer Briefs in Food, Health and Nutrition. Springer, New York, pp. 1–71.

Turner, N.W., Bramhmbhatt, H., Szabo-Vezse, M., Poma, A., Coker, R., Piletsky, S.A., 2015. Analytical methods for determination of mycotoxins: an up-date (2009-2014). Anal. Chim. Acta 901, 12–33.

Van Peteguem, C., Daeselaire, E., 2004. Residues of growth promoters. In: Nollet, L.M.L. (Ed.), Handbook of Food Analysis, second ed. Marcel Dekker Inc, New York, pp. 1037–1063.

Vermeer, I.T., Pachen, D.M., Dallinga, J.W., Kleinjans, J.C., van Maanen, J.M., 1998. Volatile N-nitrosamine formation after intake of nitrate at the ADI level in combination with an amine-rich diet. Environ. Health Perspect. 106, 459–463.

Walker, R., 1990. Nitrates, nitrites and nitrosocompounds: a review of the occurrence in food and diet and the toxicological implications. Food Addit. Contam. 7, 717–768.

Yurchenko, S., Mölder, U., 2007. The occurrence of volatile N-nitrosamines in Estonian meat products. Food Chem. 100, 1713–1721.

Zachara, A., Galkowska, D., Juszczak, I., 2017. Contamination of smoked meat and fish products from Polish market with polycyclic aromatic hydrocarbons. Food Cont. 80, 45–51.

Further reading

Brockman, R.P., Laarveld, R., 1986. Hormonal regulation of metabolism in ruminants. Rev. Livest. Prod. Sci. 14, 313–317.

EC, 2003. European Parliament and Council Regulation 2065/2003 of 10 November 2003 on Smoke Flavourings Used or Intended for Use in or on Foods.

EFSA, 2003. The effects of Nitrites/Nitrates on the microbiological safety of meat products. EFSA J. 14, 1–16.

Hill, L.H., Webb, N.B., Mongol, L.D., Adams, A.T., 1973. Changes in residual nitrite in sausages and luncheon meat products during storage. J. Milk Food Technol. 36, 515–519.

Jennings, W.G., 1990. Analysis of liquid smoke and smoked meat volatiles by headspace gas chromatography. Food Chem. 37, 135–144.

Maga, J.A., 1987. The flavour chemistry of wood smoke. Food Rev. Int. 3, 139–183.

Reig, M., Toldrá, F., 2010. Detection of chemical hazards. In: Toldrá, F. (Ed.), Handbook of Meat Processing. Wiley-Blackwell, Ames, IO, pp. 469–479.

SCF, 1995. Smoke Flavorings. Report of the Scientific Committee for Food. Opinion adopted on 23 June 1993. 34 series Food Science Techniques. European Commission.

SECTION 7

Food preservatives/additives in meat and their detection

CHAPTER 12

Use of food preservatives and additives in meat and their detection techniques

Meera Surendran Nair[1], Divek V.T. Nair[2], Anup Kollanoor Johny[2] and Kumar Venkitanarayanan[3]

[1]Department of Veterinary Population Medicine, University of Minnesota, Saint Paul, MN, United States [2]Department of Animal Science, University of Minnesota, Saint Paul, MN, United States [3]Department of Animal Science, University of Connecticut, Storrs, CT, United States

Chapter Outline
12.1 Introduction   188
12.2 Food additives   189
    12.2.1 Antioxidants   190
    12.2.2 Binders   191
    12.2.3 Emulsifiers   191
    12.2.4 Antimicrobials   192
    12.2.5 Curing agents and cure accelerators   192
    12.2.6 Flavoring agents   193
    12.2.7 Coloring agents   193
12.3 Preservatives   194
    12.3.1 Chemical preservatives   194
    12.3.2 Natural preservatives   196
12.4 Federal oversight   199
12.5 Health concerns and safety assessment   200
12.6 Analytical techniques   201
    12.6.1 Detection of sulfites   202
    12.6.2 Detection of nitrites and nitrates   203
    12.6.3 Detection of sorbic acid   203
    12.6.4 Detection of nisin and organic acids   204
    12.6.5 Detection of color additives   204
    12.6.6 Detection of synthetic phenolic antioxidants   204
12.7 Conclusion and future directions   205
References   205

12.1 Introduction

Meat is an important part of a typical diet as it contributes proteins, minerals, vitamins, and fats, thereby supporting human health. Therefore it is vital to preserve the quality of meat from factors that adversely influence its color, odor, texture, and flavor. An array of interrelated factors, such as holding time, temperature, the light intensity at storage, transportation, moisture, atmospheric oxygen level, endogenous enzymes, and microorganisms, play a critical role in causing detrimental changes to meat (Faustman and Cassens, 1990). Most importantly, the diverse and complex nutrient composition makes meat an ideal environment for the growth and propagation of spoilage and pathogenic microorganisms. Henceforth, a plethora of physical, chemical, and biological methods have been employed in the production chain to maintain and enhance the biological, physicochemical, rheological, and sensorial properties of food.

Although food, including meat products, is critical for sustaining health, it is also one of the most common sources of toxicants to which humans get exposed. Around 2500 or more chemical substances are being directly added to different types of foods globally to fortify the nutritive value, impart flavor, stabilize color and texture, as well as to make them affordable. Interestingly, about 12,000 substances also unintentionally find their way into the food that we eat (Pressman et al., 2017). Food laws have been in place from the late 1900s to curb these issues. However, mounting scientific evidence suggests that chronic exposure to many of the approved additives at low levels impacts human health adversely (Balbus et al., 2013; Maffini et al., 2017).

The quality of meat products typically diminishes from the time of harvest to consumption. This quality loss is often attributed to physical, chemical, enzymatic, and microbiological changes occurring in meat over time (Davidson et al., 2013). The reactions that deteriorate the quality of food are the principal targets for preservation (Gould, 2000). Among them, the adverse microbial changes in meat are the most serious, as they jeopardize the food quality and safety due to the growth of spoilage and pathogenic microorganisms, and the potential presence of microbial toxins (Petruzzi et al., 2017; Gould, 2000). In fact through food preservation, the physical and chemical changes that occur during storage are minimized, ensuring a better product. Therefore specific methods of preservation that incorporate the principles of microbial inhibition and minimizing other deteriorative changes such as color and oxidative changes in meat are critically important. The critical functions of preservation are to prevent deterioration of meat quality and to extend the safe time between production and consumption (shelf life) (Mukhopadhyay et al., 2017; Mustapha and Lee, 2017).

Although, there are specific roles for additives in foods (e.g., direct additives such as phosphates added in meat products to retain moisture and protect flavor), the chances of minute amounts of indirect food additives (for instance, packaging substances) finding their way into foods during processing and storage cannot be neglected. Therefore the Food and Drug Administration (FDA) requires that all materials coming in contact with food should be certified safe before they are used in the indicated manner (FSIS, 2015). Thus food additives are studied, regulated, and monitored to ensure that consumers feel safe about the foods they eat and are protected from hazards. The FDA reminds the manufacturers that it is their responsibility to ensure that the ingredients are of a food-grade purity and comply with specifications and restrictions (FDA, 2018a). In the United States, the US Department of Agriculture (USDA) shares responsibility with the FDA for regulations with food additives used in meat, poultry, and egg products. Additionally, all international trade risk assessments are conducted by an independent, international expert scientific group, the Joint FAO/WHO Expert Committee on Food Additives (JECFA) (WHO, 2018), ensuring that there are monitoring steps in place at various levels of global food production.

12.2 Food additives

The Codex Alimentarius defines a food additive as a substance not normally consumed as a food or not used as a basic food ingredient, that may have a nutritional value, whose intentional addition to the food for a technological purpose in the manufacture, processing, preparation, packaging, transportation, or storage results, or may reasonably be expected to result, in it, or its by-products becoming directly or indirectly a component of such foods (FAO, 2018). Within this broad definition, those which are chemically characterized, potentially nontoxic, and demonstrated to possess consumer and industry benefits are generally admitted to be used in the production facilities (Magnuson et al., 2013).

The first specific regulation of food additives was enacted in 1958 using the Food Additives Amendment to the Federal Food, Drug, and Cosmetic (FD&C) Act (FDA, 2018b). Currently, the WHO, in cooperation with the FAO, is responsible for assessing the safety of food additives globally along with the respective national organizations. As mentioned earlier, in the United States, the USDA shares responsibility with the FDA for regulations implicated with food additives used in meat, poultry, and egg products. The only exception to the FDA's testing and approval process of additives has been for the "generally recognized as safe (GRAS)" compounds, substances which have been historically used as food flavorants and spices with no known harmful effects. Nevertheless, since 1970 the FDA has established a comprehensive review and rulemaking procedure to affirm the GRAS status of many of such compounds that were on the list previously or those that require an affirmation status. Regulations also enforce the listing of additives on labels of meat and poultry products so that consumers will have complete information. The current decade has also witnessed the consumer-driven movement on "clean label," emphasizing increased demand for natural alternatives to synthetic chemical additives (Wang and Adhikari, 2017).

The common additives in meat include antioxidants, binders, emulsifiers, antimicrobials, curing agents and cure accelerators, flavoring agents, and coloring agents (FSIS, 2018).

12.2.1 Antioxidants

Antioxidants are synthetic or natural compounds added to meat to prevent oxidation, lipid rancidity, off-flavors, and to stabilize color. Oxidation of meat is the process of removal of electrons from the food system, which in turn generates free radicals. These highly reactive free radicals induce a chain of reactions and react with natural colors and flavors in meat (Shahidi, 2000). The oxidation of meat lipids leads to the production of hydrogen peroxide, which subsequently generates secondary volatile compounds, such as aldehydes, ketones, acids, and alcohols, which provide off-flavors in meat (Pateiro et al., 2018; Kumar et al., 2015). The undesirable brown color formation associated with the oxidation of meat pigment myoglobin to metmyoglobin also deteriorates meat quality and consumer acceptance. Moreover, metmyoglobin formation enhances the rate of lipid oxidation (Kumar et al., 2015). The oxidation of proteins affects protein functionality and consequently alters the water-holding capacity, nutrition, quality, and flavor (Lund et al., 2011).

Synthetic antioxidants such as butylated hydroxyanisole (BHA), and butylated hydroxytoluene (BHT) are used in meat as free radical scavengers or inhibitory substances to prevent lipid oxidation. These are mainly used in sausages, and the limit is 0.02% in combination with other antioxidants based on the fat content of the product (FSIS, 2018). However, the toxicity concerns of synthetic compounds and increasing health consciousness of consumers have led to extensive investigations to evaluate the efficacy of natural compounds as antioxidants in meat (Shahidi and Zhong, 2005; Sarıçoban and Yilmaz, 2014; Martins et al., 2018; Pateiro et al., 2018). For example, rosemary has been used as an antioxidant in dehydrated turkey broth powders in combination with other antimicrobials (FSIS, 2018). The aqueous extract of cinnamon bark containing phenolic and flavonoid compounds improved the oxidative stability and redness of chicken meatballs without affecting sensory properties of the meat product. Additionally, the extract had a comparable antioxidant activity to that of BHA and BHT (Chan et al., 2014). Fresh garlic, garlic powder, and garlic oil were effective in reducing lipid oxidation of raw chicken sausages during refrigerated storage at 3°C (Sallam et al., 2004). Other natural oils such as thyme or cumin essential oils at 0.05% were effective in maintaining oxidative stability, microbial safety, and shelf life stability compared to BHA or BHT in chicken patties for an extended storage period of 28 days (Sarıçoban and Yilmaz, 2014). Grape seed extract (0.02%) also possessed higher antioxidant properties than BHA and BHT in cooked, frozen pork patties for an extended storage period of 6 months (Sasse et al., 2009).

12.2.2 Binders

Binders are added to meat to maintain uniform dispersion of fat throughout the product and to preserve water during different stages of processing, heating, cooking, storage, and chilling. This process will eventually help to maintain the functional qualities of meat. For example, starch is used in meat as a binder during heating since it gelatinizes and absorbs water (Devadason et al., 2010). The gums are commonly used as binders in meat due to their high hydration ability. Xanthan gum possesses the highest water-binding capacity (232 mL/g), followed by guar gum (40 mL/g) and sodium alginate (25 mL/g) (Sánchez et al., 1995). Xanthan gum has a β-D-glucose backbone similar to cellulose, and trisaccharide side chains arising from every other glucose molecule on the backbone consisting of a glucuronic acid residue and two mannose units (Phillips and Williams, 2009). These anionic side chains provide high hydration ability to xanthan gums (Sánchez et al., 1995). In guar gum, the galactose and mannose residues present in the structure contribute to its hydration ability in cold water (Sánchez et al., 1995; Featherstone, 2015).

Carrageenan, obtained from red seaweeds, is a food-grade hydrocolloid which is used as a binder in meats. It is a polysaccharide of D-galactose and 3,6-anhydro-D-galactose that can form gels, thereby retaining water in meat (Trius et al., 1996; Wallingford and Labuza, 1983). In addition, the proteins obtained from vegetable and animal sources are used in meat as binders (FSIS, 2018). Carrageenan and modified food starch are used in turkey ham and cured pork products as binders. A combination of carrageenan, xanthan gum, and whey protein concentrate are suitable to use in sausages, cooked poultry products, beef and poultry patties, and fermented sausages.

Phosphates are also added to meat to increase water-holding capacity (Sebranek, 2015). Carboxymethyl cellulose is suitable for use in cured pork products and poultry franks. Canola proteins, carboxymethyl cellulose, and beef and pork collagens can also be used as binders in different meat and poultry products (FSIS, 2018).

12.2.3 Emulsifiers

Emulsifying agents act as an interface between the immiscible components and form stable emulsions (Santhi et al., 2017). An emulsifier increases texture and palatability, prevents separation of the food system, controls rancidity reactions, and solubilizes or disperses flavors (Eskin et al., 2001; Nash and Brickman, 1972; Martins et al., 2018; Santhi et al., 2017).

Meat proteins are good emulsifying agents. The amino acids in the proteins interact with the nonpolar fat and polar water molecules forming a continuous system in meat batter by preventing the coalescence of fat (Eskin et al., 2001). Myosin is the primary meat protein that functions as an emulsifying agent (Galluzzo and Regenstein, 1978). Plant proteins from soy, wheat, pea, and potato can be used to improve emulsification and other functional qualities of meat (Tarté, 2009). In addition, the use of collagen increases protein concentration in the emulsion and improves emulsion stability (Santana et al., 2011). Lecithin and mono- and diglycerides are also used as emulsifying agents in various poultry products due to their hydrophilic and lipophilic components (Van Nieuwenhuyzen, 1981; GPO, 2012).

12.2.4 Antimicrobials

Antimicrobials are used during meat processing to control foodborne pathogens, including *Salmonella* and *Campylobacter* (Nair and Kollanoor Johny, 2017; Nair et al., 2014a,b, 2015). Antimicrobials such as chlorine, chlorine dioxide, acidified sodium chloride (ASC, a combination of citric acid and sodium chloride), trisodium phosphate, cetylpyridinium chloride (CPC), peroxyacetic acid (PAA), and organic acids are used in wash waters in online reprocessing and chilling tanks. Additionally, during the postchill process, antimicrobial agents such as 1,3 dibromo-5,5-dimethyl hydantoin, lauramide arginine ethyl ester, and ozone can be used as a multiple hurdle approach to control *Salmonella*. Chlorine treatments such as chlorinated water or ASC can be used to treat carcass parts to prepare ground poultry products (FSIS, 2010). An aqueous solution of acidic calcium sulfate and lactic acid can be used on carcasses, parts, giblets, and ground poultry. Calcium hypochlorite, CPC, sodium hypochlorite, and electrolytically generated hypochlorous acid are used on poultry giblets (FSIS, 2018). Chemical interventions such as ASC, lactic acid, organic acids, such as lactic and citric acids, or a combination of PAA, octanoic acid, acetic acid, hydrogen peroxide, peroxyoctanoic acid, and 1-hydroxyethylidene-1,1-diphosphonic acid are used on beef carcasses, parts, trim, and organs. In addition, ozone, chlorine dioxide, electrolytically generated hypochlorous acid, and anhydrous ammonia are used on beef carcasses (Winkler and Harris, 2009; FSIS, 2018). Furthermore, a variety of natural and synthetic antimicrobial agents are used in meat products to prevent microbial contamination and quality deterioration, which are discussed under preservatives in this chapter.

12.2.5 Curing agents and cure accelerators

Curing agents provide a stable red color to meat which appeals to the consumer. The most commonly used curing agent, nitrites, provides cured meat color, flavor, antioxidant, and antimicrobial properties to meat (Sindelar et al., 2011; Sebranek, 2015). The reduction of nitrites and nitrates to nitric oxide is important for the cured meat color. The nitrosyl myoglobin, the dominant pigment in cured meat products, is formed from the interaction of nitric oxide with the heme pigment of myoglobin (Møller and Skibsted, 2006). Nitrates can be used as a source of nitrites in dry cured products since bacteria possessing nitrate reductase enzyme (e.g., nonpathogenic, coagulase-negative Staphylococci) in sausages convert nitrate to nitrites during extended curing or drying periods (Leroy et al., 2006; Sebranek, 2015; Gøtterup et al., 2008).

On the other hand, cure accelerators are added to meat to fasten the curing process, which will be beneficial in high production environments. The addition of sodium ascorbate, sodium erythorbate, ascorbic acid, or erythorbic acid fastens the curing process since they reduce nitrites to nitric oxide. Although the addition of curing agents is optional in meat, the addition of sodium erythorbate or sodium ascorbate at 550 ppm is mandatory to reduce residual nitrite in bacon to a lower level before cooking. This step reduces the risk of the formation of nitrosamines during frying since nitrosamines are considered carcinogens (Sebranek, 2015; GPO, 2012; Cantwell and Elliott, 2017; De Mey et al., 2017).

12.2.6 Flavoring agents

Spices are an integral part of food preparation in many world cuisines and used to add flavors to the meat. The spices are derived from plant parts, and the FDA has approved a list of spices that can be used in foods (FDA, 2018c). Spices contain flavor compounds such as cinnamaldehyde from cinnamon, eugenol and eugenyl acetate from clove, gingerol, shogaol, neral, and geranial from ginger, 1−8-cineole from bay leaves, piperine and beta-caryophyllene from black pepper, and capsaicin and dihydrocapsaicin from chilli pepper (Gadekar et al., 2006). In addition, the controlled burning of wood (pyrolysis) generates compounds such as acids, aldehydes, furans, and phenols, which provide flavor and color to meat products, a preservation technique known as "smoking" in meat processing (Rozum, 2009). Lactic acid, a mixture of citric, oregano, and rosemary extracts, potassium acetate (not exceeding 1.2% of product formulation), potassium citrate (not exceeding 2.25% of the product formulation), or a mixture of sodium acetate and sodium diacetate (combination should not exceed 0.8% of total formulation) are also used as flavoring agents in various poultry and meat products (FSIS, 2018).

12.2.7 Coloring agents

A variety of coloring agents are permitted in meat products. As mentioned earlier, curing agents are responsible for the stabilization of the characteristic red color to cured meat, whereas cure accelerators enhance the color fixing process. Therefore compounds such as carotene, annatto, cochineal or carmine (crimson red pigment), chlorophyll, saffron, and turmeric are used to impart color to casings or rendered fats (GPO, 2012; Kendrick, 2012). Titanium dioxide is used in ready-to-eat (RTE) and non-RTE poultry products that permit the use of coloring agents. Titanium dioxide should not exceed 0.25% of the finished product (FSIS, 2018).

12.3 Preservatives

Current meat preservation methods are designed to minimize the undesirable changes affecting the wholesomeness, sensory qualities, and nutritional composition of animal-based products. According to the FDA, the term "chemical preservative" defined in 21 CFR 101.22(a)(5), denotes "any chemical that, when added to food tends to prevent or retard deterioration thereof, but does not include common salt, sugars, vinegars, spices, or oils extracted from spices, substances added to food by direct exposure thereof to wood smoke, or chemicals applied for their insecticidal or herbicidal properties" (FDA, 1989).

As mentioned in the case of food additives, changes in consumer requirements in recent years have increased the demand for foods which are more convenient, fresher, more natural, and nutritionally healthier than those available in the past. This public perception has generated heightened interest in "biopreservation" that involves the use of naturally occurring compounds that exert a broad spectrum of antioxidant and antimicrobial activities as preservatives (Barberis et al., 2018). Subsequently, with an intention to produce minimally processed and preserved foods, new and improved technologies are also being utilized in the food industry to retain and improve the effectiveness of preservation, and ensuring the quality and safety of meat. Like food additives, general food safety legislation and other regulatory restrictions are also established for preservatives based on their history of use and toxicological information.

Preservatives in meat are broadly classified into chemical preservatives and natural preservatives. The chemical preservatives include organic acids and their derivatives, nitrites, and sulfites, whereas major natural preservatives include nisin, chitosan, plant-derived compounds, and lysozyme (FSIS, 2018; Davidson et al., 2013).

12.3.1 Chemical preservatives

12.3.1.1 Organic acids and derivatives

Organic acids such as acetic, lactic, propionic, sorbic, benzoic, and citric acids are used in meat products as preservatives. The pH of meat allows organic acid to enter the microbial cells and dissociate into proton and acid ions. The protons change the pH of microbial cytoplasm and destabilize the structural proteins of microorganisms. The anions entrapped in the microbial cytoplasm causes the denaturation of nucleic acids (Davidson et al., 2013; Theron and Lues, 2007; Mani-López et al., 2012).

Acetic acid is a commonly used organic acid due to its activity against a broad spectrum of microorganisms, including pathogens such as *Salmonella, Campylobacter*, and *Listeria* in meat (Mani-López et al., 2012; Olaimat et al., 2018; Kassem et al., 2017). Acetic acid is used in various meat products, such as dried and fermented sausages. The salts of acetic acid, such as sodium and potassium diacetate, or their combinations with other antimicrobials are used in meat products against *Listeria monocytogenes* (Porto-Fett et al., 2010; Glass et al., 2002). Potassium diacetate is used in hot dogs against pathogens at a level not exceeding 0.25% of the product formulation.

Lactic acid, a flavoring agent and acid regulator in meat and meat products, also possesses antimicrobial activity against pathogenic and spoilage organisms (Ibrahim et al., 2008; Stanojević-Nikolić et al., 2016). Combinations of lactic acid with other antimicrobials are used in RTE meat products (Davidson et al., 2013; FSIS, 2018). For example, a combination of sodium diacetate (4%), lactic acid (4%), pectin (2%), and acetic acid (0.5%) is used in cooked meat products. In addition, a mixture of sodium diacetate, salt, lactic acid, and mono- and diglycerides at a level not exceeding 0.2% of the product is used in RTE meat (FSIS, 2018).

Benzoic acid has antifungal activity (Krebs et al., 1983; López-Malo et al., 2007), which is mostly limited in foods with an acidic pH (Shahmohammadi et al., 2016). However, this organic acid exhibits synergistic antimicrobial activity with other antimicrobials against pathogenic microorganisms such as *Escherichia coli* O157:H7, *Staphylococcus aureus*, and *Bacillus mucoides* (Ceylan et al., 2004; Stanojevic et al., 2009). Benzoic acid and sodium benzoate or their combination with other antimicrobial agents such as sodium diacetate and sodium propionate are used in RTE meat and poultry products (FSIS, 2018).

Sorbic acid or sorbate has been used in foods as a preservative for several years due to their inhibitory activity against yeast and molds (Sofos and Busta, 1981; Brul and Coote, 1999). In addition, sorbate or its combination with other antimicrobial agents, including plant-derived compounds such as carvacrol, eugenol, and thymol possess inhibitory activity against pathogenic and spoilage microorganisms such as *Bacillus subtilis*, *Bacillus cereus*, *E. coli*, *Pseudomonas aeruginosa*, *S. aureus*, *Listeria innocua*, *S.* Typhimurium, and *Candida albicans* depending on the pH and water activity of the medium (Eklund, 1983; Santiesteban-López et al., 2007; Lu et al., 2011). However, they do not affect lactic acid bacteria, which make them useful as preservatives in fermented meat products (Davidson et al., 2013). In meat products such as beef snacks and jerky, potassium sorbate (0.0703% by weight) is added as a mold inhibitor. In addition, potassium sorbate can be applied to product surface and casings as an external mold inhibitor (FSIS, 2018).

12.3.1.2 Nitrites

Nitrites and nitrates are used in meat for curing purposes (Honikel, 2008). Sodium or potassium salts of these compounds are added to meat to enhance the color and flavor (Sindelar et al., 2011). The antimicrobial activity of nitrites is due to the formation of nitric oxide, nitrous oxide, or nitrous acid (Møller and Skibsted, 2002; Sindelar et al., 2011). Sodium nitrite with table salt significantly inhibits coliform growth in fresh pork sausages during a drying period of 10 days at 22°C (Bang et al., 2008). Nitrites inhibit growth, spore germination, and toxin synthesis of *Clostridium botulinum* in cured meats (Keto-Timonen et al., 2012). However, the effects of nitrites against enteropathogens such as *Salmonella* and *E. coli* are not well established (Sindelar et al., 2011; Sebranek, 2015). A combination of ascorbate and nitrite from natural sources can be used in heat-treated RTE or non-RTE meat products. A combination of either 75 and 500 ppm or 100 and 250 ppm of nitrite and ascorbate, respectively, from natural sources, should be included by the weight of the finished product to achieve antimicrobial activity (FSIS, 2018). The total amount of nitrites or nitrates or their combination should not exceed 200 ppm of nitrites in the finished product (GPO, 2012). The enhanced antimicrobial activity of nitrites is observed when they are combined with sodium chloride, EDTA, erythorbate, or ascorbate (Tompkin et al., 1978).

12.3.1.3 Sulfites

Sulfites exert antimicrobial activity against spoilage organisms, including lactic acid bacteria and molds, thereby preventing unnecessary fermentation (Taylor et al., 1986; Iammarino et al., 2017a). Health concerns such as allergic, respiratory reactions, and chronic skin symptoms in sensitive people have led to the replacement of sulfites with other antimicrobials (Vally and Misso, 2012). A higher concentration of sodium sulfite such as 50,000 and 625 ppm, are necessary to be effective against *S.* Typhimurium and *L. monocytogenes*, respectively (Lamas et al., 2016). Sodium bisulfate is generally recognized as a safe compound. However, its use in meat is limited (FDA, 2018d). It is used in poultry processing as a pH regulator in scald tanks (FSIS, 2018; GPO, 2012), and should be removed after subsequent washing operations from the carcasses (GPO, 2012).

12.3.2 Natural preservatives

12.3.2.1 Nisin

Bacteriocins are small peptide molecules produced by beneficial bacteria and exert antimicrobial activity against various pathogenic microorganisms (Cleveland et al., 2001). Nisin is a bacteriocin produced by *Lactococcus lactis* ssp. *lactis* (Cleveland et al., 2001; De Vuyst and Vandamme, 1994; Liu and Hansen, 1990; Fontes Saraiva et al., 2014; Yang et al., 2014), which is approved to be used in foods (FSIS, 2018). Nisin is a lantibiotic synthesized in ribosomes, undergoes posttranslational modifications, and contains thioether amino acids such as lanthionine and methyl-lanthionine (Yang et al., 2014). Nisin is effective against mesophilic bacteria and Gram-positive pathogens such as *L. monocytogenes*, *S. aureus*, and *B. cereus* (Benkerroum and Sandine, 1988; Piper et al., 2009; Lee et al., 2015; Le Lay et al., 2016). The antimicrobial activity of nisin is due to its binding with lipid-bound cell wall precursor lipid II, which inhibits the biosynthesis of microbial cell walls. It also leads to pore formation in cellular membranes (Li et al., 2018; Wiedemann et al., 2001). These are the active mechanisms of antimicrobial activity of nisin against Gram-positive bacteria. However, in Gram-negative bacteria, the outer membrane inhibits the direct binding of nisin to lipid II. Therefore chelators (EDTA) or processing techniques such as heating or freezing are used to enhance the activity of nisin against Gram-negative bacteria (Prudêncio et al., 2015). For example, nisin alone did not possess any antimicrobial property against *S. Enteritidis*. However, a combination of nisin (500 ppm), ascorbic acid (2000 ppm) and EDTA (250 ppm) resulted in more than 3.0 log CFU/mL reductions of *S. Enteritidis in vitro* (Sangcharoen et al., 2017). Currently, nisin preparations are used in cooked RTE meat and poultry products, meat and poultry soups, and casings. In addition, nisin is used along with rosemary extracts in frankfurters, cooked meat sausages, and cured meat products as antimicrobial agents (FSIS, 2018).

12.3.2.2 Chitosan

Chitin is a structural polysaccharide consisting of units of amino sugar glucosamine derived from the crustacean shells and fungi. The biodegradable polymer made from chitin is known as chitosan. Chitosan is a GRAS compound approved for use in foods, including poultry and meat products (FDA, 2018e). The antimicrobial activity of chitosan is pH dependent. Chitosan exists in a dissociated state at a lower pH, generates positively charged amino groups and interacts with the negatively charged carboxyl groups on bacterial cell membranes. It alters the bacterial cell permeability, disrupts cell membranes, and causes leakage of the cellular contents (Chen and Mustapha, 2008; Rabea et al., 2003; Qi et al., 2004). Chitosan exerts inhibitory effects on microorganisms such as *B. subtilis*, *E. coli*, *Pseudomonas fragi*, and *S. aureus* (Darmadji and Izumimoto, 1994). In addition, chitosan (3%) was found to be effective against *Clostridium perfringens* spore germination and growth in cooked ground beef and turkey during cooling at abuse temperatures for an extended period (Juneja et al., 2006). Chitosan shows synergistic activity against foodborne pathogens when combined with other chemical or natural antimicrobial agents. Chitosan (1.94 mg/cm^2) inhibits *L. innocua* after 24 hours storage at 10°C when used with lauric arginate ester (0.388 mg/cm^2) as edible coatings on RTE meat samples (Guo et al., 2014). It was reported that chitosan (0.01% w/w) enhances the antimicrobial activity of rutin (0.05 or 0.1% w/w) and resveratrol (0.1 or 0.2% w/w) against *E. coli* O157:H7 in undercooked hamburger patties. A combination of chitosan and these plant-derived compounds resulted in more than 5 log CFU/g reduction of *E. coli* in medium-rare (65°C) cooked beef patties compared to cooked control patties. In addition, the combination resulted in the enhanced color stability of patties (Surendran Nair et al., 2016).

12.3.2.3 Plant-derived compounds

Plant-derived compounds are traditionally used in foods as flavoring agents (Burt, 2004). These compounds are classified into four groups: herbs, botanicals, essential oils, and oleoresins, based on their biological origin, formulation, chemical description, and purity (Windisch and Kroismayr, 2007). Extensive research has been done to evaluate their efficacy as preservatives in meat products to extend shelf life. The polyphenolic extracts from cherry (0.5 g/100 g of meat) and black currant (1 g/100 g of meat) leaves increased the shelf life of vacuum packaged pork sausages by reducing mesophiles, psychrotrophs, lactic acid bacteria, and *Brochothrix* after 14 days of refrigerated storage (Nowak et al., 2016). Fresh garlic (30 g/kg), garlic powder (9 g/kg), and garlic oil (0.015 g/kg) were effective in increasing the shelf life stability of raw chicken sausages during refrigerated storage at 3°C, where fresh garlic and garlic powder increased the shelf life up to 21 days (Sallam et al., 2004). The inclusion of essential oils from bay leaf extract (0.1 g/100 g) to fresh Tuscan sausage resulted in a decrease of coliform populations during storage at 7°C for 14 days (da Silveira et al., 2014). Addition of thymol (500 ppm) or thymol in combination with modified atmosphere packaging was capable of increasing the shelf life value of reduced pork back-fat sausages to 5 days compared to control samples during refrigerated storage (Mastromatteo et al., 2011). In addition, the inclusion of marjoram (*Origanum majorana* L. (11.5 mg/g)) essential oil to fresh sausages significantly reduced aerophilic and heterophilic bacterial growth and *E. coli* populations during storage of 35 and 25 days, respectively (Busatta et al., 2008).

Plant-derived compounds target the pathogenic bacteria using a variety of mechanisms, including downregulation of virulence, affecting membrane permeability, and proton motive force, and leakage of the contents resulting in reduced infectivity or death of pathogens (Burt and Reinders, 2003; Lv et al., 2011; Surendran Nair et al., 2017; Kollanoor Johny et al., 2017; Burt, 2004; Venkitanarayanan et al., 2013). However, higher concentrations of plant-derived compounds affect the sensory quality of the meat (Hugo and Hugo, 2015). Therefore the maximum utilization of these compounds in meat can be obtained in combination with other natural antimicrobial agents such as chitosan, lysozyme, and nisin due to their cell wall perturbation effects, which cause synergistic antimicrobial effects (Surendran Nair et al., 2016; Khanjari et al., 2013; Kanatt et al., 2008; Govaris et al., 2010; FSIS, 2018).

12.3.2.4 Lysozyme

Lysozyme is one of the predominant proteins in egg albumen, which has antimicrobial activity against foodborne pathogens such as *L. monocytogenes* and *C. botulinum* (Hughey and Johnson, 1987; Yang et al., 2007). It has GRAS status for use in food products (GPO, 1998). The antimicrobial activity of lysozyme against Gram-positive cell walls is mainly due to the *N*-acetyl muramoyl hydrolase enzyme activity, which causes the hydrolysis of peptidoglycan leading to cell lysis (Masschalck and Michiels, 2003). Therefore most of the Gram-negative bacteria are insensitive to lysozyme due to their outer membranes which make the lysozyme impermeable to the cells. Therefore modifications to the lysozyme are needed to penetrate the cell membrane and to extend their spectrum of activity towards Gram-negative bacteria. In Gram-negative bacteria, lysozyme binds to cells by electrostatic binding, which leads to membrane damage and initiation of autolysis of the cells. The incorporation of fatty acids, organic acids, polysaccharides, and hydrophobic peptides has been found to enhance these interactions of lysozyme to lipopolysaccharide outer membrane. In addition, chelators such as EDTA enhance the penetration of lysozyme to the outer membrane of Gram-negative bacteria (e.g., against *E. coli* O157:H7) (Masschalck and Michiels, 2003; Davidson et al., 2013; Boland, 2003; Cegielska-Radziejewska et al., 2009).

12.4 Federal oversight

Early in the 20th century the United States federal government asserted authority over the safety and quality of foods in response to many reported industry abuses (NRC, 1998). In 1906 the initial act mentioned that "any food containing an added poisonous or other added deleterious ingredient which may render such article injurious to health would be deemed adulterated." In 1938 the legislative congress replaced this original statute with the federal FD&C act. The updated legislation retained the same basic system of regulation as in 1906, in addition to policing for adulterants in foods. In 1950 the Congress turned its attention to the growing use of chemical additives. As an initial step, the Congress amended the FD&C Act in 1954 and established tolerances for pesticide chemicals intended for use on raw agricultural commodities and created a premarket approval system for pesticide residues in foods. In 1958, the Food Additives Amendment to the FD&C Act was enacted and mandated the premarket review and approval system for all food additives. Additionally, the Congress shifted the burden of proof on the safety issue to the industry, and additives could not be used unless and until the agency certified them safe (FSIS, 2018).

The meat industry started evolving in the United States in the mid-1800s, primarily as animal slaughter and slaughter-processing industries shifted to a commercial food preservation industry in the late 19th century. The USDA Food Safety Inspection Service (FSIS) is the consumer protection agency of the USDA, which coordinates the inspection activities at meat facilities to ensure the safety, wholesomeness, and accuracy of labeling. The FSIS regulates raw beef, pork, lamb, chicken, and turkey products, as well as any product that contains 2% or more cooked poultry or 3% or more raw meat, which includes approximately 250,000 different processed meat and poultry products, including hams, sausages, soups, stews, pizzas, and frozen dinners (Clinton and Gore, 1996). Most of the regulations governing the use of various additive ingredients in processed meats are found in the Title 9 of the Code of Federal Regulations (CFR), Chapter III, Parts 317–319 (GPO, 2012). Additionally, the *USDA Processing Inspectors' Calculations Handbook* (FSIS, 1995) and the *Food Standards and Labeling Policy Book* (FSIS, 2005) describe how the regulations are applied.

12.5 Health concerns and safety assessment

Chronic exposure to food additives and the associated adverse health impacts are growing concerns globally (Balbus et al., 2013; Maffini et al., 2017). Although the mechanisms are unknown, certain additives and preservatives in meat such as monosodium glutamate have been shown to cause gastrointestinal and respiratory allergic reactions (Tarlo and Sussman, 1993). Likewise, with the evidence unveiled through epidemiological studies linking the consumption of phosphate food additives and increased cardiovascular risk, the European Food Safety Authority is currently pursuing a high-priority reevaluation of added phosphates in processed meats and other food products (Foley et al., 2009; Cancela et al., 2012; EFSA, 2013). Moreover, in 1991 the National Toxicology Program concluded that BHA is reasonably anticipated to be a human carcinogen, whereas the risk of cancer due to BHT is still controversial based on animal studies (NTP, 1991, 2016). The Center for Science in the Public Interest (CSPI), a consumer advocacy group, often disagrees with many of the FDA approval processes and cited BHA as an additive to avoid and placed BHT in its "caution list" (CSPI, 2017). There are also reports indicating that processed meat containing nitrites and nitrates increases colon and pancreatic cancer risks (Bastide et al., 2011; Larsson and Wolk, 2012).

The acceptable daily intake (ADI) levels of all food additives for human consumption have been derived from the no-observed-adverse-effect level (NOAEL) predetermined through recommended toxicological studies. These threshold values are mostly the standard limits for the safety assessment evaluations conducted by the regulatory agencies (Pressman et al., 2017). Despite the legal limits of these additives in meat products mandated by the regulatory agencies, certain food additives at their recommended concentrations have been reported to increase the risk of toxicity in humans (Pressman et al., 2017).

The existing approach to safety assessment of food additives in the United States is compiled in "the Redbook" (Toxicological Principles for the Safety Assessment of Direct Food Additives and Color Additives Used in Food) originally published in 1982 and revisions have been made in 1993 and 2007 to make the agency's testing guidelines in line with advancing international approaches (FDA, 2000). Irrespective of the food commodities, the overall approach for additive risk assessment follows four main basic principles.

1. For every food additive, toxicological information is necessary to an extent.
2. The amount of safety data required for a particular food additive is dictated by the *level of concern* (LOC).
3. LOC is defined based on the magnitude of potential human intake of an additive and its molecular structure. When available, exposure data carry more weightage than the structure.

4. If existing data suggest that there are immediate and long-term health effects associated with the ingestion of a particular additive, then the initial evaluation of testing requirements is adjusted accordingly.

The FSIS and FDA regulate the amount and type of additives and preservatives allowed in meat production, including for those produced through bioengineering or genetic modifications. However, the legal burden for proof of safety is entirely vested on the manufactures, who must satisfy the FDA's safety criteria before the marketing of a food additive. Furthermore, the most recent ruling for GRAS, published in 2016 (*Federal Register*, August 17, 2016; 81 (159), 54960–55055) mentioned that all future GRAS reviews would be "self-determinations" of GRAS status by the notifiers, and the FDA strongly encourages manufacturers to inform the agency about their GRAS conclusions on the additives used through the notification procedure finalized in the rule. Unfortunately, this law is currently being misinterpreted as allowing companies to make their safety determinations for GRAS substances they use, without notifying the FDA. Currently, this has created difficulty in obtaining vivid information about a chemical's safety and is affecting the agency's regulatory practices (Maffini et al., 2017). Additionally, the 1958 food law did not give the FDA authority to pull together any chemical hazard and exposure information from industries, or to develop a monitoring program to keep track of postmarket uses of chemicals.

12.6 Analytical techniques

Meat and meat products are complex food matrices with high concentrations of nutrients and water, thereby making them very perishable. The introduction of additives and preservatives are thus aimed at extending the shelf life and reducing the production costs. Nevertheless, the use of additives and preservatives in meat requires a stringent food safety policy as some compounds can pose adverse health risks to consumers when present in concentrations greater than the legal thresholds. Therefore different national agencies exercise their specific legislation encouraging the implementation of analytical techniques to identify and quantify various food additives in fresh meat and processed products produced during the manufacturing process. Recent advancements in technological approaches have aided the provision of more reliable, selective, sensitive, rapid, affordable, and environment-friendly techniques for quality control and food safety evaluation (Martins et al., 2018).

Alternatively, the challenging factor for chemical analysis is the high concentration of fats, oils, lipids, proteins, carbohydrates, polysaccharides, salts, surfactants, pigments, emulsions, and many other constituents present in meat, which often interfere with the analytical process. Henceforth, pretreatment of food samples is the next important step to be conducted before the chemical analysis to remove possible interfering compounds (Sun et al., 1997; Cacho et al., 2015; Martins et al., 2018). The most common pretreatment methods employed in meat additive analytics are mixing, homogenization, dilution, centrifugation, distillation, simple solvent extraction, supercritical fluid extraction, stir-bar sorptive extraction (SBSE), pressurized-fluid extraction, microwave-assisted extraction, and Soxhlet extraction (Sun et al., 1997; Cacho et al., 2015; Martins et al., 2018).

An array of analytical methods has been researched for the detection and quantification of additives and preservatives in different food commodities, including meat. The majority of these methods have been initially developed for the detection and quantification of additives, which are known to be potential toxicants to consumers. Most often, the choice of various analytics is determined based on the phase of the samples (solid, liquid, or gas), expected level of additives, and, more importantly, the levels of interfering substances present in the samples (Bahadoran et al., 2016; Martins et al., 2018). The main analytical strategies to assess the level of additives in meat include chromatography, mass spectrometry (MS), electrophoresis, electronic spin resonance, and flow injection methods, in addition to traditional enzymatic and immunoassays (Iammarino et al., 2017b; Martins et al., 2018). The following section details the detection and quantification methods used for some of the major additives and preservatives used in meat.

12.6.1 Detection of sulfites

The sulfuring treatment of fresh meat preparations utilizes the antioxidant action of sulfites. The ADI for sulfites has been assigned as 0.7 mg/kg body weight by the JECFA. The sulfite measurements are often expressed as the concentration of sulfur dioxide and its salts in meat. Traditionally, there are direct and indirect methods of SO_2 quantification. The direct methods include titrimetric, polarographic, electrometric, and colorimetric procedures, whereas the indirect approaches comprise separation by distillation in an inert atmosphere followed by absorption of the sulfur dioxide in an oxidizing agent, such as iodine or hydrogen peroxide (Wood et al., 2004). Previously, the Association of Analytical Communities (AOAC) official methods of analysis accepted Monier–Williams (AOAC 990.28) and enzymatic methods as reference methods (Edberg, 1993). The Monier–Williams method is based on the distillation of SO_2 under acid conditions, its recovery into a hydrogen peroxide solution, followed by volumetric titration of the sulfuric acid produced by SO_2 oxidation with NaOH (indirect quantification). In the enzymatic method sulfite ions are oxidized to sulfate by oxygen in the presence of sulfite oxidase, forming hydrogen peroxide, which is transformed to water by reduced nicotinamide adenine dinucleotide (NADH) in the presence of NADH peroxidase. NAD^+ formed by this reaction is proportional to the sulfite concentration, and NADH consumption is measured spectrophotometrically at 340 nm (Iammarino et al., 2017b). However, these two techniques are time-consuming and experience analytical uncertainties due to interference problems caused by other sulfurous compounds.

Several alternative methods for the determination of sulfites in foods have also been developed, and many of these methods are currently recognized as AOAC official methods of analysis for sulfites. These include differential pulse polarography (Holak and Patel, 1987), ion exclusion chromatography (IEC; Kim, 1990), flow injection analysis (FIA; Ruiz-Capillas and Jiménez-Colmenero, 2009; Sullivan et al., 1990), sequential injection analysis (Segundo, 2001), high-performance liquid chromatography (HPLC; Robbins et al., 2015; Warner et al., 1990), capillary electrophoresis methods (CE; Trenery, 1996), and vapor phase Fourier transform infrared spectrometry (Teixeira dos Santos et al., 2016). Nevertheless, most of these methods are still in the standardization stages for use in meat products owing to the fact that the determination of sulfite in meat products is difficult due to high fat content and the presence of several interfering compounds such as phosphate, sulfate, sulfide, and ascorbic acid interfering in the process (Iammarino et al., 2017b).

12.6.2 Detection of nitrites and nitrates

Nitrites and nitrates are employed in meat curing for stabilization of red meat color, retarding the oxidative rancidity, imparting flavors, and inhibiting anaerobic *C. botulinum*. As mentioned earlier, the major food groups contributing to the dietary intake of nitrites are cured and dried meat products. The ADI for nitrites, expressed as sodium nitrite, is 0.1 mg/kg body weight (Wood et al., 2004). The two AOAC official methods for the determination of nitrates and nitrites in meat are the xylenol method and colorimetric method (AOAC, 2000a). The traditional xylenol method for the determination of nitrite relies on the Griess diazotization procedure, in which an azo dye is produced by coupling a diazonium salt with an aromatic amine or phenol. The diazo compound is usually formed with sulfanilic acid or sulfanilamide, and the coupling agent is *N*-1-naphthylethylene diamine (Wood et al., 2004).

Furthermore, methods including spectroscopic determination following enzymatic reduction (Hamano et al., 1990), IEC (British Standards Institute Staff, 2005; Radisavljevic et al., 1996), FIA (Ruiz-Capillas and Jiménez-Colmenero, 2009), differential pulse voltammetry (Brunov et al., 1998), and CE (Marshall and Trenery, 1996) have been developed for the detection and quantification of nitrites in processed meat products.

12.6.3 Detection of sorbic acid

Sorbic acids are used as preservatives in various meat products to retard the growth of yeasts and molds. The ADI for sorbic acid is 25 mg/kg body weight (Wood et al., 2004). Traditionally, from carbohydrate, fat, and protein-rich foods, sorbic acid is extracted with ether and successively partitioned into aqueous NaOH and CH_2Cl_2 (AOAC, 2000b). Acids are then converted to trimethylsilyl esters and determined by gas chromatography (González et al., 1999). There are several other separation methods published for the detection of sorbic acid in foodstuffs. These include HPLC (Saad et al., 2005; FSIS, 2004), spectrophotometric (Campos et al., 1991), micellar electrokinetic chromatography (Boyce, 1999), and CE (Öztekin, 2018).

12.6.4 Detection of nisin and organic acids

Liquid chromatography-tandem mass spectrometry (MS) is a rapid and multipurpose approach capable of detecting chemical additives in meat. It provides high specificity and separation efficiency, and is capable of detecting different classes of additives in meat (Molognoni et al., 2016a,b). The preservatives such as nisin, benzoic acid, citric acid, and lactic acid in raw, cooked, dry fermented, or cured meat products could be detected using the method with acceptable recovery rates. The lower analytical limits for nisin, benzoic acid, citric acid, and lactic acid in these meat products are 1.0, 15.0, 35.0, and 25.0 mg/kg, respectively (Molognoni et al., 2018).

12.6.5 Detection of color additives

HPLC with tandem MS has been found to be an effective method for simultaneous detection of eight synthetic coloring agents in meat such as Chrysoidin, Auramine O, Sudan (I–IV), Para Red, and Rhodamine B (Li et al., 2014). The method is capable of providing quantitative data even at the trace level of these dyes ranging between 0.03–0.75 and 0.1–2.0 µg/kg depending on matrices (Li et al., 2014). The HPLC coupled with a diode-array detector and tandem MS was also validated as a sensitive method with fewer variations for the simultaneous determination of 11 coloring agents in the meat containing foodstuffs in a cost-effective manner (Qi et al., 2015). Additionally, HPLC with diode array and tandem MS is useful in the detection of sulfonate dyes such as Ponceau 4RC, Sunset yellow, Allura red, Azophloxine, Ponceau xylidine, Erythrosine, and Orange II in meat with a limit of detection in the range of 0.02–21.83 ng/mL (Zou et al., 2013). Another multiresidue meat color analysis method is microwave-assisted extraction followed by solid-phase extraction, which is a rapid, sensitive, and high-throughput color analysis method. It provides a quantification limit of 0.48–7.19 µg/kg for azo food colorants in meat sausages (Sun et al., 2013).

12.6.6 Detection of synthetic phenolic antioxidants

Synthetic phenolic antioxidants are compounds added to meat products to prevent fat degradation and rancid oxidation. The most common additives in this group are BHA, BHT, and *tert*-butyl hydroquinone. A preconcentration technique called SBSE has been used widely for the detection and quantitation of these additives. The detection and quantitative determination of these antioxidants have been accomplished using LC coupled with detectors, such as ultraviolet light (Boyce, 1999; Gao et al., 2011; González et al., 1999).

12.7 Conclusion and future directions

Meat is a popular food commodity to consumers all over the world. The nutrient-rich meat products are highly perishable. A variety of additives and preservatives are used in meat to maintain or augment the quality, safety, wholesomeness, and consumer acceptance. In the United States the meat preservatives and additives are strictly monitored and regulated by federal agencies such as the FDA and USDA. Analytics have been developed to detect and monitor the residues of many additives and preservatives for ensuring consumer health and safety in compliance with regulatory standards. However, the current analytics are time-consuming and often require pretreatments to avoid background interference. Therefore more advancement in analytics is needed, most importantly, rapid and simultaneous detection of multiple ingredients in meat at a given time.

The recent trend in human food habits is the preference for natural ingredients in foods, especially in meat. Therefore the potential for future application of natural ingredients, especially plant-derived compounds, as additives and preservatives in the meat industry is enormous. However, in-depth studies are lacking in meat products regarding the fate of these compounds during extended storage periods or the influence of processing technologies on the stability of these compounds. There are no established standards regarding the threshold level of these compounds for human consumption (e.g., ADI). Proper analytics should also be developed to determine the residue level of these compounds in finished meat products.

References

AOAC, 2000a. AOAC Official Methods 935.48. Nitrates and Nitrites in Meat. Xylenol Method. AOAC Official Method of Analysis (Chapter 39).

AOAC. 2000b. AOAC Official Method 983.16. Benzoic Acid and Sorbic Acid in Food, Gas-Chromatographic Method. NMLK−AOAC Method. AOAC Official Method of Analysis, 47.3.05, p. 9.

Bahadoran, Z., et al., 2016. Nitrate and nitrite content of vegetables, fruits, grains, legumes, dairy products, meats and processed meats. J. Food Compos. Anal. 51, 93−105.

Balbus, J.M., et al., 2013. Early-life prevention of non-communicable diseases. Lancet 381 (9860), 3−4.

Bang, W., Hanson, D.J., Drake, M.A., 2008. Effect of salt and sodium nitrite on growth and enterotoxin production of *Staphylococcus aureus* during the production of air-dried fresh pork sausage. J. Food Prot. 71 (1), 191−195.

Barberis, S., et al., 2018. Natural food preservatives against microorganisms. Food Safety and Preservation. Academic Press, pp. 621−658.

Bastide, N.M., Pierre, F.H.F., Corpet, D.E., 2011. Heme iron from meat and risk of colorectal cancer: a meta-analysis and a review of the mechanisms involved. Cancer Prev. Res. canprevres-0113.

Benkerroum, N., Sandine, W.E., 1988. Inhibitory action of nisin against *Listeria monocytogenes*. J. Dairy Sci. 71 (12), 3237−3245.

Boland, J.S., 2003. Influence of chelators on the antimicrobial activity of lysozyme against *Escherichia coli* O157:H7. University of Tennessee. Available at: <http://trace.tennessee.edu/utk_gradthes/1905> (accessed 12.10.18.).

Boyce, M.C., 1999. Simultaneous determination of antioxidants, preservatives and sweeteners permitted as additives in food by mixed micellar electrokinetic chromatography. J. Chromatogr. A. 847 (1−2), 369−375.

British Standards Institute Staff, 2005. Foodstuffs. Determination of Nitrate and/or Nitrite Content. Ion-Exchange Chromatographic (Ic) Method for the Determination of Nitrate and Nitrite Content of Meat Products. B S I Standards.

Brul, S., Coote, P., 1999. Preservative agents in foods: mode of action and microbial resistance mechanisms. Int. J. Food Microbiol. 50 (1–2), 1–17.

Brunov, A., et al., 1998. Direct determination of nitrite in food samples by electrochemical biosensor. Chem. Papers 52 (3), 156–158.

Burt, S., 2004. Essential oils: their antibacterial properties and potential applications in foods—a review. Int. J. Food Microbiol. 94 (3), 223–253.

Burt, S.A., Reinders, R.D., 2003. Antibacterial activity of selected plant essential oils against *Escherichia coli* O157:H7. Lett. Appl. Microbiol. 36 (3), 162–167.

Busatta, C., et al., 2008. Application of *Origanum majorana* L. essential oil as an antimicrobial agent in sausage. Food Microbiol. 25 (1), 207–211.

Cacho, J.I., et al., 2015. Determination of synthetic phenolic antioxidants in soft drinks by stir-bar sorptive extraction coupled to gas chromatography-mass spectrometry. Food Addit. Contam.: Part A 32 (5), 665–673.

Campos, C., et al., 1991. Determination of sorbic acid in raw beef: an improved procedure. J. Food Sci. 56 (3), 863.

Cancela, A.L., et al., 2012. Phosphorus is associated with coronary artery disease in patients with preserved renal function. PLoS one 7 (5), e36883.

Cantwell, M., Elliott, C., 2017. Nitrates, nitrites and nitrosamines from processed meat intake and colorectal cancer risk. J. Clin. Nutr. Diet. 3 (4), 27.

Cegielska-Radziejewska, R., Lesnierowski, G., Kijowski, J., 2009. Antibacterial activity of hen egg white lysozyme modified by thermochemical technique. Eur. Food Res. Technol. 228 (5), 841–845.

Ceylan, E., Fung, D.Y.C., Sabah, J.R., 2004. Antimicrobial activity and synergistic effect of cinnamon with sodium benzoate or potassium sorbate in controlling *Escherichia coli* O157: H7 in apple juice. J. Food Sci. 69 (4), FMS102–FMS106.

Chan, K.W., et al., 2014. Cinnamon bark deodorised aqueous extract as potential natural antioxidant in meat emulsion system: a comparative study with synthetic and natural food antioxidants. J. Food Sci. Technol. 51 (11), 3269–3276.

Chen, M., Mustapha, A., 2008. Natural antimicrobial compounds as meat preservatives. Food 2 (2), 102–114.

Cleveland, J., et al., 2001. Bacteriocins: safe, natural antimicrobials for food preservation. Int. J. Food Microbiol. 71 (1), 1–20.

Clinton, B., Gore, A., 1996. Reinventing Food Regulations: National Performance Review. *Diane Publishing*.

CSPI, 2017. "Clean label" - policies spurring restaurants & supermarkets to remove most of the worst additives. Available at: <https://cspinet.org/news/%E2%80%9Cclean-label%E2%80%9D-policies-spurring-restaurants-supermarkets-remove-most-worst-additives-20170112> (accessed 16.10.18.).

Darmadji, P., Izumimoto, M., 1994. Effect of chitosan in meat preservation. Meat Sci. 38 (2), 243–254.

da Silveira, S.M., et al., 2014. Chemical composition and antibacterial activity of *Laurus nobilis* essential oil towards foodborne pathogens and its application in fresh Tuscan sausage stored at 7 °C. LWT - Food Sci. Technol. 59 (1), 86–93.

Davidson, P.M., Taylor, T.M., Schmidt, S.E., 2013. Chemical preservatives and natural antimicrobial compounds. Food Microbiology. *American Society of Microbiology*, pp. 765–801.

De Mey, E., et al., 2017. Volatile N-nitrosamines in meat products: potential precursors, influence of processing, and mitigation strategies. Crit. Rev. Food Sci. Nutr. 57 (13), 2909–2923.

Devadason, I.P., Anjaneyulu, A.S.R., Babji, Y., 2010. Effect of different binders on the physico-chemical, textural, histological, and sensory qualities of retort pouched buffalo meat nuggets. J. Food Sci. 75 (1), S31–S35.

De Vuyst, L., Vandamme, E.J., 1994. Nisin, a lantibiotic produced by *Lactococcus lactis* subsp. *lactis*: properties, biosynthesis, fermentation and applications. Bacteriocins of Lactic Acid Bacteria. *Springer, Boston*, pp. 151–221.

Edberg, U., 1993. Enzymatic determination of sulfite in foods: NMKL interlaboratory study. J. AOAC Int. 76 (1), 53–58.

EFSA, 2013. Assessment of one published review on health risks associated with phosphate additives in food. EFSA J. 11 (11), 3444.

Eklund, T., 1983. The antimicrobial effect of dissociated and undissociated sorbic acid at different pH levels. J. Appl. Bacteriol. 54 (3), 383–389.

Eskin, M., Robinson, D., Clydesdale, F., Paredes-Lopez, O., et al., 2001. Food Shelf Life Stability. CRC Press, Boca Raton.

FAO, 2018. Definitions for the purposes of the Codex Alimentarius. Available at: <http://www.fao.org/docrep/005/y2200e/y2200e07.htm> (accessed 30.10.18.).

Faustman, C., Cassens, R.G., 1990. The biochemical basis for discoloration in fresh meat: a review. J. Muscle Foods 1 (3), 217–243.

FDA, 1989. Compliance Policy Guides - CPG Sec. 555.100 Alcohol; Use of synthetic alcohol in foods. Available at: <https://www.fda.gov/iceci/compliancemanuals/compliancepolicyguidancemanual/ucm074550.htm> (accessed 16.10.18.).

FDA, 2000. Redbook 2000: Guidance for industry and other stakeholders- toxicological principles for the safety assessment of food ingredients. Available at: <https://www.fda.gov/downloads/food/guidanceregulation/ucm222779.pdf> (accessed 16.10.18.).

FDA, 2018a. Food additives and ingredients: determining the regulatory status of a food ingredient. Available at: <https://www.fda.gov/food/ingredientspackaginglabeling/foodadditivesingredients/ucm228269.htm> (accessed 16.10.18.).

FDA, 2018b. Milestones in U.S. food and drug law history. Available at: <https://www.fda.gov/aboutfda/history/forgshistory/evolvingpowers/ucm2007256.htm> (accessed 26.10.18.).

FDA, 2018c. CFR - Code of Federal Regulations Title 21: Part 182, Sec. 182.10. Spices and other natural seasonings and flavorings. Available at: <https://www.accessdata.fda.gov/scripts/cdrh/cfdocs/cfcfr/CFRSearch.cfm?fr = 182.10> (accessed 16.10.18.).

FDA, 2018d. Code of Federal Regulations: Title 21-Food and drugs. Available at: <https://www.accessdata.fda.gov/scripts/cdrh/cfdocs/cfcfr/CFRSearch.cfm?fr = 182.3739> (accessed 13.10.18.).

FDA, 2018e. Food ingredients and packaging inventories: GRAS notices-GRAS No. 170. Available at: <https://www.accessdata.fda.gov/scripts/fdcc/?set = GRASNotices&id = 170&sort = FDA_s_Letter&order = ASC&startrow = 1&type = basic&search = 443> (accessed 15.10.18.).

Featherstone, S., 2015. Ingredients used in the preparation of canned foods. A Complete Course in Canning and Related Processes. *Woodhead Publishing*, pp. 147–211.

Foley, R.N., et al., 2009. Serum phosphorus levels associate with coronary atherosclerosis in young adults. J. Am. Soc. Nephrol. 20 (2), 397–404.

Fontes Saraiva, M.A., et al., 2014. Purification and characterization of a bacteriocin produced by *Lactococcus lactis* subsp. *lactis* PD6. 9. J. Microbiol. Antimicrob. 6 (5), 79–87.

FSIS, 1995. Processing inspectors' calculations handbook. Available at: <http://food-safety.guru/wp-content/uploads/2015/05/USDA-edited-cure-calculations.pdf> (accessed 30.10.18.).

FSIS, 2004. Determination of benzoic acid, sorbic acid, and methyl, ethyl, propyl, and butyl parabens by HPLC. Available at: <https://www.fsis.usda.gov/wps/wcm/connect/80689fb4-dbac-4f22-9375-96d5e2b811f4/CLG_BSP_01.pdf?MOD = AJPERES> (accessed 16.10.18.).

FSIS, 2005. Food standards and labeling policy book. Available at: <https://www.fsis.usda.gov/wps/wcm/connect/7c48be3e-e516-4ccf-a2d5-b95a128f04ae/Labeling-Policy-Book.pdf?MOD = AJPERES> (accessed 30.10.18.).

FSIS, 2010. Compliance guideline for controlling *Salmonella* and *Campylobacter* in poultry, third ed. Available at: <https://www.complianceonline.com/articlefiles/Compliance_Guide_Controling_Salmonella_Campylobacter_Poultry_0510.pdf> (accessed 16.10.18.).

FSIS, 2015. Additives in meat and poultry products. Available at: <https://www.fsis.usda.gov/wps/portal/fsis/topics/food-safety-education/get-answers/food-safety-fact-sheets/food-labeling/additives-in-meat-and-poultry-products/additives-in-meat-and-poultry-products> (accessed 16.10.18.).

FSIS, 2018. Safe and suitable ingredients used in the production of meat, poultry and egg products. Available at: <http://www.fsis.usda.gov/wps/portal/fsis/topics/regulatory-compliance/labeling> (accessed 05.10.18.).

Gadekar, Y., et al., 2006. Spices and their role in meat products: a review. Beverage Food World 33 (7), 57–60.

Galluzzo, S.J., Regenstein, J.M., 1978. Role of chicken breast muscle proteins in meat emulsion formation: myosin, actin and synthetic actomyosin. J. Food Sci. 43 (6), 1761–1765.

Gao, Y., Gu, Y., Wei, Y., 2011. Determination of polymer additives—antioxidants and ultraviolet (UV) absorbers by high-performance liquid chromatography coupled with UV photodiode array detection in food simulants. J. Agric. Food Chem. 59 (24), 12982–12989.

Glass, K.A., et al., 2002. Inhibition of *Listeria monocytogenes* by sodium diacetate and sodium lactate on wieners and cooked bratwurst. J. Food Prot. 65 (1), 116–123.

González, M., Gallego, M., Valcárcel, M., 1999. Gas chromatographic flow method for the preconcentration and simultaneous determination of antioxidant and preservative additives in fatty foods. J. Chromatogr. A 848 (1–2), 529–536.

Gøtterup, J., et al., 2008. Colour formation in fermented sausages by meat-associated Staphylococci with different nitrite- and nitrate-reductase activities. Meat Sci. 78 (4), 492–501.

Gould, G.W., 2000. Preservation: past, present and future. Br. Med. Bull. 56 (1), 84–96.

Govaris, A., et al., 2010. The antimicrobial effect of oregano essential oil, nisin and their combination against *Salmonella* Enteritidis in minced sheep meat during refrigerated storage. Int. J. Food Microbiol. 137 (2–3), 175–180.

GPO, 1998. Federal Register/Vol. 63, No. 49. Available at: <https://www.gpo.gov/fdsys/pkg/FR-1998-03-13/pdf/FR-1998-03-13.pdf> (accessed 30.10.18.).

GPO, 2012. Code of Federal Regulations: Title 9 - animals and animal products: Chapter III - Food safety and inspection service. Department of agriculture (parts 300 - 592). Available at: <https://www.gpo.gov/fdsys/pkg/CFR-2012-title9-vol2/pdf/CFR-2012-title9-vol2-chapIII.pdf> (accessed 26.10.18.).

Guo, M., et al., 2014. Antimicrobial films and coatings for inactivation of *Listeria innocua* on ready-to-eat deli turkey meat. Food Control. 40, 64–70.

Hamano, T., et al., 1990. Enzymic method for the spectrophotometric determination of aspartame in beverages. Analyst 115 (4), 435.

Holak, W., Patel, B., 1987. Differential pulse polarographic determination of sulfites in foods: collaborative study. J. Assoc. Off. Anal. Chem. 70 (3), 572–578.

Honikel, K.-O., 2008. The use and control of nitrate and nitrite for the processing of meat products. Meat Sci. 78 (1–2), 68–76.

Hughey, V.L., Johnson, E.A., 1987. Antimicrobial activity of lysozyme against bacteria involved in food spoilage and food-borne disease. Appl. Environ. Microbiol. 53 (9), 2165–2170.

Hugo, C.J., Hugo, A., 2015. Current trends in natural preservatives for fresh sausage products. Trends Food Sci. Technol. 45 (1), 12–23.

Iammarino, M., Ientile, A.R., Di Taranto, A., 2017a. Sulphur dioxide in meat products: 3-year control results of an accredited Italian laboratory. Food Addit. Contam.: Part B 10 (2), 99–104.

Iammarino, M., Marino, R., Albenzio, M., 2017b. How meaty? Detection and quantification of adulterants, foreign proteins and food additives in meat products. Int. J. Food Sci. Technol. 52 (4), 851–863.

Ibrahim, S.A., Yang, H., Seo, C.W., 2008. Antimicrobial activity of lactic acid and copper on growth of *Salmonella* and *Escherichia coli* O157:H7 in laboratory medium and carrot juice. Food Chem. 109 (1), 137–143.

Juneja, V.K., et al., 2006. Chitosan protects cooked ground beef and turkey against *Clostridium perfringens* spores during chilling. J. Food Sci. 71 (6), M236–M240.

Kanatt, S.R., Chander, R., Sharma, A., 2008. Chitosan and mint mixture: a new preservative for meat and meat products. Food Chem. 107 (2), 845–852.

Kassem, A., et al., 2017. Evaluation of chemical immersion treatments to reduce microbial populations in fresh beef. Int. J. Food Microbiol. 261, 19–24.

Kendrick, A., 2012. Natural food and beverage colourings. Natural Food Additives, Ingredients and Flavourings. *Woodhead Publishing Series in Food Science, Technology and Nutrition*, pp. 25–40.

Keto-Timonen, R., et al., 2012. Inhibition of toxigenesis of group II (nonproteolytic) *Clostridium botulinum* Type B in meat products by using a reduced level of nitrite. J. Food Prot. 75 (7), 1346–1349.

Khanjari, A., Karabagias, I.K., Kontominas, M.G., 2013. Combined effect of N, O-carboxymethyl chitosan and oregano essential oil to extend shelf life and control *Listeria monocytogenes* in raw chicken meat fillets. LWT - Food Sci. Technol. 53 (1), 94–99.

Kim, H.J., 1990. Determination of sulfite in foods and beverages by ion exclusion chromatography with electrochemical detection: collaborative study. J. Assoc. Off. Anal. Chem. 73 (2), 216–222.

Kollanoor Johny, A., et al., 2017. Gene expression response of *Salmonella enterica* serotype Enteritidis Phage Type 8 to subinhibitory concentrations of the plant-derived compounds *trans*-cinnamaldehyde and eugenol. Front. Microbiol. 8, 1828.

Krebs, H.A., et al., 1983. Studies on the mechanism of the antifungal action of benzoate. Biochem. J. 214 (3), 657–663.

Kumar, Y., et al., 2015. Recent trends in the use of natural antioxidants for meat and meat products. Compr. Rev. Food Sci. Food Safety 14 (6), 796–812.

Lamas, A., et al., 2016. An evaluation of alternatives to nitrites and sulfites to inhibit the growth of *Salmonella enterica* and *Listeria monocytogenes* in meat products. Foods 5 (4).

Larsson, S.C., Wolk, A., 2012. Red and processed meat consumption and risk of pancreatic cancer: meta-analysis of prospective studies. Br. J. Cancer 106 (3), 603–607.

Le Lay, C., et al., 2016. Nisin is an effective inhibitor of *Clostridium difficile* vegetative cells and spore germination. J. Med. Microbiol. 65 (2), 169–175.

Lee, N.-K., et al., 2015. Antimicrobial effect of nisin against *Bacillus cereus* in beef jerky during storage. Korean J. Food Sci. Anim. Resour. 35 (2), 272–276.

Leroy, F., Verluyten, J., De Vuyst, L., 2006. Functional meat starter cultures for improved sausage fermentation. Int. J. Food Microbiol. 106 (3), 270–285.

Li, J., et al., 2014. Determination of synthetic dyes in bean and meat products by liquid chromatography with tandem mass spectrometry. J. Sep. Sci. 37 (17), 2439–2445.

Li, Q., Montalban-Lopez, M., Kuipers, O.P., 2018. Increasing the antimicrobial activity of nisin-based lantibiotics against Gram-negative pathogens. Appl. Environ. Microbiol. 84 (12), AEM.00052-18.

Liu, W., Hansen, J.N., 1990. Some chemical and physical properties of nisin, a small-protein antibiotic produced by *Lactococcus lactis*. Appl. Environ. Microbiol. 56 (8), 2551–2558.

López-Malo, A., et al., 2007. Aspergillus flavus growth response to cinnamon extract and sodium benzoate mixtures. Food Control. 18 (11), 1358–1362.

Lu, H.J., et al., 2011. Antimicrobial effects of weak acids on the survival of *Escherichia coli* O157: H7 under anaerobic conditions. J. Food Prot. 74 (6), 893–898.

Lund, M.N., et al., 2011. Protein oxidation in muscle foods: a review. Mol. Nutr. Food Res. 55 (1), 83–95.

Lv, F., et al., 2011. *In vitro* antimicrobial effects and mechanism of action of selected plant essential oil combinations against four food-related microorganisms. Food Res. Int. 44 (9), 3057–3064.

Maffini, M.V., Neltner, T.G., Vogel, S., 2017. We are what we eat: regulatory gaps in the United States that put our health at risk. PLoS Biol. 15 (12), e2003578.

Magnuson, B., et al., 2013. Review of the regulation and safety assessment of food substances in various countries and jurisdictions. Food Addit. Contam.: Part A 30 (7), 1147–1220.

Mani-López, E., García, H.S., López-Malo, A., 2012. Organic acids as antimicrobials to control *Salmonella* in meat and poultry products. Food Res. Int. 45 (2), 713–721.

Marshall, P.A., Trenerry, V.C., 1996. The determination of nitrite and nitrate in foods by capillary ion electrophoresis. Food Chem. 57 (2), 339–345.

Martins, F.C.O., Sentanin, M.A., de Souza, D., 2018. Analytical methods in food additives determination: compounds with functional applications. Food Chem. 272, 732–750.

Masschalck, B., Michiels, C.W., 2003. Antimicrobial properties of lysozyme in relation to foodborne vegetative bacteria. Crit. Rev. Microbiol. 29 (3), 191–214.

Mastromatteo, M., et al., 2011. Shelf life of reduced pork back-fat content sausages as affected by antimicrobial compounds and modified atmosphere packaging. Int. J. Food Microbiol. 150 (1), 1–7.

Møller, J.K.S., Skibsted, L.H., 2002. Nitric oxide and myoglobins. Chem. Rev. 102 (4), 1167–1178.

Møller, J.K.S., Skibsted, L.H., 2006. Myoglobins: the link between discoloration and lipid oxidation in muscle and meat. Quím. Nova 29 (6), 1270–1278.

Molognoni, L., et al., 2016a. A simple and fast method for the inspection of preservatives in cheeses and cream by liquid chromatography-electrospray tandem mass spectrometry. Talanta 147, 370–382.

Molognoni, L., et al., 2016b. Development of a LC–MS/MS method for the simultaneous determination of sorbic acid, natamycin and tylosin in Dulce de leche. Food Chem. 211, 748–756.

Molognoni, L., et al., 2018. A multi-purpose tool for food inspection: simultaneous determination of various classes of preservatives and biogenic amines in meat and fish products by LC-MS. Talanta 178, 1053–1066.

Mukhopadhyay, S., et al., 2017. Principles of food preservation. Microbial Control and Food Preservation. *Springer, New York*, pp. 17–39.

Mustapha, A., Lee, J.H., 2017. Food preservation and safety. Microbial Control and Food Preservation. *Springer, New York*, pp. 17–39.

Nair, D.V.T., Kollanoor Johny, A., 2017. Food grade pimenta leaf essential oil reduces the attachment of *Salmonella enterica* Heidelberg (2011 ground turkey outbreak isolate) on to turkey skin, Front. Microbiol., 8. p. 2328.

Nair, D.V.T., Nannapaneni, R., Kiess, A., Mahmoud, B., et al., 2014a. Antimicrobial efficacy of lauric arginate against *Campylobacter jejuni* and spoilage organisms on chicken breast fillets. Poult. Sci. 93 (10), 2636–2640.

Nair, D.V.T., Nannapaneni, R., Kiess, A., Schilling, W., et al., 2014b. Reduction of *Salmonella* on turkey breast cutlets by plant-derived compounds. Foodborne Pathog. Dis. 11 (12), 981–987.

Nair, D.V.T., et al., 2015. The combined efficacy of carvacrol and modified atmosphere packaging on the survival of *Salmonella*, *Campylobacter jejuni* and lactic acid bacteria on turkey breast cutlets. Food Microbiol. 49, 134–141.

Nash, N.H., Brickman, L.M., 1972. Food emulsifiers—science and art. J. Am. Oil Chem. Soc. 49 (8), 457–461.

Nowak, A., et al., 2016. Polyphenolic extracts of cherry (*Prunus cerasus* L.) and blackcurrant (*Ribes nigrum* L.) leaves as natural preservatives in meat products. Food Microbiol. 59, 142–149.

NRC, 1998. National Research Council. Ensuring Safe Food: From Production to Consumption. *National Academies*.

NTP, 1991. National Toxicology Program. 6th Annual Report on Carcinogens. *DIANE Publishing Company*.

NTP, 2016. National Toxicology Program. Butylated hydroxyanisole. 14th Report on carcinogens, pp. 8–9. Available at: <https://ntp.niehs.nih.gov/pubhealth/roc/index-1.html> (accessed 16.10.18.).

Olaimat, A.N., et al., 2018. The use of malic and acetic acids in washing solution to control *Salmonella* spp. on chicken breast. J. Food Sci. 83 (8), 2197–2203.

Öztekin, N., 2018. Simultaneous determination of benzoic acid and sorbic acid in food products by capillary electrophoresis. Food Health 4 (3), 176–182.

Pateiro, M., et al., 2018. Essential oils as natural additives to prevent oxidation reactions in meat and meat products: a review. Food Res. Int. 113, 156–166.

Petruzzi, L., Corbo, M.R., Sinigaglia, M., Bevilacqua, A., 2017. Microbial spoilage of foods: fundamentals. The Microbiological Quality of Food. *Woodhead Publishing*, pp. 1–21.

Phillips, G.O., Williams, P.A., 2009. Handbook of Hydrocolloids. *Elsevier*.

Piper, C., et al., 2009. A comparison of the activities of lacticin 3147 and nisin against drug-resistant *Staphylococcus aureus* and *Enterococcus* species. J. Antimicrob. Chemother. 64 (3), 546–551.

Porto-Fett, A.C.S., et al., 2010. Control of *Listeria monocytogenes* on commercially-produced frankfurters prepared with and without potassium lactate and sodium diacetate and surface treated with lauric arginate using the sprayed lethality in container (SLIC®) delivery method. Meat Sci. 85 (2), 312–318.

Pressman, P., et al., 2017. Food additive safety: a review of toxicologic and regulatory issues. Toxicol. Res. Appl. Available from: https://doi.org/10.1177/2397847317723572.

Prudêncio, C.V., Dos Santos, M.T., Vanetti, M.C.D., 2015. Strategies for the use of bacteriocins in Gram-negative bacteria: relevance in food microbiology. J. Food Sci. Technol. 52 (9), 5408–5417.

Qi, L., et al., 2004. Preparation and antibacterial activity of chitosan nanoparticles. Carbohydr. Res. 339 (16), 2693–2700.

Qi, P., et al., 2015. Fast and simultaneous determination of eleven synthetic color additives in flour and meat products by liquid chromatography coupled with diode-array detector and tandem mass spectrometry. Food Chem. 181, 101–110.

Rabea, E.I., et al., 2003. Chitosan as antimicrobial agent: applications and mode of action. Biomacromolecules 4 (6), 1457–1465.

Radisavljevic, Z., et al., 1996. Determination of intracellular and extracellular nitrite and nitrate by anion chromatography. J. Liquid Chromatogr. Relat. Technol. 19 (7), 1061–1079.

Robbins, K.S., et al., 2015. Development of a liquid chromatography-tandem mass spectrometry method for the determination of sulfite in food. J. Agric. Food Chem. 63 (21), 5126–5132.

Rozum, J.J., 2009. Smoke flavor. Ingredients in Meat Products. *Springer*, New York, pp. 211–226.

Ruiz-Capillas, C., Jiménez-Colmenero, F., 2009. Application of flow injection analysis for determining sulphites in food and beverages: a review. Food Chem. 112 (2), 487–493.

Saad, B., et al., 2005. Simultaneous determination of preservatives (benzoic acid, sorbic acid, methylparaben and propylparaben) in foodstuffs using high-performance liquid chromatography. J. Chromatogr. A 1073, 393–397.

Sallam, K.I., Ishioroshi, M., Samejima, K., 2004. Antioxidant and antimicrobial effects of garlic in chicken sausage. LWT - Food Sci. Technol. 37 (8), 849–855.

Sánchez, V.E., Bartholomai, G.B., Pilosof, A.M.R., 1995. Rheological properties of food gums as related to their water binding capacity and to soy protein interaction. LWT - Food Sci. Technol. 28 (4), 380–385.

Sangcharoen, N., Klaypradit, W., Wilaipun, P., 2017. Antimicrobial activity optimization of nisin, ascorbic acid and ethylenediamine tetraacetic acid disodium salt (EDTA) against *Salmonella* Enteritidis ATCC 13076 using response surface methodology. Agric. Nat. Resour. 51 (5), 355–364.

Santana, R.C., et al., 2011. Emulsifying properties of collagen fibers: effect of pH, protein concentration and homogenization pressure. Food Hydrocolloids 25 (4), 604–612.

Santhi, D., Kalaikannan, A., Sureshkumar, S., 2017. Factors influencing meat emulsion properties and product texture: a review. Crit. Rev. Food Sci. Nutr. 57 (10), 2021–2027.

Santiesteban-López, A., Palou, E., López-Malo, A., 2007. Susceptibility of food-borne bacteria to binary combinations of antimicrobials at selected a(w) and pH. J. Appl. Microbiol. 102 (2), 486–497.

Sarıçoban, C., Yilmaz, M.T., 2014. Effect of thyme/cumin essential oils and butylated hydroxyl anisole/butylated hydroxyl toluene on physicochemical properties and oxidative/microbial stability of chicken patties. Poult. Sci. 93 (2), 456–463.

Sasse, A., Colindres, P., Brewer, M.S., 2009. Effect of natural and synthetic antioxidants on the oxidative stability of cooked, frozen pork patties. J. Food Sci. 74 (1), S30–S35.

Sebranek, J.G., 2015. An overview of functional non-meat ingredients in meat processing: the current toolbox. In: Animal Science Conference Proceedings and Presentations. 4. Available at: <http://lib.dr.iastate.edu/ans_conf/4> (accessed 14.10.18.).

Segundo, M.A., 2001. A gas diffusion sequential injection system for the determination of sulphur dioxide in wines. Anal. Chim. Acta 427, 279–286.

Shahidi, F., 2000. Antioxidants in food and food antioxidants. Food/nahrung 44 (3), 158–163.

Shahidi, F., Zhong, Y., 2005. Antioxidants: regulatory status. Bailey's Ind. Oil Fat Prod. 1, 491–512.

Shahmohammadi, M., Javadi, M., Nassiri-Asl, M., 2016. An overview on the effects of sodium benzoate as a preservative in food products. Biotechnol. Health Sci. 3 (3), 7–11.

Sindelar, J.J., Milkowski, A.L., Sindelar, J.J., 2011. Sodium nitrite in processed meat and poultry meats: A review of curing and examining the risk/benefit of its use. Am. Meat Sci. Assoc. White Paper Series 3, 1–14.

Sofos, J.N., Busta, F.F., 1981. Antimicrobial activity of sorbate. J. Food Prot. 44 (8), 614–622.

Stanojevic, D., et al., 2009. Antimicrobial effects of sodium benzoate, sodium nitrite and potassium sorbate and their synergistic action *in vitro*. Bulgarian J. Agric. Sci. 15 (4), 307–311.

Stanojević-Nikolić, S., et al., 2016. Antimicrobial activity of lactic acid against pathogen and spoilage microorganisms. J. Food Process. Preserv. 40 (5), 990–998.

Sullivan, J.J., et al., 1990. Determination of total sulfite in shrimp, potatoes, dried pineapple, and white wine by flow injection analysis: collaborative study. J. Assoc. Off. Anal. Chem. 73 (1), 35–42.

Sun, D.-h, James, K., Waters, J.K., Mawhinney, T.P., 1997. Microwave digestion with $HNO_3 - H_2O_2 - HF$ for the determination of total aluminum in seafood and meat by inductively coupled plasma atomic emission spectrometry. J. Agric. Food Chem. 45 (6), 2115–2119.

Sun, H., et al., 2013. Development of multiresidue analysis for 21 synthetic colorants in meat by microwave-assisted extraction–solid-phase extraction–reversed-phase ultrahigh performance liquid chromatography. Food Anal. Methods 6 (5), 1291–1299.

Surendran Nair, M., et al., 2016. Potentiating the heat inactivation of *Escherichia coli* O157:H7 in ground beef patties by natural antimicrobials. Front. Microbiol. 7, 15.

Surendran Nair, M., et al., 2017. Antimicrobial food additives and disinfectants: mode of action and microbial resistance mechanisms. Foodborne Pathogens and Antibiotic Resistance. *John Wiley & Sons, Inc*, pp. 275–301. Available at: <http://doi.wiley.com/10.1002/9781119139188.ch12> (accessed 23.05.17.).

Tarlo, S.M., Sussman, G.L., 1993. Asthma and anaphylactoid reactions to food additives. Can. Fam. Physician 39, 1119.

Tarté, R., 2009. Ingredients in Meat Products: Properties, Functionality and Applications. Springer Science & Business Media. Available at: <https://link.springer.com/content/pdf/10.1007/978-0-387-71327-4.pdf> (accessed 07.10.18.).

Taylor, S.L., Higley, N.A., Bush, R.K., 1986. Sulfites in foods: uses, analytical methods, residues, fate, exposure assessment, metabolism, toxicity, and hypersensitivity. Adv. Food Res. 30, 1–76.

Teixeira dos Santos, C.A., et al., 2016. Application of Fourier-transform infrared spectroscopy for the determination of chloride and sulfate in wines. LWT - Food Sci. Technol. 67, 181–186.

Theron, M.M., Lues, J.F.R., 2007. Organic acids and meat preservation: a review. Food Rev. Int. 23 (2), 141–158.

Tompkin, R.B., Christiansen, L.N., Shaparis, A.B., 1978. Enhancing nitrite inhibition of *Clostridium botulinum* with isoascorbate in perishable canned cured meat. Appl. Environ. Microbiol. 35 (1), 59–61.

Trenerry, V.C., 1996. The determination of the sulphite content of some foods and beverages by capillary electrophoresis. Food Chem. 55 (3), 299–303.

Trius, A., Sebranek, J.G., Lanier, T., 1996. Carrageenans and their use in meat products. Crit. Rev. Food Sci. Nutr. 36 (1–2), 69–85.

Vally, H., Misso, N.L.A., 2012. Adverse reactions to the sulphite additives. Gastroenterol. Hepatol. Bed Bench 5 (1), 16.

Van Nieuwenhuyzen, W., 1981. The industrial uses of special lecithins: a review. J. Am. Oil Chem. Soc. 58 (10), 886–888.

Venkitanarayanan, K., et al., 2013. Use of plant-derived antimicrobials for improving the safety of poultry products. Poult. Sci. 92 (2), 493–501.

Wallingford, L., Labuza, T.P., 1983. Evaluation of the water binding properties of food hydrocolloids by physical/chemical methods and in a low fat meat emulsion. J. Food Sci. 48 (1), 1–5.

Wang, S., Adhikari, K., 2017. Clean labeling and the 'real food' movement. University of Georgia extension. Available at: <https://secure.caes.uga.edu/extension/publications/files/pdf/B%201476_2.PDF> (accessed 16.10.18.).

Warner, C.R., et al., 1990. Determination of free and reversibly bound sulphite in foods by reverse-phase, ion-pairing high-performance liquid chromatography. Food Addit. Contam. 7 (5), 575–581.

WHO, 2018. Food additives-Facts sheets. Available at: <http://www.who.int/news-room/fact-sheets/detail/food-additives> (accessed 26.10.18.).

Wiedemann, I., et al., 2001. Specific binding of nisin to the peptidoglycan precursor lipid II combines pore formation and inhibition of cell wall biosynthesis for potent antibiotic activity. J. Biol. Chem. 276 (3), 1772–1779.

Windisch, W., Kroismayr, A., 2007. The effect of phytobiotics on performance and gut function in monogastrics. Biomin World Nutrition Forum. Available at: <https://en.engormix.com/feed-machinery/articles/phytobiotics-on-performance-gut-function-in-monogastrics-t33528.htm> (accessed 30.10.18.).

Winkler, D., Harris, K.B., 2009. Reference document: Antimicrobial interventions for beef, Texas A&M University, College Station, TX. Available at: <http://www.haccpalliance.org/sub/Antimicrobial Interventions for Beef.pdf> (accessed 08.10.18.).

Wood, R., et al., 2004. Analytical Methods for Food Additives. CRC Press.

Yang, H., et al., 2007. Enhancing antimicrobial activity of lysozyme against *Listeria monocytogenes* using immunonanoparticles. J. Food Prot. 70 (8), 1844–1849.

Yang, S.C., et al., 2014. Antibacterial activities of bacteriocins: application in foods and pharmaceuticals. Front. Microbiol. 5, 241.

Zou, T., et al., 2013. Determination of seven synthetic dyes in animal feeds and meat by high performance liquid chromatography with diode array and tandem mass detectors. Food Chem. 138 (2-3), 1742–1748.

SECTION 8

Detection and prevention of lipid oxidation products

CHAPTER 13

Analysis of lipids and lipid oxidation products

Trinidad Pérez-Palacios and Mario Estévez
IProCar Research Institute, University of Extremadura, Caceres, Spain

Chapter Outline
13.1 Introduction 217
13.2 Advances in the analysis of meat lipids 218
    13.2.1 Optimization of classic methods for total lipid quantification 218
    13.2.2 Optimization of classic methods for lipid composition 220
    13.2.3 Advanced methodologies for lipid analysis: nondestructive methods 225
13.3 Advances in the detection of lipid oxidation products 226
    13.3.1 Optimization of classic methods for lipid oxidation assessment 226
    13.3.2 Advanced methodologies for lipid oxidation assessment 230
13.4 Future perspectives: lipidomics and oxidomics 232
References 232
Further reading 239

13.1 Introduction

Pushing back the frontiers of knowledge in meat science requires a profound examination of the chemistry fundamentals of meat systems, including the precise characterization of their components and their reactivity during meat aging, storage, and processing. Lipids are key components in meat and meat products owing to the impact of their quantity and composition on assorted quality traits such as sensory properties (appearance, juiciness, etc.) and nutritional value (Gandemer, 2002). The fatty acid (FA) composition of muscle lipids and its influence on meat quality remains one of the most attractive and influential topics in the meat science field (Wood et al., 2008). Furthermore, lipids are involved in a number of reactions with the oxidation of FAs being the most salient of all, given the impact on the flavor of processed meats and the potential toxicity of particular lipid oxidation products (Gandemer, 2002; Esterbauer et al., 1993).

In recent years the analysis of muscle lipids has evolved through the optimization of classic methods for quantification and characterization of lipid and lipid oxidation products. In this regard, optimized fat extraction solvents and procedures have been proposed to guarantee accuracy in the quantification of total lipids, triglycerides, phospholipids (PLs), and their corresponding subclasses (Pérez-Palacios et al., 2008). In the same line, protocols from routine lipid oxidation methods such as the thiobarbituric acid-reactive substances (TBARS) (Turner et al., 1954) or the detection of lipid-derived volatiles (Kerler and Grosch, 1996) have evolved into faster and more precise procedures (Ganhão et al., 2011; Grotta et al., 2017). Besides the optimization of classic methodologies, the application of new and more sophisticated methodological approaches has enabled substantial advances in this field. Nondestructive methods and predictive tools have been proposed as fast and reliable procedures to assess the quantity and composition of meat lipids and their influence on meat quality. Some of the technologies applied include X-rays (de Prados et al., 2015), ultrasound (US; Ludwiczak et al., 2017), hyperspectral imaging (Kucha et al., 2018), magnetic resonance imaging (MRI) (Pérez-Palacios et al., 2017), computed tomography (Santos-Garcés et al., 2010), and near-infrared (NIR) spectrometry (Kamruzzaman et al., 2012), among others. Furthermore, the arrival of versatile mass spectrometric equipment based on exact mass technology has enabled the onset of accurate and large-scale studies of muscle lipids (*lipidomics*) including their involvement in muscle biology and redox reactions (Trivedi et al., 2016; Mi et al., 2018). This chapter reviews the most recent advances in the analysis of lipid and lipid oxidation products in meat and meat products.

13.2 Advances in the analysis of meat lipids

13.2.1 Optimization of classic methods for total lipid quantification

There is a great diversity of methodologies for the extraction of total lipids in terms of the combination and proportions of solvents, which must show adequate polarity to extract both polar and nonpolar lipids. The most frequently used methods for total lipid extraction in meat and meat products are the Soxhlet method with petroleum ether as solvent, which is the official AOAC-recommended method (Association of Official Analytical Chemist, 2000), and those described by Folch et al. (1957) and Bligh and Dyer (1959), which are based on the use of a mixture of chloroform and methanol. The two latter methods mainly differ in the proportion of chloroform:methanol and in the solvent:sample ratio. Three parts chloroform:methanol (1:2, v/v) to one part sample in the Bligh and Dyer (1959) method, and 20 parts chloroform:methanol (2:1, v/v) to one part sample in the Folch et al. (1957) method. Pérez-Palacios et al. (2008) have evaluated the efficiency of different methods for the quantification of total lipid content in meat and meat products differing in fat content and physicochemical features. Results showed that the Soxhlet with previous acid hydrolysis and the Folch et al. (1957) methods are suitable for meat and meat products with low,

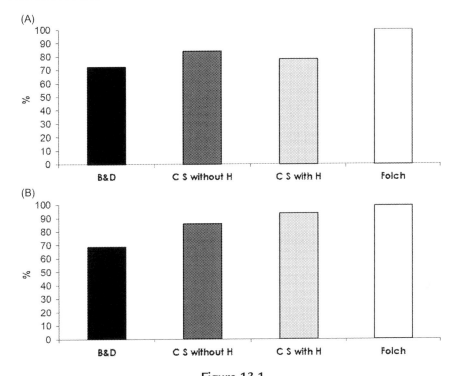

Figure 13.1
Lipid extracted by Bligh and Dyer (B&D), continuous Soxhlet (C S) with and without acid hydrolysis (H) and Folch methods in cooked breast turkey (A) and salami (B)*.
*Results are expressed as total lipid extracted in relation to the reference value provided by the label of the product (total lipids extracted (g/100 g)/total lipid labeled (g/100 g) × 100).

intermediate, high, and very high lipid content, whereas the Bligh and Dyer (1959) method underestimates the total lipid content in most meat and meat products. As an example, Fig. 13.1 exposes the total lipids extracted by these methods in cooked breast turkey (low fat content) and salami (very high fat content), expressed in relation to the reference value provided by the label of the product (total lipids extracted (g/100 g)/total lipid labeled (g/100 g) × 100). However, these authors have advised not to use the Soxhlet with acid hydrolysis method when performing further analysis of the lipid extracted, as acid hydrolysis causes lipid alterations. More recently, Pérez-Palacios et al. (2012) have found that the solvent to sample ratio influences the total lipids extracted but also the FA composition of the triacylglycerols (TG) and PL fractions in meat products. The use of a 20:1 ratio chloroform:methanol (2:1, v/v) to sample is appropriate for quantifying total lipid content in meat and meat products. However, for quantification of FA in the different lipid fractions, the volume of solvent and water used for lipid extraction should be considered.

13.2.2 Optimization of classic methods for lipid composition

13.2.2.1 Lipid classes

After lipid extraction from meat, the next stage in the analysis can involve the isolation of the lipid fractions (mainly TG and PL) and/or the determination of the FA composition of the total lipids. The separation of the lipid fractions has been traditionally carried out by means of thin-layer chromatography (TLC). However, this methodology has been largely replaced by solid-phase extraction (SPE) (Ruiz et al., 2009).

SPE is based on strong but reversible interactions of the analytes between a solid stationary phase and a mobile liquid one. Kaluzny et al. (1985) described a widely used SPE method for lipid classes separation, which is based on successive elution with chloroform/isopropanol (2:1, v-v), 2% acetic acid in diethyl ether, and methanol to elute TG, free FA, and PL, respectively. However, this method seems to be not accurate enough in separating lipid fractions in meat samples (Ruiz et al., 2004). These authors, detected a certain amount of PL in the NL fraction when fractionating lipids from pig muscle and proposed to elute TG and free FA with chloroform and diethyl ether:acetic acid, respectively, and two solvents, methanol:chloroform (6:1, v-v) and sodium acetate in methanol:chloroform, for the PL fraction. In pork loin samples, the most recent studies have also eluted the PL fraction with two extraction solvents, methanol and chloroform/methanol/water, and chloroform/2-propanol (2:1, v-v) for TG (Narvaez-Rivas et al., 2011; García-Márquez et al., 2013).

Currently, the isolation of lipid fractions can be also achieved by high-performance liquid chromatography coupled to evaporative light scattering detector (HPLC-ELSD). Perona and Ruiz-Gutierrez (2005) tried to separate cholesteryl esters, TG, free cholesterol, monoglycerides, and PL from pig muscle using hexane, 2-propanol, and methanol in gradient. However, these authors observed that the response of the ELSD was different for various lipid classes and that the response is dependent upon concentration. More recently, Habeanu et al. (2014) were able to correctly elute TG and PL of pig muscles by HPLC-ELSD with a ternary solvent gradient (isooctane:tetrahydrofurane) (99:1, v-v), acetone:dichloromethane (2/1, v-v), and 2-propanol:water + mixture acetic acid + ethanolamine (85:15, v-v). However, this technology is not so well-developed to separate TG and FL of meat lipids.

Due to the growing interest in the PL fraction, the analysis of PL classes has become a widely used analytical technique among food scientists. Nowadays, HPLC-ELSD is the most preferred method for separating and quantifying PL classes, taking over from TLC and SPE. Since ELSD is not a destructive detector, this method also allows the recovery of each PL class individually, to be further analyzed in terms of FA composition.

Some differences regarding the sample preparation, solvents and gradient used, and even the running time can be observed among the published HPLC-ELSD methods for fractionation of muscle PL classes (Pérez-Palacios et al., 2010a,b; García-Márquez et al., 2013; Ferioli and Caboni, 2010; Wang et al., 2009). Thus Pérez-Palacios et al., (2010a,b) were able to separate the four main PL classes of ham by modifying the method of Rombaut et al. (2005) for dairy products. These authors directly injected the extracted lipids (previously dissolved in chloroform:methanol) and used a linear gradient of chloroform (A), methanol (B), and triethylamine buffer (pH 3, 1 M formic acid) (C): $t = 0$ minute, 87.5% (A), 12% (B), 0.5% (C); $t = 12$ minutes, 2% (A), 90% (B), 8% (C) for 2 minutes. The mobile phase was brought back to the initial conditions at $t = 16$ minutes, and the column was allowed to equilibrate until the next injection at $t = 25$ minutes. In the methodology described by García-Márquez et al. (2013), who separated six PL classes from pork loins, the PL fraction has to be previously recovered by SPE and subsequently analyzed by HPLC. These authors used a complex gradient elution: from 0 to 5 minutes (B) was increased from 0% to 40%; from 5 to 7 minutes (B) was kept constant at 40%; from 7 to 13 minutes (B) was increased from 40% to 100%; from 13 to 20 minutes (B) was kept constant at 100%; from 20 to 25 minutes (B) was decreased from 100% to 0%; a postrun time of 5 minutes was taken to equilibrate the column, being (A) (chloroform:methanol: ammonia solution, 80:19.5:0.5, v-v:v) and (B) (chloroform:methanol:triethylamine:water, 69.53:25.58:0.49:4.40, v-v:v:v).

13.2.2.2 Fatty acids

Once the total lipids, the major lipid fractions, and/or their classes are obtained, the following step should consist of determining their FA composition. For this task, gas chromatography with flame ionization detection (GC-FID) of FA methyl esters (FAME) is the most widespread method.

The conversion of FA to FAME is most frequently carried out by acid-catalyzed esterification, mainly with methanolic hydrogen chloride or sulfuric acid in methanol solutions. Basic catalyzing transesterification methods, with potassium hydroxide in anhydrous methanol, methanolic sodium, or potassium methoxide, can be also applied. Nevertheless, the FA profile determined by CG is influenced by the esterification method. Acid-catalyzed estification methods lead to FAME from both free FA and O-acyl lipids but base-catalyzed reagents only transesterified FA of O-acyl lipids (Raes and De Smet, 2009). GC-FID is preferably used to identify and quantify the meat FA profile, by using references FAME and an internal standard. Thus the amount of FAME is visualized by the peak areas and calculated by relative response factors or calibration curves (Pérez-Palacios et al., 2009, 2010a). The current trend is to develop FA methods by using simple, fast, and one-step esterification procedures that are also accurate, safe, and environmentally friendly. In this sense, O'Fallon et al. (2007) have presented a simplified protocol to obtain FAME directly from fresh tissue, oils, or feedstuffs, without prior organic solvent extraction. With this protocol, samples are hydrolyzed for 1.5 hours at 55°C in potassium hydroxide in methanol and methylated by acid sulfuric catalysis for 1.5 hours at 55°C. Juárez et al. (2008) have also proposed an in situ method, based on homogenizing hexane and methanolic hydrochloric acid with meat samples and heated for 90 minutes in a water bath, as a good alternative for routine analysis of mixed lipid samples. Lisa et al. (2011) merely heated the sample with sodium methoxide in methanol for 10 minutes at 65°C to prepare FAME from different animal fats.

The FA profile of the outer positions of TG and PL from animal muscle may have several important consequences, both in vivo and in the derived foodstuffs. Traditional methods for determining the positional distribution of FA of TG and FL include stereospecific enzymatic hydrolysis and separation of the molecules, followed by derivatization and analysis by chiral chromatography. The determination of positional distribution of FA on TG has been performed by lipase from *Rhizopus arrhizus delemar* (Fischer et al., 1973). This enzyme removes the FA in the *sn*-1/3 positions. This technique involves incubation of the purified lipid with an aqueous suspension of the enzyme, separation of the products by TLC, and analysis FAME in each fraction. The main drawbacks of this method are related to the non-specificity of the enzyme and the impossibility to control the extent of reaction. More lately, Williams et al. (1995) tried to simplify this method by eliminating the need for TLC and determining the total enzyme activity. It is based on the fact that methanolic-NaOH only methylates FAs that are esterified to the glycerol backbone, while methanolic-HCl methylates both esterified and free FAs. Thus after lipase action, the difference in FA composition between the two methylation reactions is a quantitative measure of the FA released by the enzyme. Nevertheless, this technique does not achieve the positional distribution of *sn*-1 and *sn*-3 positions separately.

The enzymatic hydrolysis of PL fraction or PL classes is based on the specific breakage of one of the two ester bonds in the *sn*-1 and *sn*-2 positions of the glycerol backbone, yielding the FA released from the hydrolyzed position and the lyso-PL containing the FA in the remaining one. Then the two fractions are further isolated, which can be carried out by column extraction (Gładkowski et al., 2011) or by TLC (Lei et al., 2012; Simonetti et al., 2008), and subsequently transesterified and analyzed by GC, allowing the analysis of the FAs in both positions (Pérez-Palacios et al., 2006, 2007). Phospholipase A_2 from bee venom has also been used extensively for this purpose (Pacetti et al., 2005; Pérez-Palacios et al., 2006, 2007), but most recent studies have used phospholipase A_2 from porcine pancreas (Lei et al., 2012; Gładkowski et al., 2011), which hydrolyzes the ester bond in the *sn*-2 position of PL. Calcium ions are essential for the reaction to take place, which is also stimulated by diethyl ether.

Nowadays, HPLC is the more widely used technique for the analysis of FA distribution in lipid classes. In the case of TG, the combination of HPLC with MS allows unambiguous identification of some compounds, and the inclusion of atmospheric pressure chemical ionization (APCI) has provided a high suitability for this analysis. Mottram et al. (2001) have adjusted a HPLC-APCI MS technique developed for vegetable oil (Mottram et al., 1997) to carry out the regiospecific analysis of TG from beef, chicken, lamb, and pork samples. An LC-18 column (25 cm × 10 mm i.d., 5 mm particle size) was used, with propionitrile as the mobile phase. This method allowed the determination of the position of FA in TG, including isomers, that is, *cis*-18:1 was predominantly found in the 2-position of the TG and the *trans*-18:1 showed a preference for the 1/3-position.

Within HPLC techniques, nonaqueous reversed-phase (NARP) and silver ion are most widespread in the analysis of TG in natural samples. NARP-HPLC separation is based on the equivalent carbon number (ECN), increasing the retention times of TG with the increasing of ECN. In silver ion chromatography, separation of TG is governed mainly by the number of double bonds. Lisa et al. (2011) have evaluated the ability of these techniques to determine TG species in muscles from different animals and found that NARP-HPLC effectively separated individual TG species, including *trans*- and branched TGs, while silver ion mode provides the separation of TG regioisomers.

Regarding the PL, the analysis of molecular species is carried out more often on individual PL classes than in the whole PL fraction. For that, the most used methodology combines HPLC and MS detectors. It is noted that the extended procedure is usually carried out to finally achieve identification of the molecular species of the PL classes. Firstly, total lipids are extracted, then the PL fraction is eluted by SPE. Subsequently, the PL classes are separated and finally the molecular species are analyzed by HPLC. This last step has been carried out in two phases, the identification of the PL classes by HPLC-UV and the subsequent analysis of the molecular species by HPLC-ELSD-MS, as reported by Wang et al. (2009) for studying the intramuscular PL molecular species in traditional Chinese duck meat products. Nevertheless, the simultaneous determination of both the PL classes and their molecular species in meat has been achieved by coupling on-line HPLC with two detectors, ELSD and MS (Boselli et al., 2007), or a diode array detector (DAD) and MS (Narváez-Rivas et al., 2011). In addition, in some studies the MS detector has been equipped with an electrospray ionization source (ESI), the ionization technique most commonly used in the field of lipid analysis (Boselli et al., 2007; Narváez-Rivas et al., 2011). Giving a step forward in the analysis of the molecular species of PL classes, a 2D-HPLC system, which allows the use of the total lipids as a sample, has been recently improved (Takahashi et al., 2018). It consists of normal phase (NP) and reverse phase (RP) HPLC with high-pressure coupled with a charged aerosol detector (CAD) and ESI-MS. CAD is a universal HPLC detector and can be used for the detection of lipids with no chromophore, and it can detect impurities that cannot be detected under analysis of targeted compounds by MS (Libong et al., 2017). The first-dimensional-HPLC (NP-HPLC) separation was performed to separate

PL from other lipid classes in the sample by monitoring at 210 nm using a variable wavelength detectors (VWDs). The PL fraction was eluted from this first system to a second-dimensional HPLC (RP-HPLC) system to analyze the molecular species of different PL classes. This system can analyze the molecular species in one lipid class with a single analysis operation. More details about these methods for analyzing molecular species of individual PL classes in meat are summarized in Table 13.1.

Table 13.1: Summary of current methods for analyzing molecular species of individual PL classes in meat.

Technique	Column	Mobile phase	Sample	Other data[a]	Reference
HPLC-ELSD-ESI-MS	Precolumn silica (4 × 3.0 mm i.d.) Silica (150 × 4.6 mm i.d.); 3 μm particle diameter	A: chloroform:methanol:ammonia solution (30%) (70:25:1, by vol.) B: chloroform:methanol:water:ammonia solution (30%) (60:40:5.5:0.5, by vol.)	PL fraction	Pork meat PC, pPE, PE, CL, Sph, PI, PS	Boselli et al. (2007)
HPLC-ELSD-MS	C18 RP (250 × 4.6 mm i.d.); 5 μm particle diameter	A: chloroform:methanol (1/17.5, by vol.) B: acetonitrile:water (1:1, by vol.)	PC and PE recovered from HPLC-UV-ELSD analysis	Duck PC, PE	Wang et al. (2009)
HPLC-DAD-ESI-MS	Silica (250 × 6 mm i.d.); 5 μm particle diameter	A: chloroform:methanol:ammonia solution (80:19.5:0.5, by vol.) B: chloroform:methanol:triethylamine:water (69.53:25.58:0.49:4.40, by vol.)	PL fraction	Iberian pig subcutaneous fat CL, PE, PI, PS, PC, Sph, LysoPC	Narváez-Rivas et al. (2011)
2D-HPLC: first dimensional (NP-HPLC-VWD) + second dimensional (RP-HPLC-CAD-MS)	First dimensional: Silica (250 × 4.6 mm i.d.); 5 μm particle diameter Second dimensional: hybrid silica-C18 (250 × 4.6 mm i.d.); 3 μm particle diameter	First dimensional: A: hexane; B: methyl *tert*-butyl ether; C: methanol Second dimensional: acetonitrile:methanol: 20 mM ammonium acetate (25:68.5:6.5, by vol.)	Total lipids	Chicken breast PE, PC, pPE, pPC	Takahashi et al. (2018)

[a]Individual phospholipid separated (*PC*, phosphatidylcholine; *PE*, phosphatidylethanolamine; *PI*, phosphatidylinositol; *pPC*, plasmalogen phosphatidylcholine; *pPE*: plasmalogen phosphatidylethanolamine; *PS*, phosphatidylserine; *Sph*, sphingomyelin).

13.2.3 Advanced methodologies for lipid analysis: nondestructive methods

As described above, the optimized classical analytical techniques for lipid analysis are sensitive and straightforward. However, they also are time-consuming, destructive, sometimes tedious, not environment-friendly, and unsuitable for on-line and real-time measurement of fat quality. Important developments are being carried out to solve these drawbacks, by testing nondestructive methods in combination with techniques of data analysis, in order to propose reliable procedures for determining meat lipids. In this sense, researchers on computerized tomography (TC), Near-infrared reflectance spectroscopy (NIRs), hyperspectral imaging (HSI), nuclear magnetic resonance (NMR), US, and MRI, among other techniques, have been developed. For example, TC has been used to predict the fat content in animal carcasses (Vester-Christensen et al., 2009; Kongsro et al., 2009) and meat products, such as hams (Picouet et al., 2014; de Prados et al., 2015) and pork loins (Font-i-Furnols et al., 2013). Many studies have shown the ability of NIRs to predict most chemical components, mainly fat, protein, and moisture, in different meat and meat products from different animal species, as reported by Su et al. (2014). The latest studies on NIRs have been focused on development quantitative models for FA and the FA classes. Accurate results have been found in both homogenized and intact samples of adipose or meat tissues, with an average coefficient of determination ranging from 0.66 to 0.87 for major FA (palmitic acid (C16:0), stearic acid (C18:0), oleic acid (C18:1), and linoleic acid (C18:2)), while the rest of the FA obtained had lower coefficients (Kucha et al., 2018). Accurate results have also been found by using HSI to determine fat content in beef, lamb, and pork samples (Kobayashi et al., 2010; Kamruzzaman et al., 2012; Liu and Ngadi, 2014; Huang et al., 2014). However, it is necessary to seek the most sensitive wavebands to improve the speed and processing in order to implement HSI technology for rapid and nondestructive meat quality measurements (Tao and Ngadi, 2016). Classically, the lipid content of meat has also been determined by NMR (Crispilho et al., 2009), however, the great interest in NMR methods is due to their availability to quantify FA in different meat samples (Colnago et al., 2010; Siciliano et al., 2013; Stefanova et al., 2011). As for US, it has been traditionally applied for the measurement of fat content in live animals and carcasses. Nevertheless, there are some current US studies that have estimated the content of fat in different meat and meat products (Koch et al., 2011; Corona et al., 2014; Ludwiczak et al., 2017). MRI in combination with computer vision techniques have also emerged as one of the alternative methodologies to the classical analysis. Apart from being nondestructive, this technique is noninvasive, nonintrusive, nonionizing, and innocuous. There also are some recent studies showing the good ability of MRI to predict the lipid content in meat products, principally loin and hams (Monziols et al., 2006; Pérez-Palacios et al., 2014, 2017; Ávila et al., 2018).

13.3 Advances in the detection of lipid oxidation products

The relevant role played by oxidative reactions in the impairment of meat quality has challenged scientists to develop methodologies to consistently evaluate the oxidative status of muscle lipids. The selection of a unique appropriate test to accurately assess the overall extent of oxidative damage to meat lipids is complicated owing to the underlying intricate chemistry. Nevertheless, the high levels of sophistication achieved on the understanding of lipid oxidation mechanisms has enabled improved techniques for the isolation, identification, and quantification of lipid oxidation products. Together with the optimization of classic methodologies, recent years have witnessed great efforts in innovation. The application of versatile mass spectrometric methods in mechanistic studies has provided further insights into the molecular interactions of lipid oxidation products with other muscle components.

13.3.1 Optimization of classic methods for lipid oxidation assessment

Classic procedures for the assessment of lipid oxidation can be categorized into two groups: (1) those aiming at quantifying primary oxidation products, and (2) those designed for the identification and quantification of secondary oxidation products.

13.3.1.1 Assessment of primary oxidation products

Measuring the content of primary oxidation products (e.g., hydroperoxides and conjugated dienes, CDs) provides limited information on the extent of the oxidative damage, owing to the short-lived nature of these species. Hydroperoxides (or simply peroxides), are identified as the most salient early lipid oxidation products. Yet these species are not good indicators of lipid oxidation at advanced reaction stages: whenever their breakdown is as fast as or faster than their formation. Hence, the analysis of hydroperoxides is performed when the entire course of oxidation is studied and it is usually combined with the assessment of secondary oxidation products such as MDA (Kerrihard et al., 2015). Assorted blends of solvents have been used for the extraction of lipids from meat products as a required preceding stage in the analysis of lipid hydroperoxides (see methods in Section 13.2.1). The use of accelerated solvent extraction equipment minimizes hydroperoxide destruction during extraction and enables a faster and more efficient extraction procedure (Yao and Schaich, 2014). Although the classic reverse titrimetric method to determine the peroxide value is still being used in meat systems (Chauhan et al., 2018), spectrophotometric methods such as the ferrous oxidation–xylenol orange offer more consistency and objectivity (Bou et al., 2008). This method has been successfully applied directly to meat products (Dutra et al., 2017), and in studies simulating in vitro digestion and cooxidation of meat with edible oils (Martini et al., 2018). Chromatographic techniques such as gas chromatography (GC) (Leocata et al., 2016) and liquid chromatography (LC) (Ibusuki et al., 2008), commonly coupled to mass spectrometry (MS), provide evidence on the structure, concentration, and reactivity of specific hydroperoxides (Giuffrida et al., 2004). Although some applications in the field of food lipids have been reported (reviewed by Dobarganes and Velasco, 2002), the use of these technologies in meat and meat products is scarce given their complexity and the reduced association of the results with the impact of lipid oxidation on sensory properties.

The assessment of CDs is a sensitive method to follow the early stages of lipid oxidation and CD may even be indicators of free radical production (Sun et al., 2011). CD are measured by ultraviolet spectrophotometry at 234 nm for a constant mass of sample. While it has been reported to be unsuitable for determination in food matrices with a high proportion of saturated FAs, this method has been broadly applied to meat products, because it is fast, simple, requires no chemical reagents, and small samples can be processed (Pegg, 2005). Yet a prior fat extraction with chloroform/methanol (2:1) (Grau et al., 2000) or hexane/isopropanol (3:2) (Srinivasan et al., 1996) is needed. Agregán et al. (2018) and Dalle Zotte et al. (2018) among others, have recently applied the classic method, without major modifications, to diverse meat and meat products.

13.3.1.2 Assessment of secondary oxidation products

Unlike the colorless and flavorless early products of lipid oxidation, secondary oxidation products such as aldehydes, ketones, hydrocarbons, and alcohols are generally odor-active compounds (Kerrihard et al., 2015). Therefore they are more accurate indicators of the deterioration caused by lipid oxidation in meat and meat products. Among these compounds, aldehydes are considered the most salient compounds given their low odor-threshold values and their contribution to the onset of rancidity and off-flavors (Pegg, 2005).

13.3.1.2.1 Malonaldehyde

Malonaldehyde (MDA) is produced during the autoxidation of unsaturated FAs and its accretion has been for a long time used as an oxidation index in meat products. Despite being a secondary oxidation product, MDA is highly reactive and may bind to other biomolecules. This reactivity has relevant consequences in terms of toxicity (impairs protein functionality) and mutagenicity (binds to DNA), yet it enables its detection by derivatization and subsequent analysis (Estévez et al., 2017). In some cases, an acid/heat treatment of the meat matrix is required to release the bound MDA and facilitate its derivatization and detection (Ulu, 2004). The amount of MDA has been traditionally analyzed spectrophotometrically after derivatization with thiobarbituric acid (TBA) or by using chromatographic techniques.

The TBARS assay is undeniably the most common test to monitor lipid oxidation in meat and meat products (Pegg, 2005). The adduct formed by the reaction between the monoenolic form of MDA (and other unsaturated aldehydes) and the active methylene groups of TBA is a red chromogen showing a primary absorbance maximum at 532–535 nm (Fernández et al., 1997). The original method described by Turner et al. (1954) has been submitted to countless adjustments as shown in the review papers from Osawa et al. (2005) and Ghani et al. (2017). Currently three basic tests are performed depending on the treatment of the meat sample prior to derivatization with TBA, namely, lipid extraction (using organic solvents) (Younathan and Watts, 1960), acid extraction (using trichloroacetic or perchloric acid) (Witte et al., 1970), and distillation (Yu and Sinnhuber, 1967). The latter two are the most applied methods while the distillation method reported by Tarladgis et al. (1960) remains the most cited work on TBA testing among meat scientists. Most of the optimization procedures published since have been aimed at minimizing the overestimation of the concentration of MDA in the sample owing to (1) the formation of further aldehydes during extraction (Raharjo and Sofos, 1993), (2) the reaction of TBA with other food components (sugars, amino acids) (Gray and Monahan, 1992), and (3) the interference with colored pigments such as Maillard products (formed during processing or MDA extraction) and plant phenolics (added to meat products as antioxidants) (Ganhão et al., 2011). Among the strategies to solve the aforementioned problems, several may be mentioned, including (1) using SPE to remove interfering compounds (Raharjo et al., 1992); (2) optimization of sample preparation parameters (% TCA and BHT, among others) followed by a gentle extraction procedure to minimize further production of MDA (Grotta et al., 2017); (3) performing the reaction between MDA and TBA at room temperature for a longer time (15−24 hours) and under readjusted MDA concentration and pH to avoid sugar reactivity (Ganhão et al., 2011); and (4) using the distillation method to discard phenolics and other yellow pigments absorbing at 532 nm (Ganhão et al., 2011). The nitration of MDA by residual nitrite in cured muscle foods has been reported as a remarkable cause of underestimation of TBARS numbers. The addition of sulfanilamide (SA) as suggested by Zipser and Watts (1961) has been recently applied to assorted cured and processed muscle foods (Lee et al., 2017; Öztürk and Serdaroğlu, 2017). It is worth mentioning that in samples with less than 100 ppm of residual nitrite, SA may not be used given its reactivity with MDA, itself causing an underestimation (Shahidi et al., 1991).

The application of chromatographic methods enables a specific and accurate quantification of the MDA. GC analysis of total MDA is feasible upon preparation of stable pyrazole derivatives using hydrazine-based reagents (Miyake and Shibamoto, 1996). While some recent applications of GC methodologies to quantify MDA in meat products can be found in the literature (Ruan et al., 2014), LC is more commonly applied for this purpose (Wong et al., 2016). Mousa and Al-Khateeb (2017) made a quantitative analysis of MDA prior to derivatization with TBA by applying an HPLC system attached to a spectrophotometric detector. Reitznerová et al. (2017) compared the classic TBARS method with a HPLC-DAD protocol to quantify dinitrophenylhydrazine-derivatized MDA and emphasized the benefits of the chromatographic method over the routine counterpart. Jung et al. (2016) successfully recovered and quantified MDA without derivatization using an HPLC-UV method subsequent to an extraction of the reactant with acetonitrile. According to the authors, the analysis allowed the detection of MDA without interference from sodium nitrite, sodium chloride, pyrophosphates, and maltodextrin, among others.

13.3.1.2.2 Lipid-derived volatiles

The detection of volatile compounds is employed widely to assess lipid oxidation owing to the benefits of providing information on the extent of the oxidative damage and on the impact on the sensory properties of meat and meat products. Hexanal is the most abundant volatile in the headspace (HS) of most processed muscle foods and commonly used as a marker of oxidation. Yet a number of minor aldehydes, such as alkadienals, *trans*-4,5-epoxy-(*E*)-2-decenal, and 4-hydroxy-2-nonenal are typically linked to rancidity and warmed over flavor (Kerler and Grosch, 1996). The FA profile of the meat product may affect the reliability of the oxidation markers as Shahidi et al. (2017) reported that hexanal suitably indicates the extent of oxidation in products rich in ω-6 FAs while propanal may be a more consistent marker of oxidation in meats with high levels of ω-3 FAs.

The assorted methodological approaches available for the extraction and isolation of lipid-derived volatile compounds were reviewed by Ross and Smith (2006). The most common are solvent extraction, simultaneous distillation extraction, dynamic HS or purge and trap methodology, and the solid-phase microextraction (SPME). This latter technique has been extensively applied to muscle foods owing to its sensitivity, ease of use, and sound results (Iglesias and Medina, 2008). Ouyang and Pawliszyn (2008), Jeleń et al. (2012), and Xu et al. (2016) have delivered critical reviews regarding ways to improve the calibration and reproducibility of the SPME for profiling lipid-derived volatiles in food systems. Recent applications of HS-SPME in meat and meat products are found in the literature with the aims of improving the isolation and identification of assorted lipid oxidation products (Mansur et al., 2018; Wang et al., 2018a,b). While GC is usually for the separation of chemical species, MS is used for their identification. However, verifying the identity of volatile compounds has become a major issue as most scientific societies and journals are demanding on the accuracy of such identification. According to the "International Organization of the Flavor Industry" (IOFI, 2006), any identification of a flavoring substance must pass scrutiny of the latest forms of available analytical techniques. In practice, this means that any particular substance must have its identity confirmed by at least two methods, for example, comparison of chromatographic and spectrometric data, which may typically include the chromatographic linear retention index LRI and MS, with those of an authentic sample (Estévez et al., 2003). If only one method has been applied (MS data alone or retention index or Kovats index alone), the identification shall be labeled "tentative." If data on sensory properties of single compounds are reported, which is considerably frequent in studies including the assessment of lipid-derived volatiles in meat and meat products, the same organization states that the use of reference compounds is a must. In line with this, the combination of MS to olfactometry approaches seems highly recommendable in order to identify the odors imparted by particular chemical species in meat products (Zhao et al., 2017; Zhang et al., 2018).

13.3.2 Advanced methodologies for lipid oxidation assessment

13.3.2.1 Novel spectrometric methodologies

The advances in the application of sophisticated MS technology for the identification of lipid oxidation products has occurred along with the onset of alternative methods to the SPME for the previous isolation of such species. Flores et al. (2013) studied the applicability of selected ion flow tube mass spectrometry (SIFT-MS) to assess the volatile components of cooked meat and compared the outcome with that obtained from the conventional SPME-GC-MS. The alternative procedure was presented as a more suitable means to analyze volatile compounds in "real-time" conditions such as those occurring during meat mastication. Olivares et al. (2011) came to similar conclusions and calculated higher correlations between lipid oxidation-derived volatiles analyzed by SIFT-MS and the TBARS method than those obtained using SPME-GC-MS. These authors stated that IFT-MS is a fast, real-time analytical procedure for monitoring volatiles profiles in fermented sausages during processing and a valuable device to assess the oxidative status of meat products. Carrapiso et al. (2015) employed SIFT-MS to evaluate volatiles profiles in Iberian dry-cured hams. The technique enabled a fast and correct classification of the hams in accordance to the feeding background of the pigs. An inventive combination of SPMW with SIFT-MS was recently applied to for the analysis of the HS in modified atmosphere packaged poultry (Ioannidis et al., 2018). An alternative to SPME, the solid stir-bar sorptive extraction (SBSE) has been introduced to improve accuracy in the extraction of volatile and semivolatile compounds. However, the use of SBSE in meat studies has been very limited and no research has been reported in cooked ham. According to Benet et al. (2015) SBSE was useful for the detection of volatiles of assorted polarity in cooked hams. In some other cases, conventional MS detection has been replaced by more sophisticated technologies based on exact mass detection such as high-resolution time-of-flight (TOF). Wang et al. (2018a,b) recently applied two-dimensional GC coupled to TOF-MS to dry-cured hams to improve the resolution and subsequent detection of polar volatile compounds. Gamero-Negrón et al. (2015) analyzed volatile compounds in dry-cured products using proton transfer reaction TOF-MS and found an efficient discriminant capability that provided valuable analytical information.

13.3.2.2 Spectroscopic and nondestructive methodologies

The consumption of time, solvents, and the sample are some of the drawbacks of classic lipid oxidation methodologies. Nondestructive methods overcome these shortcomings while offering fast, accurate, and predictive information on the oxidative status of the food system.

Most of these methods involve spectroscopic techniques as they are highly sensitive for the detection of minor changes in the lipid fraction of meat systems. Front-face fluorescence spectroscopy (FFFS) has been applied to detect Schiff bases formed from the reaction between lipid aldehydes and amino groups in meat proteins. This technique offers a solvent-free and nondestructive measurement of the complex lipid−protein interactions during oxidative stress (Lahmar et al., 2018). Olsen et al. (2005) analyzed the early stages of lipid oxidation in porcine fat and poultry meat using FFFS and compared the results with assorted techniques including dynamic HS-GC-MS, electronic nose, and chemiluminescence and found good correlations between instrumental and sensory perceptions of rancidity. HSI has emerged as an innovative tool to nondestructively and rapidly assess the extent of lipid oxidation in muscle foods (Cheng et al., 2015). The HI system incorporates the traditional spectroscopy to computer vision. Cheng et al. (2015), in particular, acquired hyperspectral images from carp fillets by using a scanning imaging spectrograph (308−1105 nm) and a high-performance charge-couple device camera to predict TBARS numbers. The model used, calibrated with partial least-squares regression, provided satisfactory predictions of TBARS with $R^2 > 0.82$. Moreover, a model developed using multiple linear regressions enabled an interpretation of the dynamic changes caused by lipid oxidation during chilled storage of the fish muscle. Infrared spectroscopy has also been emphasized as a valuable nondestructive technology to monitor lipid degradation under oxidative environments owing to its ease of use, speed, and low cost (Barriuso et al., 2013). NIR spectroscopy is able to detect molecules at the region between 800 and 2500 nm, such as free FAs and their oxidation products including hydroperoxides and carbonyl compounds (Endo, 2018). A recent approach used NIR hyperspectral imaging to monitor carbonyl-mediated oxidation in myofibrillar proteins (Cheng et al., 2018). The data was interpreted using heterospectral two-dimensional correlations to create models to accurately reveal the extent of the oxidative damage to meat proteins. A first attempt to monitor lipid oxidation in meat samples by using Raman spectroscopy and electron spin resonance technology was recently carried out by Chen et al. (2018). The procedure enabled the detection of radical species and precise molecular structures. This outcome provides further insight into the molecular mechanisms compared to the conventional method, with which this innovative approach had high correlations. Other spectroscopic methods such as the NMR (^{1}H and ^{13}C NMR) have been used to detect lipid oxidation products in edible fats and oils. However, an accurate quantification of particular oxidation products in complex (Endo, 2018) and the applicability of such technology to meat products is still low.

13.4 Future perspectives: lipidomics and oxidomics

The understanding of the molecular basis of biochemical reactions has enabled pushing at the frontiers of knowledge and progress in the analysis of meat lipids and lipid oxidation. Untargeted MS analyses of a complex mixture of metabolites in a given biological system (the "metabolome") enables a deeper comprehension of its biology and composition. A comprehensive metabolite profiling of muscle lipids ("lipidome") has particularly been applied to meat systems for profound characterization and accurate authentication. One of the original approaches was carried out by Trivedi et al. (2016) who identified sphingolipid metabolism as a discriminating pathway between pork and beef and hence a potential marker of beef adulteration with pork. More recently, Mi et al. (2018) identified 1127 lipids in Taihe black-boned silky fowl muscles and emphasized 47 lipid compounds as potential authentication markers of this particular poultry meat.

While not fully developed for the analysis of meat products, original attempts to apply "oxidomics" approaches to food systems have recently been carried out. A complete profiling of the oxidation products occurred in a given food system may be called the "oxidome" (Hu et al., 2017). Paradiso et al. (2018) analyzed the "oxidome" of olive oil after accelerated oxidation using HPLC-ESI-MS, which enabled an in-depth insight into the reaction pathways, reactants, and oxidation products. The application of such an approach using even more sophisticated MS platforms for comprehensive "omics" studies, such as quadrupole-time of flight or orbitrap MS, would allow the deeper comprehension of oxidative reactions in meat and muscle foods, with benefits in terms of product discrimination, authenticity, quality, and safety.

References

Association of Official Analytical Chemist, 2000. seventeenth ed. Official Methods of Analysis of AOAC International, vols. 1 and 2. AOAC International, Gaithersburg, MD.

Agregán, R., Franco, D., Carballo, J., Tomasevic, I., Barba, F.J., Gómez, B., et al., 2018. Shelf life study of healthy pork liver pâté with added seaweed extracts from *Ascophyllum nodosum, Fucus vesiculosus* and *Bifurcaria bifurcate*. Food Res. Int. 112, 400−411.

Ávila, M., Caballero, D., Antequera, T., Durán, M.L., Caro, A., Pérez-Palacios, T., 2018. Applying 3D texture algorithms on MRI to evaluate quality traits of loin. J. Food Eng. 222, 258−266.

Barriuso, B., Astiasarán, I., Ansorena, D., 2013. A review of analytical methods measuring lipid oxidation status in foods: a challenging task. Eur. Food Res. Technol. 236, 1−15.

Benet, I., Ibañez, C., Guàrdia, M.D., et al., 2015. Optimisation of stir-bar sorptive extraction (SBSE), targeting medium and long-chain free fatty acids in cooked ham exudates. Food Chem. 185, 75−83.

Bligh, E.G., Dyer, E.J., 1959. A rapid method of total lipid extraction and purification. Can. J. Biochem. Physiol. 37, 911−917.

Boselli, E., Pacetti, D., Curzi, F., 2007. Determination of phospholipid molecular species in pork meat by high performance liquid chromatography−tandem mass spectrometry and evaporative light scattering detection. Meat Sci. 78, 305−313.

Bou, R., Codony, R., Tres, A., Decker, E.A., Guardiola, F., 2008. Determination of hydroperoxides in foods and biological samples by the ferrous oxidation-xylenol orange method: a review of the factors that influence the method's performance. Anal. Biochem. 377, 1−15.

Carrapiso, A.I., Noseda, B., García, C., et al., 2015. SIFT-MS analysis of Iberian hams from pigs reared under different conditions. Meat Sci. 104, 8−13.

Chauhan, P., Das, A.K., Nanda, P.K., Kumbhar, V., Yadav, J.P., 2018. Effect of *Nigella sativa* seed extract on lipid and protein oxidation in raw ground pork during refrigerated storage. Nutr. Food Sci. 48, 2−15.

Chen, Q., Xie, Y., Xi, J., et al., 2018. Characterization of lipid oxidation process of beef during repeated freeze-thaw by electron spin resonance technology and Raman spectroscopy. Food Chem. 243, 58−64.

Cheng, J.-H., Sun, D.-W., Pu, H.-B., Wang, Q.-J., Chen, Y.-N., 2015. Suitability of hyperspectral imaging for rapid evaluation of thiobarbituric acid (TBA) value in grass carp (*Ctenopharyngodon idella*) fillet. Food Chem. 171, 258−265.

Cheng, W., Sun, D.-W., Pu, H., Wei, Q., 2018. Heterospectral two-dimensional correlation analysis with near-infrared hyperspectral imaging for monitoring oxidative damage of pork myofibrils during frozen storage. Food Chem. 248, 119−127.

Colnago, M.R.M., Forato, L.A., Bouchard, D., 2010. Fast and simple nuclear magnetic resonance method to measure conjugated linoleic acid in beef. J. Agric. Food Chem. 58, 6562−6564.

Corona, E., García-Pérez, J.V., Santacatalina, J.V., Ventanas, S., Benedito, J., 2014. Ultrasonic characterization of pork fat crystallization during cold storage. J. Food Sci. 79, 828−838.

Crispilho, C., Aparecida, L., Colnago, L.A., 2009. High-throughput non-destructive nuclear magnetic resonance method to measure intramuscular fat content in beef. Anal. Bioanal. Chem. 393, 1357−1360.

Dalle Zotte, A., Cullere, M., Tasoniero, G., Gerencsér, Z., Szendrő, Z., Novelli, E., et al., 2018. Supplementing growing rabbit diets with chestnut hydrolyzable tannins: effect on meat quality and oxidative status, nutrient digestibilities, and content of tannin metabolites. Meat Sci. 146, 101−108.

de Prados, M., Fulladosa, E., Gou, P., Muñoz, I., Garcia-Pereza, J.B., Benedito, J., 2015. Non-destructive determination of fat content in green hams using ultrasound and X-rays. Meat Sci. 104, 37−43.

Dobarganes, M.C., Velasco, J., 2002. Analysis of lipid hydroperoxides. Eur. J. Lipid Sci. Technol. 104, 420−428.

Dutra, M.P., Cardoso, G.P., Fontes, P.R., Silva, D.R.G., Pereira, M.T., Ramos, A.D.L.S., et al., 2017. Combined effects of gamma radiation doses and sodium nitrite content on the lipid oxidation and color of mortadella. Food Chem. 237, 232−239.

Endo, Y., 2018. Analytical methods to evaluate the quality of edible fats and oils: the JOCS standard methods for analysis of fats, oils and related materials (2013) and advanced methods. J. Oleo Sci. 67, 1−10.

Esterbauer, H., Muskiet, F., Horrobin, D.F., 1993. Cytotoxicity and genotoxicity of lipid-oxidation products. Am. J. Clin. Nutr. 57, 779S−786S.

Estévez, M., Morcuende, D., Ventanas, S., Cava, R., 2003. Analysis of volatiles in meat from Iberian pigs and lean pigs after refrigeration and cooking by using SPME-GC-MS. J. Agric. Food Chem. 51, 3429−3435.

Estévez, M., Li, Z., Soladoye, O.P., Van-Hecke, T., 2017. Health risks of food oxidation. Adv. Food Nutr. Res. 82, 45−81.

Ferioli, F., Caboni, M.F., 2010. Composition of phospholipid fraction in raw chicken meat and pre-cooked chicken patties: influence of feeding fat sources and processing technology. Eur. Food Res. Technol. 231, 117−126.

Fernández, J., Pérez-Álvarez, J.A., Fernández-López, J.A., 1997. Thiobarbituric acid test for monitoring lipid oxidation in meat. Food Chem. 59, 345−353.

Fischer, W., Heinz, E., Zeus, M., 1973. The suitability of lipase from *Rhizopus arrhizus delemar* for analysis of fatty acid distributions in dihexosyl diglycerides, phospholipids and plant sulfolipids. Hoppe-Seyler's Z. Physiol. Chem 354, 1115−1123.

Flores, M., Olivares, A., Dryahina, K., Španěl, P., 2013. Real time detection of aroma compounds in meat and meat products by SIFT-MS and comparison to conventional techniques (SPME-GC-MS). Curr. Anal. Chem. 9, 622−630.

Folch, J., Less, M., Sloane, G.H., 1957. A simple method for the isolation and purification of total lipids from animal tissues. J. Biol. Chem. 226, 497–509.

Font-i-Furnols, M., Brun, A., Tous, N., Gispert, M., 2013. Use of linear regression and partial least square regression to predict intramuscular fat of pig loin computed tomography images. Chemom. Intell. Lab. Sys. 122, 58–64.

Gamero-Negrón, R., Sánchez del Pulgar, J., Cappellin, L., et al., 2015. Immune-spaying as an alternative to surgical spaying in Iberian × Duroc females: effect on the VOC profile of dry-cured shoulders and dry-cured loins as detected by PTR-ToF-MS. Meat Sci. 110, 169–173.

Gandemer, G., 2002. Lipids in muscles and adipose tissues, changes during processing and sensory properties of meat products. Meat Sci. 62, 309–321.

Ganhão, R., Estévez, M., Morcuende, D., 2011. Suitability of the TBA method for assessing lipid oxidation in a meat system with added phenolic-rich materials. Food Chem. 126, 772–778.

García-Márquez, I., Narvaez-Rivas, M., Gallardo, E., Cabeza, M.C., Leon-Camacho, M., 2013. Changes in the phospholipid fraction of intramuscular fat from pork loin (fresh and marinated) with different irradiation and packaging during storage. Grasas y Aceites 64, 7–14.

Ghani, M.A., Barril, C., Bedgood, D.R., Prenzler, P.D., 2017. Measurement of antioxidant activity with the thiobarbituric acid reactive substances assay. Food Chem. 230, 195–207.

Giuffrida, F., Destaillats, F., Robert, F., Skibsted, L.H., Dionisi, F., 2004. Formation and hydrolysis of triacylglycerol and sterols epoxides: role of unsaturated triacylglycerol peroxyl radicals. Free Rad. Biol. Med. 37, 104–114.

Gładkowski, W., Kiełbowicz, G., Chojnacka, A., et al., 2011. Fatty acid composition of egg yolk phospholipid fractions following feed supplementation of Lohmann Brown hens with humic-fat preparations. Food Chem. 126, 1013–1018.

Grau, A., Codony, R., Rafecas, M., Barroeta, A.G., Guardiola, F., 2000. Lipid hydroperoxide determination in dark chicken meat through a ferrous oxidation-xylenol orange method. J. Agric. Food Chem. 48, 4136–4143.

Gray, J.I., Monahan, F.J., 1992. Measurement of lipid oxidation in meat and meat products. Trends Food Sci. Tech. 3, 315–319.

Grotta, L., Castellani, F., Palazzo, F., Naceur Haouet, M., Martino, G., 2017. Treatment optimisation and sample preparation for the evaluation of lipid oxidation in various meats through TBARs assays before analysis. Food Anal. Met. 10, 1870–1880.

Habeanu, M., Thomas, A., Bispo, E., et al., 2014. Extruded linseed and rapeseed both influenced fatty acid composition of total lipids and their polar and neutral fractions in longissimus thoracis and semitendinosus muscles of finishing Normand cows. Meat Sci. 96, 99–107.

Hu, C., Wang, M., Han, X., 2017. Shotgun lipidomics in substantiating lipid peroxidation in redox biology: methods and applications. Redox Biol. 12, 946–955.

Huang, H., Liu, L., Ngadi, M.O., Gariépi, C., 2014. Rapid and non-invasive quantification of intramuscular fat content of intact pork cuts. Talanta 119, 385–395.

Ibusuki, D., Nakagawa, K., Asai, A., Oikawa, S., Masuda, Y., Suzuki, T., et al., 2008. Preparation of pure lipid hydroperoxides. J. Lipid Res. 49, 2668–2677.

Iglesias, J., Medina, I., 2008. Solid-phase microextraction method for the determination of volatile compounds associated to oxidation of fish muscle. J. Chromatogr. A 1192, 9–16.

Ioannidis, A.-G., Walgraeve, C., Vanderroost, M., et al., 2018. Non-destructive measurement of volatile organic compounds in modified atmosphere packaged poultry using SPME-SIFT-MS in tandem with headspace TD-GC-MS. Food Anal. Methods 11, 848–861.

IOFI, 2006. Statement on the identification in nature of flavouring substances, made by the working group on methods of analysis of the international organization of the flavour industry (IOFI). Flavour Fragr. J. 21, 185–188.

Jeleń, H.H., Majcher, M., Dziadas, M., 2012. Microextraction techniques in the analysis of food flavor compounds: a review. Anal. Chim. Acta 738, 13–26.

Juárez, M., Polvillo, O., Contò, M., Ficco, S., Ballico, S., Failla, S., 2008. Comparison of four extraction/methylation analytical methods to measure fatty acid composition by gas chromatography in meat. J. Chromatogr. A 1190, 327–332.

Jung, S., Nam, K.C., Jo, C., 2016. Detection of malondialdehyde in processed meat products without interference from the ingredients. Food Chem. 209, 90–94.

Kaluzny, M.A., Duncan, L.A., Merritt, M.V., Epps, D.E., 1985. Rapid separation of lipid classes in high yield and purity using bonded phase columns. J. Lipid Res. 26, 135–140.

Kamruzzaman, M., ElMasry, G., Sun, D.-W., Allen, P., 2012. Non-destructive prediction and visualization of chemical composition in lamb meat using NIR hyperspectral imaging and multivariate regression. Innov. Food Sci. Emerg. Technol. 16, 218–226.

Kerler, J., Grosch, W., 1996. Odorants contributing to warmed-over flavor (WOF) of refrigerated cooked beef. J. Food Sci. 61, 1271–1274. + 1284.

Kerrihard, A.L., Pegg, R.B., Sarkar, A., Craft, B.D., 2015. Update on the methods for monitoring UFA oxidation in food products. Eur. J. Lipid Sci. Technol. 117, 1–14.

Kobayashi, K.-I., Matsui, Y., Maebuchi, Y., Toyota, T., Nakauchi, S., 2010. Near infrared spectroscopy and hyperspectral imaging for prediction and visualisation of fat and fatty acid content in intact raw beef cuts. J. Near Infrared Spec. 18, 301–315.

Koch, T., Lakshmanan, S., Brand, S., Wicke, M., Raum, K., Mörlein, D., 2011. Ultrasound velocity and attenuation of porcine soft tissues with respect to structure and composition: II. Skin and backfat. Meat Sci. 88, 67–74.

Kongsro, J., Røe, M., Kvaal, K., Aastveit, A.H., Egelandsdal, B., 2009. Prediction of fat, muscle and value in Norwegian lamb carcasses using EUROP classification, carcass shape and length measurements, visible light reflectance and computer tomography (CT). Meat Sci. 81, 102–107.

Kucha, C., Liu, L., Ngadi, M.O., 2018. Non-destructive spectroscopic techniques and multivariate analysis for assessment of fat quality in pork and pork products: a review. Sensors 18, 377–400.

Lahmar, A., Akcan, T., Chekir-Ghedira, L., Estévez, M., 2018. Molecular interactions and redox effects of carvacrol and thymol on myofibrillar proteins using a non-destructive and solvent-free methodological approach. Food Res. Int. 106, 1042–1048.

Lee, J.H., Alford, L., Kannan, G., Kouakou, B., 2017. Curing properties of sodium nitrite in restructured goat meat (chevon) jerky. Int. J. Food Prop. 20, 526–537.

Lei, L., Li, J., Li, G.-Y., et al., 2012. Stereospecific analysis of triacylglycerol and phospholipid fractions of five wild freshwater fish from Poyang lake. J. Agric. Food Chem. 60, 1857–1864.

Leocata, S., Frank, S., Wang, Y., Calandra, M.J., Chaintreau, A., 2016. Quantification of hydroperoxides by gas chromatography-flame ionization detection and predicted response factors. Flavour Frag. J. 31, 329–335.

Libong, D., Héron, S., Tchapla, A., Chaminade, P., 2017. Lipid analysis with the corona CAD. In: Gamache, P.H. (Ed.), Charged Aerosol Detection for Liquid Chromatography and Related Separation Techniques, 2017. John Wiley & Sons, New Jersey, pp. 223–272.

Lisa, M., Netusilová, K., Franek, L., Dvořáková, H., Vrkoslav, V., Holčapek, M., 2011. Characterization of fatty acid and triacylglycerol composition in animal fats using silver-ion and non-aqueous reversed-phase high-performance liquid chromatography/mass spectrometry and gas chromatography/flame ionization detection. J. Chromatogr. A 1218, 7499–7510.

Liu, L., Ngadi, M.O., 2014. Predicting intramuscular fat content of pork using hyperspectral imaging. J. Food Eng. 134, 16–23.

Ludwiczak, A., Stanisz, M., Janiszewski, P., et al., 2017. Novel ultrasound approach for measuring marbling in pork. Meat Sci. 131, 176–182.

Mansur, A.R., Lee, H.J., Choi, H.-K., Jang, H.W., Nam, T.G., 2018. Comparison of two commercial solid-phase microextraction fibers for the headspace analysis of volatile compounds in different pork and beef cuts. J. Food Process Preserv. 42, e13746.

Martini, S., Cavalchi, M., Conte, A., Tagliazucchi, D., 2018. The paradoxical effect of extra-virgin olive oil on oxidative phenomena during in vitro co-digestion with meat. Food Res. Int. 109, 82–90.

Mi, S., Shang, K., Jia, W., Zhang, C.-H., Li, X., Fan, Y.-Q., et al., 2018. Characterization and discrimination of Taihe black-boned silky fowl (*Gallus gallus domesticus* Brisson) muscles using LC/MS-based lipidomics. Food Res. Int. 109, 187–195.

Miyake, T., Shibamoto, T., 1996. Simultaneous determination of acrolein, malonaldehyde and 4-hydroxy-2-nonenal produced from lipids oxidized with Fenton's reagent. Food Chem. Toxicol. 34, 1009–1011.

Monziols, M., Collewet, G., Bonneau, M., Mariette, F., Davenel, A., Kouba, M., 2006. Quantification of muscle, subcutaneous fat and intermuscular fat in pig carcasses and cuts by magnetic resonance imaging. Meat Sci. 72, 146–154.

Mottram, H.R., Woodbury, S.E., Evershed, R.P., 1997. Identification of triacylglycerol positional isomers present in vegetable oils by high performance liquid chromatography/atmospheric pressure chemical ionization mass spectrometry. Rapid Commun. Mass Spectrom. 11, 1240–1252.

Mottram, H.R., Crossman, Z.M., Evershed, R.P., 2001. Regiospecific characterisation of the triacylglycerols in animal fats using high performance liquid chromatography-atmospheric pressure chemical ionization mass spectrometry. Analyst 126, 1018–1024.

Mousa, R.M.A., Al-Khateeb, L.A., 2017. Influence of binary and ternary mixtures of spices on the inhibition of lipid oxidation and carcinogenic heterocyclic amines in fried hamburger patties. J. Food Process Preserv. 41, e12976.

Narváez-Rivas, M., Gallardo, E., Rios, J.J., Leon-Camacho, M., 2011. A new high-performance liquid chromatographic method with evaporative light scattering detector for the analysis of phospholipids. Application to Iberian pig subcutaneous fat. J. Chromatogr. A 1218, 3453–3458.

O'Fallon, J.V., Busboom, J.R., Nelson, M.L., Gaskins, C.T., 2007. A direct method for fatty acid methyl ester synthesis: application to wet meat tissues, oils, and feedstuffs. J. Anim. Sci 85, 1511–1521.

Olivares, A., Dryahina, K., Navarro, J.L., et al., 2011. SPME-GC-MS versus selected ion flow tube mass spectrometry (SIFT-MS) analyses for the study of volatile compound generation and oxidation status during dry fermented sausage processing. J. Agric. Food Chem. 59, 1931–1938.

Olsen, E., Vogt, G., Ekeberg, D., et al., 2005. Analysis of the early stages of lipid oxidation in freeze-stored pork back fat and mechanically recovered poultry meat. J. Agric. Food Chem. 53, 338–348.

Osawa, C.C., De Felício, P.E., Gonçalves, L., Ap, G., 2005. TBA test applied to meats and their products: traditional, modified and alternative methods. Quim. Nova 28, 655–663.

Ouyang, G., Pawliszyn, J., 2008. A critical review in calibration methods for solid-phase microextraction. Anal. Chim. Acta 627, 184–197.

Öztürk, B., Serdaroğlu, M., 2017. The effects of egg albumin incorporation on quality attributes of pale, soft, exudative (PSE-like) turkey rolls. J. Food Sci. Technol. 54, 1384–1394.

Pacetti, D., Hulan, H.W., Schreiner, M., Boselli, E., Frega, N.G., 2005. Positional analysis of egg triacylglycerols and phospholipids from hens fed diets enriched with refined seal blubber oil. J. Sci. Food Agric. 85, 1703–1714.

Paradiso, V.M., Pasqualone, A., Summo, C., Caponio, F., 2018. An "Omics" approach for lipid oxidation in foods: the case of free fatty acids in bulk purified olive oil. Eur. J. Lipid Sci. Technol. 120, 1800102.

Pegg, R.B., 2005. Measurement of primary lipid oxidation products. Handb. Food Anal. Chem. 1–2, 515–529.

Pérez-Palacios, T., Antequera, T., Muriel, E., Ruiz, J., 2006. Stereospecific analysis of phospholipid classes in rat muscle. Eur. J. Lipid Sci. Technol. 108, 835–841.

Pérez-Palacios, T., Antequera, T., Muriel, E., Martín, D., Ruiz, J., 2007. Stereospecific analysis of phospholipid classes in skeletal muscle from rats fed different fat sources. J. Agric. Food Chem. 55, 6191–6197.

Pérez-Palacios, T., Ruiz, J., Martin, D., Muriel, E., Antequera, T., 2008. Comparison of different methods for total lipids quantification in meat and meat products. Food Chem. 110, 1025–1029.

Pérez-Palacios, T., Ruiz, J., Grau, R., Flores, M., Antequera, T., 2009. Influence of pre-cure freezing of Iberian hams on lipolytic changes and lipid oxidation. Int. J. Food Sci. Technol. 44, 2287–2295.

Pérez-Palacios, T., Ruiz, J., Dewettinck, K., Trung Le, T., Antequera, T., 2010a. Individual phospholipid classes from Iberian pig meat as affected by diet. J. Agric. Food Chem. 58, 1755–1760.

Pérez-Palacios, T., Ruiz, J., Dewettinck, K., Trung Le, T., Antequera, T., 2010b. Muscle individual phospholipid classes throughout the processing of dry-cured ham: influence of pre-cure freezing. Meat Sci. 84, 431−436.

Pérez-Palacios, T., Ruiz, J., Ferreira, I., Petisca, C., Antequera, T., 2012. Effect of solvent to sample ratio on total lipid extracted and fatty acid composition in meat products within different fat content. Meat Sci. 91, 369−373.

Pérez-Palacios, T., Caballero, D., Caro, A., Rodríguez, P.G., Antequera, T., 2014. Applying data mining and Computer Vision Techniques to MRI to estimate quality traits in Iberian hams. J. Food Eng. 131, 82−88.

Pérez-Palacios, T., Caballero, D., Antequera, T., Duran, M.L., Ávila, M.M., Caro, A., 2017. Optimization of MRI acquisition and texture analysis to predict physicochemical parameters of loins by data mining. Food Bioprocess Technol. 10, 750−758.

Perona, J.S., Ruiz-Gutierrez, V., 2005. Quantitative lipid composition of Iberian pig muscle and adipose tissue by HPLC. J. Liq. Chromatogr. Relat. Technol. 28, 2445−2457.

Picouet, P.A., Muñoz, I., Fulladosa, E., Daumas, G., Gou, P., 2014. Partial scanning using computed tomography for fat weight prediction in green hams: scanning protocols and modelling. J. Food Eng. 142, 146−152.

Raes, K., De Smet, E., 2009. Fatty acids. In: Nollet, L.M.L., Toldrá, F. (Eds.), Handbook of Muscle Food Analysis. CRC Press, Boca Raton, pp. 167−786.

Raharjo, S., Sofos, J.N., 1993. Methodology for measuring malonaldehyde as a product of lipid peroxidation in muscle tissues: a review. Meat Sci. 35, 145−169.

Raharjo, S., Sofos, J.N., Schmidt, G.R., 1992. Improved speed, specificity, and limit of determination of an aqueous acid extraction thiobarbituric acid-c_{18} method for measuring lipid peroxidation in beef. J. Agric. Food Chem. 40, 2182−2185.

Reitznerová, A., Uleková, M., Nagy, J., Marcinčák, S., Semjon, B., Čertík, M., et al., 2017. Lipid peroxidation process in meat and meat products: a comparison study of malondialdehyde determination between modified 2-thiobarbituric acid spectrophotometric method and reverse-phase high-performance liquid chromatography. Molecules 22, 1988.

Rombaut, R., Camp, J.V., Dewettinck, K., 2005. Analysis of phosphor- and sphingolipids in dairy products by a new HPLC methods. J. Dairy Prod. 88, 482−488.

Ross, C.F., Smith, D.M., 2006. Use of volatiles as indicators of lipid oxidation in muscle foods. Comp. Rev. Food Sci. Food Saf. 5, 18−25.

Ruan, E.D., Aalhus, J., Juárez, M., 2014. A rapid, sensitive and solvent-less method for determination of malonaldehyde in meat by stir bar sorptive extraction coupled thermal desorption and gas chromatography/mass spectrometry with in situ derivatization. Rapid Comm. Mass Spectrom. 28, 2723−2728.

Ruiz, J., Antequera, T., Andres, A.I., Petrón, M.J., Muriel, E., 2004. Improvement of a solid phase extraction method for analysis of lipid fractions in muscle foods. Anal. Chim. Acta 520, 201−205.

Ruiz, J., Muriel, E., Perez-Palacios, T., Antequera, T., 2009. Analysis of phospholipids in muscle foods. In: Nollet, L.M.L., Toldrá, F. (Eds.), Handbook of Muscle Food Analysis, 2009. CRC Press, Boca Raton, pp. 167−786.

Santos-Garcés, E., Gou, P., Garcia-Gil, N., Arnau, J., Fulladosa, E., 2010. Non-destructive analysis of aw, salt and water in dry-cured hams during drying process by means of computed tomography. J. Food Eng. 101, 187−192.

Shahidi, F., Pegg, R.B., Harris, R., 1991. Effects of nitrite and sulfanilamide on the 2-thiobarbituric acid (TBA) values in aqueous model and cured meat systems. J. Mus. Foods 2, 1−9.

Shahidi, F., Wang, J., Wanasundara, U.N., 2017. Methods for measuring oxidative rancidity in fats and oils, Food Lipids: Chemistry, Nutrition, and Biotechnology, fourth ed. CRC Press, pp. 519−542.

Siciliano, C., Belsito, E., Marco, R., Di Gioia, M.L., Leggio, A., Liguori, A., 2013. Quantitative determination of fatty acid chain composition in pork meat products by high resolution 1H NMR spectroscopy. Food Chem. 136, 546−554.

Simonetti, M.S., Blasi, F., Bosi, A., Maurizi, A., Cossignani, L., Damiani, P., 2008. Stereospecific analysis of triacylglycerol and phospholipid fractions of four freshwater fish species: Salmotrutta, Ictaluruspunctatus, Ictalurusmelas and Micropterussalmoides. Food Chem. 110, 199–206.

Stefanova, R., Vasilev, N.V., Vassilev, N.G., 2011. 1H-NMR spectroscopy as an alternative tool for the detection of gamma-ray irradiated meat. Food Anal. Methods 4, 399–403.

Srinivasan, S., Xiong, Y.L., Decker, E.A., 1996. Inhibition of protein and lipid oxidation in beef heart surimi-like material by antioxidants and combinations of pH, NaCl, and buffer type in the washing media. J. Agric. Food Chem. 44, 119–125.

Su, H., Sha, K., Zhang, L., et al., 2014. Development of near infrared reflectance spectroscopy to predict chemical composition with a wide range of variability in beef. Meat Sci. 98, 110–114.

Sun, Y.-E., Wang, W.-D., Chen, H.-W., Li, C., 2011. Autoxidation of unsaturated lipids in food emulsion. Crit. Rev. Food Sci. Nutr. 51, 453–466.

Takahashi, R., Nakaya, M., Kotaniguchi, M., 2018. Analysis of phosphatidylethanolamine, phosphatidylcholine, and plasmalogen molecular species in food lipids using an improved 2D highperformance liquid chromatography system. J. Chromatogr. B 1077–1078, 35–43.

Tao, F., Ngadi, M., 2016. Recent advances in rapid and non-destructive assessment of meat quality using hyperspectral imaging. Proc. SPIE – Int. Soc. Opt. Eng 9860, 98600G.

Tarladgis, B.G., Watts, B.M., Younathan, M.T., Dugan Jr., L., 1960. A distillation method for the quantitative determination of malonaldehyde in rancid foods. J. Am. Oil Chem Soc. 37, 44–48.

Trivedi, D.K., Hollywood, K.A., Rattray, N.J.W., Ward, H., Trivedi, D.K., Greenwood, J., et al., 2016. Meat, the metabolites: an integrated metabolite profiling and lipidomics approach for the detection of the adulteration of beef with pork. Analyst 141, 2155–2164.

Turner, E.W., Paynter, W.D., Montie, E.J., Bessert, M.W., Struck, G.M., Olson, F.C., 1954. Use of the 2-thiobarbituric acid reagent to measure rancidity in frozen pork. Food Tech. 8, 326–330.

Ulu, H., 2004. Evaluation of three 2-thiobarbituric acid methods for the measurement of lipid oxidation in various meats and meat products. Meat Sci. 67, 683–687.

Vester-Christensen, M., Erbou, S.G.H., Hansen, M.F., et al., 2009. Virtual dissection of pig carcasses. Meat Sci. 81, 699–704.

Wang, D., Xu, W., Xu, X., et al., 2009. Determination of intramuscular phospholipid classes and molecular species in Gaoyou duck. Food Chem. 112, 150–155.

Wang, X., Zhu, L., Han, Y., et al., 2018a. Analysis of volatile compounds between raw and cooked beef by HS-SPME–GC–MS. J. Food Process. Preserv. 42, e13503.

Wang, W., Feng, X., Zhang, D., et al., 2018b. Analysis of volatile compounds in Chinese dry-cured hams by comprehensive two-dimensional gas chromatography with high-resolution time-of-flight mass spectrometry. Meat Sci. 140, 14–25.

Williams, J.P., Khan, M.U., Wong, D., 1995. A simple technique for the analysis of positional distribution of fatty acids on di- and triacylglycerols using lipase and phospholipase A2. J. Lipid Res. 36, 1407–1412.

Witte, V.C., Krause, G.F., Bailey, M.E., 1970. A new extraction method for determining 2-thiobarbituric acid values of pork and beef during storage. J. Food Sci. 35, 582–585.

Wong, D., Hu, X., Tao, N., Wang, X., Wang, M., 2016. Effect and mechanism of pyridoxamine on the lipid peroxidation and stability of polyunsaturated fatty acids in beef patties. J. Sci. Food Agric. 96, 3418–3423.

Wood, J.D., Enser, M., Fisher, A.V., Nute, G.R., Sheard, P.R., Richardson, R.I., et al., 2008. Fat deposition, fatty acid composition and meat quality: a review. Meat Sci. 78, 343–358.

Xu, C.-H., Chen, G.-S., Xiong, Z.-H., Fan, Y.-X., Wang, X.-C., Liu, Y., 2016. Applications of solid-phase microextraction in food analysis. TrAC - Trends Anal. Chem. 80, 12–29.

Yao, L., Schaich, K.M., 2014. Accelerated solvent extraction improves efficiency of lipid removal from dry pet food while limiting lipid oxidation. J. Am. Oil Chemists' Soc. 92, 141–151.

Younathan, M.T., Watts, B.M., 1960. Oxidation of tissue lipids in cooked pork. J. Food Sci. 25, 538–543.

Yu, T.C., Sinnhuber, R.O., 1967. An improved 2-thiobarbituric acid (TBA) procedure for the measurement of autoxidation in fish oils. J. Am. Oil Chem. Soc. 44, 256–258.

Zhang, M., Chen, X., Hayat, K., et al., 2018. Characterization of odor-active compounds of chicken broth and improved flavor by thermal modulation in electrical stewpots. Food Res. Int. 109, 72–81.

Zhao, J., Wang, M., Xie, J., et al., 2017. Volatile flavor constituents in the pork broth of black-pig. Food Chem. 226, 51–60.

Zipser, M.W., Watts, B.M., 1961. Lipid oxidation in heat-sterilized beef. Food Technol. 15, 445–447.

Further reading

Gamero-Negrón, R., García, C., Reina, R., Sánchez del Pulgar, J., 2018. Immune-spaying as an alternative to surgical spaying in Iberian x Duroc females: effect on the sensory traits and volatile organic compound profile of dry-cured shoulders and dry-cured loins. Meat Sci. 143, 237–241.

CHAPTER 14

Plant antioxidants, extraction strategies, and their application in meat

Zabdiel Alvarado-Martinez[1], Arpita Aditya[2] and Debabrata Biswas[3]

[1]Department of Biology-Molecular and Cellular Biology, University of Maryland, College Park, MD, United States [2]Department of Animal and Avian Sciences, University of Maryland, College Park, MD, United States [3]Center for Food Safety and Security Systems, Department of Animal and Avian Sciences, University of Maryland, College Park, MD, United States

Chapter Outline

14.1 Introduction   242
14.2 Common sources of plant antioxidants   243
14.3 Common extraction methodologies/strategies of various antioxidants from plant sources   243
    14.3.1 Extraction strategy of plant polyphenolic compounds   246
    14.3.2 Purification and fractionation   249
14.4 Application of antioxidants in animal products during the postharvest stage   251
    14.4.1 Improve the flavor and the shelf life   251
    14.4.2 Reduce microbiological contamination   251
    14.4.3 Improve the nutritional values   252
14.5 Mechanisms of action of antioxidants   252
14.6 Enzymatic antioxidants   253
    14.6.1 Superoxide dismutase   253
    14.6.2 Catalase   253
    14.6.3 Glutathione peroxidase   253
14.7 Natural nonenzymatic antioxidants   254
    14.7.1 Vitamin E   254
    14.7.2 Vitamin C   254
    14.7.3 Vitamin A   255
    14.7.4 Flavonoids   256
    14.7.5 Phenolic acids   257
14.8 Conclusion   257
References   258

14.1 Introduction

As a result of metabolic processes within eukaryotic organisms, there is the production of highly reactive molecules, or molecule fragments known as free radicals. These molecules are characterized as having one or more electrons that remain unpaired, making them highly unstable, and therefore highly reactive with other substances (Valko et al., 2006). Because of their unstable and reactive nature, free radicals can interact with different macromolecules within cells, which can lead to alterations in their chemical structures, and impairments in their normal functions. Such damage that occurs within the cell is known as oxidative stress. It is important to note that within mammalian cells, the most common free radicals are derived from reactive oxygen species (ROS) and reactive nitrogen species (RNS) (Valko et al., 2007).

Free radicals are an inevitable consequence of metabolic processes within the cell. Most ROS will originate within the mitochondria as it goes about cellular respiration. Some electrons leak from the electron transport chain and are taken up by dioxygen molecules, generating a superoxide molecule that has additional electrons and is highly oxidative as a result (Qiao et al., 2018). The generation of RNS occurs as a result of nitric oxide (NO) molecules interacting with metals from protein centers, molecular oxygen, or other free radicals (Patel et al., 1999). NO is an important cell signaler produced to regulate different processes, like inflammation and vascular pressure, but is susceptible to oxidation, which will lead to the formation of other free radicals (Van Der Vliet et al., 1997). As these excess radicals interact with other molecules within the cell and oxidize them, there will be a cascading effect by which other free radicals begin to form in increasing numbers. Such a case is that of hydroxyl and peroxylradicals that are very reactive molecules, which come from a reaction between a superoxide radical and a hydrogen peroxide molecule.

In order to balance the levels of free radicals, complex organisms have developed the use of antioxidants. Antioxidants are molecules that can neutralize free radicals, or prevent the oxidation of other molecules that could be susceptible to reacting with highly reactive radicals (Clark, 2002). Animals produce endogenous antioxidants that aid in preventing free radicals from exceeding a baseline level. However, limited production of these antioxidants, coupled with their natural half-life makes it necessary for animals to consume additional antioxidants in order to remain healthy. These can be obtained from plants and various plant products, which carry an abundance of antioxidants that can be ingested directly by consuming plant-derived products, or they can also be extracted and purified for later use (Kasote et al., 2015). Based on the available information, it is possible to visualize how the health benefits that have been correlated with consumption of antioxidants have the potential to improve the quality of meat products. Much research supports the use of antioxidants derived from natural plant products being able to be utilized as supplements that can both improve the health of farm animals, as well as improve the quality and safety of their products after they have been harvested.

14.2 Common sources of plant antioxidants

Plants are known for producing and storing a large variety and quantity of antioxidants, which makes them the main source of dietary antioxidants that are taken up by animals, including humans, on a daily basis (Pratt, 1992). The reason for producing so many antioxidants has been correlated with the plant's continuous exposure to free radicals that are formed as a result of continuous metabolic processes like cellular respiration and photosynthesis. This higher level of metabolic activity is greater than that for animals, which has led plants to evolve diverse strategies that control the formation of reactive free radicals and manage any excess stress these might cause (Kasote et al., 2015). Part of these strategies involves the development of both enzymatic and nonenzymatic antioxidants that can stabilize the large amount of reactive free radicals that are constantly being formed as a result of metabolic processes in the mitochondria and the chloroplast. The enzymatic molecules are very similar to the ones animals possess; however, nonenzymatic antioxidants are molecules of low molecular weight that are almost exclusively synthesized by plants. Even though plants are thought of as generally containing high amounts of antioxidants, there will be a considerable level of variation in the relative abundance of these molecules across different plants (Carlsen et al., 2010). Many factors can influence the amount of antioxidants that will be found in a given plant, but the most common influences will be related to the species of plant that is being harvested, the conditions in which it was grown (temperature, pH, altitude, water abundance, minerals, and the nutrients present in the soil), the part of the plant that is being harvested, the age of sample, and the method that was used to extract the product (Akula and Ravishankar, 2011).

Multiple studies have been performed in order to identify plant candidates from which antioxidants can be extracted and used as supplements to the normal dietary intake. Some of the more studied plants and plant parts that are used are listed in Table 14.1. Along with this knowledge, different extraction strategies implemented for the purpose of extracting antioxidants from plants will be discussed further.

14.3 Common extraction methodologies/strategies of various antioxidants from plant sources

Plants contain thousands of bioactive phytochemical compounds that have antioxidant properties. Among some of the more common products are vitamins, polyphenols, flavonoids, and tannins. Extracting them from natural sources is the preliminary step toward using them

Table 14.1: Sources of plant-based antioxidants.

Plant	Parts used	Extraction strategy	Antioxidants	Amount used in meat	Meat Product	Effects	Reference
Arbutus-berries (*Arbutus unedo*), common hawthorns (*Crataegus monogyna*), dog roses (*Rosa canina*) elm-leaf blackberries (*Rubus ulmifolius* Schott)		Absolute ethanol			Cooked burger patties	Reduced protein oxidation	Ganhão et al. (2010)
Black and Green tea (*Camellia sinensis*)	Leaf	Boiled water	Catechin, flava-3-ols, flavonols		Raw beef patties, Turkish dry-fermented sausage	Inhibit lipid oxidation; reduced TBARS, putrescine, histamine, and tyramine formation	Bozkurt (2006), Mitsumoto et al. (2005)
Carrot juice (*Daucus carota*)	Root	Squeezing/ evaporative concentration	Phenolics, carotenoids		Gamma irradiated raw beef sausage	Decreased lipid and protein oxidation, increase shelf life	Badr and Mahmoud (2011)
Curry (*Murraya koenigii*) and mint (*Mentha spicata*)	Leaf	Ethanol, hot water, or in combination	Phenolics		Raw ground pork in refrigeration	Reduced auto-oxidation	Biswas et al. (2012)
Defatted canola (*Brassica napus*) meal		70% Acetone	Sinapic, ferulic, *p*-hydroxybenzoic acids	15 or 100 mg gallic acid equivalents/ kg meat	Cooked beef, chicken and pork	Inhibited lipid oxidation	Brettonnet et al. (2010)
Drumstick tree (*Moringa oleifera*)	Leaf		Phenolic components		Ground pork patties	Inhibit lipid oxidation	Muthukumar et al. (2014)
Ginger (*Zingiber officinale*)	Rhizome		Phenolic and sulfur compounds		Pork	Decrease lipid oxidation	Tanabe et al. (2002)
Grape (*Vitis vinifera*)	Seed, pomace	Freeze-drying, milling, methanol, instantaneous pressure change	Phenolics mainly tannins catechins, flavonols, anthocyanidins		Raw and cooked pork patties; pork burger	Decreasing lipid oxidation; color stability	Garrido et al. (2011), Sáyago-Ayerdi et al. (2009)
Lemon balm (*Melissa officinalis*)	Leaf	Preheated (100°C) water, refluxing	Phenolics, rosmarinic acid		Meat products		de Ciriano et al. (2010)
Lotus (*Nelumbo nucifera*)	Rhizome knot and lotus leaf extract	Distill water	Phenolics, tannins, flavonoids		Ground pork and beef	Decrease lipid oxidation	Huang et al. (2011)

Plant	Part	Extraction	Active compounds	Concentration	Meat product	Effect	Reference
Mint (*Mentha spicata*)	Leaf	Reflux with distilled water	Phenolics and flavonoids		Radiation processed lamb meat	Retard lipid oxidation	Kanatt et al. (2007)
Mung bean (*Vigna radiata*), Bengal gram (*Cicer arietinum*), pigeon pea (*Cajanus cajan*)	Hull	Distilled water/ evaporative concentration	Phenolics, flavonoids			Active against common food spoilage microbes including *Escherichia coli*, *Staphylococcus aureus*	Kanatt et al. (2011)
Mustard (*Brassica juncea*)	Leaf kimchi extract	70% Ethanol			Refrigerated raw ground pork meat	Decrease lipid oxidation	Lee et al. (2010)
Oregano (*Origanum vulgare*)	Leaf, essential oil	Hydrodistillation, Diethyl ether, ethyl alcohol, and distilled water	Rosmarinic acid, phenolic acids, flavonoids		Beef and pork; irradiated beef burger	Reduce oxidation during refrigeration	Fasseas et al. (2008), Trindade et al. (2010)
Pomegranate (*Punica granatum* var *kabul*)	Seeds and rind	Grinding and filtration boiled distilled water, 70% acetone and diethyl ether	Phenolics, pro-anthocyanidins, tannins	10 mg equivalent phenolics/ 100 g meat	Chicken breast and cooked chicken patties	Inhibited protein and lipid oxidation; increase shelf life	Naveena et al. (2008), Vaithiyanathan et al. (2011)
Rosemary (*Rosmarinus officinalis*)	Leaf and secondary branches	Chloroform, ethanol, chloroform + ethanol	Carnosic acid, carnosol, rosmarinic acid rosmanol		Beef and pork	Reduce oxidation during refrigeration, color stabilization at higher doses	Djenane et al. (2003)
Sage (*Salvia officinalis*)	Leaf	Hydrodistillation	Carnosol, carnosic acid, lateolin, rosmanul, rosmarinic acid, rosmarinic acid		Beef and pork	Reduce oxidation during refrigeration	Fasseas et al. (2008)
Sansho (*Zanthoxylum piperitum* L.)	Fruit and seed				Pork	Decrease lipid oxidation	Tanabe et al. (2002)
Thuza (*Thuja occidentalis*)	Cone	Boiled sterilized distilled water	Phenolics, flavonoids		Raw ground chicken	Control lipid oxidation	Yogesh and Ali (2014)

as food additives, dietary supplements, therapeutics, and even in beauty care products (Dai and Mumper, 2010).

14.3.1 Extraction strategy of plant polyphenolic compounds

Generally, before extracting the phenolic components from plant sources, they are pretreated by grinding, milling, and homogenization. Sometimes these small pieces are dried in air, or are frozen to increase the yield. However, it is important to note that freeze-drying has been shown to have an improper preservation impact on some of the medicinal compounds (e.g., volatiles, phenolics, and carotenoids) (Abascal et al., 2005). On the other hand, some studies have found a consistently greater yield of total phenolic content when performing extractions from previously freeze-dried marionberries, strawberries, and corn that were later air-dried (Asami et al., 2003).

The yield and efficiency of extraction will depend on the type of solvent that is used during the procedure, and the product that is being extracted. Plants are the source of diverse phenolic compounds ranging from simple phenolic acids to highly complex tannins, which might require specific parameters in order to carry out a proper and effective extraction. The most commonly used solvents are aqueous solutions of methanol, ethanol, acetone, and ethyl acetate (Table 14.2). In many cases, a cocktail of these solvents is used to obtain higher yields. The solvent is one of the crucial factors for getting the desired amount of the polyphenol component (Xu and Chang, 2007), along with other important factors, such as extraction time, the temperature at which the extraction is performed, sample-to-solvent ratio, as well as the chemical and physical status of the samples. Based on the solvent composition, a mixture of soluble phenolics can be obtained from the plant source. It is also highly likely that the phenolic compounds would be mixed with many nonphenolic substances like sugar, organic acids, and fats. Further purification steps are needed to exclude those unwanted parts (Dai and Mumper, 2010).

Table 14.2: Solvents used for extracting phenolic components from plant sources.

Solvent	Extracted materials	References
Methanol	Low molecular weight polyphenols	Metivier et al. (1980)
Ethanol	Polyphenols	Shi et al. (2005)
Aqueous acetone	High molecular weight flavanols	Guyot et al. (2001)
Weak organic acids, for example, formic acid, acetic acid, citric acid, tartaric acid and phosphoric acid, and low concentrations of strong acids, for example, 0.5%–3.0% of trifluoroacetic acid <1% hydrochloric acid	Anthocyanin	Nicoué et al. (2007), Jackman et al. (1987)

14.3.1.1 Conventional methods

The effective extraction of plant nutraceuticals through conventional methods relies heavily on the solvent chosen for extraction, coupled with the use of heat and/or agitation. Existing classical techniques used to obtain nutraceuticals from plants include Soxhlet, hydrodistillation (HD), and maceration with an alcohol–water mixture or hot fat (Wang and Weller, 2006).

14.3.1.1.1 Soxhlet extraction or hot continuous extraction

Soxhlet extraction is well-established and efficient for extracting thermostable compounds. In a typical Soxhlet system, a porous bag or thimble made of strong filter paper or cellulose is filled with grounded plant material, and then placed in the thimble-holder. The solvent flask is filled with the extraction solvent (e.g., hexane, isopropanol, ethanol, hydrocarbons, water, d-limonene, etc.), and heated at the bottom. The solvent vaporizes and passes through the sample thimble as it condenses in the condenser and plops back. When the liquid level reaches the siphon arm, it is taken out of the thimble-holder and put back in the solvent flask and the process is continued until a complete extraction is achieved (Wang and Weller, 2006; Nn, 2015).

However, limiting factors such as temperature, solvent-sample ratio, and stirring speed should be considered before selecting this method. Moreover, this process generates a large amount of purified solvent, but it requires a long extraction time to achieve the final product. It works well for finely minced solids, but it cannot exclude the possibility of thermal degradation. This procedure is not considered ecofriendly, and may add to the pollution problem, compared to other advanced techniques (Amid et al., 2010; de Castro and García-Ayuso, 1998).

14.3.1.1.2 Maceration extraction

This extraction strategy is very straightforward. The plant material is placed in the maceration container (e.g., beaker or Erlenmeyer flask) with the desired solvent. The sample is kept in the dark to macerate for 72 hours at room temperature (Trusheva et al., 2007).

14.3.1.1.3 Hydrodistillation

The HD process is mainly used to extract essential oils and other volatile compounds. It is performed according to the European Pharmacopeia (1975). The plant material is subjected to HD for a couple of hours in a Clevenger-type apparatus. The plant oil extract is collected over anhydrous sodium sulfate, and stored at 0°C before use (Lucchesi et al., 2004; Sourmaghi et al., 2015).

14.3.1.2 Modern methods

14.3.1.2.1 Ultrasound-assisted extraction

Using ultrasound waves (20–1000 kHz) during solvent extraction is proven to increase the phytochemical yield. Ultrasound-assisted extraction (UAE) is a useful technology due to its simple design, which can be optimized for both small- and large-scale extraction at a lower cost (Vinatoru, 2001). In this technique, the movement of acoustic waves in the 20–800 kHz range causes mechanical vibrations in solid, liquid, and gas. While moving in a medium, the sound waves involve rhythmic expansion and compression. The expansion makes cavitation bubbles in the liquid, which in turn creates a shear force on the solid surface that collapses the bubbles (Laborde et al., 1998; Luque-García and De Castro, 2003).

Two types of ultrasound-assisted extractors are being used; ultrasonic bath and closed extractor associated with an ultrasonic horn transducer (Mason et al., 1996). The sound wave increases the penetration of solvent into the plant materials. It also causes the release of extractable cellular contents by disrupting plant cell walls, and improves mass transfer (Mason et al., 1996; Vinatoru, 2001; Vinatoru et al., 1997). The effects of ultrasound on the plant cell walls have been made evident through scanning electron microscopy. Unlike the conventional extraction, the ultrasound causes the bioactive components to diffuse through the cell walls, as well as causing the eventual rupture of the cell in a shorter period of time (Toma et al., 2001). UAE has been used to extract bioactive compounds from leaves (Albu et al., 2004), stalks (Yang and Zhang, 2008), fruits (Herrera and De Castro, 2004; Rostagno et al., 2003), and seeds (Hromádková et al., 2008).

14.3.1.2.2 Pressurized liquid extraction

This relatively new technique was first introduced by Dionex Corporation (Dionex, Sunnyvale, CA, USA) at the Pittcon Conference during the mid-1990s. They introduced it as Accelerated Solvent Extraction Technology (ASE). This technique is also known as pressurized solvent extraction, or enhanced solvent extraction. If water is used as the extraction solvent, it is named as pressurized hot water extraction, subcritical water extraction, or superheated water extraction (Mustafa and Turner, 2011).

The pressurized liquid extraction (PLE) technique uses high temperature (40°C–200°C) and pressure (3.3–20.3 MPa) to extract bioactive components from plant origins. This combination of extreme conditions requires smaller amounts of solvents, for example, 20 minutes of extraction using 10–50 mL solvent in PLE is equivalent to 10–48 hours extraction in about 200 mL of solvent during a traditional extraction process (Mendiola et al., 2007). At high temperature, the strong bonds between solutes and matrices (hydrogen bonds, van der Waals force, and dipole attractions of the solute) become weakened. Therefore the activation energy is decreased, and easily attained by the heat energy

(Richter et al., 1996). The viscosity and surface tension of the solvents are also lowered at high temperature, thus a better penetration of the solvent into the matrix is achieved, which in turn aids extraction (Möckel et al., 1987). At the same time, the pressure forces the solvent to such areas (e.g., in pores or air bubbles) that would be difficult to be contacted by solvents at normal pressures. A pressurized liquid flow helps in solubilizing air bubbles and the solvent comes in close contact with the sample matrix faster (Richter et al., 1996).

At high temperature, phenolic compounds are easily oxidized. Therefore temperature should be chosen prudently for PLE technique (Palma et al., 2001). The special conditions that the high-pressure technique requires will end up adding to the cost of production if it were to be applied at an industrial scale. This kind of trade-off may surpass the process benefits (Dai and Mumper, 2010). PLE is being used efficiently to extract antioxidants from apples (Alonso-Salces et al., 2001), barley flours (Bonoli et al., 2004), eggplants (Howard and Pandjaitan, 2008), and grape seeds and skins (Luque-Rodríguez et al., 2007; Piñeiro et al., 2006).

14.3.1.2.3 Microwave-assisted extraction

Microwave-assisted extraction (MAE) uses microwave energy to extract bioactive plant materials in an appropriate solvent. It is being used efficiently to extract small-molecules like phenolic acids (e.g., gallic acid, ellagic acid) (Li et al., 2004), isoflavone (Rostagno et al., 2003), quercetin (Huang and Zhang, 2004), anthraquinone (Hemwimon et al., 2007), and *trans*-resveratrol (Du et al., 2007). These compounds do not denature under microwave-assisted heating conditions, even at above 100°C temperature (Liazid et al., 2007). Selective extraction of the desired compounds is achieved through the application of of highly localized temperature and pressure, which is why the dipolar rotation of the polar solvent, which is influenced by the dielectric constant, will play a role in increasing its temperature significantly once it is exposed to a microwave field.

14.3.2 Purification and fractionation

Antioxidants in the crude extract are often mixed with carbohydrates and/or lipids. To obtain a concentrated fraction of antioxidants, some common purification strategies are used, such as liquid−liquid procedures, or sequential extraction and/or solid phase extraction (SPE), depending on the acidity and polarity of the polyphenols (Dai and Mumper, 2010). Usually, lipoidal materials can be excluded by washing the crude extract with nonpolar solvents such as hexane (Ramirez-Coronel et al., 2004), dichloromethane (Neergheen et al., 2006), or chloroform (Yang and Zhang, 2008).

14.3.2.1 Liquid—liquid extraction

The conventional liquid—liquid extraction strategy is now less commonly used, since it is demanding in terms of labor, time, solvent costs, and has a low recovery (Klejdus and Kubáň, 2000). A relatively advanced technique known as countercurrent chromatography (CCC) has been introduced as an alternative to liquid chromatography to purify different types of phenolic compounds (Degenhardt et al., 2000b,c). In CCC there are two immiscible liquid phases (mobile and stationary) where the phenolic compounds are partitioned and separated (Naczk and Shahidi, 2004). Solutes are partitioned between the two solvent phases according to their partition coefficient based on their hydrophobicity. As there is no solid bed, the role of two immiscible liquid phases can be switched during a run in CCC, which renders a 100% recovery of the desired components (Berthod et al., 1999). Another strategy, named high-speed countercurrent chromatography (HSCCC), has been employed in separating anthocyanins in the pigment mixture of plants based on their polarity in four different solvent systems (Degenhardt et al., 2000c). By using different solvent systems, HSCCC was also applied to the fractionation of phenolics of red wine (Vitrac et al., 2001); catechins from tea; food-related procyanidins, phenolic acids, and flavanol glycosides (Yanagida et al., 2006); and theaflavins, epitheaflavic acids, and thearubigins from black tea (Degenhardt et al., 2000a). In another strategy, a stepwise gradient elution was coupled to multilayer coil countercurrent chromatography (MCCC) to fractionate anthocyanins into glucosides and respective acetylated, coumaroylated, and caffeoylated derivatives (Vidal et al., 2004). Again, MCCC was coupled with high-performance liquid chromatography to obtain pure flavonoids from Rooibos tea (Krafczyk and Glomb, 2008).

14.3.2.2 Solid phase extraction

Sugars, organic acids, and other polar nonphenolic compounds can be removed by the SPE process. SPE is a popular purification method because it is cost-effective, fast, can be automated, and is more sensitive because of the use of cartridges and discs with various sorbents (Dai and Mumper, 2010). C_{18} Sep-Pak cartridges are most widely used to separate grape phenolics into acidic and neutral fractions. First, the aqueous sample is run through preconditioned C_{18} cartridges, which are later washed with acidified water to exclude water-soluble components like sugars, or other organic acids that might be present. After this process, the polyphenolic portion can be eluted with absolute methanol (Thimothe et al., 2007) or aqueous acetone (Ramirez-Coronel et al., 2004). Optimizing the pH of the sample renders more separation of phenolic compounds. Sorbents such as Amberlite XAD-2 (Llorach et al., 2003), XAD-7 (Salinas Moreno et al., 2005; Zhang et al., 2008), XAD-16 (Zhang et al., 2008), and Oasis HLB (Georgé et al., 2005) are also used to get purified phenolic compounds.

14.4 Application of antioxidants in animal products during the postharvest stage

14.4.1 Improve the flavor and the shelf life

Plant-based industrial by-products are being proved to be a good source of phenolic compounds which also have antioxidant properties. In countries where freezing is expensive, these plant-based extracts can be used at a chilled temperature to increase the shelf life of animal proteins. Many studies have already shown that plant-derived extracts can increase the shelf life of meat products significantly. Pomegranate peel extract at a concentration of 0.1% and 0.5% increased the shelf life of chicken meat products (chicken lollipop and chicken chilli) up to 20 days in chilled temperature (Vaithiyanathan et al., 2011). Grape seed extract has been proved to retain meat freshness in a dose-dependent manner for a longer storage time. The maximum inhibitory effect was seen at 1.6%/kg. Vacuum-packed cooked turkey breast was protected from oxidative rancidity (up to 83.72%) and volatile compounds formation (up to 89.83%) during 13 days refrigeration storage (Mielnik et al., 2006). Kulkarni et al. (2011) conducted a comparative study with grape seed extract and ascorbic acid and propyl gallate in cooked beef sausages stored at $-18°C$ for 4 months. This study reported that grape seed extract (100, 300 ppm) and propyl gallate treated beef samples were fresh, had less rancid odors, and lower thiobarbituric acid reactive substances (TBARS) values compared to the control. Green tea extracts also can increase the shelf life of beef patties by 3 days by inhibiting meat oxidation. The combination of low SO_2-vegetable extract is also effective in preserving raw meats by preventing microbial growth, loss of redness, and lipid oxidation (Bañón et al., 2007). Biswas et al. (2012) reported that uses of curry and mint leaf extract have greater potential to improve the antioxidant activity of pork products and thereby their shelf life.

Avocado peel and seed extract can reduce the color loss of pork patties in chill storage for 15 days by inhibiting protein carbonyl formation (Rodríguez-Carpena et al., 2011). Red grape pomace extract (0.06 g/100 g) also has a similar effect on pork burgers by inhibiting lipid oxidation after 6 days storage at 4°C under aerobic conditions (Garrido et al., 2011). The addition of 10% tomato paste during the manufacturing of the mortadella improved its color stability and nutritional properties, and decreased lipid oxidation during 2 months storage at 4°C (Doménech-Asensi et al., 2013). Mustard leaf kimchi can reduce microbial growth on meat and lipid oxidation as well. It helps extend the shelf life of refrigerated raw ground pork meat up to 14 days (Lee et al., 2010).

14.4.2 Reduce microbiological contamination

Besides preventing lipid and protein oxidation, plant antioxidants can contribute to preserving the meat by retarding the colonization of spoilage-causing microbes. A 0.8% v/w oregano essential oil reduced most of the lactic acid bacteria of meat by $1-2\log_{10}$ and was most effective in reducing *Salmonella* Typhimurium (Skandamis et al., 2002). Among others, clove, pimento, and thyme leaf extracts are effective against *Aeromonas hydrophila*, *Listeria monocytogenes*, and autochthonous spoilage flora at higher concentrations in refrigerated cooked beef and poultry (Hao et al., 1998a,b). One study has reported that oleoresin, rosemary, grape seed (1%), and pine bark extract (1%) can reduce some common foodborne pathogens like *Escherichia coli* O157:H7 and *S.* Typhimurium, and also retarded the growth of *L. monocytogenes* and *A. hydrophila* (Ahn et al., 2007; Fernández-López et al., 2005). A combination of chitosan and rosemary extract exerted the best effect in controlling *Enterobacteriaceae*, *Pseudomonas* spp., total viable bacteria, yeasts and molds, and lactic acid bacteria growth in refrigerated fresh pork sausage (Georgantelis et al., 2007). Hence, a combination of the plant extracts can be a powerful alternative over synthetic compounds in preventing foodborne illnesses.

14.4.3 Improve the nutritional values

When plant origin nutraceuticals are used in meat preservation it tremendously increases the nutritional value of the product because of their well-established therapeutic value in many physiological disorders and their contributions to human health (Jiang and Xiong, 2016). The addition of canola-olive oil, rice bran, and walnut extract in processed meat added more vitamin E, vitamin B, and polyphenols, which ultimately all benefit consumer health (Álvarez et al., 2011, 2012; Cofrades et al., 2004). Vitamin A content of beef patties is boosted by the addition of cooked carrot and sweet potato (Saleh and Ahmed, 1998). Meat products with added lutein are enriched with provitamin A content, which contributed to eye health (Jiang and Xiong, 2016). The dietary fibers of fruit and vegetables give additional value to meat products. Natural antioxidant-rich meat products help the endogenous antioxidants against oxidative stress and ROS-induced tissue damage. The phenolic compounds of plant extracts provide protection to the gastrointestinal tract since absorption is not needed (Halliwell et al., 2005). Naturally occurring antioxidant peptides are derived from protein hydrolysis. They are believed to have many therapeutic effects, such as antihypertensive, anticancer, antimicrobial, immunomodulatory, and opioid activities (Mine et al., 2010). These types of antioxidant are now being considered as potential food ingredients to benefit human health (Jiang and Xiong, 2016).

14.5 Mechanisms of action of antioxidants

From a biochemical standpoint, the mode of action of antioxidants is through scavenging for free radicals and reducing their high reactivity, while also reverting oxidized macromolecules to their original nonoxidized state. The high reactivity of free radicals can be greatly reduced or neutralized when exposed to a molecule that can either accept or donate an electron, like antioxidants. This exchange will stabilize free radicals that might be found in the surrounding system by completely neutralizing them, or by generating less reactive species of them. In many cases, oxidized macromolecules can also be neutralized through similar reactions, even though it might require additional reductions in order for the molecule to revert to its normal state. The exact mode of action of antioxidants will depend on whether they are enzymatic or nonenzymatic and what free radical they are interacting with.

14.6 Enzymatic antioxidants

As the name suggests, enzymatic antioxidants are enzymes produced in eukaryotic cells that are tasked with reducing the deleterious effects of oxidative stress caused by free radicals. These enzymes can be located within the mitochondrial matrix cytosol, in peroxisomes, and in the cytosol of the cell.

14.6.1 Superoxide dismutase

Superoxide dismutase (SOD) is a very common enzyme, as it converts the $O_2^{\bullet-}$ ion to the more stable O_2 molecules, while also generating an H_2O_2. SOD can be found in three forms, which are cytosolic Cu/Zn-SOD, mitochondrial Mn-SOD, and extracellular SOD (EC-SOD) (Petersen et al., 2004). Cu/Zn-SOD contains both a copper and a zinc atom in each of its subunit's binding sites. It operates by catalyzing the reaction that changes $O_2^{\bullet-}$ into oxygen and water through the reduction of its two metal ions. Mn-SOD contains a manganese atom that cycles between Mn(III) to Mn(II) and back to Mn(III) as it goes about removing $O_2^{\bullet-}$. Finally, EC-SOD is the only SOD that operates outside of the cell as it scavenges for $O_2^{\bullet-}$ in plasma, interstitial spaces, and fluids. With an affinity for heparin, EC-SOD can be distributed across tissue more effectively.

14.6.2 Catalase

After SOD eliminates $O_2^{\bullet-}$, H_2O_2 molecules remain, which are not as oxidative, but must still be properly eliminated. Catalase contains an iron(III) ion that allows it to reacts with H_2O_2, forming water and O_2. It also has the capability of acting as a peroxidase, meaning that it can interact with other hydrogen donors that contain a form of peroxide (Matés et al., 1999).

14.6.3 Glutathione peroxidase

Glutathione peroxidase works similarly to catalase; however, it contains a selenocysteine in its active site, which allows it to eliminate peroxides. This is especially important, since it can eliminate hydroperoxide radicals formed from lipids such as fatty acids, phospholipids, cholesterols, and lipoproteins (Matés et al., 1999).

14.7 Natural nonenzymatic antioxidants

Nonenzymatic antioxidants can be either natural or synthetic. Though synthetic nonenzymatic antioxidants have been used in the past, the focus of this section will be on natural nonenzymatic antioxidants and how their mode of action at a biochemical level is relevant to the discussion about improving meat products postharvest and preharvest.

14.7.1 Vitamin E

Vitamin E (tocopherols and tocotrienols) is a group of lipid-soluble antioxidants that can be found in the cell membrane. The main mode of action of vitamin E is to intercept the *proliferation* phase of lipid peroxyl radicals. It engages in scavenging for peroxyl radicals, which will protect lipids from peroxidation. The antioxidant traits of vitamin E come about by donating the hydrogen atom in the 6-hydroxyl group of the chromanol ring to the free radical, which will end up neutralizing it. However, this is a temporary state that is not reactive enough to cause any more considerable problems, as another free radical would. In addition to this, vitamin E can be renewed to its original state by reacting with ascorbic acid (vitamin C) and going through a reduction reaction. The structure of tocols consists of a chromanol ring and a long hydrophobic saturated side chain that can be either phytyl (tocopherols) or isoprenyl (tocotrienols). In addition to this, differences in the position of methyl groups in both tocopherols and tocotrienols will affect the biological activity and absorption (Schwartz et al., 2008). These methyl groups can vary in number and location within the 5, 7, and 8 positions of the phenol part of the chromanol ring, giving tocopherols and tocotrienols five main conformations designated as α-, β-, γ-, and σ-. Of these five species, α-tocopherol has been reported as the most active scavenger of oxygen species, while γ-tocopherol is a better sequesterer of free radicals derived from nitrogen (Sisein, 2014).

14.7.2 Vitamin C

Vitamin C (ascorbic acid) is different from vitamin E in that it is water soluble, but it works alongside it. As mentioned previously, vitamin E itself is oxidized after reducing a free radical, which leaves it as a radical that is stable enough to remain unreactive, but it cannot further serve as an antioxidant unless it is reverted to its original state. Vitamin C is capable of reducing the tocopherol to bring it back to its initial antioxidant form. In addition to this, it is also a scavenger of both oxygen and nitrogen reactive species, which protects DNA from endogenous oxidative damage, and lipids from peroxidation, even though it is not as effective in scavenging hydroxyl radicals. The structure of vitamin A consists of a six-carbon lactone ring, with two oxidizable hydroxyl groups in the 2 and 3 positions of the molecule. In its initial state, ascorbic acid contains two ionizable hydroxyl groups (AscH$_2$), but can donate one proton and electron to form semidehydroascorbic acid (AscH$^-$), and L-dihydroascorbic acid (DHA) after donating a second proton and electron. Ascorbic acid can react with free radicals twice, which will lead to the formation of a tricarbonyl ascorbate radical. Even though this conformation is oxidized, this molecule is very stable and can be reverted back to ascorbic acid through enzymatic or nonenzymatic processes. Oxidoreductase enzymes like the NADH-dependent semidehydroascorbatereductase or NADPH-dependent selenoenzyme thioredoxin reductase can add two protons back to DHA (Carr and Frei, 1999). Nonenzymatic reduction can occur through glutathione (Fernando et al., 2006).

14.7.3 Vitamin A

Vitamin A can be found in two main forms: retinoid (preformed vitamin) and carotenoid (provitamin). Carotenoids are metabolized by the body in order to convert them to vitamin A, which in turn will be metabolized into retinols, which are the active form of the vitamin (Hill and Johnson, 2012). Carotenoids are composed of an eight isoprenoid unit carbon chain, with 11 alternating double bonds. At each end of this chain there is a six-carbon ring with a double bond (Barua et al., 2000). The location of the double bond will determine if the ring is a β-ionone ring (carbons 5–6) or a ε-ionone ring (carbons 4–5), and will have an effect over the biochemical functions of the molecule. Carotenoids with the β-ionone ring and no functional groups are common and are called β-carotenoids. The main reaction that takes place is a one-electron reduction of the radical, thereby neutralizing it, while a positively charged vitamin A radical is formed (Martínez et al., 2008). However, carotenoids with different functional groups will be able to carry out other antioxidant functions in the cell membrane, come into contact with other antioxidants, and quench superoxide molecules directly (Hill and Johnson, 2012). In the case of retinoids, a carotenoid can be cleaved in the center, releasing two retinoid molecules (Nagao, 2004). In turn, different species of retinoids are involved in cell signaling processes, but are also capable of engaging as antioxidants (Conaway et al., 2013). At this point the vitamin A is oxidized and can continue reacting with other radicals, adding them to the other carbons of the cyclohexenyl. One of the most active forms of retinols is that of all-*trans*-retinol, which contains five conjugated double bonds located in the cyclohexenyl ring, along with a side chain that can take up different isomers (Dao et al., 2017). Vitamin A can operate as an antioxidant by scavenging for peroxyl radicals and breaking their chains, therefore preventing further propagation of radicals from lipid peroxidation. From a biochemical standpoint, vitamin A, along with other similar retinols, acts by adding the radical it scavenges to its cyclohexenyl ring (Tesoriere et al., 1997). There is also the possibility of interacting with another retinol, which would neutralize the radical. The aforementioned method is the principal pathway for free radical elimination in the system; however, there is the possibility of the retinol degrading into an alkoxyl radical and a 5,6-retinoid radical. This process would result in the elimination of a lipoperoxyl radical, but the creation of a new and reactive alkoxyl radical, which means no real net change in the number of radicals present (Tesoriere et al., 1997). These characteristics make this vitamin a proficient antioxidant. These reactions have been found to contribute to delaying meat rancidity, as well as improving other animal products, such as milk. In addition to this, it has been found to be involved in preserving the health of the animal in the areas of vision, immunology, and cardiac health.

14.7.4 Flavonoids

Flavonoids are a group of phenolic compounds found as pigments in plants. Flavonoids are composed of many bioactive compounds, such as flavonol, flavone, flavonolols, flavan-3-ols, flavonone, anthocyanidin, and isoflavone (Nimse and Pal, 2015). Flavonoids make up a large group of polyphenols, but the main characteristic that sets them apart is their benzo-γ-pyran ring, also known as the flavylium cation. This structure is modified by different functional groups, particularly hydroxyl groups, which will allow interactions with free radicals, making flavonoids good antioxidants (Kumar and Pandey, 2013). These hydroxyl groups will serve as hydrogen and electron donors to free-roaming hydroxyl, peroxyl, and peroxynitrite radicals, which will stabilize them, while the flavonoid remains relatively stable as well. Much of the activity of their scavenging functions will depend on the OH located on the third carbon, coupled with the closed aromatic ring that can be conjugated and permits resonance that lends additional stability to the molecule once it undertakes hydrogen and electron donation to a free radical (Heim et al., 2002).

Other functions of flavonoids involve protecting DNA and lipids from oxidation, as well as inhibiting ROS-generating enzymes like mitochondrial succinoxidase, NADH oxidase, glutathione S-transferase, and microsomal monooxygenase (Brown et al., 1998). Flavonoids have also been found to form complexes with ions from chelating metals like iron, copper, and zinc, but whether it will take an antioxidant or prooxidant role will depend on the ratio of metals that are present relative to the flavonoids. An excess of metals could provoke a reduction of the metal by the action of flavonoids, which will result in having ionic metals that can further engage in the production of hydroxyl radicals, through the Fenton reaction (Kasprzak et al., 2015).

However, under optimum conditions, flavonoids can chelate metal ions by binding with them at the 5-hydroxy and 4-carbonyl group, or the 3-hydroxy and 4-carbonyl group, as well as between the $3',4'$-hydroxy group in one of the rings (Symonowicz and Kolanek, 2012). The stoichiometric ratio of these reactions are usually one to one, but can also take the form of two flavonoids for one metal, or vice versa, two metal ions per flavonoid. The nature of these complexes will depend on the metal, the specific flavonoid that is interacting with them, and on the physical and chemical conditions that surround these interactions. In many cases, these complexes have been shown to increase antioxidant activity and free radical scavenging capability compared to single flavonoids, but in addition to this, part of their function is to arrest metal ions, preventing them from forming additional ROS (Kasprzak et al., 2015).

14.7.5 Phenolic acids

Phenolic acids are another type of natural antioxidant that is usually acquired through dietary intake (Heleno et al., 2015). These can be divided into two major groups known as benzoic acids and cinnamic acids. The main structure of these compounds comprises an aromatic ring that contains at least one methoxy or hydroxyl group substituting one of the hydrogens (Sánchez-Maldonado et al., 2011). It has been shown that these antioxidants prevent low-density lipoprotein oxidation (Nimse and Pal, 2015). Much of the antioxidant capabilities of phenolic acids are similar to those of flavonoids. Most of their reactions involve the donation of hydrogen and an electron to a free radical species in the medium, while resonance stabilization of the molecule keeps it from turning into a radical itself. Most of their activity is related to the functional groups that are located in the ring of these compounds, since they can act as reducing agents that can scavenge and quench metal ions, singlet oxygen, superoxide anions, and peroxynitrites (Heleno et al., 2015).

Concerns regarding the negative effects of free radicals have been extended to concerns regarding the quality of processed meats (Tomović et al., 2017). Meat products contain deposits of lipids as a result of having adipose tissue and stored triglyceride molecules between muscle fibers. As a consequence of this, the quality of these meat products can be affected by the oxidative nature of free radicals. As discussed earlier, those different forms of lipids can be susceptible to free radicals, since they contain areas in their molecular structures that are electron deficient. The highly reactive free radicals will cause a cascading effect that will lead to the degradation of saturated and unsaturated fatty acids, leading to the formation of secondary products like pentanal, hexanal, 4-hydroxynomenal, and malondialdehyde, which will give the meat an undesired flavor appearance and odor (Kumar and Pandey, 2013). Along with this effect on lipids, myoglobin also oxidizes into metmyoglobin, reducing both the esthetic and nutritional quality of the meat as it goes rancid.

14.8 Conclusion

Though synthetic antioxidants such as butylated hydroxyanisole, butylated hydroxytoluene, *tert*-butylhydroquinone, and propyl gallate can be used in preservation of meat products, the currently permitted concentrations of these products are very low since an excess of these compounds can damage the meat products as well as be unsafe for consumers. With the growing concerns over the use of various synthetic preservatives and their possible negative side effects, many have turned their attention to natural antioxidants. Not only can these compounds serve as free radical scavengers that help preserve the integrity of the meat throughout a longer period of time, but they can also serve as antimicrobials that make the food safer to ingest, while having few side effects over the consumer. Further, it is possible to implement the use of the natural antioxidants at different stages of animal farming to improve animal health, shelf life, and the nutritional value of the meat and other animal food products.

Various forms of natural antioxidants can be applied at different stages of animal farming and meat processing but the sources of raw materials and extraction methods of the natural antioxidants must be inexpensive, abundant, and easy to implement as well as consumer friendly. Given the considerations of these criteria, much more research is still needed to understand the long-term effects of antioxidant supplementation in farm animals, as well as its effect on consumers. There is also a need to standardize the concentrations and amounts of natural antioxidants that are being used to supplement animal feed, and the ones that are being used directly on the meat postharvest. In the near future, naturally extracted antioxidants could be a strategic addition to sustainable animal farming, improving the quality of the products and meeting the consumer's demands in food safety and security.

References

Abascal, K., Ganora, L., Yarnell, E., 2005. The effect of freeze drying and its implications for botanical medicine: a review. Phytother. Res. 19 (8), 655–660.

Ahn, J., Grün, I.U., Mustapha, A., 2007. Effects of plant extracts on microbial growth, color change, and lipid oxidation in cooked beef. Food Microbiol. 24 (1), 7–14.

Akula, R., Ravishankar, G.A., 2011. Influence of abiotic stress signals on secondary metabolites in plants. Plant Signal. Behav. 6 (11), 1720–1731.

Albu, S., Joyce, E., Paniwnyk, L., et al., 2004. Potential for the use of ultrasound in the extraction of antioxidants from *Rosmarinus officinalis* for the food and pharmaceutical industry. Ultrason. Sonochem. 11 (3–4), 261–265.

Alonso-Salces, R.M., Korta, E., Barranco, A., Berrueta, L.A., Gallo, B., Vicente, F., 2001. Pressurized liquid extraction for the determination of polyphenols in apple. J. Chromatogr. A 933 (1–2), 37–43.

Álvarez, D., Delles, R.M., Xiong, Y.L., et al., 2011. Influence of canola-olive oils, rice bran and walnut on functionality and emulsion stability of frankfurters. LWT-Food Sci. Technol. 44 (6), 1435–1442.

Álvarez, D., Xiong, Y.L., Castillo, M., et al., 2012. Textural and viscoelastic properties of pork frankfurters containing canola–olive oils, rice bran, and walnut. Meat Sci. 92 (1), 8–15.

Amid, A., Salim, R.J.M., Adenan, M.I., 2010. The factors affecting the extraction condition for neuroprotective activity of *Centella asiatica* evaluated by metal chelating activity assay. J. Appl. Sci. 10 (10), 837–842.

Asami, D.K., Hong, Y.J., Barrett, D.M., Mitchell, A.E., 2003. Comparison of the total phenolic and ascorbic acid content of freeze-dried and air-dried marionberry, strawberry, and corn grown using conventional, organic, and sustainable agricultural practices. J. Agric. Food Chem. 51 (5), 1237–1241.

Badr, H.M., Mahmoud, K.A., 2011. Antioxidant activity of carrot juice in gamma irradiated beef sausage during refrigerated and frozen storage. Food Chem. 127, 1119–1130. Available from: https://doi.org/10.1016/j.foodchem.2011.01.113.

Bañón, S., Díaz, P., Rodríguez, M., Garrido, M.D., Price, A., 2007. Ascorbate, green tea and grape seed extracts increase the shelf life of low sulphite beef patties. Meat Sci. 77 (4), 626–633.

Barua, A.B., Furr, H.C., Olson, et al., 2000. Vitamin A and carotenoids. Chromatogr. Sci. Series 84, 1–74.

Berthod, A., Billardello, B., Geoffroy, S., 1999. Polyphenols in counter current chromatography. An example of large scale separation. Analysis 27 (9), 750–757.

Biswas, A.K., Chatli, M.K., Sahoo, J., 2012. Antioxidant potential of curry (*Murraya koenigii* L.) and mint (*M. spicata*) leaf extracts and their effect on colour and oxidative stability of raw ground pork meat during refrigeration storage. Food Chem. 133 (2), 467–472.

Bonoli, M., Marconi, E., Caboni, M.F., 2004. Free and bound phenolic compounds in barley (*Hordeum vulgare* L.) flours: evaluation of the extraction capability of different solvent mixtures and pressurized liquid methods by micellar electrokinetic chromatography and spectrophotometry. J. Chromatogr. A 1057 (1–2), 1–12.

Bozkurt, H., 2006. Utilization of natural antioxidants: Green tea extract and Thymbra spicata oil in Turkish dry-fermented sausage. Meat Sci. 73, 442–450. Available from: https://doi.org/10.1016/j.meatsci.2006.01.005.

Brettonnet, A., Hewavitarana, A., DeJong, S., Lanari, M.C., 2010. Phenolic acids composition and antioxidant activity of canola extracts in cooked beef, chicken and pork. Food Chem. 121, 927–933. Available from: https://doi.org/10.1016/j.foodchem.2009.11.021.

Brown, E.J., Khodr, H., Hider, C.R., Rice-Evans, C.A., 1998. Structural dependence of flavonoid interactions with $Cu2+$ ions: implications for their antioxidant properties. Biochem. J. 330 (3), 1173–1178.

Carlsen, M.H., Halvorsen, B.L., Holte, K., et al., 2010. The total antioxidant content of more than 3100 foods, beverages, spices, herbs and supplements used worldwide. Nutr. J. 9 (1), 3.

Carr, A., Frei, B., 1999. Does vitamin C act as a pro-oxidant under physiological conditions? FASEB J. 13 (9), 1007–1024.

Clark, S.F., 2002. The biochemistry of antioxidants revisited. Nutr. Clin. Pract. 17 (1), 5–17.

Cofrades, S., Serrano, A., Ayo, J., Solas, M.T., Carballo, J., Colmenero, F.J., 2004. Restructured beef with different proportions of walnut as affected by meat particle size. Eur. Food Res. Technol. 218 (3), 230–236.

Conaway, H.H., Henning, P., Lerner, U.H., 2013. Vitamin A metabolism, action, and role in skeletal homeostasis. Endocr. Rev. 34 (6), 766–797.

Dai, J., Mumper, R.J., 2010. Plant phenolics: extraction, analysis and their antioxidant and anticancer properties. Molecules 15 (10), 7313–7352.

Dao, D.Q., Ngo, T.C., Thong, N.M., Nam, P.C., 2017. Is vitamin A an antioxidant or a pro-oxidant? J. Phys. Chem. B 121 (40), 9348–9357.

de Castro, M.L., García-Ayuso, L.E., 1998. Soxhlet extraction of solid materials: an outdated technique with a promising innovative future. Anal. Chim. Acta 369 (1–2), 1–10.

de Ciriano, M.G.-I., Rehecho, S., Calvo, M.I., Cavero, R.Y., Navarro, Í., Astiasarán, I., Ansorena, D., 2010. Effect of lyophilized water extracts of Melissa officinalis on the stability of algae and linseed oil-in-water emulsion to be used as a functional ingredient in meat products. Meat Sci. 85, 373–377. Available from: https://doi.org/10.1016/j.meatsci.2010.01.007.

Degenhardt, A., Engelhardt, U.H., Wendt, A.S., Winterhalter, P., 2000a. Isolation of black tea pigments using high-speed countercurrent chromatography and studies on properties of black tea polymers. J. Agric. Food Chem. 48 (11), 5200–5205.

Degenhardt, A., Knapp, H., Winterhalter, P., 2000b. Separation and purification of anthocyanins by high-speed countercurrent chromatography and screening for antioxidant activity. J. Agric. Food Chem. 48 (2), 338–343.

Degenhardt, A., Engelhardt, U.H., Lakenbrink, C., Winterhalter, P., 2000c. Preparative separation of polyphenols from tea by high-speed countercurrent chromatography. J. Agric. Food Chem. 48 (8), 3425–3430.

Djenane, D., Sánchez-Escalante, A., Beltrán, J.A., Roncalés, P., 2003. Extension of the shelf life of beef steaks packaged in a modified atmosphere by treatment with rosemary and displayed under UV-free lighting. Meat Sci. 64, 417–426. Available from: https://doi.org/10.1016/S0309-1740(02)00210-3.

Doménech-Asensi, G., García-Alonso, F.J., Martínez, E., et al., 2013. Effect of the addition of tomato paste on the nutritional and sensory properties of mortadella. Meat Sci. 93 (2), 213–219.

Du, F.Y., Xiao, X.H., Li, G.K., 2007. Application of ionic liquids in the microwave-assisted extraction of trans-resveratrol from Rhizma Polygoni Cuspidati. J. Chromatagr. 1140 (1–2), 56–62.

Fasseas, M.K., Mountzouris, K.C., Tarantilis, P.A., Polissiou, M., Zervas, G., 2008. Antioxidant activity in meat treated with oregano and sage essential oils. Food Chem. 106, 1188–1194. Available from: https://doi.org/10.1016/j.foodchem.2007.07.060.

Fernández-López, J., Zhi, N., Aleson-Carbonell, L., et al., 2005. Antioxidant and antibacterial activities of natural extracts: application in beef meatballs. Meat Sci. 69 (3), 371–380.

Fernando, M.R., Lechner, J.M., Löfgren, S., et al., 2006. Mitochondrial thioltransferase (glutaredoxin 2) has GSH-dependent and thioredoxin reductase-dependent peroxidase activities in vitro and in lens epithelial cells. FASEB J. 20 (14), 2645–2647.

Ganhão, R., Morcuende, D., Estévez, M., 2010. Protein oxidation in emulsified cooked burger patties with added fruit extracts: Influence on colour and texture deterioration during chill storage. Meat Sci. 85, 402–409. Available from: https://doi.org/10.1016/j.meatsci.2010.02.008.

Garrido, M.D., Auqui, M., Martí, N., Linares, M.B., 2011. Effect of two different red grape pomace extracts obtained under different extraction systems on meat quality of pork burgers. LWT-Food Sci. Technol. 44 (10), 2238–2243.

Georgantelis, D., Ambrosiadis, I., Katikou, P., Blekas, G., Georgakis, S.A., 2007. Effect of rosemary extract, chitosan and α-tocopherol on microbiological parameters and lipid oxidation of fresh pork sausages stored at 4 C. Meat Sci. 76 (1), 172–181.

Georgé, S., Brat, P., Alter, P., Amiot, M.J., 2005. Rapid determination of polyphenols and vitamin C in plant-derived products. J. Agric. Food Chem. 53 (5), 1370–1373.

Guyot, S., Marnet, N., Drilleau, J.-F., 2001. Thiolysis – HPLC Characterization of Apple Procyanidins Covering a Large Range of Polymerization States. J. Agric. Food Chem. 49, 14–20. Available from: https://doi.org/10.1021/jf000814z.

Halliwell, B., Rafter, J., Jenner, A., 2005. Health promotion by flavonoids, tocopherols, tocotrienols, and other phenols: direct or indirect effects? Antioxidant or not? Am. J. Clin. Nutr. 81 (1), 268S–276S.

Hao, Y.Y., Brackett, R.E., Doyle, M.P., 1998a. Efficacy of plant extracts in inhibiting *Aeromonas hydrophila* and *Listeria monocytogenes* in refrigerated, cooked poultry. Food Microbiol. 15 (4), 367–378.

Hao, Y.Y., Brackett, R.E., Doyle, M.P., 1998b. Inhibition of *Listeria monocytogenes* and *Aeromonas hydrophila* by plant extracts in refrigerated cooked beef. J. Food Prot. 61 (3), 307–312.

Heim, K.E., Tagliaferro, A.R., Bobilya, D.J., 2002. Flavonoid antioxidants: chemistry, metabolism and structure-activity relationships. J. Nutr. Biochem. 13 (10), 572–584.

Heleno, S.A., Martins, A., Queiroz, M.J.R., Ferreira, I.C., 2015. Bioactivity of phenolic acids: metabolites versus parent compounds: a review. Food Chem. 173, 501–513.

Hemwimon, S., Pavasant, P., Shotipruk, A., 2007. Microwave-assisted extraction of antioxidativeanthraquinones from roots of *Morinda citrifolia*. Sep. Purif. Technol. 54 (1), 44–50.

Herrera, M.C., De Castro, M.L., 2004. Ultrasound-assisted extraction for the analysis of phenolic compounds in strawberries. Anal. Bioanal. Chem. 379 (7–8), 1106–1112.

Hill, G.E., Johnson, J.D., 2012. The vitamin A–redox hypothesis: a biochemical basis for honest signaling via carotenoid pigmentation. Am. Nat. 180 (5), E127–E150.

Howard, L., Pandjaitan, N., 2008. Pressurized liquid extraction of flavonoids from spinach. J. Food. Sci. 73 (3), C151–C157.

Hromádková, Z., Košt'álová, Z., Ebringerová, A., 2008. Comparison of conventional and ultrasound-assisted extraction of phenolics-rich heteroxylans from wheat bran. Ultrason. Sonochem. 15 (6), 1062–1068.

Huang, B., He, J., Ban, X., Zeng, H., Yao, X., Wang, Y., 2011. Antioxidant activity of bovine and porcine meat treated with extracts from edible lotus (Nelumbo nucifera) rhizome knot and leaf. Meat Sci. 87, 46–53. Available from: https://doi.org/10.1016/j.meatsci.2010.09.001.

Huang, J., Zhang, Z., 2004. Microwave-assisted extraction of quercetin and acid degradation of its glycosides in *Psidium guajava* leaves. Anal. Sci. 20 (2), 395–397.

Jackman, R.L., Yada, R.Y., Tung, M.A., 1987. A Review: Separation and Chemical Properties of Anthocyanins Used for Their Qualitative and Quantitative Analysis. J. Food Biochem. 11, 279–308. Available from: https://doi.org/10.1111/j.1745-4514.1987.tb00128.x.

Jiang, J., Xiong, Y.L., 2016. Natural antioxidants as food and feed additives to promote health benefits and quality of meat products: a review. Meat Sci. 120, 107–117.

Kanatt, S.R., Chander, R., Sharma, A., 2007. Antioxidant potential of mint (*Mentha spicata* L.) in radiation-processed lamb meat. Food Chem. 100, 451–458. Available from: https://doi.org/10.1016/j.foodchem.2005.09.066.

Kanatt, S.R., Arjun, K., Sharma, A., 2011. Antioxidant and antimicrobial activity of legume hulls. Food Res. Int. 44, 3182–3187. Available from: https://doi.org/10.1016/j.foodres.2011.08.022.

Kasote, D.M., Katyare, S.S., Hegde, M.V., Bae, H., 2015. Significance of antioxidant potential of plants and its relevance to therapeutic applications. Int. J. Biol. Sci. 11 (8), 982.

Kasprzak, M.M., Erxleben, A., Ochocki, J., 2015. Properties and applications of flavonoid metal complexes. RSC Adv. 5 (57), 45853–45877.

Klejdus, B., Kubáň, V., 2000. High performance liquid chromatographic determination of phenolic compounds in seed exudates of *Festuca arundinacea* and *F. pratense*. Phytochem. Anal. 11 (6), 375–379.

Krafczyk, N., Glomb, M.A., 2008. Characterization of phenolic compounds in rooibos tea. J. Agric. Food Chem. 56 (9), 3368–3376.

Kulkarni, S., DeSantos, F.A., Kattamuri, S., et al., 2011. Effect of grape seed extract on oxidative, color and sensory stability of a pre-cooked, frozen, re-heated beef sausage model system. Meat Sci. 88 (1), 139–144.

Kumar, S., Pandey, A.K., 2013. Chemistry and biological activities of flavonoids: an overview. Sci. World J. 2013, 16 pp.

Laborde, J.L., Bouyer, C., Caltagirone, J.P., Gérard, A., 1998. Acoustic bubble cavitation at low frequencies. Ultrasonics 36 (1–5), 589–594.

Lee, M.A., Choi, J.H., Choi, Y.S., et al., 2010. The antioxidative properties of mustard leaf (*Brassica juncea*) kimchi extracts on refrigerated raw ground pork meat against lipid oxidation. Meat Sci. 84 (3), 498–504.

Li, H., Chen, B., Nie, L., Yao, S., 2004. Solvent effects on focused microwave assisted extraction of polyphenolic acids from *Eucommia ulmodies*. Phytochem. Anal. 15 (5), 306–312.

Liazid, A., Palma, M., Brigui, J., Barroso, C.G., 2007. Investigation on phenolic compounds stability during microwave-assisted extraction. J. Chromatagr. A 1140 (1–2), 29–34.

Llorach, R., Gil-Izquierdo, A., Ferreres, F., Tomás-Barberán, F.A., 2003. HPLC-DAD-MS/MS ESI characterization of unusual highly glycosylated acylated flavonoids from cauliflower (*Brassica oleracea* L. var. *botrytis*) agroindustrial byproducts. J. Agric. Food Chem. 51 (13), 3895–3899.

Lucchesi, M.E., Chemat, F., Smadja, J., 2004. Solvent-free microwave extraction of essential oil from aromatic herbs: comparison with conventional hydro-distillation. J. Chromatagr. A 1043 (2), 323–327.

Luque-Garcia, J.L., De Castro, M.L., 2003. Ultrasound: a powerful tool for leaching. TrAC Trends Anal. Chem. 22 (1), 41–47.

Luque-Rodríguez, J.M., Luque de Castro, M.D., Pérez-Juan, P., 2007. Dynamic superheated liquid extraction of anthocyanins and other phenolics from red grape skins of winemaking residues. Bioresour. Technol. 98, 2705–2713. Available from: https://doi.org/10.1016/j.biortech.2006.09.019.

Matés, J.M., Pérez-Gómez, C., Núñez de Castro, I., 1999. Antioxidant enzymes and human diseases. Clin. Biochem. 32, 595–603.

Martínez, A., Rodríguez-Gironés, M.A., Barbosa, A., Costas, M., 2008. Donator acceptor map for carotenoids, melatonin and vitamins. J. Phys. Chem. A 112 (38), 9037–9042.

Mason, T.J., Paniwnyk, L., Lorimer, J.P., 1996. The uses of ultrasound in food technology. Ultrason. Sonochem. 3 (3), S253–S260.

Mendiola, J.A., Herrero, M., Cifuentes, A., Ibañez, E., 2007. Use of compressed fluids for sample preparation: food applications. J. Chromatogr. A 1152 (1–2), 234–246.

Metivier, R.P., Francis, F.J., Clydesdale, F.M., 1980. Solvent Extraction of Anthocyanins from Wine Pomace. J. Food Sci. 45, 1099–1100. Available from: https://doi.org/10.1111/j.1365-2621.1980.tb07534.x.

Mielnik, M.B., Olsen, E., Vogt, G., Adeline, D., Skrede, G., 2006. Grape seed extract as antioxidant in cooked, cold stored turkey meat. LWT-Food Sci. Technol. 39 (3), 191–198.

Mine, Y., Li-Chan, E.C.Y., Jiang, B., 2010. Biologically active food proteins and peptides in health: an overview. In: Bioactive Proteins and Peptides as Functional Foods and Nutraceuticals. John Wiley & Sons, Ltd, pp. 3–11. Available from: https://doi.org/10.1002/9780813811048.ch1.

Mitsumoto, M., O'Grady, M.N., Kerry, J.P., Joe Buckley, D., 2005. Addition of tea catechins and vitamin C on sensory evaluation, colour and lipid stability during chilled storage in cooked or raw beef and chicken patties. Meat Sci. 69, 773–779. Available from: https://doi.org/10.1016/j.meatsci.2004.11.010.

Möckel, H.J., Welter, G., Melzer, H., 1987. Correlation between reversed-phase retention and solute molecular surface type and area: I. Theoretical outlines and retention of various hydrocarbon classes. J. Chromatogr. A 388, 255–266. Available from: https://doi.org/10.1016/S0021-9673(01)94487-5.

Mustafa, A., Turner, C., 2011. Pressurized liquid extraction as a green approach in food and herbal plants extraction: a review. Anal. Chim. Acta 703 (1), 8–18.

Muthukumar, M., Naveena, B.M., Vaithiyanathan, S., Sen, A.R., Sureshkumar, K., 2014. Effect of incorporation of Moringa oleifera leaves extract on quality of ground pork patties. J. Food Sci. Technol. 51, 3172–3180. Available from: https://doi.org/10.1007/s13197-012-0831-8.

Naczk, M., Shahidi, F., 2004. Extraction and analysis of phenolics in food. J. Chromatogr. A 1054 (1–2), 95–111.

Nagao, A., 2004. Oxidative conversion of carotenoids to retinoids and other products. J. Nutr. 134 (1), 237S–240S.

Naveena, B.M., Sen, A.R., Vaithiyanathan, S., Babji, Y., Kondaiah, N., 2008. Comparative efficacy of pomegranate juice, pomegranate rind powder extract and BHT as antioxidants in cooked chicken patties. Meat Sci. 80, 1304–1308. Available from: https://doi.org/10.1016/j.meatsci.2008.06.005.

Neergheen, V.S., Soobrattee, M.A., Bahorun, T., Aruoma, O.I., 2006. Characterization of the phenolic constituents in Mauritian endemic plants as determinants of their antioxidant activities in vitro. J. Plant Physiol. 163 (8), 787–799.

Nicoué, E.É., Savard, S., Belkacemi, K., 2007. Anthocyanins in Wild Blueberries of Quebec: Extraction and Identification. J. Agric. Food Chem. 55, 5626–5635. Available from: https://doi.org/10.1021/jf0703304.

Nimse, S.B., Pal, D., 2015. Free radicals, natural antioxidants, and their reaction mechanisms. RSC Adv. 5 (35), 27986–28006.

Nn, A., 2015. A Review on the Extraction Methods Use in Medicinal Plants, Principle, Strength and Limitation. Med. Aromat. Plants 4, 1–6. Available from: https://doi.org/10.4172/2167-0412.1000196.

Palma, M., Piñeiro, Z., Barroso, C.G., 2001. Stability of phenolic compounds during extraction with superheated solvents. J. Chromatogr. A 921 (2), 169–174.

Patel, R.P., McAndrew, J., Sellak, H., et al., 1999. Biological aspects of reactive nitrogen species. Biochim. Biophys. Acta 1411 (2–3), 385–400.

Petersen, S.V., Oury, T.D., Ostergaard, L., Valnickova, Z., et al., 2004. Extracellular superoxide dismutase (EC-SOD) binds to type I collagen and protects against oxidative fragmentation. J. Biol. Chem. 279 (14), 13705–13710.

Piñeiro, Z., Palma, M., Barroso, C.G., 2006. Determination of trans-resveratrol in grapes by pressurised liquid extraction and fast high-performance liquid chromatography. J. Chromatogr. A 1110 (1–2), 61–65.

Pratt, D.E., 1992. Natural antioxidants from plant material. In: ACS Symposium Series.

Qiao, J., Arthur, J.F., Gardiner, E.E., et al., 2018. Regulation of platelet activation and thrombus formation by reactive oxygen species. Redox Biol. 14, 126–130.

Ramirez-Coronel, M.A., Marnet, N., Kolli, V.K., et al., 2004. Characterization and estimation of proanthocyanidins and other phenolics in coffee pulp (*Coffea arabica*) by thiolysis − high-performance liquid chromatography. J. Agric. Food Chem. 52 (5), 1344−1349.

Richter, B.E., Jones, B.A., Ezzell, J.L., Porter, N.L., Avdalovic, N., Pohl, C., 1996. Accelerated Solvent Extraction: A Technique for Sample Preparation. Anal. Chem. 68, 1033−1039. Available from: https://doi.org/10.1021/ac9508199.

Rodríguez-Carpena, J.G., Morcuende, D., Estévez, M., 2011. Avocado by-products as inhibitors of color deterioration and lipid and protein oxidation in raw porcine patties subjected to chilled storage. Meat Sci. 89 (2), 166−173.

Rostagno, M.A., Palma, M., Barroso, C.G., 2003. Ultrasound-assisted extraction of soy isoflavones. J. Chromatogr. A 1012 (2), 119−128.

Saleh, N.T., Ahmed, Z.S., 1998. Impact of natural sources rich in provitamin A on cooking characteristics, color, texture and sensory attributes of beef patties. Meat Sci. 50 (3), 285−293.

Salinas Moreno, Y., Sánchez, G.S., Hernández, D.R., Lobato, N.R., 2005. Characterization of anthocyanin extracts from maize kernels. J. Chromatogr. Sci. 43 (9), 483−487.

Sánchez-Maldonado, A.F., Schieber, A., Gänzle, M.G., 2011. Structure−function relationships of the antibacterial activity of phenolic acids and their metabolism by lactic acid bacteria. J. Appl. Microbiol. 111 (5), 1176−1184.

Sáyago-Ayerdi, S.G., Brenes, A., Goñi, I., 2009. Effect of grape antioxidant dietary fiber on the lipid oxidation of raw and cooked chicken hamburgers. LWT-Food Sci. Technol. 42 (5), 971−976.

Schwartz, H., Ollilainen, V., Piironen, V., Lampi, A.M., 2008. Tocopherol, tocotrienol and plant sterol contents of vegetable oils and industrial fats. J. Food Comp. Anal. 21 (2), 152−161.

Sisein, E.A., 2014. Biochemistry of free radicals and antioxidants. Scholars Acad. J. Biosci. 2 (2), 110−118.

Shi, J., Nawaz, H., Pohorly, J., Mittal, G., Kakuda, Y., Jiang, Y., 2005. Extraction of Polyphenolics from Plant Material for Functional Foods—Engineering and Technology. Food Res. Int. 21, 139−166. Available from: https://doi.org/10.1081/FRI-200040606.

Skandamis, P., Tsigarida, E., Nychas, G.E., 2002. The effect of oregano essential oil on survival/death of Salmonella typhimurium in meat stored at 5 C under aerobic, VP/MAP conditions. Food Microbiol. 19 (1), 97−103.

Sourmaghi, M.H.S., Kiaee, G., Golfakhrabadi, F., Jamalifar, H., Khanavi, M., 2015. Comparison of essential oil composition and antimicrobial activity of *Coriandrum sativum* L. extracted by hydrodistillation and microwave-assisted hydrodistillation. J. Food Sci. Technol. 52 (4), 2452−2457.

Symonowicz, M., Kolanek, M., 2012. Flavonoids and their properties to form chelate complexes. Biotechnol. Food Sci. 76 (1), 35−41.

Tanabe, H., Yoshida, M., Tomita, N., 2002. Comparison of the antioxidant activities of 22 commonly used culinary herbs and spices on the lipid oxidation of pork meat. J. Anim. Sci. 73, 389−393. Available from: https://doi.org/10.1046/j.1344-3941.2002.00054.x.

Tesoriere, L., D'arpa, D., Re, R., Livrea, M.A., 1997. Antioxidant reactions of all-transretinol in phospholipid bilayers: effect of oxygen partial pressure, radical fluxes, and retinol concentration. Arch. Biochem. Biophys. 343 (1), 13−18.

Thimothe, J., Bonsi, I.A., Padilla-Zakour, O.I., Koo, H., 2007. Chemical characterization of red wine grape (*Vitis vinifera* and *Vitis* interspecific hybrids) and pomace phenolic extracts and their biological activity against *Streptococcus mutans*. J. Agric. Food Chem. 55 (25), 10200−10207.

Toma, M., Vinatoru, M., Paniwnyk, L., Mason, T.J., 2001. Investigation of the effects of ultrasound on vegetal tissues during solvent extraction. Ultrason. Sonochem. 8 (2), 137−142.

Tomović, V., Jokanović, M., Šojić, B., Škaljac, S., Ivić, M., 2017. September. Plants as natural antioxidants for meat products. In: IOP Conference Series: Earth and Environmental Science, vol. 85, No. 1. IOP Publishing, p. 012030.

Trindade, R.A., Mancini-Filho, J., Villavicencio, A.L.C.H., 2010. Natural antioxidants protecting irradiated beef burgers from lipid oxidation. LWT-Food Sci. Technol. 43, 98–104. Available from: https://doi.org/10.1016/j.lwt.2009.06.013.

Trusheva, B., Trunkova, D., Bankova, V., 2007. Different extraction methods of biologically active components from propolis: a preliminary study. Chem. Centr. J. 1 (1), 13.

Vaithiyanathan, S., Naveena, B.M., Muthukumar, M., et al., 2011. Effect of dipping in pomegranate (*Punica granatum*) fruit juice phenolic solution on the shelf life of chicken meat under refrigerated storage (4 C). Meat Sci. 88 (3), 409–414.

Valko, M., Rhodes, C., Moncol, J., et al., 2006. Free radicals, metals and antioxidants in oxidative stress-induced cancer. Chem. Biol. Interact. 160 (1), 1–40.

Valko, M., Leibfritz, D., Moncol, J., et al., 2007. Free radicals and antioxidants in normal physiological functions and human disease. Int. J. Biochem. Cell Biol. 39 (1), 44–84.

Van Der Vliet, A., Eiserich, J.P., Halliwell, B., Cross, C.E., 1997. Formation of reactive nitrogen species during peroxidase-catalyzed oxidation of nitrite a potential additional mechanism of nitric oxide-dependent toxicity. J. Biol. Chem. 272 (12), 7617–7625.

Vidal, S., Hayasaka, Y., Meudec, E., et al., 2004. Fractionation of grape anthocyanin classes using multilayer coil countercurrent chromatography with step gradient elution. J. Agric. Food Chem. 52 (4), 713–719.

Vinatoru, M., 2001. An overview of the ultrasonically assisted extraction of bioactive principles from herbs. Ultrason. Sonochem. 8 (3), 303–313.

Vinatoru, M., Toma, M., Radu, O., et al., 1997. The use of ultrasound for the extraction of bioactive principles from plant materials. Ultrason. Sonochem. 4 (2), 135–139.

Vitrac, X., Castagnino, C., Waffo-Téguo, P., et al., 2001. Polyphenols newly extracted in red wine from southwestern France by centrifugal partition chromatography. J. Agric. Food Chem. 49 (12), 5934–5938.

Wang, L., Weller, C.L., 2006. Recent advances in extraction of nutraceuticals from plants. Trends Food Sci. Technol. 17, 300–312. Available from: https://doi.org/10.1016/j.tifs.2005.12.004.

Xu, B.J., Chang, S.K.C., 2007. A comparative study on phenolic profiles and antioxidant activities of legumes as affected by extraction solvents. J. Food Sci. 72 (2), S159–S166.

Yanagida, A., Shoji, A., Shibusawa, Y., et al., 2006. Analytical separation of tea catechins and food-related polyphenols by high-speed counter-current chromatography. J. Chromatogr. A 1112 (1–2), 195–201.

Yang, Y., Zhang, F., 2008. Ultrasound-assisted extraction of rutin and quercetin from *Euonymus alatus* (Thunb.) Sieb. Ultrason. Sonochem. 15 (4), 308–313.

Yogesh, K., Ali, J., 2014. Antioxidant potential of thuja (Thuja occidentalis) cones and peach (Prunus persia) seeds in raw chicken ground meat during refrigerated (4 ± 1 °C) storage. J. Food Sci. Technol. 51, 1547–1553. Available from: https://doi.org/10.1007/s13197-012-0672-5.

Zhang, Y., Seeram, N.P., Lee, R., et al., 2008. Isolation and identification of strawberry phenolics with antioxidant and human cancer cell antiproliferative properties. J. Agric. Food Chem. 56 (3), 670–675.

SECTION 9

Strategies for elimination and detection of foodborne pathogens

CHAPTER 15

Strategies for elimination of foodborne pathogens, their influensive detection techniques and drawbacks

Sandeep Ghatak
ICAR Research Complex for North Eastern Hill Region, Umiam, India

Chapter Outline

15.1 Introduction 268
15.2 Physical methods of elimination of foodborne pathogens 268
    15.2.1 Preslaughter washing 268
    15.2.2 Removal of hair 269
    15.2.3 Spot trimming of carcasses 269
    15.2.4 Vacuum-steam/water application 269
    15.2.5 Carcass washing 270
15.3 Chemical processes for elimination of microbial pathogens 270
    15.3.1 Acidic compounds 270
    15.3.2 Chlorine and related chemicals 271
    15.3.3 Ozone 271
    15.3.4 Other chemical agents 271
15.4 Elimination of microbial pathogens by ultraviolet light 272
15.5 Irradiation of meat for eliminating microbial hazards 272
15.6 Application of low temperature 273
15.7 High-pressure processing (HPP) for elimination of pathogens 273
15.8 Other emerging approaches for elimination of microbial pathogens 273
    15.8.1 Nonthermal plasma (cold plasma) 274
    15.8.2 Dense phase carbon dioxide 274
    15.8.3 Electrolyzed oxidizing water 274
    15.8.4 Microwave and radio frequency 274
    15.8.5 Infrared heating 275
    15.8.6 Biocontrol with bacteriophage 275
15.9 Detection of microbial pathogens 275
    15.9.1 Conventional culture-based techniques 276
    15.9.2 Immunological techniques 277
    15.9.3 Nucleic acid-based techniques 278

15.9.4 Matrix-assisted laser desorption ionization-time of flight mass spectrometry 280
15.9.5 Hyperspectral imaging and analysis 280
15.9.6 Nanotechnology-based approaches 281
15.9.7 Other assays 281
References 281

15.1 Introduction

Meat occupies a central position in the human diet worldwide as a source of balanced nutrition. Meat being rich in various nutritive components, such as proteins, minerals, and vitamins, is also prone to microbial attack. Microbes associated with meat may reduce the shelf life of meat food products hurting the commercial aspects of production and/or compromising the safety of the product, often leading to foodborne illnesses. It is difficult to estimate the precise losses due to the spoilage of meat and meat products. However, a current estimate of economic costs of foodborne diseases in low- and middle-income countries alone amounted to a whopping US$110 billion (Jaffee et al., 2019). An assessment of foodborne diseases burden by the World Health Organization (WHO) in 2015 revealed that globally there were 420,000 deaths, along with a loss of 33 million healthy life years due to 31 major foodborne hazards (Foodborne Disease Burden Epidemiology Reference Group, 2015). On the other hand, rising consumer demands and awareness for quality and safe food has put increasing pressure on the food supply chain to assure the safety of all foods including highly perishable ones like meat. Technological advances in production, processing, and marketing of meat foods also pose other challenges. As a result scientists, technologists, and managers of the food industry are continuously struggling to conform and adapt to an industry scenario that is always in flux.

The transformation of the skeletal muscles of food animals into meat (and sometimes products) is a complex process involving many steps and processes. Ensuring the safety of meat and its products along these complex tortuous paths requires exercising control and supervision at every stage. While this is well-recognized in the meat food industry, the elimination of foodborne pathogens requires concerted and strategized efforts. While a detailed description of all the steps is beyond the scope of this chapter, the principles and putative applications of major aspects will be discussed.

15.2 Physical methods of elimination of foodborne pathogens

15.2.1 Preslaughter washing

Food animals and poultry harbor microbes on their hides and skins. Contamination of carcasses with microbes originating in hides or skins of animals or birds is common knowledge

(Doyle and Erickson, 2006). Therefore preslaughter washing of animals is a common practice in many slaughter premises (Sofos and Smith, 1998). Though the efficacy of preslaughter washing of animals mainly remains limited to elimination of visible contamination or dirt (Gill and Gill, 2012), the practice of washing remains commonplace in many parts of the world, including New Zealand, Australia, and the United States (Skandamis et al., 2010).

15.2.2 Removal of hair

Similar to washing of food animals, the removal of hairs also aims at minimizing visible contaminations and improving carcass appearance. Any reduction in microbiological load is thought to be consequential (Gill and Gill, 2012). For dehairing of beef carcasses chemical treatments are often resorted to (Skandamis et al., 2010), while pig carcasses are routinely put through scalding and singeing steps which considerably reduce the on-skin microbial loads (Gill and Gill, 2012) presumably due to thermal inactivation. However, the reduction in microbial loads appears to be transient as subsequent stages of slaughter operations may recontaminate the carcasses. Like pig carcasses, poultry carcasses are also scalded in high temperature water ($\approx 66°C$) causing a reduction in microbial load (Skandamis et al., 2010).

15.2.3 Spot trimming of carcasses

Following the skinning of carcasses, areas that are visibly contaminated or harboring residues of hairs and other dirt materials are trimmed off. This process, though common in many abattoir settings, is not very efficient in reducing microbial loads, rather it improves carcass appearance. Previous studies indicated that the areas with significant high load of microbial contaminant may not be apparent to the personnel involved and dirty areas may in fact harbor a lower load of bacteria (Gill and Gill, 2012; Skandamis et al., 2010).

15.2.4 Vacuum-steam/water application

Another process which is often employed at this stage involves the simultaneous application of steam or hot water under high pressure and temperature and application of vacuum suction through specially designed nozzles that aim to decontaminate the areas of carcasses polluted with hide hairs, or other contaminating objects. Compared to knife trimming of carcasses this process is believed to be more efficient as it combines the thermal properties of hot water and draining ability of the applied vacuum. Usually steam is applied at $104°C-110°C$ while the water temperature varies between $82°C$ and $94°C$ (Sofos and Smith, 1998). This popular method of elimination of carcass contamination is known to reduce microbial counts considerably (Kochevar et al., 1997).

15.2.5 Carcass washing

Carcass washing with hot or cold water is a universal step that is followed in almost all slaughter processes. Usually, brief washing with cold to lukewarm water (10°C–40°C) is given prior to evisceration. This process significantly improves carcass appearance and reduces or compensates for surface moisture loss, though it is only minimally effective in reducing microbial contamination from the meat surface (Skandamis et al., 2010). On the other hand, carcass washing with hot water with a temperature ranging between 74°C and 97°C (Skandamis et al., 2010) is reported to be an important measure in reducing carcass microbial load. Apart from the mechanical force of water to dislodge the contaminating flora, the thermal energy of the hot water contributes to the major reduction efficiency. Previous studies measuring decontamination efficiency of hot water rinsing of carcass indicated substantial reduction in microbial loads from the meat surface (Gill et al., 1997, 1999).

A variant of hot water application for carcass washing involves the application of saturated steam to achieve a pasteurization effect on the meat surface. The process is also commercially available under the trademark of "Steam Pasteurization" (Skandamis et al., 2010). Usually the process involves the holding of carcasses briefly in a commercial steam cabinet to achieve the desired effect. However, the outcome in terms of elimination of microbes is believed to be comparable with hot water washing of carcasses (Skandamis et al., 2010).

15.3 Chemical processes for elimination of microbial pathogens

Various chemicals are also used to eliminate microbial pathogens from meat. Though consumers often do not appreciate chemical treatment of food materials, prudent application of chemicals that are approved by regulatory agencies and recognized as safe may provide a desirable reduction in the microbial load of carcasses. The list of chemicals that have been studied or are in vogue is long. They can be loosely grouped into subcategories depending on their chemical nature.

15.3.1 Acidic compounds

Acidic compounds are one of the most commonly used chemical substances to achieve the decontamination of carcasses. The microbicidal effect of these compounds is chiefly attributable to the lowering of the pH creating an inhospitable environment for the contaminating microbes. Various organic acids have been used including lactic acid, acetic acid, citric acid, malic acid, gluconic acid, or various combinations of these (Gill and Gill, 2012; Skandamis et al., 2010). However, the majority of the applications included lactic acid at various concentrations ranging between 2% and 5% aqueous solution (Skandamis et al., 2010). Application of lactic acid spray is reported to reduce the microbial loads of carcasses by 2–3 log CFU (Skandamis et al., 2010). Usually acid sprays are applied prior to evisceration, postevisceration, and prior to chilling. One major advantage of the application of organic acids is the residual activity which may continue to exert antimicrobial effects on carcass surfaces (Gill and Gill, 2012; Skandamis et al., 2010).

15.3.2 Chlorine and related chemicals

Chlorine is a widely used disinfectant with industrial, household, medical, and public health applications. In the meat food industry chlorine is used for meat decontamination under various concentrations and exposure times (Skandamis et al., 2010). Factors that affect antimicrobial efficacy of chlorine include temperature, presence of organic matters, and more importantly pH, with a maximum activity in the range of 6.0–7.0. The usual mode of application is spraying chlorinated water with the required amount of free chlorine. Chlorine dioxide has also been used in the meat industry, mainly for the decontamination of poultry carcasses and is approved by the US Department of Agriculture (Trinetta et al., 2012). However, chlorine dioxide is an environmentally hazardous substance and its transportation and storage pose risks. Therefore it is produced in the place of application and is used immediately. Previous reports of the microbicidal activity of chlorine dioxide have indicated a minimum reduction of 1 log CFU in the treated chicken breasts (Ellis et al., 2006).

Another alternative to the application of chlorine for the decontamination purpose is acidified sodium chlorite (ASC). ASC is generally applied as a spray application prior to chilling and has shown considerable elimination of microbes in meat and poultry products (Lianou et al., 2012).

15.3.3 Ozone

Like chlorine, ozone is also a well-recognized sanitizer with application across industry and public health engineering. The antimicrobial activity of ozone is due to the reactive oxygen atom that is produced on dissociation of this triatomic molecule. The usual application mode of ozone in the food industry is ozonated water with varying degrees of ozone dissolved in it. Despite successful application in other foods, the application of ozone remains limited in the meat food industry primarily due to the reported lower efficacy, undesirable product modification, and reactivity with slaughterhouse equipment, especially rubber parts (Chawla et al., 2012; Kim et al., 1999, 2003; Mahapatra et al., 2005; Zweifel and Stephan, 2012).

15.3.4 Other chemical agents

A wide range of other chemical agents have been explored and are being utilized for the purpose of the elimination of harmful microbes from meat and poultry. Monochloramine, a compound derived from the reaction between ammonia and chlorine, is routinely used in the decontamination of water. Monochloramine being tasteless, odorless, and colorless with documented antimicrobial activity (Russell and Axtell, 2005) should be particularly suitable for the meat industry. Trisodium phosphate (TSP) is a bactericidal agent that has been reported to be effective by many workers (Skandamis et al., 2010). TSP is applied at a concentration of 8%–12% and has been documented to reduce *Campylobacter* contamination in carcasses (Berrang et al., 2007). Other chemical agents with demonstrated decontaminating efficacy and potential application in meat food include cetylpyridinium chloride, sodium hydroxide, bovine lactoferrin, and benzalkonium chloride (Gill and Gill, 2012; Skandamis et al., 2010; Zweifel and Stephan, 2012).

15.4 Elimination of microbial pathogens by ultraviolet light

Ultraviolet (UV) light is microbicidal and is routinely applied to decontaminate surfaces in various industrial applications including packaging materials in the food industry, medical devices, hospital settings, and for water treatment. Of the entire spectrum of UV radiation, UV-C (200–280 nm) is particularly microbicidal (Keklik et al., 2012). Antimicrobial activity of UV light is attributed to the photochemical reaction leading to free radical generation and damage to nucleic acids of microorganisms, thus arresting growth. Previous studies documented the application of UV light to eliminate meatborne pathogens in chicken meat (Isohanni and Lyhs, 2009; Zweifel and Stephan, 2012). Technological advancements in UV light generators have led to the development of pulsed UV light which in contrast to regular UV application delivers higher energy to exposed surfaces, thus producing greater microbicidal activity (Keklik et al., 2012). Nonetheless, one of the major hurdles in the application of UV light in the meat food industry is the minimal penetration, which limits the germicidal activity to the exposed surfaces only. Other concerns of UV application include the loss of sensitive nutrients such as riboflavin (Keklik et al., 2012).

15.5 Irradiation of meat for eliminating microbial hazards

Irradiation, or radiation exposure, of meat food with an intent to reduce the microbial load and enhance the shelf life is a relatively well established method (Roberts, 2014; Sommers, 2012). Microbicidal activity of radiation is derived from direct damage to DNA and RNA, and free radical damage to cell organelles. Various types of radiations have been utilized for food irradiation. Gamma rays derived from radioactive metals (Co^{60}, Cs^{137}) have been used for the purpose, as have been X-rays and electron beams (Cathode ray) (Farkas and Mohácsi-Farkas, 2011; Radomyski et al., 1994; Sommers, 2012). Microbicidal activity of food irradiation has been studied by many researchers with documented activity against foodborne bacterial pathogens, parasites, and viruses (Ahn et al., 2013; Lung et al., 2015; Monk et al., 1995; Radomyski et al., 1994; Sommers, 2012; Thayer and Boyd, 1993). Despite early approval by regulating agencies, such as the USDA, the irradiation of meat food faced consumer concerns. Major concerns relate to the loss of nutritional quality due to degradation of radiation-sensitive nutrients, loss of natural food aroma, and development of untoward flavor. Irradiated foods are also required to conform to labeling guidelines declaring the process (Farkas and Mohácsi-Farkas, 2011; O'Bryan et al., 2008; Roberts, 2014; Sommers, 2012).

15.6 Application of low temperature

Holding of meat at low temperature is an age-old practice for lengthening shelf life. Application of low temperature takes the form of chilling ($-5°C$ to $4°C$) or freezing ($\leq -18°C$) the dressed carcasses. Such application of cold temperature to meat does not necessarily eliminate any existing microbial pathogen present in meat but may arrest their proliferation. However, eukaryotic parasites are known to be susceptible to freezing (Lacour et al., 2013). Nonetheless, some evidence indicates that freezing of poultry carcasses at or below $-20°C$ may cause a reduction in count of *Campylobacter jejuni* (Bhaduri and Cottrell, 2004). Without any other decontamination approach, the efficacy of chilling or freezing in reducing the microbial load of meat remains variable and debatable (Skandamis et al., 2010).

15.7 High-pressure processing (HPP) for elimination of pathogens

Processing of meat food at high hydrostatic pressure of 100–1000 MPa is an effective method of the elimination of pathogens (Daryaei and Balasubramaniam, 2012; Gill and Gill, 2012). The process is also capable of enhancing storage quality without significant effects on organoleptic qualities. Mechanisms of microbial inactivation by HPP are multimodal and are mediated through damage of subcellular organelles, cell membranes, and inactivation of enzymes involved in DNA replication and transcription (Daryaei and Balasubramaniam, 2012). The process is particularly suitable for finished flexibly packaged products. Efficacy of HPP is reported to be in the order of 2–8 log CFU for various common bacteria associated with meat (Gill and Gill, 2012). Despite the availability of a few HPP-treated products in the market, commercial application of the process is still somewhat limited owing to the unavailability of industrial-scale apparatus capable of delivering extremely high pressure ≈ 1000 MPa (Gill and Gill, 2012).

15.8 Other emerging approaches for elimination of microbial pathogens

Due to unfettered demands of consumers for safer foods without compromising sensory qualities, researchers and technologists continuously endeavor to develop new approaches and techniques for the elimination of foodborne pathogens.

15.8.1 Nonthermal plasma (cold plasma)

Of interest among the novel applications is nonthermal plasma (cold plasma), which is an assemblage of electrically stimulated atoms, molecules, electron, photons, and ionic species in high energy state (Kong, 2012). The high energy state of plasma provides germicidal properties capable of eliminating various foodborne pathogens (Kong, 2012; Niemira, 2012; Scholtz et al., 2015). For example, a study on chicken meat revealed at least 1 log CFU reduction in *Listeria monocytogenes* (Song et al., 2009).

15.8.2 Dense phase carbon dioxide

Another promising new technology is dense phase carbon dioxide for decontamination of food including meat (Balaban et al., 2012). The process involves processing food under high-pressure carbon dioxide which permeates into the food and thus inactivates the microbial pathogens. The mechanism of inactivation is still being studied and is possibly mediated through the lowering of intracellular and extracellular pH, damage to microbial cell membranes, and extraction of important cellular functional elements (Balaban et al., 2012). While the technology is more suitable for liquid foods, experimental evidence with solid foods such as chicken meat and beef indicated 1–3 log CFU reduction in counts of *Escherichia coli*, *Staphylococcus aureus*, *L. monocytogenes*, and *Salmonella* Typhimurium (Sirisee et al., 1998; Wei et al., 1991).

15.8.3 Electrolyzed oxidizing water

Electrolyzed oxidizing water (EOW) is a powerful yet relatively simple technology for effective elimination of contaminating microbes from meat food (Cheng et al., 2012; Hricova et al., 2008). EOW is derived by electrolysis of water in a chamber divided by a membrane. The mechanism of microbial inactivation of EOW is due to various reactive ions generated during the process of electrolysis and is thus similar to the action of chlorine (Cheng et al., 2012; Gill and Gill, 2012). The technology is particularly suitable for slaughterhouse operations and commercial production units are available in the market. Previous studies on pork and chicken meat on the decontamination potential of the EOW revealed variable degrees of reduction (1–2 log CFU) in counts of foodborne pathogens (Cheng et al., 2012; Doyle and Erickson, 2006; Loretz et al., 2010).

15.8.4 Microwave and radio frequency

With rising demands for convenience food items, including meat, newer alternatives like microwave (MW) and radio frequency (RF) processing have emerged (Dev et al., 2012; Gill and Gill, 2012). MWs and RFs comprise longer wavelength parts of the electromagnetic spectrum. At the heart of the MW and RF processing of food lies the principle of dielectric heating resulting from rapid movement of polar molecules (water) in its attempt to continuously realign to the changing electromagnetic field (Dev et al., 2012). The microbicidal effect is derived from the heating that takes place inside the food. Compared to traditional thermal processing, heat transfer is more efficient and is often regarded as heating from within (Dev et al., 2012). While the process is particularly suited for meat products, the limitation of allowable heating restricts its application in fresh meat. Nevertheless, studies have documented considerable reductions in pathogenic bacteria in various meat products (Dev et al., 2012; Dinçer and Baysal, 2004).

15.8.5 Infrared heating

Similar to MW and RF heating, another alternative thermal approach to eliminate pathogens from meat is by infrared (IR) heating. The phenomenon of heating happens as IR radiation comes in contact with the food substance and delivers its energy to the substrate and the heat generated then drives the thermal inactivation of microbes (Dinçer and Baysal, 2004; Ramaswamy et al., 2012). Energy efficiency is a major advantage of the methods as IR does not heat up the air and container around the food substance. Studies have indicated a considerable reduction in log CFU of *L. monocytogenes* in ready-to-eat meat food items (Ramaswamy et al., 2012).

15.8.6 Biocontrol with bacteriophage

Apart from the physicochemical approach, an emerging biological approach is application of bacteriophage for targeted elimination of foodborne pathogens from meat. The advantages of this approach include high specificity, minimal negative impact on food sensory qualities, and negligible environmental footprint. A number of products are available commercially although there is a need to validate and standardize their application in various food matrices. Experimental evidence to date has indicated comparable efficacy with other decontamination procedures (Cooper, 2016; Moye et al., 2018; Sabouri et al., 2017; Sulakvelidze, 2013; Wang et al., 2017).

15.9 Detection of microbial pathogens

Foodborne illnesses caused by unsafe food continue to be a global public health challenge. Foodborne diseases take major toll on global health and economy (Foodborne Disease Burden Epidemiology Reference Group, 2015; Jaffee et al., 2019). On the other hand, there are consumers who are increasingly more aware and who demand safe products. Therefore ensuring the safety of foods, including meat and meat products, has become a social responsibility, economic imperative, and commercial necessity to maintain the continual and sustainable growth of the meat food sector.

The ability to detect microbial pathogens in meat and meat products is integral to ensuring safety. Early and reliable detection minimizes untoward effects on health and business. Since the early days of the development of the science of meat and meat products, considerable efforts have been devoted toward the development of new techniques and fine-tuning of the existing techniques for the detection of various foodborne pathogens, including those associated with meat and its products. While a detailed description of all the techniques will be beyond the scope of current discourse, influential detection methods will be discussed.

15.9.1 Conventional culture-based techniques

These techniques are the oldest, most studied, widely evaluated and reevaluated, and still constitute the mainstay of universally accepted detection methodologies. Essentially these are bacteriological techniques developed for microbiological research and eventually adopted for the detection of foodborne bacterial pathogens and occasionally other pathogens. Classical bacteriological techniques provide visually verifiable results indicating the isolation of a viable pathogen in pure form and are highly reproducible across laboratories. Commensurate with the diversity of foodborne bacteriological pathogens, a wide variety of specific media have been developed for selective detection of target organisms, although general assessment techniques (aerobic plate count, total coliform count) intended for quantitative estimation of overall sanitary quality of meat and meat products are also widely used (Priyanka et al., 2016). These techniques often form the basis for comparing other assays and enjoy legal standing all over the world. Advantages notwithstanding, the conventional detection techniques suffer serious shortcomings too. Major limitations of these methodologies include a lack of rapidity with the process often stretching from days to weeks, inability to detect viable but nonculturable forms of bacteria, and the difficulty in subtyping the pathogen for tracing foodborne outbreaks (Priyanka et al., 2016). Rising cost of manpower and the labor-intensive nature of conventional techniques also add to the disadvantages of these techniques. Despite these shortcomings, it is believed that overwhelming majority of laboratories still use culture-based techniques.

15.9.1.1 Automated microbial identification systems

Considering the labor-intensive nature and limited throughput of culture-based identification approaches, automation was natural and inevitable. As a result, a number of automated microbial identification systems were developed and marketed. Automated microbiological identification instruments incorporate miniaturized systems for microbial growth in the presence of various substrates, followed by detection of growth from changes (chromogenic, fluorogenic, redox potential, pH) in substrate media, and matching the pattern of growth with an internal database to report an identification (O'Hara, 2005; Stager and Davis, 1992). The systems are usually computerized with the necessary software for analysis of results. A number of comparable systems are available in the market, for example, Phoenix 100, VITEK, MicroScan WalkAway, Sensititre, ARIS, OmniLog, etc. (O'Hara, 2005; Odumeru et al., 1999; Stager and Davis, 1992). While the systems were initially developed to cater to clinical microbiology laboratories, they are increasingly being used for pathogen identification from food. The automated systems offer many advantages, including speed, reduced labor requirements, cheaper per sample processing cost, and standardized reports, although they do require high initial investment. When applied to foodborne pathogens these systems are reported to perform well (Odumeru et al., 1999; Wiedmann et al., 2000).

15.9.2 Immunological techniques

By harnessing the ability of the specific interaction between an antigen and its antibody, various techniques and assays were developed for the rapid detection of meatborne and other foodborne pathogens (Mangal et al., 2016; Umesha and Manukumar, 2018). Relative ease of execution and fairly straightforward interpretation, made these assays popular. Many commercial assays are available in the market for ready detection of pathogens. Various formats were developed with enzyme-linked immunosorbent assay (ELISA) and the latex agglutination test (LAT) being most popular.

15.9.2.1 Enzyme-linked immunosorbent assay (ELISA)

ELISA is perhaps the most popular immunological technique applied for the detection of meat and other foodborne pathogens. The strength of the assay lies in the amplification of the antigen−antibody binding signal through the application of an enzyme-coupled secondary antibody and substrate reaction, thus increasing the assay sensitivity manifold (Priyanka et al., 2016). Other advantages are high throughput with the standardized 96-well format and the quantitative nature of the assay. Many commercial assays for a diverse range of pathogens are available in the market. However, high initial investment costs coupled with the perishable nature of the reagents, and the need for specialized plate readers keep them beyond reach for a large number of laboratories (Alahi and Mukhopadhyay, 2017; Law et al., 2014).

15.9.2.2 Latex agglutination test (LAT)

When antibody-coated minute latex beads are allowed to interact with the specific antigen (foodborne pathogen), they produce agglutinating clumps visible to the naked eye against a dark background (D'Aoust et al., 1991). The ease of performance and relative simplicity are reasons for the popularity of LAT. Commercial assays for a number of foodborne pathogens are available. Weaknesses include qualitative results, short shelf life of latex beads, and scope for the subjective interpretation of assay results (D'Aoust et al., 1991; Mangal et al., 2016; Miller et al., 2008).

15.9.3 Nucleic acid-based techniques

Since the discovery of the specific binding principles of DNA to its complementary sequence, the property of complementary binding has been utilized to devise a host of techniques to detect a specific pathogen by identifying its DNA (or RNA). Initial developments were based on simple hybridization techniques, but were rapidly replaced by PCR (polymerase chain reaction)-based techniques, though many newer varieties have also been developed subsequently (Kuchta et al., 2014; Souii et al., 2016).

15.9.3.1 Polymerase chain reaction (PCR)

PCR for detection of foodborne pathogens involves enzymatic and exponential amplification of a suspect DNA (or RNA) string to enable easy visualization of the amplified product under agarose gel electrophoresis in case of a positive match. The specificity of PCR is derived from the complementary binding of a known primer (short DNA sequence) with the suspected target (pathogen) sequence, while the sensitivity of the assay is due to exponential nature of amplification (Kuchta et al., 2014; Law et al., 2014; Souii et al., 2016; Umesha and Manukumar, 2018). Over the last few decades, PCR has evolved to be fairly robust and increasingly cheaper in terms of operating costs. Commercial assays or PCR-based detection services for the detection of foodborne pathogens are available. However, the qualitative nature of the assay, inability to selectively detect live pathogens, requirement for sophisticated instrument (thermocycler) are areas of concerns (Law et al., 2014; Mangal et al., 2016; Souii et al., 2016; Zhao et al., 2014).

15.9.3.2 Real-time polymerase chain reaction

Also known as quantitative PCR (qPCR) real-time PCR had been a significant improvement towards the quantification of foodborne pathogens in samples. The principle of real-time PCR hinges on the traditional PCR technique coupled with continuous monitoring of the amplification process by measuring the increase in fluorescence activity of either a DNA binding dye or specialized dye-tagged probes. Quantification of the target DNA of a suspect pathogen is calculated from the amount of fluorescence following comparison with experimental controls. The assay has been employed for the detection of many pathogens including those associated with foodborne illnesses. While the principle of the assay is straightforward, it is a complex process to perform requiring expertise. Moreover, instrumentation is costly and thus largely remains out of bounds for most laboratories (Kuchta et al., 2014; Law et al., 2014; Mangal et al., 2016; Souii et al., 2016; Umesha and Manukumar, 2018; Valderrama et al., 2016; Zhao et al., 2014).

15.9.3.3 Nucleic acid sequence based amplification

First reported in 1991, nucleic acid sequence based amplification (NASBA) is an isothermal transcription-based amplification methodology for the detection of RNA. The technique depends on sequential interplay of three key enzymes—T7 RNA polymerase, RNase H, and AMV (avian myeloblastosis virus) reverse transcriptase (Kuchta et al., 2014; Law et al., 2014; Mangal et al., 2016; Pilla and Rickeb, 1995; Souii et al., 2016; Umesha and Manukumar, 2018; Zhao et al., 2014). NASBA being isothermal offers the advantage of doing away with a costly thermocycler. Refinement of NASBA by incorporating fluorescent labeled probes paved way for development of real-time NASBA which has been successfully used for detecting a number of foodborne pathogens including *Salmonella*, *Vibrio cholerae*, *Campylobacter*, and *S. aureus* (Kuchta et al., 2014; Law et al., 2014; Zhao et al., 2014). NASBA offers the advantages of high throughput and shorter incubation time and assessment of viability of foodborne pathogens (Kuchta et al., 2014). Commercial assays are available in the market, such as Nuclisens EasyQ (Kuchta et al., 2014).

15.9.3.4 Loop-mediated isothermal amplification assay

To overcome the variable temperature amplification constraint of PCR requiring a thermocycler, loop-mediated isothermal amplification was developed with the amplification of target DNA occurring under isothermal conditions (Li et al., 2017a; Niessen et al., 2013). This was a major breakthrough as the need for precise temperature control was no longer needed. The end product in case of positive detection can be easily ascertained form agarose gel electrophoresis or from naked eye inspection as a DNA binding dye (e.g., SyBR Green) is incorporated in the reaction mix (Li et al., 2017a; Mangal et al., 2016; Umesha and Manukumar, 2018; Zhao et al., 2014). Researchers developed and optimized many assays targeting various foodborne pathogens including *E. coli*, *L. monocytogenes*, *Campylobacter*, and *Salmonella* (Law et al., 2014; Li et al., 2017b, 2018; Niessen et al., 2013).

15.9.3.5 DNA microarray

In order to enable detection of multiple pathogens and to scale up the throughput, the DNA microarray was devised so that multiple targets (suspected foodborne pathogens) can be checked for simultaneously. A DNA microarray is a collection of immobilized hybridization probes specific for multiple pathogens (though this may vary) which on reaction with dye-labeled sample DNA bind and emit fluorescence. From the location of the fluorescence and the corresponding identity of the immobilized probe the suspect pathogen is identified (Kostić and Sessitsch, 2011; López-Campos et al., 2012; Rasooly and Herold, 2008). Like qPCR the theoretical framework is quite simple, yet the method requires a high degree of expertise in fabrication of the chip (solid support onto which probes are immobilized and fixed), probe design, and assay technique. Moreover, instrumentation costs are high with the requirement for a specialized chip reader and software for result interpretation (López-

Campos et al., 2012; Rasooly and Herold, 2008). Nonetheless, several researchers have designed and successfully employed this technique for the detection of common foodborne pathogens (Kostić and Sessitsch, 2011; López-Campos et al., 2012; Rasooly and Herold, 2008).

15.9.4 Matrix-assisted laser desorption ionization-time of flight mass spectrometry

Matrix-assisted laser desorption ionization-time of flight mass spectrometry (MALDI-TOF MS) is a developing technique for identifying foodborne microbial contaminants (Pavlovic et al., 2013; Sauget et al., 2017; Singhal et al., 2015). The technique involves laser bombardment of an analyte (foodborne pathogen) embedded and crystallized in a suitable matrix leading to ionization and vaporization of the analyte and subsequent flight of the ions in an electrical field. By measuring the ratio of the mass to charge (m/z) of the charged vaporized particle and its time of flight along the electrical field (often called flight tube), a peptide mass fingerprint (PMF) is generated. The final analysis and identification of the analyte (foodborne pathogen) is achieved by comparing the PMF with a standard database (Mazzeo et al., 2006; Pavlovic et al., 2013; Singhal et al., 2015). This technology has been successfully applied for the laboratory identification of microbes and is known to be extremely rapid, accurate, and cost-effective, although initial investments are high, and it requires expertise in mass-spectral analysis (Pavlovic et al., 2013; Tran et al., 2015; Urwyler and Glaubitz, 2016). Evaluation of MALDI-TOF for detection of foodborne pathogens indicated excellent performance and the technique has been successfully applied for the detection of many foodborne pathogens such as *Campylobacter*, *Salmonella*, and *Staphylococcus* (Carbonnelle et al., 2007; Dieckmann and Malorny, 2011; Kumar and Thakur, 2018; Mandrell et al., 2005; Mazzeo et al., 2006).

15.9.5 Hyperspectral imaging and analysis

Advancements in terrestrial remote sensing technology combining the strengths of conventional imaging, spectral analysis, and radiometry yielded a unique imaging system that has become popularly known as hyperspectral imaging (Park et al., 2015; Xiong et al., 2015). The principle of hyperspectral imaging involves accurate radiometric measurement of individual pixels of an image over a range of spectrum. Resultant data contain spatial (two-dimensional) information as well as spectral information, thus adding another dimension to the spatial data (hence the name hyperspectral imaging) (Elmasry et al., 2012). Hyperspectral analysis of images can provide important information on the physicochemical properties of the imaged object that are otherwise not possible with conventional imaging techniques. The technique has been applied for various purposes—from meat quality analysis to the identification and classification of foodborne pathogens. Specific studies of *E. coli* and *S. aureus* have yielded encouraging results (Elmasry et al., 2012; Park et al., 2014, 2015; Xiong et al., 2015).

15.9.6 Nanotechnology-based approaches

Rapid advances in nanoscale science have ushered in new opportunities for the development of new kind of assays based on molecular interaction at the submicroscopic scale, thereby allowing more efficient detection of foodborne microbes (Krishna et al., 2018; Rowland et al., 2016). Moreover, the application of nanotechnology, particularly nanofabrication techniques, resulted in the significant improvement and considerable miniaturization of existing technologies offering greater portability. Biomolecules behave differently at nanoscale environments and quantum effects take precedence over classical physical laws (Rowland et al., 2016). Taking advantage of these principles, a wide variety of assays are based on nanoparticles (usually gold and silver), quantum dots, quantum beads, quantum rods, fluorescent polymeric nanoparticles, magnetic particles, fiber optic-based sensors, cantilever assays, etc. (Billington et al., 2014; Krishna et al., 2018; Pérez-López and Merkoçi, 2011; Rowland et al., 2016). These assays were targeted toward important foodborne pathogens such as *E. coli* O157:H7, *Salmonella*, *L. monocytogenes*, and *Vibrio parahaemolyticus*, among others (Stephen Inbaraj and Chen, 2016). Despite offering potential benefits including portability and enhanced sensitivity, currently the application of nanotechnology-based assays for the detection of foodborne pathogens is in a growing stage and requires further refinements and validations. For a detailed review on the subject, please see the excellent articles available (Billington et al., 2014; Kaittanis et al., 2010; Krishna et al., 2018; Pérez-López and Merkoçi, 2011; Rowland et al., 2016; Stephen Inbaraj and Chen, 2016).

15.9.7 Other assays

In addition to the described assays a large variety of detection platforms/techniques have been developed, notably biosensors based on optical, mass-action, electrical activities, and surface plasmon resonance; microfluidic devices and lateral flow assays based on immunological principles; next-generation sequencing (NGS)-based multigene panel platforms; digital PCR; and impedimetric sensors. These novel methods are under intense development and offer many advantages over the traditional approaches. For a detailed description of them please see other relevant chapter(s) of this book.

References

Ahn, D.U., Kim, I.S., Lee, E.J., 2013. Irradiation and additive combinations on the pathogen reduction and quality of poultry meat. Poult. Sci. 92, 534–545.

Alahi, M.E.E., Mukhopadhyay, S.C., 2017. Detection methodologies for pathogen and toxins: a review. Sensors 17, 1–20.

Balaban, M.O., Ferrentino, G., Spilimbergo, S., 2012. Dense phase CO_2 (DPCD) for microbial decontamination of food. In: Demirci, A., Ngadi, M.O. (Eds.), Microbial Decontamination in the Food Industry. Woodhead Publishing, pp. 665–697.

Berrang, M.E., Bailey, J.S., Altekruse, S.F., Patel, B., et al., 2007. Prevalence and numbers of *Campylobacter* on broiler carcasses collected at Rehang and Postchill in 20 U.S. processing plants. J. Food Prot. 70, 1556–1560.

Bhaduri, S., Cottrell, B., 2004. Survival of cold-stressed *Campylobacter jejuni* on ground chicken and chicken skin during frozen storage. Appl. Environ. Microbiol. 70, 7103–7109.

Billington, C., Hudson, J.A., D'Sa, E., 2014. Prevention of bacterial foodborne disease using nanobiotechnology. Nanotechnol. Sci. Appl. 7, 73–83.

Carbonnelle, E., Beretti, J.-L., Cottyn, S., Quesne, G., Berche, P., Nassif, X., et al., 2007. Rapid identification of Staphylococci isolated in clinical microbiology laboratories by matrix-assisted laser desorption ionization-time of flight mass spectrometry. J. Clin. Microbiol. 45, 2156–2161.

Chawla, A.S., Kasler, D.R., Sastry, S.K., Yousef, A.E., 2012. Microbial decontamination of food using ozone. In: Demirci, A., Ngadi, M.O. (Eds.), Microbial Decontamination in the Food Industry. Woodhead Publishing, pp. 495–532.

Cheng, K.-C., Dev, S.R.S., Bialka, K.L., Demirci, A., 2012. Electrolyzed oxidizing water for microbial decontamination of food. In: Demirci, A., Ngadi, M.O. (Eds.), Microbial Decontamination in the Food Industry. Woodhead Publishing, pp. 563–591.

Cooper, I.R., 2016. A review of current methods using bacteriophages in live animals, food and animal products intended for human consumption. J. Microbiol. Methods 130, 38–47.

D'Aoust, J.-Y., Sewell, A.M., Greco, P., 1991. Commercial latex agglutination kits for the detection of foodborne *Salmonella*. J. Food Prot. 54, 725–730.

Daryaei, H., Balasubramaniam, V.M., 2012. Microbial decontamination of food by high pressure processing. In: Demirci, A., Ngadi, M.O. (Eds.), Microbial Decontamination in the Food Industry. Woodhead Publishing, pp. 370–406.

Dev, S.R.S., Birla, S.L., Raghavan, G.S.V., Subbiah, J., 2012. Microbial decontamination of food by microwave (MW) and radio frequency (RF). In: Demirci, A., Ngadi, M.O. (Eds.), Microbial Decontamination in the Food Industry. Woodhead Publishing, pp. 274–299.

Dieckmann, R., Malorny, B., 2011. Rapid screening of epidemiologically important Salmonella enterica subsp. enterica serovars by whole-cell matrix-assisted laser desorption ionization-time of flight mass spectrometry. Appl. Environ. Microbiol. 77, 4136–4146.

Dinçer, A.H., Baysal, T., 2004. Decontamination techniques of pathogen bacteria in meat and poultry. Crit. Rev. Microbiol. 30, 197–204.

Doyle, M.P., Erickson, M.C., 2006. Reducing the carriage of foodborne pathogens in livestock and poultry Poultry Science Oxford Academic. Poult. Sci. 85, 960–973.

Ellis, M., Cooksey, K., Dawson, P., Han, I., Vergano, P., 2006. Quality of fresh chicken breasts using a combination of modified atmosphere packaging and chlorine dioxide sachets. J. Food Prot. 69, 1991–1996.

Elmasry, G., Barbin, D.F., Sun, D.W., Allen, P., 2012. Meat quality evaluation by hyperspectral imaging technique: an overview. Crit. Rev. Food Sci. Nutr. 52, 689–711.

Farkas, J., Mohácsi-Farkas, C., 2011. History and future of food irradiation. Trends Food Sci. Technol. 22, 121–126.

Foodborne Disease Burden Epidemiology Reference Group, 2015. WHO Estimates of the Global Burden of Foodborne Diseases. Switzerland. https://doi.org/10.1016/j.fm.2014.07.009.

Gill, A., Gill, C.O., 2012. Microbial decontamination of raw and ready-to-eat meats. In: Demirci, A., Ngadi, M.O. (Eds.), Microbial Decontamination in the Food Industry. Woodhead Publishing, pp. 30–59.

Gill, C., Bedard, D., Jones, T., 1997. The decontaminating performance of a commercial apparatus for pasteurizing polished pig carcasses. Food Microbiol. 14, 71–79.

Gill, C., Bryant, J., Bedard, D., 1999. The effects of hot water pasteurizing treatments on the appearances and microbiological conditions of beef carcass sides. Food Microbiol. 16, 281–289.

Hricova, D., Stephan, R., Zweifel, C., 2008. Electrolyzed water and its application in the food industry. J. Food Prot. 71, 1934–1947.

Isohanni, P.M.I., Lyhs, U., 2009. Use of ultraviolet irradiation to reduce *Campylobacter jejuni* on broiler meat. Poult. Sci. 88, 661–668.

Jaffee, S., Henson, S., Unnevehr, L., Grace, D., Cassou, E., 2019. The Safe Food Imperative: Accelerating Progress in Low- and Middle-Income Countries. World Bank, Washington, DC. Available from: https://doi.org/10.1596/978-1-4648-1345-0.

Kaittanis, C., Santra, S., Perez, J.M., 2010. Emerging nanotechnology-based strategies for the identification of microbial pathogenesis. Adv. Drug Deliv. Rev. 62, 408–423.

Keklik, N.M., Krishnamurthy, K., Demirci, A., 2012. Microbial decontamination of food by ultraviolet (UV) and pulsed UV light. In: Demirci, A., Ngadi, M.O. (Eds.), Microbial Decontamination in the Food Industry. Woodhead Publishing, pp. 344–369.

Kim, J.-G., Yousef, A.E., Dave, S., 1999. Application of ozone for enhancing the microbiological safety and quality of foods: a review. J. Food Prot. 62, 1071–1087.

Kim, J.-G., Yousef, A.E., Khadre, M.A., 2003. Ozone and its current and future application in the food industry. Adv. Food Nutr. Res. 45, 167–218.

Kochevar, S.L., Sofos, J.N., LeValley, S.B., Smith, G.C., 1997. Effect of water temperature, pressure and chemical solution on removal of fecal material and bacteria from lamb adipose tissue by spray-washing. Meat Sci. 45, 377–388.

Kong, M.G., 2012. Microbial decontamination of food by non-thermal plasmas. In: Demirci, A., Ngadi, M.O. (Eds.), Microbial Decontamination in the Food Industry. Woodhead Publishing, pp. 472–492.

Kostić, T., Sessitsch, A., 2011. Microbial diagnostic microarrays for the detection and typing of food- and water-borne (bacterial) pathogens. Microarrays 1, 3–24.

Krishna, V.D., Wu, K., Su, D., Cheeran, M.C.J., Wang, J.P., Perez, A., 2018. Nanotechnology: review of concepts and potential application of sensing platforms in food safety. Food Microbiol. 75, 47–54.

Kuchta, T., Knutsson, R., Fiore, A., Kudirkiene, E., Höhl, A., Horvatek Tomic, D., et al., 2014. A decade with nucleic acid-based microbiological methods in safety control of foods. Lett. Appl. Microbiol. 59, 263–271.

Kumar, D., Thakur, S., 2018. Molecular tools to study preharvest food safety challenges. Microbiol. Spectr. 6, 1–16.

Lacour, S.A., Heckmann, A., Macé, P., Grasset-Chevillot, A., Zanella, G., Vallée, I., et al., 2013. Freeze-tolerance of Trichinella muscle larvae in experimentally infected wild boars. Vet. Parasitol. 194, 175–178.

Law, J.W.-F., Ab Mutalib, N.-S., Chan, K.-G., Lee, L.-H., 2014. Rapid methods for the detection of foodborne bacterial pathogens: principles, applications, advantages and limitations. Front. Microbiol. 5, 1–19.

Li, Y., Fan, P., Zhou, S., Zhang, L., 2017a. Loop-mediated isothermal amplification (LAMP): a novel rapid detection platform for pathogens. Microb. Pathog. 107, 54–61.

Li, Y., Yang, L., Fu, J., Yan, M., Chen, D., Zhang, L., 2017b. The novel loop-mediated isothermal amplification based confirmation methodology on the bacteria in Viable but Non-Culturable (VBNC) state. Microb. Pathog. 111, 280–284.

Li, Y., Bai, C., Yang, L., Fu, J., Yan, M., Chen, D., et al., 2018. High flux isothermal assays on pathogenic, virulent and toxic genetics from various pathogens. Microb. Pathog. 116, 68–72.

Lianou, A., Koutsoumanis, K.P., Sofos, J.N., 2012. Organic acids and other chemical treatments for microbial decontamination of food. In: Demirci, A., Ngadi, M.O. (Eds.), Microbial Decontamination in the Food Industry. Woodhead Publishing, pp. 592–664.

López-Campos, G., Martínez-Suárez, J.V., Aguado-Urda, M., López-Alonso, V., 2012. Microarray detection and characterization of bacterial foodborne pathogens. Food, Health, and Nutrition. Springer, pp. 13–33.

Loretz, M., Stephan, R., Zweifel, C., 2010. Antimicrobial activity of decontamination treatments for poultry carcasses: a literature survey. Food Control 21, 791–804.

Lung, H.-M., Cheng, Y.-C., Chang, Y.-H., Huang, H.-W., Yang, B.B., Wang, C.-Y., 2015. Microbial decontamination of food by electron beam irradiation. Trends Food Sci. Technol. 44, 66–78.

Mahapatra, A.K., Muthukumarappan, K., Julson, J.L., 2005. Applications of ozone, bacteriocins and irradiation in food processing: a review. Crit. Rev. Food Sci. Nutr. 45, 447−461.

Mandrell, R.E., Harden, L.A., Bates, A., Miller, W.G., Haddon, W.F., Fagerquist, C.K., 2005. Speciation of *Campylobacter coli*, *C. jejuni*, *C. helveticus*, *C. lari*, *C. sputorum*, and *C. upsaliensis* by matrix-assisted laser desorption ionization-time of flight mass spectrometry. Appl. Environ. Microbiol. 71, 6292−6307.

Mangal, M., Bansal, S., Sharma, S.K., Gupta, R.K., 2016. Molecular detection of foodborne pathogens: a rapid and accurate answer to food safety. Crit. Rev. Food Sci. Nutr. 56, 1568−1584.

Mazzeo, M.F., Sorrentino, A., Gaita, M., Cacace, G., Stasio, M.D., et al., 2006. Matrix-assisted laser desorption ionization − time of flight mass spectrometry for the discrimination of food-borne microorganisms. Appl. Environ. Microbiol. 72, 1180−1189.

Miller, R.S., Speegle, L., Oyarzabal, O.A., Lastovica, A.J., 2008. Evaluation of three commercial latex agglutination tests for identification of *Campylobacter* spp. J. Clin. Microbiol. 46, 3546−3547.

Monk, J.D., Beuchat, L.R., Doyle, M.P., 1995. Irradiation inactivation of food-borne microorganisms. J. Food Prot. 58, 197−208.

Moye, Z.D., Woolston, J., Sulakvelidze, A., 2018. Bacteriophage applications for food production and processing. Viruses 10, 1−22.

Niemira, B.A., 2012. Cold plasma decontamination of foods. Annu. Rev. Food Sci. Technol. 3, 125−142.

Niessen, L., Luo, J., Denschlag, C., Vogel, R.F., 2013. The application of loop-mediated isothermal amplification (LAMP) in food testing for bacterial pathogens and fungal contaminants. Food Microbiol. 36, 191−206.

O'Bryan, C.A., Crandall, P.G., Ricke, S.C., Olson, D.G., 2008. Impact of irradiation on the safety and quality of poultry and meat products: a review. Crit. Rev. Food Sci. Nutr. 48, 442−457.

Odumeru, J.A., Steele, M., Fruhner, L., Larkin, C., et al., 1999. Evaluation of accuracy and repeatability of identification of food- borne pathogens by automated bacterial identification systems. J. Clin. Microbiol. 37, 944−949.

O'Hara, C.M., 2005. Manual and automated instrumentation for identification of Enterobacteriaceae and other aerobic gram-negative bacilli. Clin. Microbiol. Rev. 18, 147−162.

Park, B., Eady, M., Choi, S., et al., 2014. Rapid identification of *Salmonella* serotypes with stereo and hyperspectral microscope imaging methods. IAFP's European Symposium on Food Safety. International Association of Food Protection, Budapest.

Park, B., Seo, Y., Yoon, S.C., Windham, W.R., Lawrence, K.C., 2015. Hyperspectral microscope imaging methods to classify gram-positive and gram-negative foodborne pathogenic bacteria. Trans. ASABE 58, 5−16.

Pavlovic, M., Huber, I., Konrad, R., Busch, U., 2013. Application of MALDI-TOF MS for the identification of food borne bacteria. Open Microbiol. J. 7, 135−141.

Pérez-López, B., Merkoçi, A., 2011. Nanomaterials based biosensors for food analysis applications. Trends Food Sci. Technol. 22, 625−639.

Pilla, S.D., Rickeb, S.C., 1995. Strategies to accelerate the applicability of gene amplification protocols for pathogen detection in meat and meat products. Crit. Rev. Microbiol. 21, 239−261.

Priyanka, B., Patil, R.K., Dwarakanath, S., 2016. A review on detection methods used for foodborne pathogens. Indian J. Med. Res. 144, 327−338.

Radomyski, T., Murano, E.A., Olson, D.G., Murano, P.S., 1994. Elimination of pathogens of significance in food by low-dose irradiation: a review. J. Food Prot. 57, 73−86.

Ramaswamy, R., Krishnamurthy, K., Jun, S., 2012. Microbial decontamination of food by infrared (IR) heating. In: Demirci, A., Ngadi, M.O. (Eds.), Microbial Decontamination in the Food Industry. Woodhead Publishing, pp. 450−471.

Rasooly, A., Herold, K.E., 2008. Food microbial pathogen detection and analysis using DNA microarray technologies. Foodborne Pathog. Dis. 5, 531−550.

Roberts, P.B., 2014. Food irradiation is safe: half a century of studies. Radiat. Phys. Chem. 105, 78−82.

Rowland, C.E., Brown, C.W., Delehanty, J.B., Medintz, I.L., 2016. Nanomaterial-based sensors for the detection of biological threat agents. Mater. Today 19, 464−477.

Russell, S.M., Axtell, S.P., 2005. Monochloramine versus sodium hypochlorite as antimicrobial agents for reducing populations of bacteria on broiler chicken carcasses. J. Food Prot. 68, 758–763.

Sabouri, S., Sepehrizadeh, Z., Amirpour-Rostami, S., Skurnik, M., 2017. A minireview on the in vitro and in vivo experiments with anti-*Escherichia coli* O157:H7 phages as potential biocontrol and phage therapy agents. Int. J. Food Microbiol. 243, 52–57.

Sauget, M., Valot, B., Bertrand, X., Hocquet, D., 2017. Can MALDI-TOF mass spectrometry reasonably type bacteria? Trends Microbiol. 25, 447–455.

Scholtz, V., Pazlarova, J., Souskova, H., Khun, J., Julak, J., 2015. Nonthermal plasma—a tool for decontamination and disinfection. Biotechnol. Adv. 33, 1108–1119.

Singhal, N., Kumar, M., Kanaujia, P.K., Virdi, J.S., 2015. MALDI-TOF mass spectrometry: an emerging technology for microbial identification and diagnosis. Front. Microbiol. 6, Article 791.

Sirisee, U., Hsieh, F., Huff, H.E., 1998. Microbial safety of supercritical carbon dioxide processes. J. Food Process. Preserv. 22, 387–403.

Skandamis, P.N., Nychas, G.-J.E., Sofos, J.N., 2010. Meat decontamination. In: Toldrá, F. (Ed.), Handbook of Meat Processing. John Wiley & Sons, Ltd, pp. 43–85.

Sofos, J.N., Smith, G.C., 1998. Nonacid meat decontamination technologies: model studies and commercial applications. Int. J. Food Microbiol. 44, 171–188.

Sommers, C.H., 2012. Microbial decontamination of food by irradiation. In: Demirci, A., Ngadi, M.O. (Eds.), Microbial Decontamination in the Food Industry. Woodhead Publishing, pp. 322–343.

Song, H.P., Kim, B., Choe, J.H., Jung, S., Moon, S.Y., Choe, W., et al., 2009. Evaluation of atmospheric pressure plasma to improve the safety of sliced cheese and ham inoculated by 3-strain cocktail *Listeria monocytogenes*. Food Microbiol. 26, 432–436.

Souii, A., M'hadheb-Gharbi, M.B., Gharbi, J., 2016. Nucleic acid-based biotechnologies for food-borne pathogen detection using routine time-intensive culture-based methods and fast molecular diagnostics. Food Sci. Biotechnol. 25, 11–20.

Stager, C.E., Davis, J.R., 1992. Automated systems for identification of microorganisms. Clin. Microbiol. Rev. 5, 302–327.

Stephen Inbaraj, B., Chen, B.H., 2016. Nanomaterial-based sensors for detection of foodborne bacterial pathogens and toxins as well as pork adulteration in meat products. J. Food Drug Anal. 24, 15–28.

Sulakvelidze, A., 2013. Using lytic bacteriophages to eliminate or significantly reduce contamination of food by foodborne bacterial pathogens. J. Sci. Food Agric. 93, 3137–3146.

Thayer, D.W., Boyd, G., 1993. Elimination of *Escherichia coli* O157:H7 in meats by gamma irradiation. Appl. Environ. Microbiol. 59, 1030–1034.

Tran, A., Alby, K., Kerr, A., Jones, M., Gilligan, P.H., 2015. Cost savings realized by implementation of routine microbiological identification by matrix-assisted laser desorption ionization-time of flight mass spectrometry. J. Clin. Microbiol. 53, 2473–2479.

Trinetta, V., Morgan, M., Linton, R., 2012. Chlorine dioxide for microbial decontamination of food. In: Demirci, A., Ngadi, M.O. (Eds.), Microbial Decontamination in the Food Industry. Woodhead Publishing, pp. 533–562.

Umesha, S., Manukumar, H.M., 2018. Advanced molecular diagnostic techniques for detection of food-borne pathogens: current applications and future challenges. Crit. Rev. Food Sci. Nutr. 58, 84–104.

Urwyler, S.K., Glaubitz, J., 2016. Advantage of MALDI-TOF-MS over biochemical-based phenotyping for microbial identification illustrated on industrial applications. Lett. Appl. Microbiol. 62, 130–137.

Valderrama, W.B., Dudley, E.G., Doores, S., Cutter, C.N., 2016. Commercially Available Rapid Methods for Detection of Selected Food-borne Pathogens. Crit. Rev. Food Sci. Nutr. 56, 1519–1531.

Wang, L., Qu, K., Li, X., Cao, Z., Wang, X., Li, Z., et al., 2017. Use of bacteriophages to control *Escherichia coli* O157:H7 in domestic ruminants, meat products, and fruits and vegetables. Foodborne Pathog. Dis. 2266. Available from: https://doi.org/10.1089/fpd.2016.2266. XX, fpd. 2016.

Wei, C.I., Balaban, M.O., Fernando, S.Y., Peplow, A.J., 1991. Bacterial effect of high pressure CO_2 treatment on foods spiked with *Listeria* or *Salmonella*. J. Food Prot. 54, 189–193.

Wiedmann, M., Weilmeier, D., Dineen, S.S., Ralyea, R., Boor, K.J., 2000. Molecular and phenotypic characterization of Pseudomonas spp. isolated from milk. Appl. Environ. Microbiol. 66, 2085–2095.

Xiong, Z., Xie, A., Sun, D.W., Zeng, X.A., Liu, D., 2015. Applications of hyperspectral imaging in chicken meat safety and quality detection and evaluation: a review. Crit. Rev. Food Sci. Nutr. 55, 1287–1301.

Zhao, X., Lin, C.W., Wang, J., Oh, D.H., 2014. Advances in rapid detection methods for foodborne pathogens. J. Microbiol. Biotechnol. 24, 297–312.

Zweifel, C., Stephan, R., 2012. Microbial decontamination of poultry carcasses. In: Demirci, A., Ngadi, M.O. (Eds.), Microbial Decontamination in the Food Industry. Woodhead Publishing, pp. 60–95.

CHAPTER 16

Modern techniques for rapid detection of meatborne pathogens

Prabhat Kumar Mandal[1] and Ashim Kumar Biswas[2]

[1]*Department of Livestock Products Technology, Rajiv Gandhi Institute of Veterinary Education and Research, Puducherry, India* [2]*Division of Post-Harvest Technology, ICAR-Central Avian Research Institute, Izatnagar, Bareilly, India*

Chapter Outline
16.1 Introduction 288
16.2 Need for rapid detection method 288
    16.2.1 Trends in rapid detection of meatborne pathogens 289
16.3 Biosensors-based detection techniques 290
    16.3.1 Bioluminescence sensors 291
    16.3.2 Fiber optic biosensor 291
    16.3.3 Surface plasmon resonance biosensor 293
    16.3.4 Electrical impedance biosensor 294
    16.3.5 Impedance-based biochip sensor 294
    16.3.6 Piezoelectric biosensors 295
    16.3.7 Cell-based sensor 295
    16.3.8 Fourier transform infrared spectroscopy 295
    16.3.9 Flow cytometry 296
    16.3.10 Solid phase cytometry 296
16.4 Nucleic acid-based assays 297
    16.4.1 DNA hybridization 297
    16.4.2 Polymerase chain reaction 298
    16.4.3 DNA microarrays (gene chip technology) 298
    16.4.4 Loop-mediated isothermal amplification 300
16.5 Requirements for rapid detection methods 300
16.6 Limitations of rapid detection methods 301
References 301

16.1 Introduction

The safety of meat and meat products is a key issue in the modern food trade. In the post-WTO era, meat entrepreneurs have increased productivity, but have few safeguards on product quality. The creation of World Trade Organization (WTO) in 1995 resulted in significantly increased trade in foods of animal origin between countries. But the emergence and reemergence of foodborne diseases are the key concern of the new food trades. According to a CDC report (Center for Disease Control and Prevention) the annual cost due to foodborne illness in the United States is nearly US$10 billion.

The bacterial pathogens most frequently identified in illnesses associated with beef products are *Salmonella* spp., *Clostridium perfringens*, and *Staphylococcus aureus*. The interest in *Escherichia coli* O157:H7 increased after a highly publicized outbreak of food poisoning associated with undercooked beef patties in the United States in 1993. Likewise, multidrug-resistant *Salmonella typhimurium* DT-104 has spread widely since it was first detected in the United Kingdom (JECFA 2002). The incidence of *Salmonella* was recorded to be up to 9% in red meat in India (Rao and Mahendrakar, 2003). Another foodborne emerging disease includes *Listeriosis*, which spread throughout France and Canada, where meat and meat products were implicated as a source of *Listeria monocytogenes* (Borch and Arinder, 2002). Similarly, *Staphylococcal* food poisoning or food intoxication was first reported in 1894, it is now a global problem in the meat industry.

Bacterial pathogens that have caused human illness in the last decades have been through the consumption of undercooked or minimally processed ready-to-eat meats (hotdogs, sliced luncheon meats, and salami). However, the presence of pathogens in ready-to-eat products is a serious concern since these products generally do not receive any further treatment before consumption. Food animals and poultry are the most important reservoirs for many of the foodborne pathogens (Biswas et al., 2008), while animal by-products used as feed supplements may also transmit pathogens to other animals. Further, consumers have become much more aware of food safety issues as a result of publicity given to foodborne diseases in the media (Griffiths, 1993; Chang, 2000; Park, 2001). Hence it is an emerging need to implement programs such as HACCP as a part of Good Manufacturing Practices and Sanitary and Phytosanitary measures to monitor the presence of pathogens and microbial toxins (APHA, 1987).

16.2 Need for rapid detection method

The effective testing of bacteria requires methods of analysis that can meet a number of challenging criterions. Time and sensitivity of analysis (Table 16.1) are the most important limitations related to the usefulness of microbial testing. The food industry

Table 16.1: Characteristics of some alternative and rapid methods.

Method	Detection limit (cfu/mL or g)	Time before result	Specificity
Plating technique	1	1–3 days	Good
Bioluminescence	10^4	30 min	No
Flow cytometry	10^2–10^3	30 min	Good
DEFT	10^3–10^4	30 min	No
Impedimetry	1	6–24 h	Moderate/good
Immunological methods	10^4	1–2 h	Moderate/good
Nucleic acid-based assays	10^3	6–12 h	Excellent

is in need of more rapid methods which are sensitive for the following reasons (APHA, 1987):

1. To provide immediate information on the possible presence of pathogen in raw material and finished products.
2. Low numbers of pathogenic bacteria are often present in complex biological environments along with many other nonpathogenic organisms.
3. The presence of even a single pathogenic organism in the food may be an infectious dose.
4. For monitoring of process control, cleaning, and hygienic practices during manufacture.
5. To reduce human errors and to save time and labor cost.

16.2.1 Trends in rapid detection of meatborne pathogens

The analysis of meat for the presence of both pathogenic and spoilage bacteria is a standard practice for ensuring food safety and quality (Doyle, 2001). However, the advent of biotechnology has greatly altered food testing methods and there are numerous companies that are actively developing assays that are specific, faster, and more sensitive than conventional methods in testing microbial contaminants in food (Boer and Beumer, 1999).

A rapid method can be an assay that gives instant or real-time results, but on the other hand it can also be a simple modification of a procedure that reduces the assay time. These rapid methods not only deal with the early detection and enumeration of microorganisms, but also with the characterization of isolates by use of microbiological, chemical, biochemical, biophysical, molecular biological, immunological, and serological methods (Boening and Tarr, 1995; Yongsheng et al., 1996; Westerman et al., 1997; Groisman and Ochman, 2000; Shah et al., 2003; Naravaneni and Jamil, 2005; Biswas et al., 2008). The degree to which rapid methods and automation are accepted and used for microbiological analysis is determined by the range and type of testing required, the volume throughput of samples to be tested, the availability of trained laboratory staff, and the nature of manufacturing practices (Vasavada, 1993). Rapid methods, such as online monitoring systems, can be useful to quickly screen a large number of samples and thereby enhance the processing efficiency.

16.3 Biosensors-based detection techniques

The science of biosensor is a multidisciplinary area. Biosensors (Fig. 16.1) are defined as indicators of biological compound that can be as simple as temperature sensitive paints or as complex as DNA−RNA probes. The potential application of biosensor technology to food testing offers several attractive features. Many of the systems are portable, can be used on site in time, are rapid, and are capable of undertaking multiple analyses simultaneously. Biosensing methods for pathogen detection are based on four basic physiological or genetic properties of microorganisms: metabolic patterns of substrate utilization, phenotypic expression analysis of signature molecules by antibodies, nucleic acid analysis, and the analysis of the interaction of pathogens with eukaryotic cells. Many commercially available rapid methods use culture-based methods coupled with automated or semiautomated nucleic acids-, antibody-, or substrate utilization-based methods to obtain results in 24−72 hours. Interestingly, many of the modern-day biosensor-based methods being developed utilize one of the above four principles or their combinations. However, antibody-based methods are the most popular because of their versatility, convenience, and relative ease in interpretation of the data. Hence the majority of the biosensors use antibodies for the capture and detection of the target analyte (Ritcher, 1993).

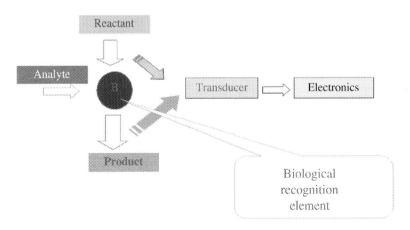

Figure 16.1
Schematic sketch of a biosensor.

16.3.1 Bioluminescence sensors

Recent advances in bioanalytical sensors have led to the utilization of the ability of certain enzymes to emit photons as a by-product of their reaction. This phenomenon is known as bioluminescence and may be used to detect the presence and biological condition of the cells. Among the emerging technologies for rapid microbiological analysis, this technique gives results in a short time. Two distinct areas of bioluminescence are in use in the food industry:

16.3.1.1 ATP bioluminescence

All living cells contain the molecule ATP. This molecule may be analyzed simply using an enzyme and coenzyme complex (Luciferase–Luciferin) found in the tail of the firefly (*Photinus pyralis*). The total light output of the sample is directly proportional to the amount of ATP present and can be quantified by luminometers. At least 10^4 cells are required to produce a signal. This system lacks specificity, but because of the rapid response time for obtaining results, this system is very suitable for online monitoring of HACCP programs. This technique has a detection limit of 1 pg ATP which is equivalent to 1000 bacterial cells. ATP is present in both nonmicrobial and microbial cells. To determine microbial ATP selective extraction is used. First, nonmicrobial ATP is extracted with nonionic detergents and then destroyed with high levels of potato ATPase for 5 minutes. Subsequently, microbial ATP is extracted using either trichloroacetic acid (5%) or an organic solvent (ethanol, acetone, or chloroform).

16.3.1.2 Bacterial bioluminescence

The gene responsible for bacterial bioluminescence (lux gene) has been identified and cloned. The DNA carrying this gene can be introduced into host-specific phages. These phages do not possess the intracellular biochemistry necessary to express this gene, hence they remain dark. However, the transfer of the lux gene to the host bacterium during infection results in light emission that can be easily detected by luminometers. This technique can detect 1×10^2 cells in 60 minutes. The specificity of this assay depends on phage specificity, e.g., Bacteriophage p22 is specific for *S. typhimurium* (Fig. 16.2).

16.3.2 Fiber optic biosensor

The fiber optic biosensor was one of the first commercially available optical biosensors, marketed by Research International (Monroe, WA) for the detection of foodborne pathogens. The basic principle of the fiber optic sensor is that when light propagates through the core of the optical fiber, that is, waveguide, it generates an evanescent field outside the surface of the waveguide. The waveguides are generally made up of polystyrene fibers or glass

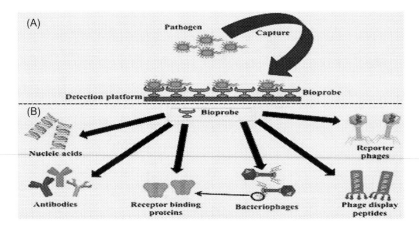

Figure 16.2
Capturing the pathogen on the sensing platform. (A) Pathogen detection platform (B) Molecules as bio-probe. Source: Adapted from Alahi, M.E.E., Mukhopadhyay, S.C., 2017. Detection methodologies for pathogen and toxins: a review. Sensors 17, 1885. doi:10.3390/s17081885 (Alahi and Mukhopadhyay, 2017).

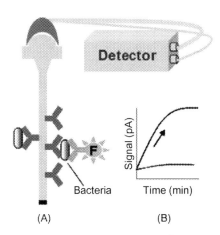

Figure 16.3
Schematic representation of fiber optic sensor. (A) Waveguide with captured antibody and fluorophor-labeled antibody, (B) signal acquisition over time. Source: Adapted from Bhunia, A.K., Banada, P., Banerjee, P., et al., 2007. Light scattering, fiber optic and cell-based sensors for sensitive detection of foodborne pathogens. J. Rapid Method Autom. Microbiol. 15, 121–145 (Bhunia et al., 2007).

slides. When fluorescent-labeled analytes such as pathogens or toxins bound to the surface of the waveguide are excited by the evanescent wave generated by a laser (635 nm), they emit a fluorescent signal (Bhunia, 2008; Taitt et al., 2005) and the signal travels back through the waveguide in high order mode to be detected by a fluorescence detector in real time (Fig. 16.3).

16.3.3 Surface plasmon resonance biosensor

Surface plasmon resonance (SPR) is a phenomenon that occurs during optical illumination of a metal surface, and it can be used for biomolecular interaction analysis. Receptors or antibodies immobilized on the surface of a thin film of a precious metal (gold) deposited on the reflecting surface of an optically transparent waveguide are used to capture the target analyte (Fig. 16.4). The sensing surface is located above or below a high index-resonant layer and a low index coupling layer. When visible or near-infrared (IR) radiation is passed through the waveguide in such a way, it causes an internal total reflection on the surface of the waveguide. At a certain wavelength in the red or near-IR region, the light interacts with a plasma or cloud of electrons on the high-index metal surface, and the resonance effect causes a strong absorbance. The exact wavelength of this absorption depends on the angle of incidence, the metal, the amount of capture molecules immobilized on the surface, and the surrounding material. The presence of ligands or antigens interacting with the receptor or antibody causes a shift in the resonance to longer wavelengths, and the amount of shift can be related to the concentration of the bound molecules.

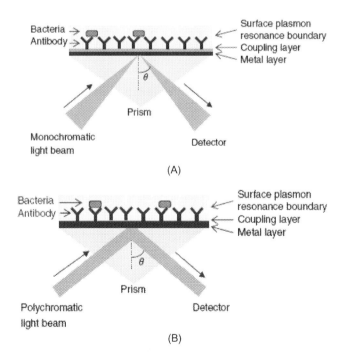

Figure 16.4
Schematic drawing of SPR sensor with angular (A) and wavelength (B) modulation. *Source: Adapted from Bhunia, A.K., 2008. Biosensors and bio-based methods for the separation and detection of foodborne pathogens. Adv. Food Nutr. Res. 54, 1–44.*

SPR-based sensors are governed by two basic principles: wavelength interrogation and angle interrogation. Wavelength interrogation uses a fixed angle of incidence but measures spectral changes, while in angle interrogation a fixed wavelength is used but the angle of reflectance is monitored. Most of the commercial SPR systems are operated based on the angle interrogation mode. SPR-based sensors allow real-time or near real-time detection of binding events between two molecules. The detection system is label-free, thus eliminating the need for additional reagents, assay steps, and time. The sensor can be reused for the same analyte repeatedly. It is highly sensitive and it can detect molecules in the femtomolar range (Bhunia, 2008; Rasooly and Herold, 2006).

16.3.4 Electrical impedance biosensor

Impedance microbiology detects microbes either directly due to production of ions from metabolic end products or indirectly from the liberation of CO_2. Microbial metabolism usually results in an increase in both conductance and capacitance, causing a decrease in impedance. A bridge circuit usually measures impedance. This method is well-suited for detection of bacteria in clinical samples and to monitor quality and detect specific food pathogens.

In this method, a population of microbes is provided with nutrients (nonelectrolyte) like lactose and microbes may utilize that nutrient and convert it to lactic acid (ionic form) thus changing the impedance. This impedance is measured over a period of 20 hours after inoculation in specific media. Since this does not involve serial dilution, this technique is simple to perform and faster than agar plate count. This system is capable of analyzing hundreds of sample at the same time since the instrument (Bactometer) is computer driven and automated to enable continuous monitoring. Typically most impedance analysis of food samples can be completed in 24 hours. This technique is not suited for testing samples with a low number of microorganisms and the food matrix may interfere with the analysis.

16.3.5 Impedance-based biochip sensor

The concept of this detection method, although old, is now getting wider popularity. Impedance is based on the changes in conductance in a medium due to the microbial breakdown of inert substrates into electrically charged ionic compounds and acidic by-products. The principle of all impedance-based systems is that they measure the relative or absolute changes in conductance, impedance, or capacitance at regular intervals. So the threshold value for the detection of target pathogens mainly depends on initial inoculums and the physiological state of the cells. In media-based impedance methods, bacterial metabolism results in an increased conductance and capacitance, with decreased impedance (Ivnitski et al., 1999). The major advantage of this system is that it allows the detection of only the viable cells, which is the major concern in food safety. The basic technical equipment required for performing impedance microbiology consists of special incubators and their culture vessels and an evaluation unit with computer, printer, and appropriate software.

16.3.6 Piezoelectric biosensors

This system is very attractive and offers a real-time output, simplicity of use, and cost-effectiveness. The general principle is based on coating the surface of piezoelectric sensor with a selective binding substance, for example, antibodies to bacteria, and then placing it in a solution containing bacteria. The bacteria will bind to the antibodies and the mass of the crystal will increase while the resonance frequency of oscillation will decrease proportionally.

16.3.7 Cell-based sensor

Cell-based assays (CBAs) continue to serve as a reliable method for the detection of pathogens in food samples. The CBA systems can report perturbations in the "normal" physiological activities of mammalian cells as a result of exposure to an "external" or environmental challenge. For this, mammalian cells are used as electrical capacitors. Electrical impedance (EI) uses the inherent electrical properties of cells to measure the parameters related to the tissue environment. The mechanical contact between cell—cell and cell—substrates is measured via conductivity or EI. The cell can be equated to a simple circuit since it is nothing more than conductive fluid encapsulated by a membrane surrounded by another conductive fluid. The conductive fluids make up the resistance elements of the circuit, while the membrane acts as a capacitor. Changes in impedance were able to detect changes in cell density, growth, or cellular behavior. These biosensors are able to provide detailed information about the growth characteristics of the tissue culture, including information on spreading, attachment, and cellular morphology. Mammalian cells have been widely used for the analysis of the pathogenic potential of foodborne bacteria (Bhunia and Wampler, 2005; Gray, 2004).

16.3.8 Fourier transform infrared spectroscopy

Fourier transform infrared spectroscopy (FTIR) is used to generate bacterial spectral scans based on the molecular composition of a sample, and mainly consists of the infrared source, the sample, and the detector. It is a nondestructive rapid method and sample identification depends on the available spectral library. When IR is absorbed or transmitted through the sample to the detector, it generates a "scan" or "fingerprint" profile. A library of spectral scans can be generated for different bacterial species and strains, which can be used for future comparison. This method requires the transfer of cells (biomass) from the growth media to an IR-reflecting substrate for spectral collection. FTIR has been used for the classification or identification of several foodborne pathogens: *Yersinia, Staphylococcus, Salmonella, Listeria, Klebsiella, Escherichia, Enterobacter, Citrobacter*, etc. (Gupta et al., 2005; Mossoba et al., 2005; Sivakesava et al., 2004). FTIR photoacoustic spectroscopy was used for the identification of spores of several *Bacillus* species with 100% accuracy (Thompson et al., 2003).

16.3.9 Flow cytometry

This may be considered as the form of automated fluorescence microscopy in which instead of the sample being fixed to a slide, it is injected into a fluid (dye), which passes through a sensing medium of flow cell. In a flow cytometer the cells are carried by laminar flow of water through a focus of light, the wavelength of which matches the absorption spectrum of the dye with which the cells have been stained. On passing through the focus each cell emits a pulse of fluorescence and the scattered light is collected by lenses and directed on to selective detectors (photomultiplier tubes). These detectors transform the light pulses into an equivalent electrical signal. The light scattering of the cells gives information on their size, shape, and structure. This system is a highly effective means for rapid analysis of individual cells at the rate of a thousand cells per second.

16.3.10 Solid phase cytometry

Solid phase cytometry (SPC) is a novel technique that allows the rapid detection of bacteria at the single cell level, without the need for a growth phase (Haese and Nclis, 2002). The short time detection inherent in this approach is of considerable advantage over conventional plating techniques, especially for slow growing bacteria.

SPC combines aspects of flow cytometry and epifluorescence microscopy. The microbes are isolated from their matrix by the membrane filter, fluorescent-labeled with argon laser excitable dye, and automatically counted by a laser scanning device. During the 3-minute scanning process the entire membrane filter surface is scanned yielding a theoretical detection limit of one cell per membrane filter. During scanning two photomultiplier tubes with wavelengths of 500–530 nm (green) and 540–585 nm (amber) detect the fluorescent light emitted by the labeled cells. The signals are processed with software which differentiates between viable signals (target cells) and background noises (electronic noise and fluorescent panicles). Scanned results are displayed as primary and secondary maps. The actual nature of each fluorescent spot can be further examined by epifluorescent microscope.

16.4 Nucleic acid-based assays

Advances in biotechnology have led to the development of a diverse array of assays for the detection of foodborne pathogens. Rapid analysis using nucleic acid hybridization and nucleic acid amplification techniques offer more sensitivity and specificity than culture-based methods as well as dramatic reduction in the time required. Although molecular techniques have improved food microbiology to a great extent, they are not wonder techniques. Certain techniques and methods look good and work well if used in research laboratories by skillful technician, but are not useful for routine testing of food pathogens (Rijpens and Herman, 2002).

Many methods have also achieved a high level of automation, facilitating their application as routine sample screening assays (Wang et al., 1997). The essential principle of nucleic acid-based assays is the specific formation of double-stranded nucleic acid molecules from two complementary single-stranded molecules under defined physical and chemical conditions. There are many nucleic acid-based assays but only DNA probe and polymerase chain reaction (PCR) have been developed commercially for detecting food pathogens (Wang, 2002). Recently, a number of DNA-based molecular typing methods, including pulse field gel electrophoresis, restriction fragment length polymorphism, and ribotyping have also been developed.

16.4.1 DNA hybridization

The identification of bacteria by DNA probe hybridization is based on the presence or absence of particular genes. A gene probe is composed of nucleic acid molecules—most often double-stranded DNA. It consists of either an entire gene or a fragment of a gene with a known function. Alternatively, short pieces of single-stranded DNA can be synthesized, based on the nucleotide sequence of the known gene (Laizard et al., 1991). Double-stranded DNA probes must be denatured before the hybridization reaction, whereas oligonucleotide and RNA probes, which are single-stranded, need not be denatured.

Probes can be labeled with radioactive substances by two methods: nick translation and random priming technique. Oligonucleotide probes are usually labeled at 5″ with 32p, using bacteriophage T_4 polynucleotide kinase and gamma AT 32p. Although radioactive probes seem to have the greatest sensibility in the hybridization process, they are potential hazards and the disposal of radioactive wastes can be expensive. Currently, the labeling of probes with nonradioactive substances such as alkaline phosphatase has been used without effecting the kinetics or specificity of the hybridization (Pitchur et al, 1989).

Target nucleic acids are denatured by high temperature (above 95°C) or high pH (above 12) and then the labeled gene probe is added. If the target nucleic acid in the sample contains the same nucleotide sequence as that of the gene probe, the probe will form a hydrogen bond with the target. The unreacted, labeled probe is removed by washing the solid support and the presence of probe target complexes is signaled by the bound label and detected by autoradiography (Laizard et al., 1991).

16.4.2 Polymerase chain reaction

The PCR is an in vitro method used to increase the number of specific DNA sequences in a sample. PCR is used increasingly in research in food microbiology because of its high sensibility or specificity. By this method, a specific DNA fragment is amplified during a cyclic three-step process (Olsen et al., 1995).

1. The target DNA is denatured at high temperature.
2. Two synthetic oligonucleotides (primers) are annealed to opposite strands at a temperature that only allows hybridization to the correct target.
3. Polymerization is performed with the oligonucleotide as primers for the enzymes and the target DNA as template.

When this is performed over and over with synthesized DNA as the template in addition to the original target DNA, an exponential amplification of the DNA fragment between two primers is obtained (Biswas et al, 2008). Theoretically, PCR can amplify a single copy of DNA by a million-fold in less than 2 hours, hence it has the potential to disseminate or greatly reduce the dependence on cultural enrichment. In a PCR system, assuming a sensitivity of one cell/reaction tube, an approximately 10^3 bacteria/mL sample is required to ensure a reliable and repeatable amplification (Wang et al., 1997).

The limitations of this technique are that, although PCR is a powerful technology, the reactions can be dramatically affected by the presence of inhibitory compounds in foods and selective microbiological media, such as bile salts and acriflavin. A problem with the routine use of PCR in a food testing lab is that the procedures are rather complicated and a very clean environment is needed to perform the tests. Further, PCR cannot distinguish between live and dead cells and hence provides more false-negative results (Biswas et al, 2008).

16.4.3 DNA microarrays (gene chip technology)

This technology uses photolithography, which was developed by computer chip makers (Fig. 16.5). A gene chip can be made of glass or nylon membrane and there are two basic variants. In one format, target DNA is amplified by PCR and spread on to a membrane, which is then probed either singly or simultaneously by hundreds of labeled probes to determine specific hybridization (Fig. 16.6). In the other format, an array of oligonucleotides are synthesized directly on a glass chip and then exposed to labeled target DNA. DNA chip

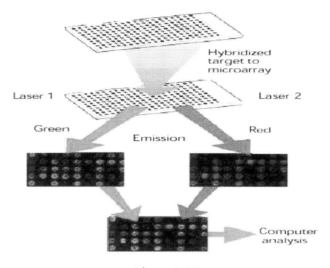

Figure 16.5
Microarray hybridization and scanning.

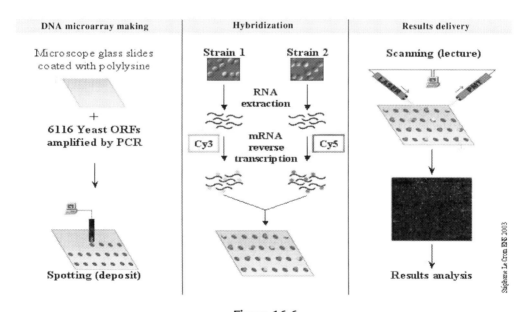

Figure 16.6
DNA microarray analysis.

technology also makes it possible to detect diverse individual sequences in complex DNA samples (Pitchur et al., 1989). The development of this approach is continuing at a rapid pace and for microbiologists, this technology will be one of the major tools for the future.

16.4.4 Loop-mediated isothermal amplification

This is an isothermal nucleic acid amplification technique in which the reaction is carried out at constant temperature using two or three sets of primers along with a DNA polymerase with high displacement strand activity, instead of a series of alternating temperature cycles as used in PCR techniques (Notomi et al., 2000). In this technique usually four different primers are used to identify six distinct regions on the target gene with high specificity, and the inclusion of additional pair of primers (loop primer) further enhances the reaction with the production of a higher amount of DNA than is usually produced in PCR-based amplification. This helps in the visualization of amplified products for a larger reaction volume. It has been further updated over the years, and now can be applied in combination with the other molecular approaches, such as reverse transcription and multiplex amplification for the detection of microbes. Further, although the technique started with a high volume of reaction mixture, it can now also be performed for small volumes using simple colorimetric and fluorescent detection techniques, and for real-time monitoring using a turbidity meter or lateral flow assay (Wong et al., 2018). Indeed loop-mediated isothermal amplification (LMAP) is a newly developed gene amplification method, that combines rapidity, simplicity, and high specificity. The only drawback is that it cannot be used for cloning or multiplexing approaches since LMAP-based techniques are less developed than for PCR.

16.5 Requirements for rapid detection methods

There are several factors which must be considered before adapting a new alternative or rapid method.

1. *Accuracy*: false-positive and false-negative results must be minimal or preferably zero. The method must be as sensitive as possible and the detection limit as low as possible. In many cases, the demand is less than one cell per 25 g of food, as small numbers of some pathogens may cause disease. Analytical tests for these agents need only be qualitative (presence/absence). For rapid screening methods, a higher false-positive frequency may be acceptable, as positive screening tests are followed by confirmation tests.
2. *Validation*: the alternative test should be validated against standard tests and evaluated by collaborative studies. In these studies preference should be given to naturally contaminated food specimens; the tests are then performed under conditions in which users will apply them. Results obtained with samples containing a low contamination level should be emphasized, since there is sufficient evidence that in most cases high numbers of target cells will lead to positive results.
3. *Speed*: rapid tests for the detection of pathogens or toxins should give an accurate result within hours. However, many detection systems need an overnight enrichment for resuscitation and amplification of the target pathogens, as they rely on the presence of at least 10^4-10^5 organism/mL for results to be reliable.

4. *Automaton and computerization*: the ability to test many samples at the same time. Many systems utilizing the microtiter plate format can handle 96 samples at one time. However, for smaller laboratories, the availability of single unit tests is also very important.
5. *Sample matrix*: new systems should give a good performance of the matrices to be tested. Baseline extinction values may depend on the type of food being tested. Background flora, natural substances or debris can interfere with the method and invalidate the test result.
6. *Costs*: purchasing reagents, supply, operational costs, upkeep. The initial financial investment for rapid methods may be high, because many systems require expensive instruments. The operating costs of many commercial rapid test kits are also high.
7. *Miscellaneous*: methods should be user-friendly, which means easy to operate and manipulate. Reagents and supply should be rapidly available. Training, technical service, and company support is essential.

16.6 Limitations of rapid detection methods

A major disadvantage of rapid methods over culturing methods is that most methods need the damaging of the cells and therefore viable cells for confirmation and further characterization can only be obtained by repeat analysis using standard culturing procedures. Moreover, rapid methods usually detect only one specific pathogen, while culturing methods may simultaneously detect and isolate many pathogens by including several types of numerous microbiological examinations or samples, selective media in the analysis. The use of several rapid assays to do multipathogen analyses on a food makes this analysis unacceptably expensive.

AOAC international approved rapid methods are mostly designed for preliminary screening; negative results are regarded definitive, but positive results are considered presumptive and must be confirmed. The evaluation of rapid methods shows that the same methods may perform better in some foods. This may be attributed mostly to interference by normal microbiota or inhibitors in food. In the case of an illness investigation, the food implicated may be suspected to contain a particular pathogen based on clear symptoms, but the actual pathogen is unknown. In these situations, when multiple pathogen analysis may be needed it makes the procedure complex and costly.

References

Alahi, M.E.E., Mukhopadhyay, S.C., 2017. Detection methodologies for pathogen and toxins: a review. Sensors 17, 1885. Available from: https://doi.org/10.3390/s17081885.
APHA, 1987. Compeodium of Methods for Microbiological Examination of Food, third ed. American Public Health Association, New York.
Bhunia, A.K., 2008. Biosensors and bio-based methods for the separation and detection of food-borne pathogens. Adv. Food Nutr. Res. 54, 1–44.

Bhunia, A.K., Wampler, J.L., 2005. Animal and cell culture models for foodborne bacterial pathogens. In: Fratamico, P., Bhunia, A.K., Smith, J.L. (Eds.), Foodborne Pathogens: Microbiology and Molecular Biology. Caister Academic Press, Norfolk, pp. 15–32.

Bhunia, A.K., Banada, P., Banerjee, P., et al., 2007. Light scattering, fiber optic and cell-based sensors for sensitive detection of foodborne pathogens. J. Rapid Method Autom. Microbiol. 15, 121–145.

Biswas, A.K., Kondaiah, N., Anjaneyulu, A.S.R., et al., 2008. Microbial profiles of frozen trimmings and silver sides prepared at Indian buffalo meatpacking plants. Meat Sci. 80, 418–422.

Boening, D.W., Tarr, P.I., 1995. Proposed method for isolation of *Escherichia coli* O157:H7 from environmental samples. J. Environ. Health 57, 9–21.

Boer, E.D., Beumer, R.R., 1999. Methodology for typing and detection of food home microorganisms. Int. J. Food Microbiol. 50, 119–130.

Borch, E., Arinder, P., 2002. Bacteriological safety issues in red meat and ready-to-eat meat products, as well as control measures. Meat Sci. 62, 381–390.

Chang, Y.H., 2000. Prevalence of *Salmonella* spp. in poultry broilers and shell eggs in Korea. J. Food Prot. 63 (5), 655–658.

Doyle, M.P., 2001. Food Microbiology: Fundamentals and Frontiers, second ed. ASM press, Washillgton, DC.

Gray, K.M., 2004. Cytotoxicity and cell-based sensors for detection of *Listeria monocytogenes* and *Bacillus* species (Ph.D. dissertation). Purdue University, West Lafayette, p. 130.

Griffiths, M.W., 1993. Application of bioluminescence in the dairy industry. J. Dairy Sci. 76, 3118–3125.

Groisman, E.A., Ochman, H., 2000. The path to *Salmonella*. ASM News 66, 21–26.

Gupta, M.J., Irudayaraj, J.M., Debroy, C., et al., 2005. Differentiation of food pathogens using FTIR and artificial neural networks. Trans. ASAE 48, 1889–1892.

Haese, E.D., Nclis, H.J., 2002. Rapid detection of single cell bacteria as a novel approach in food microbiology. J. AOAC Int. 85, 979–983.

Ivnitski, D., Harnid, L.A., Atanasov, P., Wilkins, E., 1999. Biosensors for detection of pathogenic bacteria. Biosens. Bioelectron. 14, 599–624.

JECFA, 2002. Future trends in veterinary public health. In: Fifty-Eight Report of the Joint FAO/WHO Expert Committee on Food Additives. Report of a WHO study group. WHO Technical Report Series. No. 907.

Laizard, P.W., Zijderveld, A., Linders, K., et al., 1991. Simplified marnmalian DNA isolation procedure. Nuclcic Acid Res. 19, 4293.

Mossoba, M.M., Al-Khaldi, S.F., Kirkwood, J., et al., 2005. Printing microarrays of bacteria for identification by infrared micro spectroscopy. Vib. Spectrosc. 38, 229–235.

Naravaneni, R., Jamil, K., 2005. Rapid detection of food-borne pathogens by using molecular techniques. J. Med. Microbiol. 54, 51–54.

Notomi, T., Okayama, H., Masubuchi, H., et al., 2000. Loop-mediated isothermal amplification of DNA. Nucleic Acids Res. 28 (12), E63.

Olsen, J.E., Aabo, S., Hill, W., Olsvik, O., 1995. Probes and PCR for detection of foodborne bacterial pathogens. Int. J. Food Microbiol. 28, 1–78.

Park, Y.H., 2001. Application of biotechnology in animal products safety. In: Jinju, S. (Ed.), Proc. 3rd CJK Symposium on Biotechnology and Animal Production. Jinju National University, Korea, pp. 33–69.

Pitchur, D.G., Saunders, N.A., Owens, R.J., 1989. Rapid extraction of bacterial genomic DNA with guanidium thiocyanate. Let. Appl. Microbiol. 8, 151–156.

Rao, D.N., Mahendrakar, N.S., 2003. Challenges of the quality control of Indian meat and meat products in view of global market. In: First convention of Indian Meat Science Association and National Symposium on Impact of Globalization on Indian Meat Industry. 11th-12th December, 2003, pp. 18–26.

Rasooly, A., Herold, K.E., 2006. Biosensors for the analysis of food and waterborne pathogens and their toxins. J. AOAC Int. 89, 873–883.

Rijpens, N.P., Herman, L.M.F., 2002. Molecular methods for identification and detection of bacterial food pathogens. J. AOAC Int. 85, 984–995.

Ritcher, E.R., 1993. Biosensors: application for dairy food industry. J. Dairy Sci. 76 (3), 1143117.

Shah, J., Chemburu, S., Wilkins, E., Abdel-Hamidb, I., 2003. Rapid amperometric immunoassay for *Escherichia coli* based on graphite coated nylon membranes. Electroanalysis 15, 23–24.

Sivakesava, S., Irudayaraj, J., DebRoy, C., 2004. Differentiation of microorganisms by FTIR-ATR and NIR spectroscopy. Trans. ASAE. 47, 951–957.

Taitt, C.R., Anderson, G.P., Ligler, E.S., 2005. Evanescent wave fluorescence biosensors. Biosens. Bioelectron. 20, 2470–2487.

Thompson, S.E., Foster, N.S., Johnson, T.J., et al., 2003. Identification of bacterial spores using statistical analysis of Fourier transform infrared photoacoustic spectroscopy data. Appl. Spectrosc. 57, 893–899.

Vasavada, P.C., 1993. Rapid methods and automation in dairy microbiology. J. Dairy Sci. 76, 3101–3113.

Wang, H., 2002. Rapid method for detection and enumeration of *Campylobacter* spp. in food. J. AOAC Int. 85, 996–1002.

Wang, R.F., Cao, W.W., Cerniglia, C.E., 1997. A universal protocol for PCR detection of 13 species of food borne pathogens in food. J. Appl. Microbiol. 83, 727–736.

Westerman, R.B., Yongsheng, H.E., Keen, J.E., et al., 1997. Production and characterization of monoclonal antibodies specific for the lipopolysaccharide of *Escheriria coli* O157:H7. J. Clin. Microbiol. 35, 679–684.

Wong, Y.P., Othman, S., Lau, Y.L., et al., 2018. Loop-mediated isothermal amplification (LAMP): a versatile technique for detection of micro-organisms. J. Appl. Microbiol. 124 (3), 626–643.

Yongsheng, H.E., Keen, J.E., Westerman, R.B., et al., 1996. Monoclonal antibodies for detection of the H7 antigen of *Escheriria coli* O157:H7. Appl. Environ. Microbiol. 35, 679–684.

SECTION 10

Modern biological concept of meat deterioration and its detection

CHAPTER 17

Spoilage bacteria and meat quality

Abraham Joseph Pellissery, Poonam Gopika Vinayamohan,
Mary Anne Roshni Amalaradjou and Kumar Venkitanarayanan

Department of Animal Science, University of Connecticut, Storrs, CT, United States

Chapter Outline

17.1 Introduction 308
17.2 Causes of meat spoilage 308
17.3 Microbiome of spoiled meat 309
    17.3.1 Microflora of fresh meat 309
    17.3.2 Spoilage microflora associated with aerobically packaged meat 310
    17.3.3 Spoilage microflora in vacuum-packaged meat 312
    17.3.4 Spoilage microflora in modified atmosphere packaged meat 313
17.4 Chemistry of meat spoilage 314
    17.4.1 Nonmicrobial/biochemical spoilage of meat 314
    17.4.2 Microbiological spoilage of meat 314
17.5 Characteristics of spoiled meat 317
    17.5.1 Discoloration 317
    17.5.2 Off-odors and off-flavors 318
    17.5.3 Gas production 318
    17.5.4 Filaments and slime formation 318
17.6 Factors affecting microbial meat spoilage 319
    17.6.1 Intrinsic factors 320
    17.6.2 Extrinsic factors 321
    17.6.3 Implicit factors 322
17.7 Indicators of microbial spoilage 322
17.8 Detection of microbial spoilage in meat 323
    17.8.1 Enumeration 323
    17.8.2 Detection of bacterial metabolites 324
    17.8.3 Molecular methods 326
17.9 Conclusion 328
References 328
Further reading 334

17.1 Introduction

Meat and meat-based products constitute a vital component of our daily diet. In fact, the demand for meat as a source of high-value animal protein has continued to increase over the last 50 years. Toward this, global meat production has increased tremendously by four- to fivefold since 1961. According to the Food and Agricultural Organization, over 300 million tons of meat were produced worldwide in 2014 with a per capita consumption of 95 lbs. (Ritchie and Roser, 2018). Unfortunately the high nutritive nature of meat also makes it a highly perishable food commodity susceptible to degradation and spoilage. Spoilage can be defined as any change that renders a product unsuitable for human consumption (Huis in't Veld, 1996). In the case of meat, spoilage is a complex event involving biological, physical, and chemical activities that lead to product deterioration (Casaburi et al., 2015). Therefore spoilage has an adverse effect on consumer acceptance and correlates strongly with food wastage. The FAO estimates that one-third of foods produced for human consumption is wasted each year (FAO, 2011).

Based on the susceptibility to spoilage, foods can be classified as (1) highly perishable such as meat, poultry, milk, and eggs that have a quality retention time of 1–3 days, (2) semiperishable, including pasteurized milk, eggs and refrigerated foods with a shelf life of 1–3 weeks, (3) semishelf stable products such as vacuum-packed meat and fermented sausages possessing a shelf life of 1–4 months, and (4) shelf-stable foods that have been subject to processing including high heat, sterilization, and hermetic sealing. These foods can remain edible and retain their quality for up to 1 year. Examples include dried food products, flour, and sugar (Erkmen and Bozoglu, 2016).

Meat and meat products fall into the highly perishable category due to the following inherent characteristics:

1. High nutritive content that can support rapid microbial growth
2. High water activity of 0.85 or greater
3. pH of 4.6 or greater
4. Presence of autolytic enzymes
5. Presence of a significant amount of fat and lipids favoring oxidation and spoilage (Erkmen and Bozoglu, 2016; Dave and Ghaly, 2011; Iulietto et al., 2015)

17.2 Causes of meat spoilage

Meat spoilage can be caused by physical, chemical and biological agents, including (1) microorganisms-bacteria, yeast and mold, (2) action of enzymes in meat such as lipases and proteases, (3) chemical reactions in foods such as browning and oxidation, and (4) physical changes introduced by freezing, drying, and application of pressure

(Iulietto et al., 2015; Erkmen and Bozoglu, 2016). Although several agents are implicated in meat spoilage, microorganisms are the most common cause of quality deterioration in foods of animal origin. The spoilage organisms break down fat, carbohydrate, and protein in meat resulting in the development of off-flavors, slime formation, and discoloration, thereby rendering the meat disagreeable for consumption (Ercolini et al., 2006; Nychas et al., 2008). It is estimated that microbial spoilage is responsible for 25% of postharvest food loss globally (Cenci-Goga et al., 2014).

17.3 Microbiome of spoiled meat

Microbial spoilage can be defined as the biochemical changes in meat brought about by dominant microorganisms that make up a significantly higher proportion of the microbial community associated with meat (Pothakos et al., 2015). The overall composition of spoilage microflora is diverse and primarily determined by the environment in which the animals are raised, and the postharvest and processing environment of meat (Hultman et al., 2015). These spoilage organisms are conventionally grouped as Gram-negative rods, Gram-positive spore formers, lactic acid bacteria (LAB), other Gram-positive bacteria, yeast, and molds (Huis in't Veld, 1996; Jay et al., 2005). In general, meat products are not commonly degraded by yeast due to their inability to produce extracellular proteases. Some exceptions to this include *Yarrowia lipolytica*, *Rhodoturola*, *Cryptococcus*, *Pichia*, and *Saccharomyces* in fresh and refrigerated meat and poultry (Dave and Ghaly, 2011; Petruzzi et al., 2017). Similarly, mold found on meat that could play a role in spoilage includes *Alternaria*, *Aspergillus*, *Fusarium*, *Rhizopus*, and *Cladosporium* (Jay et al., 2005). Given that the environment has a predominant role in determining the spoilage microbiome and the propensity of bacteria to cause spoilage, this section will discuss meat-associated spoilage bacteria along the meat processing continuum.

17.3.1 Microflora of fresh meat

The muscle tissue in healthy living animals is essentially sterile. Thus the initial microbial load and composition of fresh meat are primarily influenced by the physiological status of the animal at the time of slaughter, the spread of microbes during slaughter and the slaughterhouse environment (Erkmen and Bozoglu, 2016). Following sacrifice, the main contamination of meat occurs when the carcass is opened and the offals are removed. For instance, bacteria from the intestines, lymph nodes, skin, hide, handlers, cutting knives, and the processing facility can potentially contaminate meat. These microorganisms acquired by meat can be termed as the slaughterhouse microbiome, which is a combination of the microbial population in the facility and the animal's gut (Koutsoumanis and Sofos, 2004). Toward this, Mills and coworkers demonstrated that *Carnobacterium* spp. identified on lamb carcasses were traced back to the meat processing environment (Mills et al., 2018). Further, investigations of microbial prevalence revealed that the core microbiota at the slaughterhouse consisted of *Staphylococcus* spp., *Streptococcus* spp., *Brocothrix* spp., *Psychrobacter* spp., *Acinetobacter* spp., and LAB. On the other hand, Proteobacteria especially *Pseudomonas* spp. and members of *Enterobacteriaceae* were found to dominate the carcass microflora (Stellato et al., 2016).

Overall, the initial microflora on the carcass is diverse and contains more mesophiles than psychrotrophs (Schaefer-Seidler et al., 1984). The total microbial load on fresh meat ranges from 10^2 to 10^5 CFU/cm^2 (Bell, 1997; Nychas et al., 1988). This initial microbiome is usually composed of *Achromobacter, Acinetobacter, Moraxella, Aeromonas, Bacillus* sp., *Brochothrix thermosphacta, Clostridium, Flavobacterium, Cytophaga, Enterobacteriaceae* including *Escherichia coli*, LAB, *Micrococcus, Pseudomonas, Staphylococcus,* and *Streptococcus* (Bell, 1997; Newton and Gill, 1978). The microbiome that eventually develops on meat resulting in spoilage will depend on the method of packaging used for the product. Fresh meat is packed in one of three different ways, namely aerobic, vacuum, and modified atmosphere packaging (MAP). The difference in the microenvironment surrounding packed meat, specifically, the presence or absence of oxygen, drives the eventual composition of the meat microbiome and determines its shelf-stability (Farkas, 2007; Silva et al., 2011).

17.3.2 Spoilage microflora associated with aerobically packaged meat

Under aerobic conditions, the initial signs of spoilage, including off-odor are evident when the bacterial density reaches 10^8 cells/cm^2 (Nychas et al., 1988). This increase in bacterial number is associated with a concomitant reduction in available glucose levels in meat. Therefore at this stage, the microbes switch to using protein as the source of energy (Gill, 1976). This change in substrate availability favors the growth of proteolytic organisms and coincides with organoleptic spoilage (Nychas et al., 2008). Given their ability to break down protein, *Pseudomonas* species including *P. fragi, P. putida, P. lundensis,* and *P. fluorescens* are the predominant flora associated with spoilage in aerobically packaged meat (Doulgeraki and Nychas, 2013). Under humid conditions, the characteristic slime formation in spoilt meat appears when the Pseudomonad population reaches 10^7-10^8 CFU/g. In effect, *Pseudomonas* spp. are the predominant flora on beef carcasses, hung beef cuts, and high pH beef (Simard et al., 1984). Several investigators have reported that in some cases *Pseudomonas* spp. can comprise up to 96% of the spoilage microflora (Bailey et al., 1979a; Enfors et al., 1979; Asensio et al., 1988). More specifically, among the different Pseudomonads, higher proportions of *P. fluorescens* were associated with meat before spoilage, while *P. fragi* was predominant in spoiled meat (Shaw and Latty, 1982). Further, *P. fragi* was observed to become dominant following extended storage of meat due to its ability to survive under different packing and storage conditions (Ercolini et al., 2006).

Similarly, fresh pork and poultry meat stored under aerobic conditions had a higher proportion of *P. putida* than *P. fluorescens*, while *P. putida* and *P. fragi* were identified in minced beef (Doulgeraki and Nychas, 2013). In the case of spoiled poultry meat, a significant proportion of the spoilage microflora was found to consist of *Pseudomonas putrefaciens* along with other pigmented and nonpigmented *Pseudomonas* spp.

The prevalence of Pseudomonads is followed by other Gram-negative rods, including *Acinetobacter*, *Moraxella*, and *Flavobacterium* (Ercolini et al., 2006). These bacterial species have been shown to promote the growth of Pseudomonads by limiting the oxygen availability to competing organisms (Gill, 1986). In addition to packaging conditions, storage temperature also plays an important role in defining the spoilage microbiome of meat. When the meat was stored at 15°C–25°C, large numbers of *Enterobacteriaceae* including *Kurthia*, *Enterobacter*, and *Hafnia* were prevalent in stored pork (Gardner et al., 1967). Similarly, when meat was held at a higher temperature of 30°C, *Acinetobacter* and *Enterobacteriaceae* were the dominant microflora (Gill and Newton, 1980). This transition in microbial population favoring facultative anaerobes can be attributed to the oxygen-limiting conditions encountered in meat following the rapid proliferation and predominance of Pseudomonads (Enfors and Molin, 1984).

In addition to Gram-negative organisms, Gram-positive bacteria also play a role in meat spoilage. Most commonly in aerobically packaged meat the predominant Gram-positive organism is *Brocothrix* spp., particularly *B. thermospacta*. Members of this genus are known to break down meat components to produce acid, including acetic acid, formic acid, isobutyric acid, and isovaleric acid (Huis in't Veld, 1996). Besides *Brocothrix*, meat stored aerobically was also shown to harbor *Micrococcus*, *Achromobacter*, and *Staphylococcus* (Barlow and Kitchell, 1966). However, *Brocothrix* was found to rapidly multiply and become the dominant flora following prolonged storage of meat at 5°C (Barlow and Kitchell, 1966). Further, *Brocothrix* was identified to be part of the natural microflora of lamb surface and pork fat (Barlow and Kitchell, 1966; Blickstad and Molin, 1983). In the case of cured meat products, *Micrococcus* was found to be more commonly associated with spoilage due to their ability to withstand high levels of salt (Huis in't Veld, 1996). In the case of beef, LAB were found to be commonly associated with spoilage in addition to *Brocothrix* and *Pseudomonas*. LAB constitutes a large group of Gram-positive organisms that are usually associated with fresh and cooked meat products (Pothakos et al., 2015). The growth of LAB on meat is associated with the formation of lactic acid, slime, and carbon dioxide, leading to a reduction in pH and off-flavors (Huis in't Veld, 1996; Pothakos et al., 2015). However, these organisms are slow growing under refrigeration temperatures and are often outcompeted by Pseudomonads under aerobic conditions. Hence LAB, although present in low numbers, are rarely associated with spoilage of fresh and aerobically stored meat (Borch et al., 1996; Doulgeraki et al., 2010). On the other hand, LAB have been implicated in the spoilage of meat stored under vacuum or MAP (Pothakos et al., 2015).

17.3.3 Spoilage microflora in vacuum-packaged meat

As discussed previously, an aerobic environment contributes to spoilage by accelerating the growth of Pseudomonads. As a means to counter spoilage and increase the shelf life of meat, vacuum packaging was developed. Meat storage under vacuum is accomplished by wrapping the meat in an oxygen impermeable packaging material following the removal of air from the pack. This is done by flushing the package with carbon dioxide prior to the application of vacuum and leaving residual levels of the gas within the package. The film serves as a barrier between the outside environment and the microenvironment within the container by minimizing oxygen diffusion into the meat/meat product (Church and Parsons, 1995; Asensio et al., 1988). Although this limits the growth of aerobic organisms, facultative and obligate anaerobes can survive in this environment and predominate the spoilage microflora of vacuum-packed meats (Hodges et al., 1974).

As with aerobically packed meat, the anaerobic environment of vacuum-packed meats also supports a diverse microflora. The total microbial counts in vacuum-packed meats often do not exceed 10^8 CFU/cm^2 (Nychas et al., 1988). When low oxygen transmissible films are used in packaging, the number of Pseudomonads is significantly less when compared to aerobically packed meat. On the other hand, when vacuum-packed meats were wrapped with high oxygen transmission polyethylene, 78% of the microflora on spoiled pork was *Pseudomonas* while *B. thermosphacta* constituted the remaining 22% (Asensio et al., 1988). Although the products were vacuum-packed, the diffusion of oxygen across the permeable film failed to support the anaerobic microenvironment, thus favoring the growth of these aerobic organisms (Shay and Egan, 1987). In most cases, however, the initial microbial population that is associated with vacuum-packed meat included Gram-negative facultative anaerobes such as *Hafnia*, *Serratia liquefaciens*, and *Enterobacter* (Dainty and Mackey, 1992).

In the case of vacuum-packed meat, the predominant populations associated with spoilage are Gram-positive organisms including LAB and *B. thermosphacta* (Enfors et al., 1979; Bailey et al., 1979a). In effect, Bailey et al. (1979b) observed that in vacuum-packed chicken meat, 60% of the microflora on day 5 of storage comprised Gram-positive bacteria with their proportion eventually comprising 95% of the microbial population. Similarly, an increase in Gram-positive population was also observed in other vacuum-packed meats following 7–14 days of storage (Christopher et al., 1980). This increase in Gram-positive facultative and obligate anaerobes such as LAB is due to the anaerobic condition that develops in vacuum-packed meat. Seman et al. (1988) observed that LAB numbers in vacuum-packed meat increased from 10^1 on week 1 to 10^7 CFU/g after 18 weeks of storage. Although predominantly associated with spoilage in vacuum-packed meat, LAB constitute a controversial and highly diverse group of microbes that can either contribute to spoilage or serve as bioprotectants in the control of spoilage and pathogenic bacteria (Pothakos et al.,

2015). For instance, development of certain *Lactobacillus*, *Carnobacterium*, and *Leuconostoc* species in foods is linked to the production of bacteriocins that limit the growth of other bacterial species (Newton and Gill, 1978; Jack et al., 1996). Alternatively, certain members of the genus *Lactobacillus* (*L. sakei*, *L. curvatus*, *L. algidus*, *L. fuchiensis*, *L. oligofermentas*) have been implicated in meat spoilage associated with severe acidification, the production of off-odors, and ropy slime (Doulgeraki et al., 2010). Similarly, the genus *Leuconostoc* (*L. gelidum*, *L. carnosum*, *L. mesentroides*) is associated with the production of buttery odor, slime formation, and green discoloration in meats (Diez et al., 2009; Nieminen et al., 2011). *Carnobacterium*, specifically *C. divergens* and *C. maltaromaticum*, are often associated with spoilage in beef, poultry, and pork when stored under low oxygen packaging (Casaburi et al., 2011). Similar to *Lactobacillus*, the genera *Lactococcus*, although primarily associated with the production of fermented foods, also includes species (*L. piscium*, *L. raffinolactis*) that are associated with spoilage mainly in beef stored under vacuum (Rahkila et al., 2012). Also certain species of *Weisella* (*W. viridescens*) are often associated with gas production and bulging of vacuum-packed poultry and minced meat products (Diez et al., 2009; Zhang et al., 2012).

17.3.4 Spoilage microflora in modified atmosphere packaged meat

As opposed to vacuum packaging, in MAP the microenvironment within the container is altered by the inclusion of specific gas mixes to slow food respiration rates, reduce microbial growth, and inhibit enzymatic spoilage/autolysis (Young et al., 1988). The gases primarily used for MAP are nitrogen, carbon dioxide, and oxygen. These gases are mixed in varying proportions to generate a modified atmosphere within the package. For instance, the use of higher levels of carbon dioxide in the gas mix enhances the growth of LAB while inhibiting *Enterobacteriaceae* and *Pseudomonas* (Gill, 1996). Use of 20%–60% carbon dioxide was found to be effective against controlling *Acinetobacter*, *Pseudomonas*, and *Moraxella* while slowing the growth of *B. thermosphacta* and LAB. A gas mixture consisting of 20% carbon dioxide, 78% nitrogen, and 2% oxygen led to an increase in *B. thermosphacta* population, which was higher than LAB (Farkas, 2007). Contrary to LAB that requires 10^8 CFU/g to cause spoilage, *Brocothrix* can bring about spoilage when appearing in low numbers. This is because *Brocothrix* ferments glucose to produce lactate, acetate, and formate that can render foods organoleptically unacceptable. Another group of organisms that are of concern in MAP meats are the *Enterobacteriaceae*. Members of this family are the only Gram-negative bacteria of concern in MAP, particularly in the presence of high carbon dioxide levels, due to their facultatively anaerobic nature (Gill and Penney, 1988). Although they can survive and grow in the presence of carbon dioxide, these organisms experience a prolonged lag phase. Hence *Enterobacteriaceae* can increase in numbers toward the end of storage. Facultative anaerobes such as *Serratia*, *Enterobacter*, *Proteus*, and *Hafnia* metabolize amino acids in meat to produce amines, ammonia, methylsulfides, and mercaptans resulting in putrefaction, off-odor, and pinkish discoloration of meat.

17.4 Chemistry of meat spoilage

Postmortem conversion of muscle to meat generates a conducive environment for natural biochemical reactions after slaughter or for microbes to invade and utilize the soluble low molecular weight substrates. These processes can be categorized as nonmicrobial and microbial reactions favoring meat spoilage.

17.4.1 Nonmicrobial/biochemical spoilage of meat

Generally, the intrinsic biochemical changes are categorized as lipid and protein oxidation as well as autolytic enzymatic spoilage of meat. Nonmicrobial oxidative deterioration of lipids, proteins, and off-flavor development in meat are a consequence of natural processes that occur following autooxidation and production of free radicals after slaughter. The cessation of blood circulation, oxygen supply, and metabolic processes following slaughter promotes fatty acid and protein oxidation in tissues (Papuc et al., 2017). Meat handling processes such as cutting, mincing, irradiation, handling, packaging, storage, and cooking are known to induce chemical (nonenzymatic) and enzymatic reactions in proteins and lipids (Papuc et al., 2017). Nonmicrobial reactions that favor meat spoilage include oxidation of lipid and protein in raw meat, generation of toxic metabolites from lipid peroxidation, reduced bioavailability of essential fatty acids, and undesirable changes in flavor and color (Papuc et al., 2017).

17.4.2 Microbiological spoilage of meat

Meat and meat products are ideal growth media for animalborne as well as environmental sources of microbes. In general, the skin and intestinal contents are the primary sources of animalborne microbes in meat (Jay et al., 2005). Muscle glycogen-derived lactic acid from anaerobic glycolysis along with minor quantities of glucose and glucose-6-phosphate are some of the molecules available for microbial utilization. Glucose is the first source of energy, which is metabolized more rapidly by obligate aerobic pseudomonads than by facultative anaerobes such as *B. thermosphacta* and oxidative strains of *Shewanella putrefaciens* (Dainty et al., 1985; Tsigarida et al., 2003). Pseudomonads are predominantly seen during spoilage as a result of their faster growth rate along with a higher affinity for oxygen. Once glucose reserves are depleted, lactate is the next energy source utilized both under aerobic and anaerobic conditions, followed by amino acids. In general, sensorial meat spoilage development is due to the metabolic activity of meat surface microbiota on nutrient substrates such as sugars, fatty acids, and free amino acids favoring the release of undesirable volatile organic compounds (VOCs), including alcohol, aldehydes, ketones, esters, and volatile fatty acids. Aerobic bacteria such as pseudomonads oxidize glucose and glucose-6-phosphate to form D-gluconate, pyruvate and 6-phosphogluconate. Odoriferous metabolites derived from amino acids such as sulfides, methyl esters, and ammonia are usually the first manifestation of spoilage of chilled meat and poultry. Some of the microorganisms commonly involved in putrefaction include *P. fragi, S. putrefaciens, Proteus, Citrobacter, Hafnia,* and *Serratia* (Dainty et al., 1985).

17.4.2.1 Alcohols

Alcohols are produced by spoilage microbes during the chilling of fresh meat when stored aerobically, and under vacuum packaging and MAP. Microbial metabolism favors the breakdown of proteins and amino acids, reduction of ketones, and aldehydes derived from lipid peroxidation to produce a variety of alcohols. Alcohols associated with spoilage of meat stored aerobically and in vacuum packaging include methyl-1-butanol, 1-octen-3-ol, 2-ethyl-1-hexanol, 2, 3-butanediol, butanol, 1-heptanol, 1-hexanol, and 3-phenoxy-1-propanol, whereas 1-octen-3-ol is associated with MAP meats. Among the different spoilage organisms, *Pseudomonas* spp. and *Carnobacterium* spp. are predominantly involved in the production of alcohols, and some of the compounds generated are indicative of possible off-odor in meat (Casaburi et al., 2015; Smit et al., 2009).

17.4.2.2 Aldehydes

The production of aldehydes by spoilage organisms is known to impart sharp acidic to fatty flavor in meat. Acidic flavors are commonly attributed to short-chain aldehydes, whereas an increase in aldehyde chain length with varying degrees of unsaturation contributes to fattiness. These compounds are derived from triglyceride hydrolysis, oxidation of unsaturated fatty acids, or lipid autooxidation. Moreover, aldehydes can also be generated from imide intermediates of amino acid transamination reactions. The species mainly contributing to off-flavors by aldehyde production include *Pseudomonas* spp., *Carnobacterium* spp., and *Enterobacteriaceae* spp. (Soncin et al., 2007). For example, hexanal, nonanal, benzaldehyde, and 3-methylbutanal are aldehydes seen in naturally spoiled meat, which at detectable threshold levels are known to generate fresh green fatty aldehydic grass leafy, fruity sweaty odor, fatty and green herbal odor, volatile almond oil and burning aromatic taste, and cheese and pungent apple-like odor, respectively (Calkins and Hodgen, 2007). Exceedingly higher concentrations than detectable odor threshold values are known to produce very unpleasant and rancid aromas in meat. Although aldehydes are known to produce off-flavors in fresh as well as spoiled meat, correlating their presence with spoilage bacteria is difficult due to their low concentration and oxidation to acids during the early storage phase (Casaburi et al., 2015).

17.4.2.3 Ketones

Ketones are generated either via chemical or by microbial spoilage, and they are produced in fresh meat stored under varying atmospheric conditions. Lipolysis and microbial alkane degradation or dehydrogenation of secondary alcohols are some of the putative routes for ketone production in fresh meat. As with aldehydes, *Pseudomonas* spp., *Carnobacterium* spp., and *Enterobacteriaceae* are also known to be primarily associated with volatile ketones from spoiled meat. Acetoin and diacetyl are major ketones that contribute to cheesy odor and butter, sweet, creamy, and pungent caramel flavor, respectively. Acetoin is known to be generated from glucose catabolism by *B. thermosphacta*, *Carnobacterium* spp., and *Lactobacillus* spp. and also by the microbial breakdown of aspartate (Casaburi et al., 2015).

17.4.2.4 Esters

Esters are predominantly seen in fresh meat stored aerobically and their production is attributed to *P. fragi*, which is considered the major ester producer. Microbial esterase activity favors the esterification of alcohols and carboxylates found in meat resulting in a fruity off-flavor. Some of the volatile esters produced from naturally spoiled meat or an inoculated model meat system include ethyl acetate, ethyl butanoate, ethyl-3-methylbutanoate, ethyloctanoate, ethyl hexanoate, and ethyl decanoate (Dainty and Mackey, 1992; Toldra, 1998).

17.4.2.5 Volatile fatty acids

Volatile fatty acids are another group of compounds that originate from fresh meat following the hydrolysis of triglycerides and phospholipids. Amino acid degradation or the oxidation of ketones, esters, and aldehydes are other plausible reaction pathways for their production (Toldra, 1998). *B. thermosphacta* and *Carnobacterium* spp. are associated with the production of volatile fatty acids in fresh meat. *B. thermosphacta* are known to produce 2- and 3-methylbutanoic acid from aerobically stored fresh meat, wherein isoleucine, leucine, and valine act as precursors for amino acid degradation. These acids provide a pungent, acid, and roquefort cheese odor and a sour, stinky, feet, sweaty, and cheese odor, respectively, in aerobically stored fresh meat. Butanoic acid is produced by LAB via breakdown of amino acids through Stickland reaction, or by *Clostridia* through butyric fermentative metabolism in vacuum-packaged meats (Martín et al., 2010). Butanoic acid is known to produce a rancid, sharp, acid, cheesy, butter, and fruity odor in spoiled meat (Weimer, 2007).

17.4.2.6 Sulfur compounds

Volatile sulfur compounds are produced by spoilage microbes as a result of degradation of sulfur-containing amino acids (methionine and cysteine) producing compounds such as dimethylsulfide, dimethyldisulfide, dimethyltrisulfide, and methyl thioacetate. Pseudomonads are commonly associated with the production of volatile sulfur compounds which generate a wide variety of odors providing a sulfurous, cooked onion, vegetable, radish-like, and savory meaty odors (Yvon and Rijnen, 2001; Blank, 2002). Biogenic amines are also a consequence of meat spoilage by bacteria producing amino acid decarboxylases. The primary end product of bacterial amino acid metabolism in meat includes putrescine and cadaverine. Production of these amines leads to the development of putrefying odors associated with spoiled fresh meat (Nowak and Czyzowska, 2011).

17.5 Characteristics of spoiled meat

Given the diverse nature of microorganisms involved in meat spoilage, their predilection for specific nutrient resource and the factors influencing their growth and survival, the observed effects are variable, including slime formation (visible growth), textural changes (degradation), development of off-odors and off-flavors, and discoloration (Borch et al., 1996; Nychas et al., 2008). In general, the changes observed in spoiled meat are driven by the available nutrients (glucose, lactic acid, protein, free amino acids, and other nitrogenous compounds) that serve as precursors of microbial metabolites responsible for spoilage (Nychas et al., 2008; Iulietto et al., 2015).

17.5.1 Discoloration

A significant criterion for visual appraisal of meat is color. Some color defects seen in meat arise as a result of chemical modification of myoglobin without the involvement of spoilage microbes. Chemical modification leads to the alteration of meat myoglobin, wherein innate oxidative processes can lead to the formation of brown myoglobin. Microbial etiologies are known to destroy myoglobin to cause greening/graying of meat by combining with hydrogen sulfide of microbial origin or be broken down to form yellow or green bile pigments by microbial hydrogen peroxide. Greening/graying of meat can be attributed to the presence of LAB. In addition, *L. sakei*, *Hafnia alvei*, and *S. putrefaciens* are also involved in green sulfmyoglobin formation. Hydrogen peroxide produced by *Leuconostoc* spp. and *W. viridescens* contributes to meat greening by oxidizing nitrosomyochromogen in aerobically stored meat (Borch et al., 1996; Dušková et al., 2013). Microbial pigment production is also visually appreciable on meat where Pseudomonads, molds of the genera *Cladosporium*, *Sporotrichium*, and *Penicillium* produce a variety of blue, green, yellow, black, and white pigments. Yellow fluorescent pigment associated with *P. fluorescens* is a siderophore produced by the bacteria for the utilization of iron in meat (Cornelis, 2010; Lawrie, 2006). Blue pigment production by *P. fluorescens* leads to "blue pork or beef" in spoiled meat.

17.5.2 Off-odors and off-flavors

The production of volatile end products such as sulfur compounds, ketones, aldehydes, acids, esters, alcohols, and ammonia following bacterial metabolism is responsible for the off-odors associated with microbially spoiled meat (Casaburi et al., 2015). In the case of fresh and aerobically packaged meat, off-odors are noticed when the bacterial population reaches a density of 10^7 CFU/g. At this population density and under aerobic conditions, carbohydrates are depleted and the bacterial communities start utilizing amino acids as their source of energy. The predominant microflora under this environment includes *Pseudomonas*, *Moraxella*, *Alcaligens*, *Aeromonas*, *Serratia*, and *Pantoea* (Nychas et al., 2008). The breakdown of amino acids and other nitrogenous compounds results in the formation of offensive odors. Catabolism of sulfur-containing compounds by *Enterobacteriaceae* and *Pseudomonas* spp. leads to the production of hydrogen sulfide and dimethyl sulfide, respectively, resulting in sulfuric odor in meat. Similarly, carbohydrate metabolism by homofermentative *Lactobacillus*, *Enterobacteriaceae*, and *B. thermosphacta* results in the formation of a cheesy odor due to the production of acetoin, diacetyl, and 3-methylbutanol (Casaburi et al., 2015). In the case of vacuum and MAP meat, the oxygen-limiting environment results in less strong off-odor and mainly consists of sour, acid aroma produced by LAB (Pin et al., 2002).

17.5.3 Gas production

Gas production is commonly seen in vacuum-packaged meat and is primarily associated with the growth of *Clostridium* spp. (Iulietto et al., 2015). These anaerobes lead to extensive proteolysis in meat resulting in accumulation of a large amount of gases (hydrogen and carbon dioxide), putrid odors, exudate formation, changes in pH and color. Similar changes in refrigerated, vacuum-packed meat can also be produced by certain psychrotrophic bacteria, including LAB (Yang et al., 2014). In addition to *Clostridium*, LAB also play a role in the production of volatile, organic acids detected in the headspace of spoiled meat. Besides acid, heterofermentative Lactobacilli and Leuconostocs also produce carbon dioxide and off-odors (Iulietto et al., 2015).

17.5.4 Filaments and slime formation

Ropy slime appearance in meat is another visual defect of meat that affects consumer choice. This defect is commonly seen in vacuum-packed, cooked meat products due to the presence of *Lactobacillus* spp. and *Leuconostoc* spp. The ropy nature of the slime is attributed to the bacterial polysaccharide polymers extending from the meat surface to the casing or between slices of meat (Iulietto et al., 2015). Slime formation may also be associated with green discoloration such as in the case of *W. viridescens*. The growth of *W. viridescens*

Table 17.1: Microbial spoilage characteristics and associated causative agents.

Characteristic feature	Microbial cause	Associated chemical changes	References
Yellow discoloration	Pseudomonas fluorescence	Utilization of iron by yellow fluorescent siderophore pigment	Cornelis (2010), Andreani et al. (2015), Borch et al. (1996), Dušková et al. (2013), Doulgeraki et al. (2012) Peirson et al. (2003)
Blue discoloration	P. fluorescence	Pigment gene associated with tryptophan production pathway	
Green discoloration	Lactic acid bacteria, Hafnia alvei, Shewanella putrefaciencs, Weisella viridescens	Myoglobin discoloration by hydrosulfide or hydrogen peroxide	
Ropy slime appearance	Lactobacillus sakei, Leuconostoc spp. and other lactic acid bacteria	Extracellular polysaccharide production by microbes (glucan production)	Björkroth and Korkeala (1997) Notararigo et al. (2013), Soultos et al. (2009)
Foul smelling/ putrid odor	Pseudomonas spp., Brochothrix thermosphacta Bacteria harboring amino acid decarboxylases	Acetoin, acetic acid Tyramine and histamine	Koutsoumanis et al. (2006) Casaburi et al. (2015) Nowak and Czyzowska (2011)
Sulfuric odors	Enterobacteriaceae Pseudomonas spp.	Hydrogen sulfide Dimethyl sulfide	Casaburi et al. (2015)
Cheesy odors	Enterobacteriaceae, B. thermosphacta, homofermentative Lactobacillus spp.	Acetoin/diacetyl and 3-methylbutanol	Casaburi et al. (2015)
Gas formation (Gassiness)— Blow pack defect	Enterobacteriaceae, Clostridium estertheticum, Clostridium gasigenes	Gas production in vacuum-packaged meat due to bacterial metabolism	Brightwell et al. (2007), Remenant et al. (2015)

on the meat surface in high numbers results in the formation of a greenish slime on the surface of the product (Björkroth and Korkeala, 1997; Dušková et al., 2013). Table 17.1 summarizes the major qualitative changes associated with the microbial spoilage of meat.

17.6 Factors affecting microbial meat spoilage

Spoilage of meat is principally caused by the growth and degradation of the nutrients in the product by a diverse group of microorganisms. The composition of this microflora is dependent on the product itself and the processing and storage conditions it is subjected to (Huis in't Veld, 1996; Iulietto et al., 2015). In general, the factors that influence microbial proliferation on meat are grouped into three categories (Dave and Ghaly, 2011; Cenci-Goga et al., 2012):

1. Intrinsic parameters: These include the physical and chemical composition of the substrate, water activity, pH, nutrient availability, initial microflora and presence of natural antimicrobial substances.
2. Extrinsic parameters: The storage and handling environment specifically temperature, humidity, and atmosphere condition (aerobic, anaerobic, and MAP).
3. Implicit parameters: These constitute the synergistic and antagonistic effects of the factors mentioned above on the development and establishment of the spoilage microflora.

17.6.1 Intrinsic factors

17.6.1.1 Meat composition and antimicrobial hurdles

Like higher animals, microorganisms also require energy for their growth and survival, essential nutrients and components for the constitution of cells. They acquire these molecules from their substrate or surrounding food environment (Iulietto et al., 2015). In this regard, meat and muscle foods, in general, are rich in proteins, lipids, minerals, and vitamins, but poor sources of carbohydrates. This nutrient composition and availability select for the growth and survival of certain groups of microbes (initial microflora) over the others (Cenci-Goga et al., 2012). Further, the initial break down of these macromolecules to simpler molecules paves the way for microbial succession by organisms that in turn feed on these metabolites. Beyond nutrient availability, the presence of growth factors, natural and added inhibitors select for specific strains (Ray and Bhunia, 2013). These antimicrobial hurdles include food additives, preservatives, natural antimicrobials, and bioprotective cultures that are incorporated in food to improve shelf life and promote food safety (Burt, 2004).

17.6.1.2 pH

Postmortem pH of meat is determined by the amount of lactic acid produced from glycogen during anaerobic glycolysis, and is an essential determinant for the growth of spoilage microbes. After slaughter, muscle pH reduces typically to 5.4–5.8, which can inhibit spoilage microbes to a certain extent. Meat from stressed animals produces a pH greater than or equal to 6.0 (dark, firm, and dry meat), and this makes it an ideal environment for microbes to multiply, eventually resulting in spoilage. The presence of lipid (adipose tissue) and high pH favor rapid bacterial proliferation, utilization of nutrients, and eventual spoilage of meat (Ray and Bhunia, 2013; Lawrie, 2006).

17.6.1.3 Water activity

High moisture content and low solute concentrations tend to provide a favorable environment for microbial growth on meat. Water activity (a_W) of a solution is defined as the ratio of its vapor pressure to that of pure water at the same temperature, and it is inversely proportional to the number of solute molecules present. Spoilage molds and yeast are more tolerant to higher osmotic pressures than bacteria. Bacteria tend to grow at a a_W ranging from 0.75 to 1.0, whereas yeast and molds grow slowly at an a_W of 0.62. Dried products (a_W of less than 0.85), which are stored and distributed at ambient temperatures do not support growth and toxin production bacteria such as *Staphylococcus aureus* and *Clostridium botulinum* (Iulietto et al., 2015; Scott, 1953; Scott, 1957). The microbe population in curing salt solutions such as bacon brines has a shift in population toward osmotolerant and halotolerant organisms. For instance, certain *Lactobacillus* spp. can tolerate high sugar concentration generally used in ham-curing brine. They are capable of growing on cured unprocessed hams and produce polysaccharides with associated deterioration in flavor and appearance (Lawrie, 2006).

17.6.2 Extrinsic factors

17.6.2.1 Temperature

Temperature is a major factor that controls bacterial growth. An understanding of time and temperature management to control spoilage microbes is essential to improve the shelf life of a product (Iulietto et al., 2015). Based on the survivability of microbes at different temperatures, they can be classified as psychrotrophs, mesophiles, and thermophiles, whose tolerability includes the following temperature ranges: 2°C–7°C, 10°C–40°C, and 43°C–66°C, respectively (Lawrie, 2006). Aerobic spoilage microflora at chilling temperatures consists predominantly of pseudomonads, while LAB are the primary organisms of concern under anaerobic conditions or MAP (Gill and Newton, 1978). The nutrient content at certain storage temperatures in meat is another factor that influences microbial growth. An inverse relation has been observed with temperature and amino acid utilization by *Lactobacillus arabinosus*, wherein the bacterium requires phenylalanine, tyrosine, and aspartate for growth at 39°C, phenylalanine and tyrosine at 37°C, and none of these amino acids at 26°C (Borek and Waelsch, 1951). Also, a high microbial load before freezing can contribute to the persistence of microbial enzymes such as lipases even at freezing temperatures. Although the microbial growth process is arrested by freezing, microbial enzymes may continue to produce deleterious changes in meat quality even at temperatures as low as −30°C (Sulzbacher and Gaddis, 1968).

17.6.2.2 Packaging and gaseous atmosphere

The gaseous atmosphere within a packed meat product has a significant impact on the spoilage microbiome (Borch et al., 1996). *Pseudomonas* spp., *Acinetobacter* spp. and *Moraxella* spp. are predominant bacterial genera involved in aerobically stored meat products within a temperature range of $-1°C$ to $25°C$. Specifically, *P. fluorescens, P. fragi, P. ludensis*, and *P. putida* are the significant species commonly isolated from aerobically packaged meat (Ercolini et al., 2006, 2007). In vacuum-packed and MAP meat, there is a shift from aerobic bacteria to the overgrowth and prevalence of facultative and strict anaerobic spoilage microbes (Yost and Nattress, 2000). *Shewanella* spp., *Brochothrix* spp. (*B. thermosphacta* and *B. campestris*), *Serratia* spp., and LAB are the major groups involved in spoilage of vacuum and/or MAP meat products. *S. putrefaciens* is a predominant spoilage bacterium found in chilled and vacuum-packaged meat (Doulgeraki et al., 2012). Reduced water activity along with microaerophilic conditions inhibits gram-negative spoilage microbes and favors the growth and establishment of LAB (Yost and Nattress, 2000; Borch et al., 1996).

17.6.3 Implicit factors

Implicit factors influencing spoilage develop as a result of microbial succession that occurs in meat through the production continuum (Mossel et al., 1995). The factors previously described can either have a synergistic or antagonistic effect on strain selection and eventual composition of the spoilage microflora. Synergistic effects include the breakdown of macromolecules in meat by the initial microflora, thereby providing easily accessible nutrients for a subsequent group of microorganisms that would otherwise be unable to sustain themselves in the food environment. Similarly, changes in acidity or buffering capacity of meat and water activity help select for strains that are tolerant to the altered conditions thereby establishing the secondary spoilage microflora on meat. While these conditions may serve to support a certain group of organisms, they are antagonistic to other species that are sensitive to this food environment (Huis in't Veld, 1996).

17.7 Indicators of microbial spoilage

Microbial spoilage of meat leads to food wastage and significant economic loss. To reduce the costs associated with microbial spoilage, it is important to determine the storage life and predict the shelf-stability of foods (Erkmen and Bozoglu, 2016). Several spoilage indicators are employed to ascertain the shelf-stability of foods. Since one microbial criterion may not be effective in comprehensively predicting the shelf life and spoilage status of meats, the following factors must be considered when identifying indicators of microbial spoilage: (1) indicators must be present in low number in the fresh product, (2) along the farm to fork continuum these microbes should increase to very high numbers, (3) at the onset of spoilage these organisms must be the predominant cause of spoilage, and (4) spoilage characteristics attributed to these organisms must be easily and rapidly detectable (Erkmen and Bozoglu, 2016).

Based on these criteria, some of the microbial spoilage indicators identified for different foods include:

- Gram-negative rods, most importantly *Pseuodomonas* spp. for refrigerated, aerobically stored fresh (raw) meat
- Psychrotrophic LAB, psychrotrophic *Enterobacteriaceae*, psychrotrophic *Clostridium* spp. (*C. laramie*) for refrigerated, vacuum-packed raw meats and refrigerated, low-heat processed, vacuum-packed meats
- Aerobic plate count for investigating the efficacy of sanitary procedures in processing, handling and storage environment

17.8 Detection of microbial spoilage in meat

Microbial spoilage of meat and in turn the associated microorganisms can be detected using direct and indirect enumeration, identification of specific metabolites, and by molecular techniques (Ellis and Goodacre, 2001; Doulgeraki et al., 2012).

17.8.1 Enumeration

Direct methods conventionally employed to detect microbial spoilage of foods are based on quantifying bacterial populations such as total viable count, specific microbial count, direct microbial count, and yeast and mold count (Erkmen and Bozoglu, 2016). These procedures, in general, are time-consuming and only aid in estimating the viable bacterial load in a food at the time of sampling. However, over the last two decades, rapid detection tests have been developed to facilitate bacterial quantification. One such method is the ATP bioluminescence assay that measures ATP levels in bacterial cells in order to indirectly quantify the number of cells in a food sample (de Boer and Beumer, 1999). However, since all living cells generate and utilize ATP, the bioluminescence measured is inclusive of ATP that originates from the muscle tissue and associated microorganisms (D'Souza, 2001; Siragusa et al., 1996). Hence this approach is commonly employed to validate sanitization procedures employed in the processing environment. Other methods include the measurement of changes in electric current or conductivity associated with bacterial growth. The increased electrical conductivity is due to bacterial metabolism of uncharged particles in the culture medium (Betts, 1999).

17.8.2 Detection of bacterial metabolites

17.8.2.1 Spectroscopic methods

These methods are based on the interaction between food components and electromagnetic radiation from a light source, which leads to the production of visible or infrared spectral formation. These are further analyzed by multivariate analysis tools to establish the relationship between spectra and microbial values. Spectroscopic techniques that are commonly used for rapid microbial detection include Fourier transform infrared (FTIR) spectroscopy, Raman spectroscopy, and hyperspectral imaging (HSI) (ur Rahman et al., 2016). FTIR spectroscopy is used to measure the biochemical changes occurring inside the food material and provides data on the formation of distinct microbial metabolites in a food. The resulting absorbance spectrum can be used to discriminate between closely related bacterial species (Rodriguez-Saona et al., 2001). FTIR was exploited to measure microbial spoilage in chicken breasts, which also suggests proteolysis as the major indicator for the onset of spoilage when microorganisms reach 10^7 CFU/g (Ellis et al., 2002). FTIR can also be combined with chemometrics to detect meat spoilage by measuring bacterial counts, pH, and sensorial qualities at various temperatures and storage conditions (Ammor et al., 2009). Raman spectroscopy is used to provide vibrational and rotational spectra especially of covalent bonds present in food molecules (Schmitt and Popp, 2006). These are less influenced by water and hence make it more suitable for analyzing complex heterogeneous food substances. They also provide information about the secondary and tertiary structure of proteins (Tuma, 2005). HSI has been used for the analyses of food products by measuring color, pH, water-holding capacity, and tenderness, and obtains the spectrum using reflectance, scattering, and fluorescence modes (Elmasry et al., 2012). Applications of HSI include the determination of chemical composition and microbial contamination in meat products (Peng et al., 2011).

17.8.2.2 Gas chromatography and mass spectroscopy

Meat spoilage is often accompanied by both chemical and physical changes including the release of VOCs, many of which are odorous (Dainty, 1996). These VOCs give spoiled meat its characteristic pungent, sour or sulfury odor. Characterizing meat spoilage based on the released VOCs can be performed using gas chromatography and mass spectroscopy (GC-MS) methods (Franke and Beauchamp, 2017; Bhattacharjee et al., 2011). Conventional methods for detecting VOCs involved periodic sampling and provide limited information on the kinetics of VOC production and release. Proton-transfer-reaction mass spectrometry is an on-line technique that was used for the detection of VOCs in chicken breast fillets inoculated with *B. thermosphacta* and stored under MAP conditions at 4°C for 1 week (Rattanasomboon et al., 1999). Other miniaturized techniques like GC-DMS (gas chromatography-differential mobility spectroscopy) along with GC-MS have been used for the detection of O-nitrophenol and indole released by the coliforms, including *E. coli* (Saptalena et al., 2012).

17.8.2.3 Biosensors

Biosensors for bacterial detection are based on the interaction between a biological recognition component, such as receptors, nucleic acids, or antibodies, and an appropriate transducer (Ali et al., 2017). Therefore biosensors can be used to detect any compound of interest, including bacterial antigen, toxin, a byproduct of microbial contamination, or spoilage precursor (Schaertel and Firstenberg-Eden, 1988). Depending on the method of signal transduction, biosensors may be divided into four basic groups: optical, mass, electrochemical, and thermal sensors (Sethi, 1994). Optical biosensors can identify microbes in food by their ability to detect minute changes in the refractive index or thickness which occur when bacterial cells bind to receptors on the transducer surface (Ivnitski et al., 1999). Electrochemical biosensors detect changes observed in the main electrical parameters due to the interaction between the sensor and sample. The optical biosensor contains a polymer degradable by lytic enzymes secreted by microorganisms during the decay of the natural product. As the bacterial populations increase, there is an increased release of enzymes causing deterioration of food, which will be visible as the polymer is degraded. Thermometric biosensors exploit the fundamental property of absorption or evolution of heat during biological reactions (Spink and Wadsö, 1976). Thus biosensors can be used as new tools for the rapid analysis of food deterioration and food quality (Ibrišimović et al., 2015).

17.8.2.4 Electronic nose

This instrument comprises an array of chemical sensors with signature specificity and pattern recognition systems capable of recognizing simple and complex odors (Gardner and Bartlett, 1994). In order to classify samples, the electronic nose combines the response profiles of various sensors, which react to various types of compounds in the odor or headspace of the meat product (El Barbri et al., 2008). For instance, the electronic nose was trained to differentiate between unspoiled and spoiled meat. Data obtained from this study demonstrated that the electronic nose was able to classify samples accurately with a success rate of 98.81% and 96.43% for spoiled and unspoiled meat, respectively (El Barbri et al., 2008). Further, a good correlation was obtained between the electronic nose and conventional bacteriological analysis indicating that this approach could be employed for the rapid detection of spoilage in meat.

17.8.3 Molecular methods

17.8.3.1 Immunological methods

Enzyme-linked immunosorbent assay (ELISA), enzyme-linked fluorescent assay, and immunomagnetic separation are the common immunological methods employed in microbial detection. These techniques are based on the binding of antibodies raised to capture specific bacterial antigens (Ellis and Goodacre, 2001). A competitive indirect ELISA using a polyclonal antibody was developed to detect tyramine which is mainly formed as a result of microbial decarboxylation of amino acids in meat and seafood (Sheng et al., 2016). Similar assays have been developed for the detection of proteases produced by the food spoilage genus *Pseudomonas* (Jabbar and Joishy, 1999).

Besides ELISA, flow cytometry has been widely used for studying cell size and structure, for quantifying cellular components, cell physiological activity estimation, as well as for the characterization and detection of microorganisms (Allman et al., 1992). Flow cytometry has been employed by the food industry to detect and quantify foodborne pathogens and other microbial contaminants at population levels as low as 10^2 cells/mL (Tortorello et al., 1997; Jericho et al., 1996). For instance, flow cytometry was found to be useful in quantifying individual cells of *B. thermosphacta* exhibiting complex morphologies (Rattanasomboon et al., 1999).

17.8.3.2 Nucleic acid-based methods

Application of these molecular methods have enabled the rapid and reliable identification and typing of microorganisms, thereby provide a comprehensive picture of the microbial community associated with food in a particular environment at a given point in time. Further, these techniques have also helped to provide valuable information on microbial succession that occurs in food over time (Ercolini et al., 2011). In general, these methods utilize short complementary nucleic acid probes that bind to target sequences in the test samples.

17.8.3.2.1 Polymerase chain reaction -based molecular typing

Polymerase chain reaction (PCR) is the most widely applied nucleic acid-based technique for bacterial identification (Mullis and Faloona, 1987). Traditional and quantitative PCR (qPCR)-based identification of meat spoilage bacteria were developed using primer pairs designed from *Pseudomonas aeruginosa* 23S rDNA sequence (Venkitanarayanan et al., 1996, 1997). Both techniques amplified a unique 207-base-pair DNA product capable of identifying nine major meat spoilage bacteria. Electrochemiluminescence-based qPCR product quantification indicated a high correlation coefficient with regards to cycle number and initial number of bacteria present in the meat sample (Venkitanarayanan et al., 1997). Furthermore, qPCR-ELISA was also developed for the enumeration of spoilage bacteria in refrigerated raw meat (Gutiérrez et al., 1998). This technique is based on 16S rRNA gene amplification utilizing digoxigenin-labeled primers that were biotinylated in situ during the PCR reaction. The amplified biotin–digoxigenin hybrids were quantified by a streptavidin-based ELISA. The captured biotinylated–digoxigenin PCR products in the ELISA were detected with a peroxidase-linked antidigoxigenin conjugate. The enzymatic-based substrate conversion provided absorbance readings that linearly correlated with bacterial counts ranging from 10^2 to 10^7 cm^{-2} in spoiled meat samples (Gutiérrez et al., 1998).

Depending on the stringency of the PCR conditions and primer specificity, it can be employed to differentiate bacteria at the species and intraspecies level (Welsh and McClelland, 1992; Seal et al., 1992). Also strain differences can be detected by the use of primers that anneal to various regions on the genome, thereby producing a banding that is strain-specific. This is often accomplished through the use of PCR variants such as randomly amplified polymorphic DNA-PCR (RAPD-PCR) and repetitive extragenic palindromic-PCR (rep-PCR) (Cocolin et al., 2008). RAPD-PCR has been used to characterize mesophilic and psychrotrophic bacteria isolated from meat (Casaburi et al., 2011). Along the same lines, rep-PCR was used to differentiate closely related bacterial species and spoilage bacteria of artisan-type cooked ham (Vasilopoulos et al., 2010). In terms of PCR-based techniques, PCR-denaturing gradient gel electrophoresis (PCR-DGGE) of ribosomal RNA genes is the most commonly used culture-independent technique for microbial fingerprinting (Ercolini, 2004). This technique is widely used to analyze DNA associated with meat and meat products (Pennacchia et al., 2011). More recently, high-throughput sequencing techniques such as pyrosequencing can be a valuable tool to elucidate the microbial diversity associated with meat and meat products. Ercolini et al. (2011) demonstrated that pyrosequencing of 16s rRNA gene amplicons from meat stored under different packaging conditions was more sensitive and accurate than PCR-DGGE in the detection of microbial succession during storage (Nieminen et al., 2011; Doulgeraki et al., 2012).

17.8.3.2.2 Non-polymerase chain reaction based molecular typing

Although based on the detection of target genetic material, these techniques do not require the amplification of bacterial DNA for detection. For example, restriction enzyme analysis coupled with pulsed-field gel electrophoresis (REA-PFGE) is routinely used for strain characterization (Doulgeraki et al., 2012). This technique involves the digestion of nucleic acid from different bacterial strains using specific restriction enzymes and fragment separation using PFGE resulting in strain-specific banding patterns (Cocolin et al., 2008). PFGE has been intensively used to monitor bacterial succession during storage of meat at the species and strain-level (Doulgeraki et al., 2010, 2011). These studies identified that different LAB and enterobacterial strains within the same species were found depending on the storage conditions. Besides REA-PFGE, terminal restriction fragment length polymorphism has been applied to study bacterial composition and succession in foods like fish, milk, and meat (Reynisson et al., 2009; Rasolofo et al., 2011; Nieminen et al., 2011).

17.9 Conclusion

Given the significant role that microorganisms play in the spoilage of meat and meat products, it is critical to develop effective and feasible approaches to prevent and curtail the growth of spoilage microorganisms. However, in order to develop practical antimicrobial hurdles, it is important to identify, characterize, and understand the predisposing factors in a food system that promote bacterial growth and spoilage. Furthermore, the elucidation of the microbial signature associated with different foods, and various handling and storage conditions will help to develop intervention strategies that are product specific and can be applied along the food production continuum.

References

Ali, J., Najeeb, J., Ali, M.A., Aslam, M.F., Raza, A., 2017. Biosensors: their fundamentals, designs, types and most recent impactful applications: a review. J. Biosens. Bioelectron. 8 (235), 2.

Allman, R., Hann, A.C., Manchee, R., Lloyd, D., 1992. Characterization of bacteria by multiparameter flow cytometry. J. Appl. Bacteriol. 73 (5), 438−444.

Ammor, M.S., Argyri, A., Nychas, G.J.E., 2009. Rapid monitoring of the spoilage of minced beef stored under conventionally and active packaging conditions using Fourier transform infrared spectroscopy in tandem with chemometrics. Meat Sci. 81 (3), 507−514.

Andreani, N.A., Carraro, L., Martino, M.E., Fondi, M., et al., 2015. A genomic and transcriptomic approach to investigate the blue pigment phenotype in *Pseudomonas fluorescens*. Int. J. Food Microbiol. 213, 88−98.

Asensio, M.A., Ordoñez, J.A., Sanz, B., 1988. Effect of carbon dioxide and oxygen enriched atmospheres on the shelf-life of refrigerated pork packed in plastic bags. J. Food Prot. 51 (5), 356−360.

Bailey, J.S., Reagan, J.O., Carpenter, J.A., Schuler, G.A., Thomson, J.E., 1979a. Types of bacteria and shelf-life of evacuated carbon dioxide-injected and ice-packed broilers. J. Food Prot. 42 (3), 218−221.

Bailey, J.S., Reagan, J.O., Carpenter, J.A., Schuler, G.A., 1979b. Microbiological condition of broilers as influenced by vacuum and carbon dioxide in bulk shipping packs. J. Food Sci. 44 (1), 134−137.

Barlow, J., Kitchell, A.G., 1966. A note on the spoilage of prepacked lamb chops by *Microbacterium thermosphactum*. J. Appl. Bacteriol. 29 (1), 185−188.

Bell, R.G., 1997. Distribution and sources of microbial contamination on beef carcasses. J. Appl. Microbiol. 82 (3), 292−300.

Betts, R., 1999. Analytical microbiology—into the next millenium. New Food 2, 9−16.

Bhattacharjee, P., Panigrahi, S., Lin, D., Logue, C.M., et al., 2011. A comparative qualitative study of the profile of volatile organic compounds associated with *Salmonella* contamination of packaged aged and fresh beef by HS-SPME/GC-MS. J. Food Sci. Technol. 48 (1), 1−13.

Björkroth, J., Korkeala, H., 1997. Ropy slime-producing *Lactobacillus sakei* strains possess a strong competitive ability against a commercial biopreservative. Int. J. Food Microbiol. 38 (2−3), 117−123.

Blank, I., 2002. Sensory relevance of volatile organic sulfur compounds in food, *ACS Symposium Series* 2002, 1999. American Chemical Society, Washington, DC, pp. 25−53.

Blickstad, E., Molin, G., 1983. Carbon dioxide as a controller of the spoilage flora of pork, with special reference to temperature and sodium chloride. J. Food Prot. 46 (9), 756−763.

Borch, E., Kant-Muermans, M.L., Blixt, Y., 1996. Bacterial spoilage of meat and cured meat products. Int. J. Food Microbiol. 33 (1), 103−120.

Borek, E., Waelsch, H., 1951. The effect of temperature on the nutritional requirement of microorganisms. J. Biol. Chem. 190 (1), 191−196.

Brightwell, G., Clemens, R., Urlich, S., Boerema, J., 2007. Possible involvement of psychrotolerant *Enterobacteriaceae* in blown pack spoilage of vacuum-packaged raw meats. Int. J. Food Microbiol. 119 (3), 334−339.

Burt, S., 2004. Essential oils: their antibacterial properties and potential applications in foods—a review. Int. J. Food Microbiol. 94 (3), 223−253.

Calkins, C.R., Hodgen, J.M., 2007. A fresh look at meat flavor. Meat Sci. 77 (1), 63−80.

Casaburi, A., Nasi, A., Ferrocino, I., Di Monaco, R., et al., 2011. Spoilage-related activity of *Carnobacterium maltaromaticum* strains in air-stored and vacuum-packed meat. Appl. Environ. Microbiol. 77 (20), 7382−7393.

Casaburi, A., Piombino, P., Nychas, G.J., Villani, F., Ercolini, D., 2015. Bacterial populations and the volatilome associated to meat spoilage. Food Microbiol. 45, 83−102.

Cenci-Goga, B.T., Rossitto, P.V., Sechi, P., Parmegiani, S., et al., 2012. Effect of selected dairy starter cultures on microbiological, chemical and sensory characteristics of swine and venison (*Dama dama*) nitrite-free dry-cured sausages. Meat Sci. 90 (3), 599−606.

Cenci-Goga, B.T., Karama, M., Sechi, P., Iulietto, M.F., Novelli, S., Mattei, S., 2014. Evolution under different storage conditions of anomalous blue coloration of Mozzarella cheese intentionally contaminated with a pigment-producing strain of *Pseudomonas fluorescens*. J. Dairy Sci. 97 (11), 6708−6718.

Christopher, F.M., Carpenter, Z.L., Dill, C.W., Smith, G.C., Vanderzant, C., 1980. Microbiology of beef, pork and lamb stored in vacuum or modified gas atmospheres. J. Food Prot. 43 (4), 259−264.

Church, I.J., Parsons, A.L., 1995. Modified atmosphere packaging technology: a review. J. Sci. Food Agric. 67 (2), 143−152.

Cocolin, L., Dolci, P., Rantsiou, K., 2008. Molecular methods for identification of microorganisms in traditional meat products. Meat Biotechnology. Springer, New York, NY, pp. 91−127.

Cornelis, P., 2010. Iron uptake and metabolism in pseudomonads. Appl. Microbiol. Biotechnol. 86 (6), 1637−1645.

Dainty, R., 1996. Chemical/biochemical detection of spoilage. Int. J. Food Microbiol. 33 (1), 19−33.

Dainty, R.H., Mackey, B.M., 1992. The relationship between the phenotypic properties of bacteria from chill (4 ± 1°C) stored meat and spoilage processes. J. Appl. Bacteriol. 73, 103s−114s.

Dainty, R., Edwards, R., Hibbard, C., 1985. Time course of volatile compound formation during refrigerated storage of naturally contaminated beef in air. J. Appl. Bacteriol. 59 (4), 303−309.

Dave, D., Ghaly, A.E., 2011. Meat spoilage mechanisms and preservation techniques: a critical review. Am. J. Agric. Biol. Sci. 6 (4), 486−510.

de Boer, E., Beumer, R.R., 1999. Methodology for detection and typing of foodborne microorganisms. Int. J. Food Microbiol. 50 (1−2), 119−130.

Diez, A.M., Björkroth, J., Jaime, I., Rovira, J., 2009. Microbial, sensory and volatile changes during the anaerobic cold storage of morcilla de Burgos previously inoculated with *Weissella viridescens* and *Leuconostoc mesenteroides*. Int. J. Food Microbiol. 131 (2-3), 168−177.

Doulgeraki, A.I., Nychas, G.J.E., 2013. Monitoring the succession of the biota grown on a selective medium for pseudomonads during storage of minced beef with molecular-based methods. Food Microbiol. 34 (1), 62−69.

Doulgeraki, A.I., Paramithiotis, S., Kagkli, D.M., Nychas, G.J.E., 2010. Lactic acid bacteria population dynamics during minced beef storage under aerobic or modified atmosphere packaging conditions. Food Microbiol. 27 (8), 1028−1034.

Doulgeraki, A.I., Paramithiotis, S., Nychas, G.J.E., 2011. Characterization of the *Enterobacteriaceae* community that developed during storage of minced beef under aerobic or modified atmosphere packaging conditions. Int. J. Food Microbiol. 145 (1), 77−83.

Doulgeraki, A.I., Ercolini, D., Villani, F., Nychas, G.J.E., 2012. Spoilage microbiota associated to the storage of raw meat in different conditions. Int. J. Food Microbiol. 157 (2), 130–141.

D'Souza, S.F., 2001. Microbial biosensors. Biosens. Bioelectron. 16 (6), 337–353.

Dušková, M., Kameník, J., Karpíšková, R., 2013. *Weissella viridescens* in meat products—a review. Acta Vet. Brno 82 (3), 237–241.

El Barbri, N., Llobet, E., El Bari, N., Correig, X., Bouchikhi, B., 2008. Electronic nose based on metal oxide semiconductor sensors as an alternative technique for the spoilage classification of red meat. Sensors 8 (1), 142–156.

Ellis, D.I., Goodacre, R., 2001. Rapid and quantitative detection of the microbial spoilage of muscle foods: current status and future trends. Trends Food Sci. Technol. 12 (11), 414–424.

Ellis, D.I., Broadhurst, D., Kell, D.B., Rowland, J.J., Goodacre, R., 2002. Rapid and quantitative detection of the microbial spoilage of meat by Fourier transform infrared spectroscopy and machine learning. Appl. Environ. Microbiol. 68 (6), 2822–2828.

Elmasry, G., Barbin, D.F., Sun, D., Allen, P., 2012. Meat quality evaluation by hyperspectral imaging technique: an overview. Crit. Rev. Food Sci. Nutr. 52 (8), 689–711.

Enfors, S.O., Molin, G., 1984. Carbon dioxide evolution of refrigerated meat. Meat Sci. 10, 197–206.

Enfors, S.O., Molin, G., Ternström, A., 1979. Effect of packaging under carbon dioxide, nitrogen or air on the microbial flora of pork stored at 4 °C. J. Appl. Bacteriol. 47 (2), 197–208.

Ercolini, D., 2004. PCR-DGGE fingerprinting: novel strategies for detection of microbes in food. J. Microbiol. Methods 56 (3), 297–314.

Ercolini, D., Russo, F., Torrieri, E., Masi, P., Villani, F., 2006. Changes in the spoilage-related microbiota of beef during refrigerated storage under different packaging conditions. Appl. Environ. Microbiol. 72 (7), 4663–4671.

Ercolini, D., Russo, F., Blaiotta, G., Pepe, O., Mauriello, G., Villani, F., 2007. Simultaneous detection of *Pseudomonas fragi, P. lundensis*, and *P. putida* from meat by use of a multiplex PCR assay targeting the carA gene. Appl. Environ. Microbiol. 73 (7), 2354–2359.

Ercolini, D., Ferrocino, I., Nasi, A., Ndagijimana, M., et al., 2011. Microbial metabolites and bacterial diversity in beef stored in different packaging conditions monitored by pyrosequencing, PCR-DGGE, SPME-GC/MS and 1HNMR. Appl. Environ. Microbiol. 77 (20), 7372–7381.

Erkmen, O., Bozoglu, T.F., 2016. Spoilage of meat and meat products. In: Erkmen, O., Bozoglu, T.F. (Eds.), Food Microbiology: Principles into Practice. John Wiley & Sons. Available from: http://dx.doi.org/10.1002/9781119237860.ch16.

FAO G., 2011. Global Food Losses and Food Waste—Extent, Causes and Prevention. Save Food: An Initiative on Food Loss and Waste Reduction.

Farkas, J., 2007. Physical methods of food preservation, Food Microbiology: Fundamentals and Frontiers, *third ed.* American Society of Microbiology, pp. 685–712.

Franke, C., Beauchamp, J., 2017. Real-time detection of volatiles released during meat spoilage: a case study of modified atmosphere-packaged chicken breast fillets inoculated with *Br. thermosphacta*. Food Anal. Methods 10 (2), 310–319.

Gardner, J.W., Bartlett, P.N., 1994. A brief history of electronic noses. Sensors Actuators B: Chem. 18 (1–3), 210–211.

Gardner, G.A., Carson, A.W., Patton, J., 1967. Bacteriology of prepacked pork with reference to the gas composition within the pack. J. Appl. Bacteriol. 30 (2), 321–332.

Gill, C.O., 1986. The control of microbial spoilage in fresh meat. In: Pearson, A.M., Dutson, T.R. (Eds.), Meat and Poultry Microbiology. Macmillan Publishers, Basingstoke, pp. 49–88.

Gill, C., 1996. Extending the storage life of raw chilled meats. Meat Sci. 43, 99–109.

Gill, C.O., 1976. Substrate limitation of bacterial growth at meat surfaces. J. Appl. Bacteriol. 41 (3), 401–410.

Gill, C.O., Newton, K.G., 1978. The ecology of bacterial spoilage of fresh meat at chill temperatures. Meat Sci. 2 (3), 207–217.

Gill, C.O., Newton, K.G., 1980. Growth of bacteria on meat at room temperatures. J. Appl. Bacteriol. 49 (2), 315–323.

Gill, C.O., Penney, N., 1988. The effect of the initial gas volume to meat weight ratio on the storage life of chilled beef packaged under carbon dioxide. Meat Sci. 22 (1), 53–63.

Gutiérrez, R., Garcia, T., Gonzalez, I., Sanz, B., et al., 1998. Quantitative detection of meat spoilage bacteria by using the polymerase chain reaction (PCR) and an enzyme linked immunosorbent assay (ELISA). Lett. Appl. Microbiol. 26 (5), 372–376.

Hodges, J.H., Cahill, V.R., Ockerman, H.W., 1974. Effect of vacuum packaging on weight loss, microbial growth and palatability of fresh beef wholesale cuts. J. Food Sci. 39 (1), 143–146.

Huis in't Veld, J.H., 1996. Microbial and biochemical spoilage of foods: an overview. Int. J. Food Microbiol. 3 (1), 1–18.

Hultman, J., Rahkila, R., Ali, J., Rousu, J., Björkroth, K.J., 2015. Meat processing plant microbiome and contamination patterns of cold-tolerant bacteria causing food safety and spoilage risks in the manufacture of vacuum-packaged, cooked sausages. Appl. Environ. Microbiol. 81, 7088–7097. AEM-02228.

Ibrišimović, N., Ibrišimović, M., Kesić, A., Pittner, F., 2015. Microbial biosensor: a new trend in the detection of bacterial contamination. Monatshefte Chem.-Chem. Monthly 146 (8), 1363–1370.

Iulietto, M.F., Sechi, P., Borgogni, E., Cenci-Goga, B.T., 2015. Meat spoilage: a critical review of a neglected alteration due to ropy slime producing bacteria. Italian J. Anim. Sci. 14 (3), 4011.

Ivnitski, D., Abdel-Hamid, I., Atanasov, P., Wilkins, E., 1999. Biosensors for detection of pathogenic bacteria. Biosens. Bioelectron. 14 (7), 599–624.

Jabbar, H., Joishy, K.N., 1999. Rapid detection of *Pseudomonas* in seafoods using protease indicator. J. Food Sci. 64 (3), 547–549.

Jack, R.W., Wan, J., Gordon, J., Harmark, K., et al., 1996. Characterization of the chemical and antimicrobial properties of piscicolin 126, a bacteriocin produced by *Carnobacterium piscicola* JG126. Appl. Environ. Microbiol. 62 (8), 2897–2903.

Jay, J.M., Loessner, M.J., Golden, D.A., 2005. Modern Food Microbiology, seventh ed. Springer Science and Business Media, NY, pp. 63–101, ISBN: 0387231803.

Jericho, K., Kozub, G., Loewen, K., Ho, J., 1996. Comparison of methods to determine the microbiological contamination of surfaces of beef carcasses by hydrophobic grid membrane filters, standard pour plates or flow cytometry. Food Microbiol. 13 (4), 303–309.

Koutsoumanis, K., Sofos, J., 2004. Microbial contamination of carcasses and cuts. Encyclopedia of Meat Sciences. Elsevier, pp. 727–737.

Koutsoumanis, K., Stamatiou, A., Skandamis, P., Nychas, G.J., 2006. Development of a microbial model for the combined effect of temperature and pH on spoilage of ground meat, and validation of the model under dynamic temperature conditions. Appl. Environ. Microbiol. 72 (1), 124–134.

Lawrie, R., 2006. In: Ledward, D. (Ed.), Lawrie's Meat Science, *seventh ed.* Woodhead Publishing Ltd, Abington, pp. 157–188.

Martín, A., Benito, M.J., Aranda, E., Ruiz-Moyano, S., et al., 2010. Characterization by volatile compounds of microbial deep spoilage in iberian dry-cured ham. J. Food Sci. 75 (6), M360–M365.

Mills, J., Horváth, K.M., Reynolds, A.D., Brightwell, G., 2018. Farm and abattoir sources of *Carnobacterium* species and implications for lamb meat spoilage. J. Appl. Microbiol. 125 (1), 142–147.

Mossel, D.A.A., Corry, J.E.L., Struijk, C.B., Baird, R.M., 1995. Essentials of the Microbiology of Foods: A Textbook for Advanced Studies. Wiley, England, pp. 175–214.

Mullis, K.B., Faloona, F.A., 1987. Specific synthesis of DNA in vitro via a polymerase-catalyzed chain reaction. *Methods in Enzymology, 155.* Academic Press, pp. 335–350.

Newton, K.G., Gill, C.O., 1978. The development of the anaerobic spoilage flora of meat stored at chill temperatures. J. Appl. Bacteriol. 44 (1), 91–95.

Nieminen, T.T., Vihavainen, E., Paloranta, A., Lehto, J., et al., 2011. Characterization of psychrotrophic bacterial communities in modified atmosphere-packed meat with terminal restriction fragment length polymorphism. Int. J. Food Microbiol. 144 (3), 360–366.

Notararigo, S., Nácher-vázquez, M., Ibarburu, I., et al., 2013. Comparative analysis of production and purification of homo-and hetero-polysaccharides produced by lactic acid bacteria. Carbohydr. Polym. 93 (1), 57–64.

Nowak, A., Czyzowska, A., 2011. In vitro synthesis of biogenic amines by *Brochothrix thermosphacta* isolates from meat and meat products and the influence of other microorganisms. Meat Sci. 88 (3), 571–574.

Nychas, G.J., Dillon, V., Board, R.G., 1988. Glucose, the key substrate in the microbiological changes occurring in meat and certain meat products. Biotechnol. Appl. Biochem. 10 (3), 203–231.

Nychas, G.J.E., Skandamis, P.N., Tassou, C.C., Koutsoumanis, K.P., 2008. Meat spoilage during distribution. Meat Sci. 78 (1–2), 77–89.

Papuc, C., Goran, G.V., Predescu, C.N., Nicorescu, V., 2017. Mechanisms of oxidative processes in meat and toxicity induced by postprandial degradation products: a review. Comp. Rev. Food Sci. Food Saf. 16 (1), 96–123.

Peirson, M.D., Guan, T.Y., Holley, R.A., 2003. Aerococci and carnobacteria cause discoloration in cooked cured bologna. Food Microbiol. 20 (2), 149–158.

Peng, Y., Zhang, J., Wang, W., Li, Y., et al., 2011. Potential prediction of the microbial spoilage of beef using spatially resolved hyperspectral scattering profiles. J. Food Eng. 102 (2), 163–169.

Pennacchia, C., Ercolini, D., Villani, F., 2011. Spoilage-related microbiota associated with chilled beef stored in air or vacuum pack. Food Microbiol. 28 (1), 84–93.

Petruzzi, L., Corbo, M.R., Sinigaglia, M., Bevilacqua, A., 2017. Microbial spoilage of foods: fundamentals. The Microbiological Quality of Food. Woodhead Publishing, pp. 1–21.

Pin, C., de Fernando, G.D.G., Ordóñez, J.A., 2002. Effect of modified atmosphere composition on the metabolism of glucose by *Brochothrix thermosphacta*. Appl. Environ. Microbiol. 68 (9), 4441–4447.

Pothakos, V., Devlieghere, F., Villani, F., Björkroth, J., Ercolini, D., 2015. Lactic acid bacteria and their controversial role in fresh meat spoilage. Meat Sci. 109, 66–74.

Rahkila, R., Nieminen, T., Johansson, P., Säde, E., Björkroth, J., 2012. Characterization and evaluation of the spoilage potential of *Lactococcus piscium* isolates from modified atmosphere packaged meat. Int. J. Food Microbiol. 156 (1), 50–59.

Rasolofo, E.A., LaPointe, G., Roy, D., 2011. Assessment of the bacterial diversity of treated and untreated milk during cold storage by T-RFLP and PCR-DGGE methods. Dairy Sci. Technol. 91 (5), 573–597.

Rattanasomboon, N., Bellara, S., Harding, C., et al., 1999. Growth and enumeration of the meat spoilage bacterium *Brochothrix thermosphacta*. Int. J. Food Microbiol. 51 (2-3), 145–158.

Ray, B., Bhunia, A., 2013. Microbial food spoilage, Fundamental Food Microbiology, fifth ed. CRC Press, Boca Raton, FL, pp. 245–304.

Remenant, B., Jaffres, E., Dousset, X., Pilet, M.F., Zagorec, M., 2015. Bacterial spoilers of food: behavior, fitness and functional properties. Food Microbiol. 45, 45–53.

Reynisson, E., Lauzon, H.L., Magnússon, H., et al., 2009. Bacterial composition and succession during storage of North-Atlantic cod (*Gadus morhua*) at superchilled temperatures. BMC Microbiol. 9 (1), 250.

Ritchie, H., Roser, M., 2018. Meat and Seafood Production and Consumption, 12. Our World in Data, p. 2018, Retrieved September.

Rodriguez-Saona, L.E., Khambaty, F.M., Fry, F.S., Calvey, E.M., 2001. Rapid detection and identification of bacterial strains by Fourier transform near-infrared spectroscopy. J. Agric. Food Chem. 49 (2), 574–579.

Saptalena, L.G., Kerpen, K., Kuklya, A., Telgheder, U., 2012. Rapid detection of synthetic biomarkers of *Escherichia coli* in water using microAnalyzer: a field dependence study. Int. J. Ion Mob. Spectr. 15 (2), 47–53.

Schaertel, B., Firstenberg-Eden, R., 1988. Biosensors in the food industry: present and future. J. Food Prot. 51 (10), 811–820.

Schaefer-Seidler, C.E., Judge, M.D., Cousin, M.A., Aberle, E.D., 1984. Microbiological contamination and primal cut yields of skinned and scalded pork carcasses. J. Food Sci. 49 (2), 356–358.

Schmitt, M., Popp, J., 2006. Raman spectroscopy at the beginning of the twenty-first century. J. Raman Spectr. 37 (1-3), 20–28.

Scott, W., 1953. Water relations of *Staphylococcus aureus* at 30 C. Aus. J. Biol. Sci. 6 (4), 549–564.

Scott, W., 1957. Water relations of food spoilage microorganisms. Advances in Food Research. Elsevier, pp. 83–127.

Seal, S.E., Jackson, L.A., Daniels, M.J., 1992. Use of tRNA consensus primers to indicate subgroups of *Pseudomonas solanacearum* by polymerase chain reaction amplification. Appl. Environ. Microbiol. 58 (11), 3759–3761.

Seman, D.L., Drew, K.R., Clarken, P.A., Littlejohn, R.P., 1988. Influence of packaging method and length of chilled storage on microflora, tenderness and colour stability of venison loins. Meat Sci. 22 (4), 267–282.

Sethi, R.S., 1994. Transducer aspects of biosensors. Biosens. Bioelectron. 9 (3), 243–264.

Shaw, B.G., Latty, J.B., 1982. A numerical taxonomic study of *Pseudomonas* strains from spoiled meat. J. Appl. Bacteriol. 52 (2), 219–228.

Shay, B.J., Egan, A.F., 1987. The packaging of chilled red meats. Food Technol. Aus. 39, 283–285.

Sheng, W., Sun, C., Fang, G., et al., 2016. Development of an enzyme-linked immunosorbent assay for the detection of tyramine as an index of freshness in meat and seafood. J. Agric. Food Chem. 64 (46), 8944–8949.

Silva, A.R., Paulo, É.N., Sant'Ana, A.S., et al., 2011. Involvement of *Clostridium gasigenes* and *C. algidicarnis* in 'blown pack' spoilage of Brazilian vacuum-packed beef. Int. J. Food Microbiol 148 (3), 156–163.

Simard, R.E., Zee, J., L'heureux, L., 1984. Microbial growth in carcasses and boxed beef during storage. J. Food Prot. 47 (10), 773–777.

Siragusa, G.R., Dorsa, W.J., Cutter, C.N., Perino, L.J., Koohmaraie, M., 1996. Use of a newly developed rapid microbial ATP bioluminescence assay to detect microbial contamination on poultry carcasses. J. Bioluminescence Chemiluminescence 11 (6), 297–301.

Smit, B.A., Engels, W.J., Smit, G., 2009. Branched chain aldehydes: production and breakdown pathways and relevance for flavour in foods. Appl. Microbiol. Biotechnol. 81 (6), 987–999.

Soncin, S., Chiesa, L., Cantoni, C., Biondi, P., 2007. Preliminary study of the volatile fraction in the raw meat of pork, duck and goose. J. Food Comp. Anal. 20 (5), 436–439.

Soultos, N., Tzikas, Z., Christaki, E., Papageorgiou, K., Steris, V., 2009. The effect of dietary oregano essential oil on microbial growth of rabbit carcasses during refrigerated storage. Meat Sci. 81 (3), 474–478.

Spink, C., Wadsö, I., 1976. Calorimetry as an analytical tool in biochemistry and biology. Methods Biochem. Anal. 23 (1), 1–160.

Stellato, G., La Storia, A., De Filippis, F., 2016. Overlap of spoilage microbiota between meat and meat processing environment in small-scale vs large-scale retail distribution. Appl. Environ. Microbiol. AEM-00793.

Sulzbacher, W., Gaddis, A., 1968. Meats. Preservation of quality by frozen storage. The Freezing Preservation of Foods. AVI Publishing Co., Inc, Westport, CT, p. 159.

Toldra, F., 1998. Proteolysis and lipolysis in flavour development of dry-cured meat products. Meat Sci. 49, S101–S110.

Tortorello, M.L., Stewart, D.S., Raybourne, R.B., 1997. Quantitative analysis and isolation of *Escherichia coli* O157: H7 in a food matrix using flow cytometry and cell sorting. FEMS Immunol. Med. Microbiol. 19 (4), 267–274.

Tsigarida, E., Boziaris, I.S., Nychas, G.J., 2003. Bacterial synergism or antagonism in a gel cassette system. Appl. Environ. Microbiol. 69 (12), 7204–7209.

Tuma, R., 2005. Raman spectroscopy of proteins: from peptides to large assemblies. J. Raman Spectr. 36 (4), 307–319.

ur Rahman, U., Shahzad, T., Sahar, A., Ishaq, A., 2016. Recapitulating the competence of novel and rapid monitoring tools for microbial documentation in food systems. LWT-Food Sci. Technol. 67, 62–66.

Vasilopoulos, C., De Maere, H., De Mey, E., Paelinck, H., 2010. Technology-induced selection towards the spoilage microbiota of artisan-type cooked ham packed under modified atmosphere. Food Microbiol. 27 (1), 77–84.

Venkitanarayanan, K.S., Faustman, C., Crivello, J.F., 1997. Rapid estimation of spoilage bacterial load in aerobically stored meat by a quantitative polymerase chain reaction. J. Appl. Microbiol. 82 (3), 359–364.

Venkitanarayanan, K.S., Khan, M.I., Faustman, C., Berry, B.W., 1996. Detection of meat spoilage bacteria by using the polymerase chain reaction. J. Food Prot. 59 (8), 845–848.

Weimer, B.C., 2007. Improving the Flavour of Cheese. Elsevier, pp. 102–120.

Welsh, J., McClelland, M., 1992. PCR-amplified length polymorphisms in tRNA intergenic spacers for categorizing *Staphylococci*. Mol. Microbiol. 6 (12), 1673–1680.

Yang, X., Youssef, M.K., Gill, C.O., Badoni, M., López-Campos, Ó., 2014. Effects of meat pH on growth of 11 species of psychrotolerant clostridia on vacuum packaged beef and blown pack spoilage of the product. Food Microbiol. 39, 13–18.

Yost, C., Nattress, F., 2000. The use of multiplex PCR reactions to characterize populations of lactic acid bacteria associated with meat spoilage. Lett. Appl. Microbiol. 31 (2), 129–133.

Young, L.L., Reviere, R.D., Cole, A.B., 1988. Fresh Red Meats: A Place to Apply Modified Atmospheres. Food Technology, USA.

Yvon, M., Rijnen, L., 2001. Cheese flavour formation by amino acid catabolism. Int. Dairy J. 11 (4–7), 185–201.

Zhang, Q.Q., Han, Y.Q., Cao, J.X., et al., 2012. The spoilage of air-packaged broiler meat during storage at normal and fluctuating storage temperatures. Poult. Sci. 91 (1), 208–214.

Further reading

Casla, D., Requena, T., Gómez, R., 1996. Antimicrobial activity of lactic acid bacteria isolated from goat's milk and artisanal cheeses: characteristics of a bacteriocin produced by *Lactobacillus curvatus* IFPL105. J. Appl. Bacteriol. 81 (1), 35–41.

Hanna, M.O., Savell, J.W., Smith, G.C., Purser, D.E., Gardner, F.A., Vanderzant, C., 1983. Effect of growth of individual meat bacteria on pH, color and odor of aseptically prepared vacuum-packaged round steaks. J. Food Prot. 46 (3), 216–221.

McEvoy, J.M., Sheridan, J.J., Blair, I.S., McDowell, D.A., 2004. Microbial contamination on beef in relation to hygiene assessment based on criteria used in EU Decision 2001/471/EC. Int. J. Food Microbiol. 92 (2), 217–225.

Newton, K.G., Harrison, J.C.L., Wauters, A.M., 1978. Sources of psychrotrophic bacteria on meat at the abattoir. J. Appl. Bacteriol. 45 (1), 75–82.

CHAPTER 18

Modern concept and detection of spoilage in meat and meat products

V.J. Ajaykumar[1] and Prabhat Kumar Mandal[2]

[1]Department of Veterinary Public Health, Rajiv Gandhi Institute of Veterinary Education and Research, Puducherry, India [2]Department of Livestock Products Technology, Rajiv Gandhi Institute of Veterinary Education and Research, Puducherry, India

Chapter Outline

18.1 Introduction 335
18.2 Microbial spoilage of meat 336
18.3 Microbial metabolites 337
18.4 Detection of spoilage bacteria 338
    18.4.1 Enumeration methods 338
    18.4.2 Detection methods 339
    18.4.3 Detection by molecular methods 339
18.5 Detection of microbial metabolites 340
18.6 Modern trends in the spoilage detection of meat 340
    18.6.1 Odor sensors and electronic nose technology 340
    18.6.2 Conducting organic polymers 341
    18.6.3 Metal oxide semiconductor 341
    18.6.4 Spectroscopy in advanced forms 342
    18.6.5 Use of lasers in detecting food quality 344
    18.6.6 Smartphone-based food diagnostic technologies 345
    18.6.7 Chemical and biological sensors for food-quality monitoring 345
18.7 Conclusion 346
References 346
Further reading 349

18.1 Introduction

Spoilage of food is a challenge faced by humankind from the time they started living in communities and perhaps one of the first experiments humans might have done would have been in the storage of excess food or storing food for lean seasons. Foods are not only nutritious to consumers, but are also excellent sources of nutrients for microbial growth.

Depending upon the microorganisms present, foods may spoil or be preserved by fermentation. Meat and meat products are integral parts of the diet for human beings and any issue pertaining to the acceptability of this is of great concern. Nutrient-rich matrix meat is the first-choice source of animal protein for many people all over the world and consumption is increasing steadily. On the other hand, a significant portion of meat and meat products are spoiled every year and approximately 3.5 billion kg of poultry and meat were wasted at the consumer, retailer, and foodservice levels, which have a substantial economic and environmental impact. Significant portion of this loss is due to microbial spoilage (Dave and Ghaly, 2011). Microorganisms can not only be useful to transform raw foods into fermented delights, including yoghurt, cheese, sausages, tempeh, pickles, wine, beers, and other alcoholic products, but also can act as a reservoir for disease transmission, and thus the detection and control of pathogens and spoilage organisms are important areas of food microbiology.

During the entire sequence of food handling from the producer to the final consumer, microorganisms can affect food quality and human health. Muscle foods, which include both meat and poultry, are an integral part of the human diet and have been so for several thousand years (Ellis et al., 2002). The spoilage can be also due to autolytic, oxidative, and other chemical changes which can happen to meat during storage and supply. At the level of the consumer, the detection of spoilage can be really challenging. Most of the earlier methods focused on spoilage detection directly on meat, whereas now in most of the cases the consumer is not getting raw meat but mostly ready-to-eat products. In this situation the detection of spoilage is becoming more challenging as most of the ready-to-eat foods have contents other than meat. As the spoilage is practically the result of microbial growth and autolytic changes, a background on microbial spoilage and the microbial metabolites responsible for spoilage is given and the methods to detect spoilage are discussed.

18.2 Microbial spoilage of meat

Meat and meat products provide excellent growth media for a variety of microflora (bacteria, yeasts, and molds) some of which are pathogens (Jay et al., 2005). The intestinal tract and the skin of the animal are the main sources of these microorganisms. The composition of microflora in meat depends on various factors: (1) preslaughter husbandry practices (free range vs intensive rearing), (2) age of the animal at the time of slaughtering, (3) handling during slaughtering, evisceration, and processing, (4) temperature controls during slaughtering, processing, and distribution, (5) preservation methods, (6) type of packaging, and (7) handling and storage by consumer (Cerveny et al., 2009). These factors may determine the type of microbe present in the raw meat, and the microbial picture in the product will be entirely different as many other ingredients maybe added to meat. However, the modern spoilage microbe detection methods rarely look for specific organisms, rather those methods are looking for the presence of microbial metabolic products which are common for many organisms.

Muscle foods are described as spoiled if organoleptic changes make them unacceptable to the consumer (Mandal et al., 2018). These organoleptic characteristics can include changes in appearance (i.e., discoloration), the development of odors, slime formation, or any other characteristic which makes the food undesirable for human consumption (Jay et al., 2005). It is generally accepted that detectable organoleptic spoilage is a result of decomposition and the formation of metabolites by the growth of microorganisms (Braun et al., 1999). The organoleptic changes vary according to the species of microflora present, the characteristics of the meat, processing methods, product composition, and the storage conditions (Jackson et al., 1997).

The first stage of colonization and growth of microorganisms is the attachment of bacterial cells on meat surfaces. The second and irreversible stage involves the production of a glycocalyx by the bacterium (Costerson et al., 1981). The main spoilage organisms belong to the genus *Pseudomonas* and other major spoilage flora of meat stored aerobically under refrigeration include the *Moraxella*, *Psychrobacter*, and *Acinetobacter*.

Fresh meats generally have a pH range between 5.5 and 5.9 and contain sufficient glucose and other simple carbohydrates to support approximately 10^9 cfu/cm^2. The pseudomonads grow fastest and utilize glucose at refrigeration temperatures (Seymour et al., 1994). At levels of 10^7 cfu/cm^2 odors may become evident in the form of a faint "dairy" type aroma and once it has reached 10^8 cfu/cm^2 the recognizable off-odors develop leading to "sensory" spoilage (Stanbridge and Davies, 1998). The development of off-odors depends upon the free amino acid utilization. These odors are described as dairy/buttery/fatty/cheesy at 10^7 cfu/cm^2; a sickly sweet/fruity at 10^8 cfu/cm^2 and finally putrid odor at 10^9 cfu/cm^2 (Adams and Moss, 2000). The surface of the meat will be tacky, indicating the first stages of slime formation due to bacteria growth and synthesis of polysaccharides. Deterioration in the color of meat is due to a fall in the partial pressure of oxygen. Once the population of bacteria approaches 10^8 cfu/cm^2, the nitrogenous compounds lead to the formation of malodorous substances such as ammonia (NH_3), dimethylsulfide (C_2H_6S), and diacetyl ($C_4H_6O_2$) (Stanbridge and Davies, 1998).

18.3 Microbial metabolites

Numerous attempts have been made to associate metabolites with the microbial spoilage of meat and to provide information about spoilage and possibly determine remaining shelf life (Alomirah et al., 1998; Braun et al., 1999). Once the surface levels of glucose have been depleted bacteria will metabolize secondary substrates such as free amino acids and lactate. Borch et al. (1991) concluded that glucose limitation caused a switch from a saccharolytic to an amino acid-degrading metabolism in at least some bacterial species. The physicochemical changes during the spoilage process occur within the aqueous phase of meat and this phase contains low molecular weight compounds, such as glucose, lactic acid, certain amino acids, nucleotides, urea, and water-soluble proteins that are catabolized by the vast majority of the meat microflora (Drosinos and Board, 1994; Nychas et al., 1998).

Many bacteria secrete proteases and in general Gram-negative bacteria in chilled meat predominantly secrete aminopeptidases (Nychas et al., 1998). The utilization of free amino acids by bacteria leads to an increase in levels of ammonia and it has been observed that the switch from a saccharolytic to an amino acid-degrading metabolism occurs whilst considerable levels of glucose are still present deep within the muscle tissue (Seymour et al., 1994). In addition to ammonia, the by-products of amino acid utilization include sulfides, indole, scatole, and amines, such as the diamines putrescine and cadaverine (Adams and Moss, 2000). It is the production of these compounds, amongst others, that lead to the characteristic changes associated with spoiled meat, such as malodors and the increase in pH.

18.4 Detection of spoilage bacteria

The conventional microbiological approach to food sampling has changed little over the last half century and it has been estimated that there are currently in excess of 40 methods to measure and detect bacterial spoilage in meats (Betts, 1999; Nychas et al., 1998). The development of rapid microbiological test procedures over the last two decades can be divided into two main groups: enumeration and presence/absence tests (Mandal et al., 2018)

18.4.1 Enumeration methods

Enumeration methods are generally based on microscopy, ATP bioluminescence, or the measurement of electrical phenomena. In the case of microscopic methods sophisticated techniques have been developed where microorganisms are stained with fluorescent dyes and viewed with an epifluorescent microscope. The problems such as staining of both viable and nonviable cells were overcome with the introduction of the direct epifluorescent filter technique, but the procedure is time-consuming and laborious (Pyle et al., 1999). Though fully automated systems and the use of flow cytometry have been developed (Rattanasomboon et al., 1999), the results from low levels of microorganisms can still take 18–20 hours (Betts, 1999) and disaggregation of the spoilage organism from the meat is difficult.

ATP bioluminescence acts by measuring ATP levels in bacterial cells in culture in order to calculate the number of cells present in that culture (Champiat et al., 2001; D'Souza, 2001).

The problem with this method is that ATP present in meat has to be destroyed before microbial ATP can be measured. Electrical measuring methods are based on the detection of electrical current during microbial growth, as changes are caused by bacteria that metabolize uncharged particles in any growth medium, thereby increasing the conductivity of that medium. Commercially available instruments include the Bactometer, Malthus Analyzer, Rabit, and Bactrac (Betts, 1999).

18.4.2 Detection methods

Current detection methods are based on immunological or nucleic acid-based procedures. Immunological methods employ antibodies that are raised to react to surface antigens of specific microorganisms (Betts, 1999). The most common form of these methods is the enzyme-linked immunosorbent assays and these are based on the use of an enzyme label. Those in use are currently aimed at the detection of foodborne pathogens such as *Salmonella*, *Listeria*, *Escherichia coli* O157:H7 as well as toxins produced by *Staphylococcus aureus* and proteases from the food spoilage genus *Pseudomonas* (Jabbar and Joishy, 1999). Nucleic acid-based procedures utilize probes that are small segments of single-stranded complementary nucleic acid that are used to detect specific genetic sequences in test samples. Nucleic acid probes can be used to detect either DNA or RNA sequences in order to identify accurately a specific microorganism (Alexandre et al., 2001; Mandal et al., 2011).

The most widely applied nucleic acid detection method at present utilizes the polymerase chain reaction (PCR) (Mullis and Faloona, 1987). This method has been reported to allow for rapid and selective identification and/or detection of microorganisms in different matrices by amplifying specific gene fragments and detecting the PCR amplicons by gel electrophoresis (Cloak et al., 2001; Yost and Nattress, 2000). Thus like for nucleic acid probes, the DNA sequence of the target organism must be known prior to the analysis. This method has limitations, for as long as intact nucleic acid sequences are present in a sample they will be amplified by PCR. Therefore DNA from nonviable microorganisms can lead to false positive results. The final limitation of PCR is yet again the time factor, as this can be a time-consuming method especially as regards large-scale testing and the tedious and exacting nature of the reaction setup (Barbour and Tice, 1997). However, PCR is at present one of the most rapid procedures available for the detection of pathogens in foods, with test times for *Salmonella* spp., for example, of approximately 18 hours (Warneck, 2001; Mandal et al., 2011).

18.4.3 Detection by molecular methods

PCR has been used in detecting bacteria that can spoil vacuum-packaged meats. These bacteria are able to grow in temperatures of $-1.5°C$. As this temperature is the optimal storage temperature for the transport of red meat, the early screening of these bacteria in the meat is important, particularly *Brochothrix campestris* and *Brochothrix thermosphacta*, which can cause spoilage within 9 weeks of storage (Gribble and Brightwell, 2013). PCR has been used to identify and differentiate the two bacteria accurately. Similarly Reid et al. (2017) developed a set of real-time PCR methods for the detection of various *Clostridium* species (*C. estertheticum, C. gasigenes,* and *C. ruminantium*), the causative agents of blown pack spoilage in vacuum-packaged beef. The 16SrRNA gene sequencing analysis identified the predominant potential spoilage organism as *Enterobacter* and *Acinetobacter* at room temperature, *Pseudomonas* and *Aeromonas* at refrigerated storage, and *Aeromonas* and *Enterococcus* at ice storage (Don et al., 2018).

Very recently, Illikoud and collaborators evaluated the genotypic and phenotypic diversity of *B. thermosphacta*, which is one of the major bacterial species involved in meat and seafood spoilage of meat and seafood. The authors collected 161 *B. thermosphacta* strains isolated from different foods, spoiled and otherwise, and from a slaughterhouse environment. Using the rpoB gene, a PCR test was thus developed for a fast screening of *B. thermosphacta* isolates.

18.5 Detection of microbial metabolites

Instead of detecting the spoilage organisms many researchers have indirectly targeted the metabolites released by the spoilage organisms. This has enabled them to detect spoilage by a group of bacteria which actually releases the same metabolite, thus making the test more attainable than looking for a specific microbe. The use of solid phase microextraction gas chromatography and mass spectroscopy of dimethyl sulfate ethyloctonate, 2-ethyl 1 hexonate, etc. has recently been attempted by researchers as a method of the detection of spoilage indirectly. However, these methods are not very popular as the spoilage to some extent has to be detected more quickly than the time taken for most of the tests. But many times the results are superior to that of mere detection of the organisms, as not always does the presence of organisms translate into spoilage. Glucose, gluconate, total lactate, D-lactate, ethanol, free amino acids, ammonia, acetone, sulfur, dimethyl sulfide, dimethyl disulfide, hydrogen sulfide, diacetyl, acetoin biogenic amines, diamines, and proteolysis products, such as amides, amines, etc., are the common metabolites looked for (Nychas et al., 2007)

18.6 Modern trends in the spoilage detection of meat

18.6.1 Odor sensors and electronic nose technology

These types of sensors are actually based on human olfaction. The electronic nose (or e-nose) is modeled on the chemical interactions between odor compounds and primary neurons found in the human nasal cavity (Ghasemi-Varnamkhastim et al., 2009). The e-nose technology consists of many electrochemical sensors and pattern classification algorithms which can detect odors, that is, there is some resemblance between the primary neurons of the nasal cavity and the chemical sensors of the e-nose (Hasan et al., 2012). These algorithms can actually analyze a large quantity of chemical data in a short period of time and give an indication about the quality of the meat or any food product. There are many advancements in sensor technology and many types of chemosensors are readily available. Of these products, metal oxide semiconductor (MOS) and conducting organic polymer (CP) will be briefly reviewed in this section (Hasan et al., 2012).

18.6.2 Conducting organic polymers

Conducting polymers are a recent addition in the rapid detection of meat spoilage. They involve a set of active mechanisms which detect odors and convert chemical vapors into electrical signals (Casalinuovo et al., 2006). Furthermore, CP have vital mechanical properties that allow for the construction of food spoilage sensors (Bai and Shi, 2007). The conducting polymer characteristics depend on doping levels, ion size of the dopant, water content, and protonation levels. The use of diagnostic medical reagents and distinguishable chemical memory are the primary materials of industrial sensors. These sensors may be classified into the mode of transduction and application (Gupta et al., 2006). Despite these advantages, the research output involving CP-based e-noses has been relatively limited. As early as 2001, work by Neely and collaborators looked at developing semiconductor-based e-noses to distinguish between alpaca and llama meats in Peru and Bolivia (Neely et al., 2001). With the aid of linear discriminant analysis of data generated, the authors were able to apply their device to the examination of meat and reported some success, as well as problems experienced in analyzing the data relating to sample size (Neely et al., 2001). Since this work, few reports of e-noses based on CPs applied toward meat examination have emerged, with the bulk of the e-nose sensors based on MOS-based modifications (Fletcher et al., 2018).

18.6.3 Metal oxide semiconductor

This technology is in tune with the trending electric noses. The MOS is most commonly used in e-noses (Majchrzak et al., 2018). This is a kind of gas sensor which detects early signs of meat spoilage. For example, Timsorn and coworkers have proposed an e-nose based on eight MOS sensors that was used for the evaluation of chicken meat freshness and bacterial population on chicken meat stored at 4.0°C and 30.0°C for up to 5 days (Timsorn et al., 2016). The authors demonstrated their results to illustrate the classification of chicken meat freshness corresponding to different storage days and temperatures. Furthermore, the e-nose exhibited a good correlation (0.94) of the bacterial population on chicken, suggesting that the developed e-nose system can be used as a rapid and alternative way for the evaluation of bacterial populations on meats and offers several advantages including being fast, portable, low cost, and allowing nondestructive measurement with high relative accuracy (Timsorn et al., 2016). The findings provided valuable points on improving aspects of vacuum packaging, for example, headspace and refrigeration conditions as well as duration. Zhang and collaborators have similarly developed an e-nose based on MOS sensors for characterizing microbial colonies developed during the refrigeration of vacuum-packed Yao meat (Zhang et al., 2018). Their findings provided new insights into packaging methods and areas for improvements. Zou and coworkers assessed the effects of ultrasonic-assisted cooking on the chemical profiles of spiced beef taste and flavor using an e-nose system equipped with 18 MOS sensors (Zou et al., 2018). They found that ultrasonic treatment could significantly increase the content of sodium, some essential amino acid content, and the essential amino acid/nonessential ratios in beef. They concluded that the application of ultrasound during cooking has a positive effect on chemical profiles of spiced beef taste and Wang and coworkers evaluated the sensory qualities of the fresh chilled pork (spiked with *Salmonella*) using a portable e-nose containing a detector unit composed of an array of 10 different MOS-type chemical sensors (Wang et al., 2016). The authors found that the e-nose was able to assist in the detection and minimization of odor.

18.6.4 Spectroscopy in advanced forms

Spectroscopy was one of the earlier methods to detect many compounds. But the drawbacks of the earlier methods have been overcome by the development of new technological additions to good old spectroscopy. Many emerging techniques are designed to detect and quantify the spoilage as early as possible (Lohumi et al., 2015). Oto et al. (2013) have investigated the potential of fluorescence spectroscopy for the nondestructive evaluation of adenosine triphosphate (ATP) content and plate count on pork meat surface stored aerobically at 15°C for 3 days. Shirai et al. (2016) have similarly applied excitation−emission matrix spectroscopy in the rapid, nondestructive evaluation of cleanliness in meat processing plants using ATP as an indicator of microbial contamination. The ATP content and plate count were quantified by them. The field of spectroscopy itself is moving toward more advanced methods of spectral imaging, such as surfaced-enhanced Raman spectroscopy and hyperspectral imaging (HIS) (Hagen and Kudenov, 2014; Wu and Sun, 2013). These techniques are still relatively new, but can provide greater precision, accuracy, clarity, and identification of sampled materials (Schlücker, 2014). Through these new techniques, the advancement of computing software and the minimization of technology through branches of science (such as nanoscience), spectroscopy machines are predicted to become more readily available to businesses and households in the foreseeable future

18.6.4.1 Fourier transform infrared spectroscopy

Fourier transform infrared spectroscopy (FTIR) has been used by many workers for detection of meat spoilage. In a study conducted on beef the following conclusions were made. The study demonstrated the effectiveness of the detection approach based on FTIR spectroscopy which in combination with an appropriate machine learning strategy could become an effective tool for monitoring meat spoilage during aerobic storage at various temperatures. The collected spectra could be considered as biochemical "signatures" containing information for the discrimination of meat samples in quality classes corresponding to different spoilage levels, whereas at the same time they could be used to predict satisfactorily the microbial load directly from the sample surface. The research however revealed two open problems. The use of any machine learning method cannot be considered as a panacea to problems that include sensorial devices. It is well-known that PLS regression models do have problems in modeling high nonlinear dynamics problems. There is a need to explore further the use of advanced intelligent systems, and this paper has attempted for the first time to associate FTIR spectra with such systems. Research work is in progress to develop algorithms based on fuzzy logic that will generate "virtual" spectral data from limited experimental spectral meat samples.

18.6.4.2 Electrical impedance spectroscopy

Electrical impedance spectroscopy (EIS) is a method to analyze electrical properties of materials and systems by inducing alternating electrical signals at different frequencies into them and measuring the responding signals. A function of impedance according to frequencies is established and further correlated with physical parameters or properties of materials and systems, for the aim of analysis and evaluation. EIS was originally applied in research on electrochemical system, and from the 1920s it began to be used for biological systems. Until now EIS has had an extensive application in biological research. According to the biological objects, the application of EIS can be divided into three aspects, that is, electrical impedance tomography in medical imaging, quality and safety assessment in the food industry, and phytophysiology in agronomy. The challenges of EIS lie in the impact factors of impedance measurement, including electrode polarization, materials and structure of electrodes, and the measurement setup configuration, and also lie in the difficulty in data processing and interpretation for diverse and complex tested meat or fish tissues. These challenges still need to be carefully considered before moving this technology from the laboratories to industrial real-time detection.

18.6.4.3 Near-infrared spectroscopy

Near-infrared (NIR) spectroscopy has shown great potential for the estimation of quality and safety attributes in meat and meat products in recent years, and has been considered as one of the most effective and progressive techniques (Warriss, 2004). It is a potential analytical tool for sensitive and fast analysis with a simplicity in sample preparation, which allows a simultaneous assessment of numerous meat properties. NIR has shown enormous potential to predict quality attributes, such as protein, fat, moisture, ash, myoglobin, pH value, water-holding capacity, color, marbling, tenderness, and safety attributes (freshness, total bacterial count, adulteration) (Alomar et al., 2003). The existing researches indicated that NIR has been successfully applied to the quantitative determination of such attributes in meat with high accuracy and the coefficients of determination (R) of some indicators were up to 0.90 between predicted and reference values (Peng and Wang, 2015).

All of the technologies are fast, nondestructive, and suitable for the development of on-line detecting instruments. However, NIR spectroscopy and HSI demand expensive equipment, and electronic nose technology requires specific environmental conditions for measurement. Compared with other new technologies, EIS shows outstanding advantages of being inexpensive and having operating low requirements.

18.6.5 Use of lasers in detecting food quality

Laser speckle imaging has been introduced to monitor moving particles in optically inhomogeneous media by analyzing time-varying laser speckle patterns (Dunn et al., 2001). Light impinging on turbid media such as biological tissues experiences multiple light scattering, and scattered light produces laser speckle patterns by light interference. Because of the deterministic nature of multiple light scattering, scattered light from a static turbid medium generates a constant laser speckle pattern (Mosk et al., 2012). However, if scatters are spontaneously moving inside a turbid medium, time-varying speckle patterns are produced, from which information of the moving scatters can be retrieved. First, this method is noncontact and noninvasive. Unlike other conventional chemical or molecule methods which inevitably involve invasive and contact procedures, this method is based on the analysis of dynamic laser speckles which can be obtained by simply measuring the reflectance of a laser beam from a sample. Meats sealed with transparent plastic wraps can also be examined with this method because the present method only detects time-varying signals in reflected laser beams and a transparent plastic wrap does not cause time-varying signals. Second, this method can provide rapid assessment. The activity of live bacteria can be identified within a few seconds. Third, the technique is extremely simple and cost-effective. Without precise optical alignment, using a simple instrument consisting of a coherent laser source and an image sensor can achieve the noninvasive assessment of bacterial activity in food. The simplicity of the instrumentation allows significant flexibility in its applications. However, the method is not able to identify different pathogenic bacterial strains (Yoon et al., 2016).

18.6.6 Smartphone-based food diagnostic technologies

Mobile diagnostics is gaining more and more attention in healthcare, environmental monitoring, and agrofood sectors, allowing rapid and on-site analysis for preliminary and meaningful information extraction. Nevertheless, smartphones cannot function alone as laboratory instruments (Rateni et al., 2017). Rather, they need to be augmented by other accessories. Such augmented devices have great potential as mobile diagnostic platforms for food analysis. In recent years, many external sensor modules have been designed and integrated with smartphones to extend their capabilities for extracting more-sophisticated diagnostic information. These portable, low-cost devices have the potential to run routine tests, which are currently performed by trained personnel using laboratory instrumentation, rapidly and on-site, thanks to the global widespread use of smartphones. Basically most of these devices are not using any new technology but are combining existing technology with the ability of smartphones, thus leaving the core ability of perceiving, processing, and analyzing of data with the smartphone (Rateni et al., 2017). This reduces the cost of equipment significantly. Thus now we have companies already developing biosensors, fluorescence imaging, calorimetric readers, electroanalytical readers, and smartphone spectroscopy. In future these technologies are going to revolutionize the field of food testing.

18.6.7 Chemical and biological sensors for food-quality monitoring

When meat, fish, or poultry undergo degradation, different spoilage indicators can be found indicating lipid decay, protein breakdown, and ATP decay. The speed of degradation is dependent on the type of product, storage temperature, feeding habits, and harvesting methods. Traditional methods to assess freshness rely on human senses; although they are essential, they provide no quantitative data of spoiled food. Methods that can quantitatively measure markers of degradation through chemical or biological reactions can provide the means to more precisely assess the status and quality of food. In fish products, for example, one of the main freshness indicators is hypoxanthine, which is produced by the metabolic degradation of ATP (Ashie et al., 1996). An electrode modified with gold nanoparticles was reported and tested on chicken and meat samples, with a limit of detection of 2.2×10^{-7} M hypoxanthine (Agüí et al., 2006). The use of nanomaterials in both sensing and packaging technologies is also growing and has demonstrated promising potential. However, toxicity concerns and safety need to be evaluated for promoting the further use of nanotechnologies in the food industry. Moreover, ensuring connectivity of sensing devices and the development of wireless, independently operated sensors are needed to facilitate the rapid monitoring of a large number of samples and to provide real-time status during shipping and long-term storage (Mustafa et al., 2017).

18.7 Conclusion

Current methods for the rapid detection of spoilage in meats are inadequate and all have the same recurring theme in that they are time-consuming, labor-intensive, and therefore give retrospective information. The processes involved in the microbial spoilage of meats are well-established and for three decades microbial metabolites have been put forward as potential indicators of organoleptic spoilage and remaining shelf life. Despite this knowledge, the ability to correlate biochemical change with microbial biomass is a complex problem. Achieving multifunctionality through the combined use of various detection methods is another growing field of research. The integration of these technologies is a challenge, but the rapid development of new technologies and their functional use is expected to positively impact the development of new strategies. Inspection authorities need reliable methods for quality control purposes. Retailers and wholesalers demand these valid methods to ensure the freshness and safety of their products and in case of disputes between buyers and sellers. A reliable indication of the safety and quality status of meat in retail and until consumed is always desirable. It is therefore crucial to have valid methods to monitor freshness and safety to be able to ensure what the quality is, regardless of whose perspective we take, that is, that of the consumer, the industry, the inspection authority, or the scientist. There is immense scope for research in this field and the opportunities are unlimited in developing a reliable, flexible, and cost-effective method for the detection of muscle food deterioration.

References

Adams, M.R., Moss, M.O., 2000. Food Microbiology. The Royal Society of Chemistry, Cambridge.

Agüí, L., Manso, J., Yáñez-Sedeño, P., Pingarrón, J.M., 2006. Amperometric biosensor for hypoxanthine based on immobilized xanthine oxidase on nanocrystal gold-carbon paste electrodes. Sens. Actuators B Chem. 113, 272–280.

Alexandre, M., Prado, V., Ulloa, M.T., Arellano, C., Rios, M., 2001. Detection of enterohemorrhagic *Escherichia coli* in meat foods using DNA probes, enzyme-linked immunosorbent assay and polymerase chain reaction. J. Vet. Med. (B) 48, 321–330.

Alomar, D., Gallo, C., Castaneda, M., Fuchslocher, R., 2003. Chemical and discriminant analysis of bovine meat by near infrared reflectance spectroscopy (NIRS). Meat Sci. 63, 441–450.

Alomirah, H.F., Alli, I., Gibbs, B.F., Konishi, Y., 1998. Identification of proteolytic products as indicators of quality in ground and whole meat. J. Food Quality 21, 299–316.

Ashie, I., Smith, J., Simpson, B., Haard, N.F., 1996. Spoilage and shelf life extension of fresh fish and shellfish. Crit. Rev. Food Sci. Nutr. 36, 87–121.

Bai, H., Shi, G., 2007. Gas sensors based on conducting polymers. Sensors 7, 267–307.

Barbour, W.M., Tice, G., 1997. Genetic and immunologic techniques for detecting foodborne pathogens and toxins. In: Doyle, M.C., Beuchat, L.R., Montville, T.J. (Eds.), Food Microbiology: Fundamentals and Frontiers. ASM Press, Washington, DC.

Betts, R., 1999. Analytical microbiology—into the next millenium. New Food 2, 9–16.

Borch, E., Berg, H., Holst, O., 1991. Heterolactic fermentation by a homofermentative *Lactobacillus* sp during glucose limitation in anaerobic continuous culture with complete cell recycle. J. Appl. Bacteriol. 71, 265–269.

Braun, P., Fehlhaber, K., Klug, C., Kopp, K., 1999. Investigations into the activity of enzymes produced by spoilage-causing bac-teria: a possible basis for improved shelf-life estimation. Food Microbiol. 16, 531–540.

Casalinuovo, I.A., Di Pierro, D., Coletta, M., Di Francesco, P., 2006. Application of electronic noses for disease diagnosis and food spoilage detection. Sensors 6, 1428–1439.

Cerveny, J., Meyer, J.D., Hall, P.A., 2009. Microbiological spoilage of meat and poultry products. In: Sperber, W.H., Doyle, M.P. (Eds.), Compendium of the Microbiological Spoilage, of Foods and Beverages. Food Microbiology and Food Safety. Springer Science and Business Media, New York, pp. 69–868.

Champiat, D., Matas, N., Monfort, B., Fraass, H., 2001. Applications of biochemiluminescence to HACCP. Luminescence 16, 193–198.

Cloak, O.M., Duy, G., Sheridan, J.J., Blair, I.S., McDowell, D.A., 2001. A survey on the incidence of *Campylobacter* spp. and the development of a surface adhesion polymerase chain reaction (SA–PCR) assay for the detection of *Campylobacter jejuni* in retail meat products. Food Microbiol. 18, 287–298.

Costerson, J.W., Irvin, R.T., Cheng, K.J., 1981. The bacterial glycocalyx in nature and disease. Annu. Rev. Microbiol. 35, 299–324.

Dave, D., Ghaly, A.E., 2011. Meat spoilage mechanisms and preservation techniques: a critical review. Am. J. Agric. Biol. Sci. 6 (4), 486–510.

Don, S., Xavier, K.A.M., Devi, S.T., Nayak, B.B., Kannuchamy, N., 2018. Identification of potential spoilage bacteria in farmed shrimp (*Litopenaeus vannamei*): application of relative rate of spoilage models in shelf life-prediction. LWT Food Sci. Technol. 97, 295–301.

Drosinos, E.H., Board, R.G., 1994. Metabolic activities of pseudomonads in batch cultures of extract of minced lamb. J. Appl. Bacteriol. 77, 613–620.

D'Souza, S.F., 2001. Microbial biosensors. Biosens. Bioelectron. 16, 337–353.

Dunn, A.K., Bolay, T., Moskowitz, M.A., Boas, D.A., 2001. Dynamic imaging of cerebral blood flow using laser speckle. J. Cerebr. Blood Flow Metab. 21, 195–201.

Ellis, D.I., Broadhurst, D., Kell, D.B., Rowland, J.J., Goodacre, R., 2002. Rapid and quantitative detection of the microbial spoilage of meat by fourier transform infrared spectroscopy and machine learning. Appl. Environ. Microbiol. 68 (6), 2822–2828.

Fletcher, B., Mullane, K., Platts, P., Todd, E., Power, A., Roberts, J., et al., 2018. Advances in meat spoilage detection: a short focus on rapid methods and technologies. J. Food 16 (1), 1037–1044.

Ghasemi-Varnamkhastim, M., Mohtasebi, S.S., Siadat, M., Balasubramanian, S., 2009. Meat quality assessment by electronic nose (Machine olfactory technology). Sensors 9, 6058–6083.

Gribble, A., Brightwell, G., 2013. Spoilage characteristics of *Brochothrix thermosphacta* and *campestris* in chilled vacuum packaged lamb, and their detection and identification by real time PCR. Meat Sci. 94 (3), 361–368.

Gupta, N., Sharma, S., Mir, I.A., Kumar, D., 2006. Advances in sensors based on conducting polymers. J. Sci. Ind. Res. 65, 549–557.

Hagen, N.K., Kudenov, M.W., 2014. Review of snapshot spectral imaging technologies. Opt. Eng. 52 (9), 1–23.

Hasan, N.U., Ejaz, N., Ejaz, W., Kim, H.S., 2012. Meat and fish freshness inspection system based on odor sensing. Sensors 12, 15542–15557.

Jay, J.M., Loessner, M.J., Golden, D.A., 2005. Modern Food Microbiology, seventh ed. Springer Science and Business Media, New York0387231803, pp. 63–101, ISBN.

Jabbar, H., Joishy, K.N., 1999. Rapid detection of *Pseudomonas* in seafoods using protease indicator. J. Food Sci. 64, 547–549.

Jackson, T.C., Acu, G.R., Dickson, J.S., 1997. Meat, poultry, and seafood. In: Doyle, M.P., Beuchat, L.R., Montville, T.J. (Eds.), Food Microbiology: Fundamentals and Frontiers. ASM Press, Washington, DC, pp. 83–100.

Lohumi, S., Lee, S., Lee, H., Cho, B.K., 2015. A review of vibrational spectroscopic techniques for the detection of food authenticity and adulteration. Trends Food Sci. Technol. 1 (46), 85–98.

Majchrzak, T., Wojnowski, W., Dymerski, T., Gębicki, J., Namieśnik, J., 2018. Electronic noses in classification and quality control of edible oils: a review. Food Chem. 246, 192–201.

Mandal, P.K., Biswas, A.K., Choi, K., Pal, U.K., 2011. Methods for rapid detection of food-borne pathogens: an overview. Am. J. Food Technol. 6 (2), 87–102.

Mandal, P.K., Pal, U.K., Kasthuri, S., 2018. Advances in rapid detection of microbial spoilage of muscle foods. In: Compendium of 8th Conference of Indian Meat Science Association and International Symposium on "Technological Innovations in Muscle Food Processing for Nutritional Security, Quality and Safety" Kolkata from 22 - 24 November, 2018, pp. 263–266.

Mullis, K.B., Faloona, F.A., 1987. Specific synthesis of DNA in vitro via a polymerase–catalysed chain reaction. Methods Enzymol. 155, 335–350.

Mosk, A.P., Lagendijk, A., Lerosey, G., Fink, M., 2012. Controlling waves in space and time for imaging and focusing in complex media. Nat. Photonics 6, 283–292.

Mustafa, F., Hassan, R.Y., Andreescu, S., 2017. Multifunctional nanotechnology-enabled sensors for rapid capture and detection of pathogens. Sensors 17, 2121.

Neely, K., Taylor, C., Prosser, O., Hamlyn, P.F., 2001. Assessment of cooked alpaca and llama meats from the statistical analysis of data collected using an 'electronic nose'. Meat Sci. 58 (1), 53–58.

Nychas, G.J.E., Drosinos, E.H., Board, R.G., 1998. Chemical changes in stored meat. In: Davies, A., Board, R. (Eds.), The Microbiology of Meat and Poultry. Blackie Academic & Professional, London, pp. 288–320.

Nychas, G.J.E., Marshall, D., Sofos, J., 2007. Food microbiology fundamentals and frontiers. In: Doyle, M.P., Beuchat, L.R., Montville, T.J. (Eds.), Meat, Poultry and Seafood. ASM Press.

Oto, N., Oshita, S., Makino, Y., Kawagoe, Y., 2013. Non-destructive evaluation of ATP content and plate count on pork meat surface by fluorescence spectroscopy. Meat Sci. 93 (3), 579–585.

Peng, Y., Wang, W., 2015 Application of Near-infrared Spectroscopy for Assessing Meat Quality and Safety. https://doi.org/10.5772/58912.

Pyle, B.H., Broadaway, S.C., McFeters, G.A., 1999. Sensitive detection of *Escherichia coli* O157: H7 in food and water by immunomagnetic separation and solid-phase laser cytometry. Appl. Environ. Microbiol. 65, 1966–1972.

Rateni, G., Dario, P., Cavallo, F., 2017. Smartphone-based food diagnostic technologies: a review. Sensors 17, 1453. Available from: https://doi.org/10.3390/s17061453.

Rattanasomboon, N., Bellara, S.R., Harding, C.L., Fryer, P.J., Thomas, C.R., Al-Rubeai, M., et al., 1999. Growth and enumeration of the meat spoilage bacterium *Brochothrix thermosphacta*. Int. J. Food Microbiol. 51, 145–158.

Reid, R., Burgess, C.M., McCabe, E., Fanning, S., Whyte, P., Kerry, J., et al., 2017. Real-time PCR methods for the detection of blown pack spoilage causing *Clostridium* species; *C. estertheticum*, *C. gasigenes* and *C. ruminantium*. Meat Sci. 133, 56–60.

Schlücker, S., 2014. Surface-enhanced Raman spectroscopy: concepts and chemical applications. Angew. Chem. 53 (9), 4756–4795.

Seymour, I.J., Cole, M.B., Coote, P.J., 1994. A substrate mediated assay of bacterial proton efflux/influx to predict the degree of spoilage of beef mince stored at chill temperatures. J. Appl. Bacteriol. 76, 608–615.

Shirai, H., Oshita, S., Makino, Y., 2016. Detection of fluorescence signals from ATP in the second derivative excitation–emission matrix of a pork meat surface for cleanliness evaluation. J. Food Eng. 168, 173–179.

Stanbridge, L.H., Davies, A.R., 1998. The microbiology of chill-stored meat. In: Davies, A., Board, R. (Eds.), The Microbiology of Meat and Poultry. Blackie Academic & Professional, London, pp. 174–219.

Timsorn, K., Thoopboochagorn, T., Lertwattanasakul, N., Wongchoosuk, C., 2016. Evaluation of bacterial population on chicken meats using a briefcase electronic nose. Biosyst. Eng. 151, 116–125.

Wang, S., He, Y., Wang, Y., Tao, N., Wu, X., Wang, X., et al., 2016. Comparison of flavour qualities of three sourced *Eriocheir sinensis*. Food Chem. 200, 24–31.

Warneck, H.W., 2001. Method for detection of salmonella species within 18 hours. Fleischwirtschaft 81, 79–81.

Warriss, P.D., 2004. Meat Science. An Introductory Text. CABI Publishing, Wallingford, Oxon.

Wu, D., Sun, D.W., 2013. Advanced applications of hyperspectral imaging technology for food quality and safety analysis and assessment: a review-Part II: Applications. Innov. Food Sci. Emerg. Technol. 19, 15–28.

Yoon, J., Lee, K., Park, Y., 2016. A simple and rapid method for detecting living microorganisms in food using laser speckle decorrelation. https://www.researchgate.net/publication/301836645.

Yost, C.K., Nattress, F.M., 2000. The use of multiplex PCR reactions to characterize populations of lactic acid bacteria associated with meat spoilage. Lett. Appl. Microbiol. 31, 129–133.

Zhang, Y., Yao, Y., Gao, L., Wang, Z., Xu, B., 2018. Characterization of a microbial community developing during refrigerated storage of vacuum packed Yao meat, a Chinese traditional food. LWT Food Sci. Technol. 90, 562–569.

Zou, Y., Kang, D., Liu, R., Qi, J., Zhou, G., Zhang, W., 2018. Effects of ultrasonic assisted cooking on the chemical profiles of taste and flavor of spiced beef. Ultrason. Sonochem. 46, 36–45.

Further reading

Ellis, D.I., Goodacre, R., 2001. Rapid and quantitative detection of the microbial spoilage of muscle foods: current status and future trends. Trends Food Sci. Technol. 12, 414–424.

Illikoud, N., Rossero, A., Chauvet, R., Courcoux, P., Pilet, M.-F., Charrier, T., et al., 2018. Genotypic and phenotypic characterization of the food spoilage bacterium *Brochothrix thermosphacta*. Food Microbiol. 2018, 1–10.

Miyasaki, T., Hamaguchi, M., Yokoyama, S., 2011. Change of volatile compounds in fresh fish meat during ice storage. J. Food Sci. 76 (9), C1319–C1325.

Nychas, G.J.E., Skandamis, P.N., Tassou, C.C., Koutsoumanis, K.P., 2008. Meat spoilage during distribution. Meat Sci. 78, 77–89.

Zhao, X., Zhuang, H., Yoon, S., Dong, Y., Wang, W., Zhao, W., 2017. Electrical impedance spectroscopy for quality assessment of meat and fish: a review on basic principles, measurement methods, and recent advances. Hindawi J. Food Quality 2017, ID 6370739.

SECTION 11

Proteomic and genomic tools in meat quality evaluation

… CHAPTER 19

Application of proteomic tools in meat quality evaluation

M.N. Nair and C. Zhai

Department of Animal Sciences, Colorado State University, Fort Collins, CO, United States

Chapter Outline

19.1 Introduction 353
19.2 Proteomics 354
    19.2.1 Gel-based approaches 355
    19.2.2 Gel-free approaches 357
19.3 Proteomic approaches to meat quality 358
    19.3.1 Meat color 359
    19.3.2 Tenderness 361
    19.3.3 Water-holding capacity 363
19.4 Conclusion 365
References 365

19.1 Introduction

Meat quality attributes such as color, tenderness, and juiciness are highly variable and are often dictated by the functionality of proteins present in meat. Quality defects and inconsistencies in quality are major challenges for the meat industry. Although extensively researched, some of the fundamental mechanisms contributing to these quality differences are not completely understood, and developing a clear understanding of these processes at a biochemical and molecular level is critical to improving meat quality consistently.

Proteins constitute around 20% of meat and serve as a connection between the genetic profile of the animals and meat quality. In general, meat proteins could be categorized into three broad categories based on their solubility, namely (1) sarcoplasmic, (2) myofibrillar, and (3) stromal (connective tissue) proteins. Among these groups, sarcoplasmic proteins are located in the sarcoplasm (cytoplasm) of the muscle cells and constitute about 25%−30% of the total muscle proteins. These proteins are water-soluble and are generally globular proteins. There are hundreds of proteins present in the sarcoplasm, and their functionalities can influence the meat quality. For example, the glycolytic enzymes present in the sarcoplasm can influence postmortem metabolism, which in turn can influence meat quality. Myoglobin is another sarcoplasmic protein that critically influences meat quality, especially meat color. Although the physiological function of myoglobin is to store oxygen for muscle metabolism, the redox chemistry of myoglobin postmortem determines meat color.

Myofibrillar proteins constitute around 50%−60% of total meat proteins and are salt soluble. They are largely responsible for textural and structural properties of meat and meat products. Myosin (thick filament) and actin (thin filament) are two major myofibrillar proteins that play a critical role in the process of muscle contraction and postmortem rigor. These proteins dictate functionalities such as gelation and emulsification during meat processing. Stromal proteins are typically insoluble in water and require a strong acid/alkali solution for solubilizing them (e.g., collagen, elastin, reticulin). They are part of the extracellular matrix that provides structure and organization to the muscles. However, stromal proteins can critically influence meat quality parameters, particularly meat tenderness. Among the stromal proteins, collagen has been studied extensively in relation to meat tenderness. The large number and varied nature of the meat proteins make it challenging to examine the interactions between them using traditional meat quality analysis approaches. Moreover, the protein profile could be very dynamic in comparison to the static genome. Hence the application of novel tools such as proteomics has gathered a lot of interest from the meat scientific community in the past decade. This chapter will focus on some of the commonly used proteomic tools and their applications for meat quality evaluation.

19.2 Proteomics

The term "proteome" is defined as the protein complement of the genome comprising the total amount of proteins expressed at a certain time point in an animal (Wilkins et al., 1996). Proteomics is the systematic analyses of the proteome, which could include protein identification, quantification, and functional characterization (Liebler, 2002). When compared to the static genome, the proteome is dynamic and influenced by various factors related to protein synthesis or degradation. In eukaryotes, the alternative splicing of genes and various posttranslational modifications (PTMs) that a protein undergoes, like acetylation, phosphorylation, and glycosylation, result in the proteome being larger than the genome.

From a meat science perspective the proteome can be considered as the molecular linkage between the genome and meat quality traits (Hollung et al., 2007). Although cutting-edge tools in mass spectrometry and proteomics have been extensively utilized in many disciplines within agricultural sciences, the application of proteomics in meat research grew significantly only in the last decade as it enables researchers to investigate postmortem protein changes at a molecular level. Chromatography-based techniques such as ion-exchange chromatography, size exclusion chromatography, and affinity chromatography are conventionally used in the purification of proteins, whereas techniques such as enzyme-linked immunosorbent assay and Western blotting are used for the analysis of selective proteins (not discussed in this chapter). These techniques are restricted to the analysis of few individual proteins and might be poor in defining protein expression levels. However, proteomics enables the characterization of hundreds of proteins in a single analysis by identifying cleavage sites and protein modifications such as protein oxidation and phosphorylation. Since proteins are effectors of biological function whose levels are not only dependent on corresponding mRNA levels but also on host translational control and regulation, proteomics could be considered as the most relevant data set to characterize a biological system (Huang and Lametsch, 2013).

The proteomic approaches can be broadly classified into two categories: (1) gel-based approaches, and (2) gel-free approaches. Each of these approaches will be discussed briefly in this chapter, whereas Baggerman et al. (2005) have comprehensively reviewed the topic.

19.2.1 Gel-based approaches

Sodium dodecyl sulfate-polyacrylamide gel electrophoresis (SDS-PAGE), often referred to simply as one-dimensional electrophoresis (1-DE), is a technique for the separation of proteins according to their size, which enables approximation of the molecular weight. Proteins are neutrally charged at their isoelectric point (pI). When an electric field is applied in a medium having pH different from their pI, the proteins migrate on the gels (usually on polyacrylamide gels) and form bands, which are then visualized through staining techniques. The rate/velocity of migration depends on the ratio between the protein's charge and mass. However, the addition of sodium dodecyl sulfate denatures the proteins and therefore separates them absolutely according to molecular weight (Aslam et al., 2017). Although 1-DE is very useful for the separation of proteins at a higher level, the resolution of the gels is very low, with the possibility of multiple proteins with similar molecular weight appearing in a single protein band.

The two-dimensional PAGE often referred to simply as two-dimensional electrophoresis (2-DE), enables the separation of proteins by their mass and charge. The 2-DE is capable of resolving ~5000 different proteins successively, depending on the size of the gel. In this technique, proteins are separated by charge (pI) by isoelectric focusing in the first dimension, and then according to their molecular weight in the second dimension. Since these two parameters are unrelated, an almost uniform distribution of protein spots can be obtained across a two-dimensional gel, and the resulting map of protein spots can be considered as the protein fingerprint of that sample. As in SDS-PAGE gels, the protein spots are visualized through staining methods. The most popular protein staining method is using Coomassie blue, as it is inexpensive, easy to use, and has a wide linear range that makes relative quantification easy. Moreover, Coomassie staining is compatible with downstream analysis by mass spectrometry. However, due to its medium sensitivity, not all proteins can be visualized by Coomassie blue. Another option is to use silver staining which has a 20−50 times higher detection limit (highest sensitivity). The disadvantage of silver staining is that it is not very compatible with mass spectrometry and extensive destaining methods have to be used before the identification of the protein spot. Moreover, due to its limited dynamic range, the reproducibility of the spot intensities is low while using silver staining. This problem can be somewhat circumvented by differentially labeling different protein samples and separating them on the same gel.

One such method is the two-dimensional differential gel electrophoresis (2D-DIGE), which is performed by covalently tagging two protein samples (control and treatment) with two distinct N-hydroxysuccinamide derivatives with different fluorescent emission spectra, but identical masses and electrophoretic mobility. The labeled extracts are pooled and separated by 2-DE. An imager then scans the gel at the two Cy-dye emission wavelengths, and the image analysis software can identify spots of different intensity by superimposing the images. The major advantage of the 2D-DIGE is that it allows the comparison of two protein samples on the same gel. However, in larger studies where multiple conditions have to be compared, there could be issues with gel-to-gel variation. A novel approach of using a third dye was suggested by Alban et al. (2003) to solve this problem. These researchers used a third dye to label a third sample that will function as an internal standard. The standard sample comprises equal amounts of each sample to be compared. This standard is then mixed with two of the protein samples that have to be compared and subjected to 2-DE, which will enable the relative quantification easier.

However, 2-DE gels cannot visualize all proteins in a complex sample due to different technical limitations. A typical 2-DE gel can visualize only 30%−50% of the entire proteome (depending on the type of tissue), although some prefractionation methods can be used to overcome this drawback to some extent. Especially, proteins present in extremely low concentrations or proteins that cannot be separated on 2-DE gels due to their physicochemical properties (molecular weight, pI, hydrophobicity) will not be typically detected in 2-DE gels. Proteins with high (>150 kDa) and low (<10 kDa) molecular weight, and proteins with extreme pIs are usually outside the detection limit in a standard 2-DE. Additionally, the hydrophobic proteins are not soluble in the typical buffers used for sample preparation or could precipitate during the electrophoretic process (Lescuyer et al., 2004). This means that the membrane proteins and small proteins (peptides) cannot be examined in most proteomic studies. Strong detergents that will disrupt the double layer of phospholipids and release the embedded hydrophobic protein can be used for extraction of the membrane proteins. However, these detergents are not compatible with the first dimension separation (isoelectric focusing) in 2-DE. Briefly, the 2-DE process is time-consuming, labor-intensive, and requires significant technical expertise to generate reproducible gels (Rabilloud, 2002).

To overcome some of these challenges, several gel-free high-throughput technologies for proteome analysis have been developed in recent years.

19.2.2 Gel-free approaches

Instead of using gel-based approaches, multidimensional capillary liquid chromatography (LC) coupled to tandem mass spectrometry (MS/MS) can be used to separate and identify the peptides obtained from the enzymatic digest of an entire protein extract. This bottom-up (shotgun) proteomics technique (multidimensional protein identification technology; MudPIT) allows analysis of hydrophobic proteins as well as peptides. Basically, complex protein mixtures are digested to peptides using proteolytic enzymes, fractionated using chromatographic columns according to different chemical properties, and subsequently analyzed by MS/MS resulting in protein identification. Several drawbacks associated with 2-DE such as underrepresentation of extreme acid/basic proteins and the poor sensitivity for lowly expressed proteins can be avoided through gel-free approaches. Moreover, the MudPIT method simplifies sample handling, avoids sample loss in the gel matrix and increases throughput and data acquisition (Bantscheff et al., 2007; Lewis et al., 2012). However, it is important to realize that proteins are not examined in their intact state, but instead as peptides obtained through proteolytic cleavage. Since it is easier to separate peptides by LC compared to proteins, a peptide-based proteomic analysis can be performed much faster and more cheaply than a gel-based analysis. Although the MudPIT technology is fast and sensitive with good reproducibility, it lacks the ability to provide quantification.

In gel-free proteomic approaches, there are two basic possibilities of quantification: (1) a relative quantification of proteins in compared samples (e.g., control vs treatment) or (2) an absolute quantification (Yates et al., 2009). One of the most popular methods for relative quantification is stable isotope labeling of proteins in samples before analysis. Proteins are labeled with heavy isotopologues of C, H, N, or O through chemical derivatization processes before mass spectrometric (MS) analysis. Isotope-coded affinity tags (ICATs), dimethyl labeling, and isobaric mass tags are some of the common labeling techniques for protein quantification. Except for isobaric mass tags, stable isotope derivatization methods introduce a small mass difference to identical peptides from two or more samples which can be distinguished in the MS1 spectrum. The relative abundance ratios of peptides are measured by comparing heavy/light peptide pairs, and then protein levels are inferred from a statistical evaluation of the peptide ratios.

The ICAT reagents generally comprise an affinity tag for the isolation of labeled peptides, an isotopically coded linker, and a reactive group (Shiio and Aebersold, 2006). In an ICAT experiment, either light or heavy ICAT reagents are used to label protein samples. The mixtures of labeled proteins are then digested by trypsin and separated through a multistep chromatographic separation, which can be then identified with MS/MS. Integrated LC peak areas of the heavy and light versions of the ICAT-labeled peptides can be then used to infer the relative quantities of the peptide. Stable isotopic labeling with amino acids in cell culture (SILAC) is another gel-free mass spectrometry-based approach for quantitative proteomics. The whole proteome of different cells grown in cell culture are labeled with light or heavy forms of amino acids (metabolic labeling) and are differentiated through mass spectrometry. The SILAC was primarily developed as a technique to study the regulation of gene expression, cell signaling, and PTMs, but is not widely used for meat science applications.

Isobaric tag for relative and absolute quantitation (iTRAQ) is also a gel-free multiplex protein labeling technique for protein quantification based on tandem mass spectrometry. In iTRAQ, each sample is derivatized with a different isotopic variant of an isobaric mass tag, and then the samples are pooled and analyzed simultaneously in the mass spectrometer. The N-terminus and side chain amine groups of proteins are usually labeled and are fractionated through LC columns before mass spectrometric analysis. The iTRAQ method has several advantages such as the ability to multiplex several samples, easier quantification, simplified analysis, and increased analytical precision and accuracy (Aggarwal et al., 2006; Zieske, 2006; Lund et al., 2007). Since the tags are isobaric, the differentially labeled peptides appear as a single composite peak at the same m/z value in the MS1 scan. When the iTRAQ tagged peptides are subjected to MS/MS it generates two types of product ions: (1) reporter ion peaks and (2) peptide fragment ion peaks. The reporter ions provide relative quantitative information of proteins, whereas the original peptide fragments are used to obtain the identity of the proteins. The quantification is accomplished by directly correlating the relative intensity of reporter ions to that of the peptide selected for MS/MS fragmentation. Since every tryptic peptide can be labeled in an isobaric labeling method, more than one peptide representing the same protein could be identified, which in turn increases the confidence in both the identification and quantification of the protein (Rauniyar and Yates, 2014). Additionally, techniques such as Edman degradation can be used to determine the amino acid sequence of a particular protein (Smith, 2001). X-ray crystallography and nuclear magnetic resonance spectroscopy are also major high-throughput techniques that provide three-dimensional structure of a protein that might be helpful to understand its biological function (Smyth and Martin, 2000).

19.3 Proteomic approaches to meat quality

Proteomic tools can be applied to investigate preharvest as well as postharvest aspects of meat production (Bendixen, 2005; Mullen et al., 2006). Preharvest applications explain the biochemistry of food animal growth (Doherty et al., 2004) and muscle biology (Okumura et al., 2005), whereas postharvest aspects primarily focus on the fundamental aspects of meat quality, such as color (Nair et al., 2017), tenderness (Picard and Gagaoua, 2017), and water-holding capacity (WHC; Di Luca et al., 2011).

19.3.1 Meat color

Meat color is an important quality attribute that critically influences consumer purchase decisions. As mentioned earlier in this chapter, sarcoplasmic proteins in meat play a critical role in meat color, especially through the interactions with myoglobin. The changes in the skeletal muscle proteome continue during the postmortem period (Hollung et al., 2007) and can critically influence meat color (Nair et al., 2018a,b). Earlier meat color research using proteomic tools focused on the interaction between myoglobin and small biomolecules, especially the effect of lipid oxidation products on myoglobin redox chemistry using 4-hydroxy-2-nonenal (HNE) as a model aldehyde. This research demonstrated that the covalent modification of histidine residues (via Michael addition) in myoglobins of horse (Faustman et al., 1999), pork (Lee et al., 2003), beef (Alderton et al., 2003), emu (Nair et al., 2014), and ostrich (Nair et al., 2014) by reactive aldehydes is responsible for lipid oxidation-induced meat discoloration. Further, Yin et al. (2011) compared lipid oxidation-induced oxidation in various livestock and poultry myoglobins. These authors reported that the effect of HNE on myoglobin oxidation was correlated with a number of histidine residues in myoglobins, with a greater oxidation rate observed in myoglobins containing a greater number of histidine residues.

Suman et al. (2006) reported mono- and diadducts between HNE and beef myoglobin, whereas only monoadducts were present in pork myoglobin at typical meat conditions (pH 5.6, 4°C). While tandem mass spectrometry revealed four histidine adduction sites (36, 81, 88, and 152) in beef myoglobin, only two histidines (24 and 36) were found to be adducted in pork myoglobin, which indicated that the effect of lipid oxidation on myoglobin redox stability and meat color are species-specific. Further studies (Suman et al., 2007) revealed that histidine 36 was preferentially adducted in pork myoglobin, whereas histidine 81 and 88 were the major sites of early HNE adduction in beef myoglobin. These authors also concluded that the preferential adduction of HNE at proximal histidine (93) observed exclusively in beef myoglobin was responsible for increased lipid oxidation-induced oxidation in beef myoglobin compared to pork myoglobin. These studies (Suman et al., 2006, 2007) also explained why vitamin E is effective in stabilizing color in beef, but not in pork.

Proteomic tools are also used to examine the fundamental basis of muscle-specificity in meat color. Joseph et al. (2012) compared the sarcoplasmic proteome of color-stable longissimus lumborum and color-labile psoas major using 2-DE and tandem mass spectrometry and reported differential abundance of several proteins, including metabolic enzymes, antioxidant proteins, and chaperones. Further, Wu et al. (2015, 2016) reported differentially abundant sarcoplasmic proteins in longissimus lumborum, psoas major, and semitendinosus muscle from Luxi yellow cattle during postmortem storage for 0, 5, 10, and 15 days, indicating that the variation in the sarcoplasmic proteins contributes to the muscle-specificity in meat color. Clerens et al. (2016) performed proteomic and peptidomic analysis of four muscles (semitendinosus, longissimus thoraciset lumborum, psoas major, and infraspinatus) from New Zealand-raised Angus steers. Although the muscles exhibited similar 2-DE profile, there was significant intensity difference between many proteins, including hemoglobin subunit beta, carbonic anhydrase 3, triosephosphate isomerase, phosphoglycerate mutase 2, serum albumin, and β-enolase. Nair et al. (2018a,b) performed comprehensive proteome analysis of differentially color-stable beef muscles, longissimus lumborum, psoas major, semitendinosus, and semimembranosus during postmortem aging and reported significant changes in proteins associated with glycolysis and energy metabolism in relation to meat color. Yu et al. (2017) utilized label-free mass spectrometry to characterize the effect of postmortem storage time (0, 4, and 9 days) on the proteome changes of semitendinosus muscle in Holstein cattle, and correlations between differentially abundant proteins and meat color traits. A total of 118 proteins with significant changes (fold change >1.5; $P < .05$) were identified by comparisons of day 4 versus day 0, day 9 versus day 0, and day 9 versus day 4. Bioinformatics analyses revealed that most of these proteins were involved in glycolysis and energy metabolism, electron-transfer processes, and antioxidative function, which implied an underlying connection between meat discoloration and these biological processes.

Beef semimembranosus is a large muscle in beef hindquarter that exhibits intramuscular differences in color stability and could be separated into the color-stable outside (OSM) and color-labile inside (ISM) regions. The variations in temperature decline and pH drop during carcass chilling are considered to be partly responsible for this intramuscular color difference. Nair et al. (2016) investigated the proteome basis of the color difference between OSM and ISM steaks using 2-DE and tandem mass spectrometry which revealed that ISM steaks had a greater abundance of glycolytic enzymes (fructose-bisphosphate aldolase A, phosphoglycerate mutase 2, and β-enolase) than their OSM counterparts. A combination of rapid pH decline (due to possible rapid glycolysis as a result of increased enzyme levels) and the high temperature (due to the location within the carcass) in ISM during the immediate postmortem period could have an adverse effect on myoglobin redox stability (Faustman et al., 2010; Suman and Joseph, 2013), thus compromising the meat color stability.

Li et al. (2018) used TMT labeling in combination with TiO_2 phosphopeptide enrichment to perform a quantitative analysis of protein phosphorylation in ovine longissimus muscles with different color stability. These researchers performed informatics analysis and reported that among the differentially phosphorylated proteins, 27 phosphoproteins were key color-related proteins, including glycolytic enzymes and myoglobin. Sayd et al. (2006) characterized the sarcoplasmic proteome of pale versus dark pork meat (semimembranosus muscle) using 2-DE and tandem mass spectrometry, along with the correlation of protein expression to color-related attributes. These researchers reported an overexpression of oxidative enzymes related to mitochondrial metabolism, hemoglobin, and chaperone/regulatory proteins in darker meat. On the other hand, the pale meat revealed greater expression of glycolytic enzymes and glutathione S-transferase. The authors correlated such differential abundance in porcine meat with faster postmortem metabolism, possibly with accelerated ATP depletion and subsequent pH decline, which could lead to protein denaturation and thereby resulting in discoloration.

19.3.2 Tenderness

Tenderness is an important quality attribute of meat, which is considered to be critical for consumer eating satisfaction and repurchase decisions. Postmortem degradation of several structural proteins has been implicated in the development of meat tenderness. However, the fundamental mechanisms for meat tenderization and biochemical basis of variation in tenderness are yet to be completely understood (Huff-Lonergan et al., 2010). Most proteomic studies on tenderness have been conducted by comparing extreme groups (tender vs tough) using different proteomic methods.

Zapata et al. (2009) used SDS-PAGE and functional proteomics to associate electrophoretic bands from the myofibrillar proteins to meat tenderness to understand the mechanisms of beef tenderness. Six significant electrophoretic bands were identified by electrophoretic and statistical analysis and were sequenced by nano-LC-MS/MS. These authors reported that the shear values were associated with the structural proteins, myosin heavy chains, myosin light chains, actin, desmin, and tubulin or their fragments. Marino et al. (2015) investigated postmortem proteolysis in psoas major, longissimus dorsi, and semitendinosus muscle from Podolian young bulls aged 1, 7, 14, and 21 days using SDS-PAGE, Western blotting, and 2-DE. Throughout postmortem aging, some structural proteins changed in intensity in all the muscles analyzed. The blotting profile highlighted that desmin and troponin-T bands were affected by both muscle and aging effects. Moreover, postmortem aging of semitendinosus muscle did not result in the same improvement in tenderness observed in longissimus and psoas muscles during aging, which was supported by proteolysis analysis, particularly troponin-T degradation.

Proteome analysis of troponin-T degradation in beef longissimus muscle, utilizing 2-DE demonstrated that several troponin-T isoforms were fragmented by 14 days postmortem (Muroya et al., 2007a). Further mass spectrometric analyses of peptides revealed that all the isoforms were cleaved exclusively at the glutamic acid-rich amino terminal region, and the troponin-T fragments yielded a conventional 30 kDa band in the gel which could be utilized as biomarkers for monitoring postmortem beef tenderization. Morzel et al. (2008) examined the proteome of longissimus thoracis of Blonde d'Aquitaine beef animals to identify early predictors of tenderness using 2-DE and LC-ESI-MS. Although succinate dehydrogenase was suggested as an excellent candidate protein for predicting initial and overall tenderness, HSP-27 (Heat Shock Protein-27 kDa) and its fragments correlated well with the sensory scores, indicating the possibility of HSP-27-related cellular mechanisms influencing tenderness as well as the suitability of HSP-27 as a potential marker for beef tenderness. Variations in tenderness of longissimus thoracis within Charolais young bulls were examined by Laville et al. (2009) utilizing proteome analyses of tough and tender muscles. Mass spectrometric analyses revealed the presence of a greater quantity of actin fragments and proteins from inner and outer mitochondrial membranes in the tender group on day 0 postmortem. Mitochondrial fragmentation occurs during apoptosis, and the findings of this study suggested a possible role for cell apoptotic process in meat tenderization. Muroya et al. (2007b) and Jia et al. (2009) utilized the 2D-DIGE approach to assess the changes in the myosin light chains of longissimus myofibril proteins during postmortem aging and for identification of protein markers of meat tenderness in longissimus muscle respectively.

iTRAQ and 2-DE were utilized by Bjarnadóttir et al. (2012) to find potential biomarkers for meat tenderness in bovine longissimus thoracis muscle and to compare the two methods. Although the overlap among significantly changed proteins was relatively low between iTRAQ and 2-DE analysis, certain proteins predicted to have the same function were found in both analyses and showed similar changes between the groups, such as structural proteins and proteins related to apoptosis and energy metabolism. The iTRAQ approach was also utilized by Mao et al. (2016) for the identification of the differentially expressed proteins involved in intramuscular fat deposition.

The molecular basis of meat tenderization in pork during the first 72 hours postmortem was investigated by Lametsch et al. (2003) through proteome analyses using 2-DE. The researchers identified 27 proteins with pronounced changes, including fragments of actin, myosin heavy chain, titin, myosin light chain I, myosin light chain II, Cap Z, and cofilin. Statistical analysis revealed a significant correlation between myosin light chain II, and several actin and myosin heavy chain fragments with shear force (objective measurement of tenderness), indicating that postmortem degradation of actin and myosin heavy chain is critical in meat tenderization. Further research by Lametsch et al. (2004) characterized the proteolytic changes in pork myofibrils after incubation with μ-calpain at 4°C for 4 days using electrophoresis and mass spectrometry. In contrast to the previous reports, these authors suggested actin and myosin heavy chain as substrates for μ-calpain. Also, several proteins including desmin, troponin-T, tropomyosin α-1, thioredoxin, and Cap Z were degraded by μ-calpain.

One of the limitations of proteomic approaches for studying meat tenderness is the difficulty of extracting proteins from extracellular matrices that are not solubilized by typical buffers. Hence connective tissue proteins known to be important for meat tenderness are typically not considered in these studies. However, these approaches provide opportunities to study

PTMs such as phosphorylation, glycosylation/glycation, oxidation, and ubiquitination. These PTMs play major roles in the postmortem process in muscle/meat science (Huang et al., 2011; D'Alessandro and Zolla, 2013). For example, phosphorylation is a reversible protein modification that can affect the protein structure and activity of many enzymes in vivo, and hence have a potential role in meat tenderization through regulation of the activities of glycolytic enzymes. Muroya et al. (2007b) reported that myosin light chain (MyLC2) was doubly phosphorylated during rigor mortis in bovine longissimus muscle. Furthermore, the appearance of the doubly phosphorylated protein after 8 hours postmortem correlated with the fast phase of rigor mortis, suggesting its potential influence on muscle shortening.

19.3.3 Water-holding capacity

WHC refers to the ability of meat to hold on to water during the postmortem period and processing. It is not only important for visual and sensory acceptability and economic reasons, but also because of its role in molding muscle structure and the consequent effects on quality (Hughes et al., 2014).

Marcos and Mullen (2014) utilized 2-DE to examine the relationship between pressure-induced changes on individual proteins and quality parameters of bovine longissimus thoracis and reported that the solubilization of myofibrillar proteins and insolubilization of sarcoplasmic proteins due to pressure resulted in paler meat with decreased WHC. These results indicated that the sarcoplasmic proteins play a critical role in determining the WHC of meat. Similarly, other authors have highlighted the role of sarcoplasmic proteins such as aldehyde dehydrogenase, glycerol-3-P-dehydrogenase, protein DJ-1, serotransferrin, β-enolase, creatine kinase M-type, and heat shock protein 70 kDa on meat color (L^*; lightness) and drip loss (Hwang et al., 2004; Sayd et al., 2006; van de Wiel and Zhang, 2007; Kwasiborski et al., 2008).

Zuo et al. (2016) used proteomic tools to identify differentially expressed proteins during postmortem aging of yak longissimus lumborum muscle which was classified into high and low drip loss groups. Heat shock protein, myosin light chain, and triosephosphate isomerase were identified as differentially expressed between the groups. Further research by Zuo et al. (2018) compared proteome profile of longissimus thoracis of yak classified into low cooking loss and high cooking loss (HCL) groups. The results showed that cooking loss could be attributed to structural proteins, metabolic enzymes, stress-related proteins, and transport protein. There was a greater expression in the level of desmin, troponin-T, and L-lactate dehydrogenase in the HCL group.

Di Luca et al. (2011) compared protein abundance of diverse WHC phenotypes in pork using SDS-PAGE across time-points postmortem and identified several significant associations between the protein/fragment band volumes and WHC. Their results indicated that proteins such as HSP-70 could have the potential for inclusion in biomarker panels for the early prediction of meat quality in an industrial setting. Further, Di Luca et al. (2016) used 2-D DIGE and mass spectrometry to investigate the changes in metabolic proteins that occur over 7 days (day 1, 3, and 7) of postmortem aging using centrifugal exudate from pigs with divergent WHC. These researchers used a machine-learning algorithm (L1-regularized logistic regression), to derive a model with the ability to discriminate between high and low drip phenotypes using a subset of 25 proteins with an accuracy of 63%.

Phongpa-Ngan et al. (2011) compared the pectoralis proteome of chicken with different growth rates and WHC within the same genotype. The differentially expressed proteins included creatine kinase, pyruvate kinase, triosephosphate isomerase, ubiquitin, heat shock proteins, as well as several structural and contractile proteins. Many of these proteins were proposed as markers of WHC and growth rate, demonstrating the potential of proteomics for the selection of quality and production traits. Further, Zhang et al. (2019) used label-free quantitative mass spectrometry to understand the mechanisms underlying drip loss and to identify the protein markers associated with WHC of goose meat. They identified 21 differentially abundant proteins between high and low drip loss groups, which generally fell into the structural proteins, metabolic enzymes, antioxidant enzymes, and stress response proteins.

Desai et al. (2016) examined the whole muscle proteome of normal and pale, soft, exudative (PSE) broiler breast meat and identified 15 differentially abundant proteins. Actin alpha, myosin heavy chain, phosphoglycerate kinase, creatine kinase M-type, β-enolase, carbonic anhydrase 2, proteasome subunit alpha, pyruvate kinase, and malate dehydrogenase were overabundant in PSE broiler breast, whereas phosphoglycerate mutase-1, α-enolase, ATP-dependent 6-phosphofructokinase, and fructose 1,6-bisphosphatase were overabundant in normal meat. These results indicated that the overabundance of proteins involved in glycolytic pathways, muscle contraction, proteolysis, ATP regeneration, and energy metabolism in PSE breast could be related to the quality differences between normal and PSE meat.

Woody breast is a quality defect in poultry breast that is characterized by hardened areas and pale ridge-like bulges at both the caudal and cranial regions of the breast, and can be classified as slight, moderate, and severe (Owens, 2016; Tijare et al., 2016). Proteomic tools were utilized to understand the biochemical basis of this quality defect (Cai et al., 2018). Whole muscle proteome analysis using 2-DE reported that eight proteins were differentially expressed between normal and woody breast meat samples, and indicated increased oxidative stress in woody breast meat when compared to normal meat.

19.4 Conclusion

Proteomic investigations have expanded the understanding of the cellular and biochemical mechanisms governing the quality of fresh muscle foods, and results of these studies will aid the food industry's efforts to engineer novel processing strategies to improve the quality of muscle foods. Furthermore, proteomics studies have implicated that biological functions such as apoptosis, oxidative stress, and autophagy play a critical role in postmortem metabolism and in turn on meat quality. The technological advances in the field of mass spectrometry and the development of novel proteomic tools will further facilitate the application of these tools to address meat quality issues.

References

Aggarwal, K., Choe, L.H., Lee, K.H., 2006. Shotgun proteomics using the iTRAQ isobaric tags. Brief Funct. Genomics 5, 112–120.

Alban, A., David, S.O., Bjorkesten, L., et al., 2003. A novel experimental design for comparative two-dimensional gel analysis: two-dimensional difference gel electrophoresis incorporating a pooled internal standard. Proteomics 3, 36–44.

Alderton, A.L., Faustman, C., Liebler, D.C., Hill, D.W., 2003. Induction of redox instability of bovine myoglobin by adduction with 4-hydroxy-2-nonenal. Biochemistry 42, 4398–4405.

Aslam, B., Basit, M., Nisar, M.A., Khurshid, M., Rasool, M.H., 2017. Proteomics: technologies and their applications. J. Chromatogr. Sci. 55, 182–196.

Baggerman, G., Vierstraete, E., De Loof, A., Schoofs, L., 2005. Gel-based versus gel-free proteomics: a review. Comb. Chem. High Throughput Screen. 8, 669–677.

Bantscheff, M., Schirle, M., Sweetman, G., Rick, J., Kuster, B., 2007. Quantitative mass spectrometry in proteomics: a critical review. Anal. Bioanal. Chem. 389, 1017–1031.

Bendixen, E., 2005. The use of proteomics in meat science. Meat Sci. 71, 138–149.

Bjarnadóttir, S.G., Hollung, K., Høy, M., et al., 2012. Changes in protein abundance between tender and tough meat from bovine Longissimus thoracis muscle assessed by isobaric Tag for Relative and Absolute Quantitation (iTRAQ) and 2-dimensional gel electrophoresis analysis. J. Anim. Sci. 90, 2035–2043.

Cai, K., Shao, W., Chen, X., et al., 2018. Meat quality traits and proteome profile of woody broiler breast (pectoralis major) meat. Poultry Sci. 97, 337–346.

Clerens, S., Thomas, A., Gathercole, J., et al., 2016. Proteomic and peptidomic differences and similarities between four muscle types from New Zealand raised Angus steers. Meat Sci. 121, 53–63.

D'Alessandro, A., Zolla, L., 2013. Meat science: from proteomics to integrated omics towards system biology. J. Proteomics 78, 558–577.

Desai, M.A., Jackson, V., Zhai, W., et al., 2016. Proteome basis of pale, soft, and exudative-like (PSE-like) broiler breast (Pectoralis major) meat. Poultry Sci. 95, 2696–2706.

Di Luca, A., Mullen, A.M., Elia, G., Davey, G., Hamill, R.M., 2011. Centrifugal drip is an accessible source for protein indicators of pork ageing and water-holding capacity. Meat Sci. 88, 261–270.

Di Luca, A., Hamill, R.M., Mullen, A.M., Slavov, N., Elia, G., 2016. Comparative proteomic profiling of divergent phenotypes for water holding capacity across the post mortem ageing period in porcine muscle exudate. PLoS One 11, e0150605.

Doherty, M.K., McLean, L., Hayter, J.R., et al., 2004. The proteome of chicken skeletal muscle: changes in soluble protein expression during growth in a layer strain. Proteomics 4, 2082–2093.

Faustman, C., Liebler, D.C., McClure, T.D., Sun, Q., 1999. α,β-Unsaturated aldehydes accelerate oxymyoglobin oxidation. J. Agric. Food Chem. 47, 3140–3144.

Faustman, C., Sun, Q., Mancini, R., Suman, S.P., 2010. Myoglobin and lipid oxidation interactions: mechanistic bases and control. Meat Sci. 86, 86−94.

Hollung, K., Veiseth, E., Jia, X., Færgestad, E.M., Hildrum, K.I., 2007. Application of proteomics to understand the molecular mechanisms behind meat quality. Meat Sci. 77, 97−104.

Huang, H., Lametsch, R., 2013. Challenges and applications of proteomics for analysis of changes in early postmortem meat. In: Toldrá, F., Nollet, L.M.L. (Eds.), Proteomics in Foods: Principles and Applications. Springer US, Boston, MA, pp. 103−109.

Huang, H., Larsen, M.R., Karlsson, A.H., et al., 2011. Gel-based phosphoproteomics analysis of sarcoplasmic proteins in postmortem porcine muscle with pH decline rate and time differences. Proteomics 11, 4063−4076.

Huff-Lonergan, E., Zhang, W., Lonergan, S.M., 2010. Biochemistry of postmortem muscle—lessons on mechanisms of meat tenderization. Meat Sci. 86, 184−195.

Hughes, J.M., Oiseth, S.K., Purslow, P.P., Warner, R.D., 2014. A structural approach to understanding the interactions between colour, water-holding capacity and tenderness. Meat Sci. 98, 520−532.

Hwang, I.H., Park, B.Y., Cho, S.H., 2004. Identification of muscle proteins related to objective meat quality in Korean native black pig. Asian Austral. J. Anim. 17, 1599−1607.

Jia, X., Veiseth-Kent, E., Grove, H., et al., 2009. Peroxiredoxin-6—a potential protein marker for meat tenderness in bovine longissimus thoracis muscle. J. Anim. Sci. 87, 2391−2399.

Joseph, P., Suman, S.P., Rentfrow, G., Li, S., Beach, C.M., 2012. Proteomics of muscle-specific beef color stability. J. Agric. Food Chem. 60, 3196−3203.

Kwasiborski, A., Sayd, T., Chambon, C., et al., 2008. Pig Longissimus lumborum proteome: Part II: Relationships between protein content and meat quality. Meat Sci. 80, 982−996.

Lametsch, R., Karlsson, A., Rosenvold, K., et al., 2003. Postmortem proteome changes of porcine muscle related to tenderness. J. Agric. Food Chem. 51, 6992−6997.

Lametsch, R., Roepstorff, P., Møller, H.S., Bendixen, E., 2004. Identification of myofibrillar substrates for μ-calpain. Meat Sci. 68, 515−521.

Laville, E., Sayd, T., Morzel, M., et al., 2009. Proteome changes during meat aging in tough and tender beef suggest the importance of apoptosis and protein solubility for beef aging and tenderization. J. Agric. Food Chem. 57, 10755−10764.

Lee, S., Phillips, A.L., Liebler, D.C., Faustman, C., 2003. Porcine oxymyoglobin and lipid oxidation in vitro. Meat Sci. 63, 241−247.

Lescuyer, P., Hochstrasser, D.F., Sanchez, J.C., 2004. Comprehensive proteome analysis by chromatographic protein prefractionation. Electrophoresis 25, 1125−1135.

Lewis, C., Doran, P., Ohlendieck, K., 2012. Proteomic analysis of dystrophic muscle. In: DiMario, J.X. (Ed.), Myogenesis: Methods and Protocols. Humana Press, Totowa, NJ, pp. 357−369.

Li, Z., Li, M., Li, X., et al., 2018. Quantitative phosphoproteomic analysis among muscles of different color stability using tandem mass tag labeling. Food Chem 249, 8−15.

Liebler, D.C., 2002. Proteomics and the new biology. In: Liebler, D.C. (Ed.), Introduction to Proteomics: Tools for the New Biology. Humana Press, Totowa, NJ, pp. 3−13.

Lund, T.C., Anderson, L.B., McCullar, V., et al., 2007. iTRAQ is a useful method to screen for membrane-bound proteins differentially expressed in human natural killer cell types. J. Proteome Res. 6, 644−653.

Mao, Y., Hopkins, D.L., Zhang, Y., et al., 2016. Beef quality with different intramuscular fat content and proteomic analysis using isobaric tag for relative and absolute quantitation of differentially expressed proteins. Meat Sci. 118, 96−102.

Marcos, B., Mullen, A.M., 2014. High pressure induced changes in beef muscle proteome: correlation with quality parameters. Meat Sci. 97, 11−20.

Marino, R., dellaMalva, A., Albenzio, M., 2015. Proteolytic changes of myofibrillar proteins in Podolian meat during aging: focusing on tenderness. J. Anim. Sci. 93, 1376−1387.

Morzel, M., Terlouw, C., Chambon, C., Micol, D., Picard, B., 2008. Muscle proteome and meat eating qualities of Longissimus thoracis of "Blonde d'Aquitaine" young bulls: a central role of HSP27 isoforms. Meat Sci. 78, 297−304.

Mullen, A.M., Stapleton, P.C., Corcoran, D., Hamill, R.M., White, A., 2006. Understanding meat quality through the application of genomic and proteomic approaches. Meat Sci. 74, 3–16.

Muroya, S., Ohnishi-Kameyama, M., Oe, M., Nakajima, I., Chikuni, K., 2007a. Postmortem changes in bovine troponin T isoforms on two-dimensional electrophoretic gel analyzed using mass spectrometry and western blotting: the limited fragmentation into basic polypeptides. Meat Sci. 75, 506–514.

Muroya, S., Ohnishi-Kameyama, M., Oe, M., et al., 2007b. Double phosphorylation of the myosin regulatory light chain during rigor mortis of bovine longissimus muscle. J. Agric. Food Chem. 55, 3998–4004.

Nair, M.N., Suman, S.P., Li, S., Joseph, P., Beach, C.M., 2014. Lipid oxidation–induced oxidation in emu and ostrich myoglobins. Meat Sci. 96, 984–993.

Nair, M.N., Suman, S.P., Chatli, M.K., et al., 2016. Proteome basis for intramuscular variation in color stability of beef semimembranosus. Meat Sci. 113, 9–16.

Nair, M.N., Costa-Lima, B.R.C., Wes Schilling, M., Suman, S.P., 2017. Chapter 10 - Proteomics of color in fresh muscle foods. In: Colgrave, M.L. (Ed.), Proteomics in Food Science. Academic Press, pp. 163–175.

Nair, M.N., Li, S., Beach, C.M., Rentfrow, G., Suman, S.P., 2018a. Changes in the sarcoplasmic proteome of beef muscles with differential color stability during postmortem aging. Meat Muscle Biol. 2, 1–17.

Nair, M.N., Li, S., Beach, C., Rentfrow, G., Suman, S.P., 2018b. Intramuscular variations in color and sarcoplasmic proteome of beef semimembranosus during postmortem aging. Meat Muscle Biol. 2, 92–101.

Okumura, N., Hashida-Okumura, A., Kita, K., et al., 2005. Proteomic analysis of slow- and fast-twitch skeletal muscles. Proteomics 5, 2896–2906.

Owens, C.M., 2016. Woody breast meat. In: 69th Reciprocal Meat Conference Proceedings. June 18th-23rd, San Angelo, Texas.

Phongpa-Ngan, P., Grider, A., Mulligan, J.H., Aggrey, S.E., Wicker, L., 2011. Proteomic analysis and differential expression in protein extracted from chicken with a varying growth rate and water-holding capacity. J. Agric. Food Chem. 59, 13181–13187.

Picard, B., Gagaoua, M., 2017. Chapter 11 - Proteomic investigations of beef tenderness. In: Colgrave, M.L. (Ed.), Proteomics in Food Science. Academic Press, pp. 177–197.

Rabilloud, T., 2002. Two-dimensional gel electrophoresis in proteomics: old, old fashioned, but it still climbs up the mountains. Proteomics 2, 3–10.

Rauniyar, N., Yates, J.R., 2014. Isobaric labeling-based relative quantification in shotgun proteomics. J. Proteome Res. 13, 5293–5309.

Sayd, T., Morzel, M., Chambon, C., et al., 2006. Proteome analysis of the sarcoplasmic fraction of pig semimembranosus muscle: implications on meat color development. J. Agric. Food Chem. 54, 2732–2737.

Shiio, Y., Aebersold, R., 2006. Quantitative proteome analysis using isotope-coded affinity tags and mass spectrometry. Nat. Protoc. 1, 139–145.

Smith, J.B., 2001. Peptide Sequencing by Edman Degradation. eLS. American Cancer Society.

Smyth, M.S., Martin, J.H.J., 2000. x-ray crystallography. Mol. Pathol. 53, 8–14.

Suman, S.P., Joseph, P., 2013. Myoglobin chemistry and meat color. Annu. Rev. Food Sci. Technol. 4, 79–99.

Suman, S.P., Faustman, C., Stamer, S.L., Liebler, D.C., 2006. Redox instability induced by 4-hydroxy-2-nonenal in porcine and bovine myoglobins at pH 5.6 and 4 °C. J. Agric. Food Chem. 54, 3402–3408.

Suman, S.P., Faustman, C., Stamer, S.L., Liebler, D.C., 2007. Proteomics of lipid oxidation-induced oxidation of porcine and bovine oxymyoglobins. Proteomics 7, 628–640.

Tijare, V.V., Yang, F.L., Kuttappan, V.A., 2016. Meat quality of broiler breast fillets with white striping and woody breast muscle myopathies. Poultry Sci. 95, 2167–2173.

van de Wiel, D.F.M., Zhang, W.L., 2007. Identification of pork quality parameters by proteomics. Meat Sci. 77, 46–54.

Wilkins, M.R., Pasquali, C., Appel, R.D., et al., 1996. From proteins to proteomes: large scale protein identification by two-dimensional electrophoresis and amino acid analysis. Nat. Biotechnol. 14, 61–65.

Wu, W., Gao, X., Dai, Y., et al., 2015. Post-mortem changes in sarcoplasmic proteome and its relationship to meat color traits in M. semitendinosus of Chinese Luxi yellow cattle. Food Res. Int. 72, 98–105.

Wu, W., Yu, Q., Fu, Y., et al., 2016. Towards muscle-specific meat color stability of Chinese Luxi yellow cattle: a proteomic insight into post-mortem storage. J. Proteomics 147, 108–118.

Yates, J.R., Ruse, C.I., Nakorchevsky, A., 2009. Proteomics by mass spectrometry: approaches, advances, and applications. Annu. Rev. Biomed. Eng. 11, 49–79.

Yin, S., Faustman, C., Tatiyaborworntham, N., et al., 2011. Species-specific myoglobin oxidation. J. Agric. Food Chem. 59, 12198–12203.

Yu, Q., Wu, W., Tian, X., et al., 2017. Unraveling proteome changes of Holstein beef M. semitendinosus and its relationship to meat discoloration during post-mortem storage analyzed by label-free mass spectrometry. J. Proteomics 154, 85–93.

Zapata, I., Zerby, H.N., Wick, M., 2009. Functional proteomic analysis predicts beef tenderness and the tenderness differential. J. Agric. Food Chem. 57, 4956–4963.

Zhang, M., Wang, D., Xu, X., Xu, W., 2019. Comparative proteomic analysis of proteins associated with water holding capacity in goose muscles. Food Res. Int 116, 354–361.

Zieske, L.R., 2006. A perspective on the use of iTRAQTM reagent technology for protein complex and profiling studies. J. Exp. Bot. 57, 1501–1508.

Zuo, H., Han, L., Yu, Q., et al., 2016. Proteome changes on water-holding capacity of yak longissimus lumborum during postmortem aging. Meat Sci. 121, 409–419.

Zuo, H., Han, L., Yu, Q., et al., 2018. Proteomic and bioinformatic analysis of proteins on cooking loss in yak longissimus thoracis. Eur. Food Res. Technol. 244, 1211–1223.

CHAPTER 20

Application of genomics tools in meat quality evaluation

T.K. Bhattacharya
ICAR-Directorate of Poultry Research, Hyderabad, India

Chapter Outline
20.1 Introduction 369
20.2 Genomic tools for meat quality assessment 370
    20.2.1 Isolation of nucleic acid (DNA and RNA) 370
    20.2.2 Quantification and purity measurement of DNA 372
    20.2.3 Polymerase chain reaction 372
    20.2.4 Real-time PCR 374
    20.2.5 Microsatellite analysis 375
    20.2.6 Agarose gel electrophoresis 384
    20.2.7 Polyacrylamide gel electrophoresis 385
    20.2.8 SDS-polyacrylamide gel electrophoresis 386
    20.2.9 Western blot 386
    20.2.10 Two-dimensional gel electrophoresis 387
20.3 Some important requisites 388
    20.3.1 Sterilization and disinfection of laboratory wares 388
    20.3.2 Sterilization of glasswares 388
    20.3.3 Dry heat sterilization 388
    20.3.4 Moist heat sterilization 388
    20.3.5 Disinfection and sterilization 388
20.4 Conclusion 389
References 389

20.1 Introduction

Meat is consumed by nonvegetarians as the prime source of protein and for other macro- as well as micronutrients. The quality of meat is very important as it should be tasty, tender, the desired color, devoid of undesired flavor, and free from pathogenic organisms. Today, one of the major concerns is to have clean and safe foods, which is most important for maintaining good and sound health. Raw foods are obtained in the market from different food channels, which sometimes make the food contaminated due to lack of practice of proper standard procedures by manufacturers/handlers/suppliers. But all this contamination of food should be properly diagnosed so that they can be properly treated in order to make it edible or be discarded. Just by looking at contaminated foods, it is sometimes impossible to detect contamination, and consequently it may be undiagnosed putting our health at risk of getting diseases by handling or consuming those foods. Hence the diagnosis of healthy foods including meat is our primary concern for which several laboratory procedures are available. Of the many quality assessment techniques, genomic approaches have added advantages for the diagnosis of the good healthy meats, which make our life healthy and safe from various diseases and health hazards (Bhattacharya et al., 2008). Genomic tools are very much confirmatory techniques to determine food safety in terms of contamination and hazards. In this chapter some important techniques of genomics and proteomics have been described one by one. The most important genomic tools for meat quality assessment are polymerase chain reaction (PCR), real-time PCR, DNA sequencing, and microsatellite analysis. Each technique has been described briefly for easy understanding.

20.2 Genomic tools for meat quality assessment

20.2.1 Isolation of nucleic acid (DNA and RNA)

The nucleus and mitochondria in animal cells contain genetic materials, which are mostly packaged with histone and nonhistone proteins besides other metallic substances. Genomic materials include nucleic acids, which are of two types, namely deoxyribonucleic acid (DNA) and ribonucleic acid (RNA). DNA molecules of more than 150 kb are prone to breakage by forces generated during their isolation. Getting longer DNA is desirable in experiments. Hence proper care should be taken during its isolation. DNA can be extracted from fresh or frozen whole blood, blood stains, semen, feather follicles, tissues, etc. The basic extraction procedure remains the same and slight modifications are required depending on the type of biological material used for isolation. The phenol−chloroform extraction method is followed for the isolation of genomic DNA. The genome contains not only DNA but also different categories of RNA to exert biological activity by synthesizing proteins and peptides. A typical cell contains mainly three classes of RNA: rRNA, mRNA, and tRNA, of which 80%−85% of total RNA is rRNA more specifically 28S, 18S, and 5S, 15%−20% is tRNA, and 1%−5% is mRNA. Besides these, small nuclear RNAs (snRNAs) and other noncoding RNAs like micro-RNA (miRNA), small interfering RNA (siRNA), and Piwi-interacting RNA (piRNA) are also found in animal cells. snRNA is a small RNA molecule found within the nucleus of eukaryotic cells and is involved in several processes such as RNA splicing, regulation of transcription factors and RNA polymerase II, and maintaining telomeres. miRNA is 21−22 bases in size and is found in the eukaryotic cell. It forms an effector miRNA complex, which can either cleave mRNA or block mRNA from being translated or accelerate mRNA degradation, and thus inhibit translation. siRNA either cleaves mRNA or methylates target genes to interfere with gene expression. Animals also have piRNA which is 29–30 bases in length and found in germline cells. It protects the functional genome from transposons and aids in gametogenesis.

20.2.1.1 Isolation of genomic DNA

Tissue samples are either freshly collected or stored at −20°C are allowed to attain room temperature before starting the experiment. A small portion of tissue is to be taken into a 1.5 mL eppendorf tube using sterile forceps and is homogenized/ground using a micropestle. A volume of 500 μL phosphate buffer saline (PBS) solution is added, mixed well, and centrifuged at 2000 rpm for 5 minutes. The supernatant is decanted. About 100 μL of suspension is taken from the homogenized tissue. A volume of 900 μL tissue lysis buffer, 15 μL proteinase K, and 100 μL SDS are added to the tissue suspension and incubated at 50°C–60°C overnight. After overnight incubation, 900 μL of Tris-saturated phenol is added and mixed gently. The mixture is centrifuged at 10,000 rpm for 10 minutes. The supernatant is transferred into another tube and 900 μL phenol:chloroform:isoamyl alcohol is added. It is mixed gently and centrifuged at 10,000 rpm for 10 minutes. The supernatant is collected into another tube and 900 μL chloroform:isoamyl alcohol is added followed by spinning at 10,000 rpm for 10 minutes. The supernatant is transferred into another tube and isopropanol (equal volume) is added. It is mixed gently and DNA will precipitate. The tube is centrifuged at 5000 rpm for 1 minute and a DNA pellet will settle at the bottom. Decant the supernatant and add 800 μL 70% ethanol. It is spun at 5000 rpm for 1 minute and the supernatant is decanted. Add 1 mL 70% ethanol, spin at 5000 rpm for 1 minute, and decant the supernatant. Dry the DNA pellet at 37°C for 2 hours. Add 200 μL TE to dissolve the pellet, incubate at 65°C for 1 hour, and later DNA is stored at 4°C for further experiments.

20.2.1.2 Isolation of RNA

Before the onset of the experiment, all the labwares are treated with 0.1% DEPC-treated water at 37°C overnight and autoclaved. An amount of 1 g tissues is taken in a tube containing PBS and triturated in pestle and mortar. Equal volumes of RNA extraction buffer and proteinase K digestion buffer are added. It is mixed by vortexing. The proteinase K (20 mg/mL) to a final concentration of 200 μg/mL is added in it, mixed well, and incubated for 45 minutes at 37°C temperature. An equal volume of phenol:chloroform is added and centrifuged for 10 minutes at room temperature. The upper aqueous phase is collected to a fresh tube and 2.5 volume of ice-cold ethanol is added. It is mixed well and chilled to 0°C for an hour. Then ethanol mixed solution is centrifuged at 5000 rpm for 10 minutes at 0°C and supernatant is discarded. The pellet is washed with 70% ethanol containing 0.1 M sodium acetate (pH 5.2). Ethanol is decanted and the pellet is allowed to dry in the open air. The pellet is dissolved with 200 μL TE buffer. The $MgCl_2$ and DTT is added to a final concentration of 10 and 0.1 mM, respectively, and then, RNAase inhibitor or vanadylribonucleoside complexes is added to a final concentration of 1000 U/mL or 10 mM, respectively. The DNAase I is added to a final concentration of 2 μg/mL and incubated for an hour at 37°C. The EDTA and SDS are added to final concentrations of 10 mM and 0.2%, respectively. An equal volume of phenol:chloroform is included in it and it is centrifuged at 5000 rpm for 10 minutes at room temperature. The upper aqueous phase is transferred to a fresh tube containing 3 M sodium acetate (pH 5.2) to a final concentration of 0.3 M. About 2.5 volume of ice-cold ethanol is added, mixed well, and chilled for 1 hour on ice. Then it is centrifuged at 12,000g for 5 minutes at 4°C. The ethanol is decanted and the pellet is allowed to dry. The pellet is dissolved in 200 μL of TE buffer (pH 7.6) and RNA is ready for further experiments.

20.2.2 Quantification and purity measurement of DNA

Checking the quantity and quality of DNA are essential steps before carrying out further experiments or works. Two methods are mostly used to estimate quantity and check quality of DNA and they are the spectrophotometric method and the ethidium bromide fluorescence method in agarose gel. Samples should be used to take spectrophotometric readings at 260 and 280 nm. The ratio of OD_{260} and OD_{280} reflects the purity of DNA. If the ratio value falls in the range of 1.7 and 2.0, the DNA is expected to be pure. If the value is less than the range, it indicates contamination with protein or phenol. The concentration of DNA is estimated with the optical density value at 260 nm by following the formula:

$$\text{DNA concentration } (\mu g/\mu L) = \frac{OD_{260} \times \text{Dilution factor} \times 50}{1000}$$

$$(1 \text{ OD value at 260 nm is equivalent to 50 ng DNA}/\mu L)$$

For the quantitation of DNA, this method is of a subjective type and needs some expertise. For the quality of DNA, the types of bands found in the gel are the indicators of the quality of DNA. By comparing with a known quantity of different concentrations of λ DNA bands, the quantity of the unknown DNA band can be determined.

20.2.3 Polymerase chain reaction

PCR is an in vitro technique to amplify a specific region of the DNA, generating thousands to millions of copies of a particular DNA sequence. The technique was developed by Kary B. Mullis in 1983. For this invention, Kary Mullis was awarded the Nobel prize in Chemistry. Currently, there are two methods available for amplifying and making copies of DNA and they are cloning and PCR. Cloning takes a long time to generate the sufficient number of clones, while PCR quickly amplifies DNA to generate millions of copies. For PCR we need a small quantity of cells or tissues like blood, tissue, hair, feather, etc.

On account of these advantages, PCR technology has been used as the new gold standard for detecting a wide variety of templates in molecular biology research.

The PCR technique is based on thermal cycling conditions. It involves DNA replication. It consists of cycles of repeated heating and cooling of the reaction mixture involving DNA denaturation and enzymatic replication of DNA. Primers, the short length DNA fragments containing sequences complementary to the target region along with DNA polymerase enzyme, help to initiate and engage selective and repeated amplification. As PCR progresses, the amplified DNA is itself used as a template for replication. Consequently, a chain of reaction is contemplated in which the DNA template gets amplified exponentially. The PCR is basically used for the detection of mutations, DNA cloning, diagnosis of pathogens, DNA fingerprinting, prenatal diagnosis, mapping of genome, etc.

PCR is a simple and highly versatile technique. It has been modified in a variety of ways to suit specific applications.

- *Inverse PCR*: inverse PCR is especially useful in identifying flanking sequences of various genomic inserts. In this PCR the amplification of DNA of an unknown sequence is carried out from a known sequence.
- *Anchored PCR*: in this PCR a small sequence of nucleotides can be attached or tagged to the target DNA. The anchor is frequently a poly G to which a poly C primer is used.
- *Reverse transcriptase PCR*: this type of PCR is done on RNA using reverse transcriptase enzymes to prepare cDNA. This is widely used in expression profiling, gene expression, or to identify the sequence of an RNA transcript.
- *Asymmetric PCR*: this type of PCR is used in sequencing applications which require only one of the complementary strands of DNA.
- *Allele specific polymerase PCR*: this PCR uses primers specific to a particular allele, where amplification is found in the presence of a specific allele and absent in the presence of other alleles. It is used as a diagnostic technique to identify single nucleotide polymorphisms.
- *Hot start PCR*: it reduces nonspecific amplification during the initial stages of PCR. It increases the specificity of products and can be used in cloning experiments.
- *Multiplex PCR*: it uses multiple primer sets within a single PCR mixture to produce PCR products of varying sizes. The PCR products are specific to different DNA sequences.
- *Real-time PCR*: real-time PCR measures the quantity of DNA using highly precise DNA polymerase enzymes with specific chemistry, such as SYBR Green or Taqman for the detection of molecules. Real-time PCR involves the fluorescence reporter molecule, for example, SYBR Green, binding to the double-stranded DNA. As the PCR product accumulates in each cycle of amplification, it emits fluorescence which is detected by the real-time PCR machine.

- *Methylation specific PCR*: it rapidly assesses the methylation states of any group of CpG sites within a CpG island, independent of the use of methylation sensitive restriction enzymes. Methylation is determined by the ability of the specific primers to achieve amplification. This kind of PCR is used in disease diagnosis.

20.2.4 Real-time PCR

Real-time PCR or quantitative PCR or qPCR is an in vitro technique to quantify the presence of DNA templates (Dhanasekaran et al., 2014). It is used to amplify and simultaneously quantify a targeted DNA molecule. The conventional PCR suffers from major disadvantages like time-consuming post-PCR handling. The qPCR has overcome the shortcomings of conventional PCR. qPCR is highly sensitive and specific. qPCR monitors the progress of the PCR reaction in real time to detect and quantify the presence of nucleic acid sequences. Real-time PCR is based on the detection of the increasing fluorescence emitted by a fluorescence reporter molecule at a particular wavelength in every cycle. This occurs due to the accumulation of the PCR product with each cycle of amplification. These fluorescent reporter molecules include dyes that bind to the double-stranded DNA, for example, SYBR Green or sequence-specific probes, such as molecular beacons or Taqman probes. The reaction can start with the minimal amounts of nucleic acid and quantify the end product accurately. Moreover, there is no need for the post-PCR processing through agarose gel electrophoresis (AGE). These advantages of the fluorescence-based real-time PCR technique have completely revolutionized the approach to PCR-based quantification of DNA and RNA. Real-time PCR has wide applications such as for testing DNA damage, in vivo imaging of cellular process, detection of inactivation at x chromosomes, prenatal diagnosis of hemoglobinopathies etc., in allelic discrimination, detection of gene mutations and genome instabilities, quantification of gene expression, to study the microbial agents causing infectious disease, monitoring of chemotherapy, cytokine quantification, etc.

Real-time PCR uses a fluorescent dye or reporter molecule, which is used to monitor the amplification of nucleic acid. The fluorescence emitted by the reporter molecule multiplies as the PCR product accumulates with each cycle of amplification (Guru Vishnu et al., 2017). Based on the molecule used for the detection, the real-time PCR techniques can be grouped into two classes: nonspecific detection using DNA binding dyes, and specific detection with target specific probes.

In real-time PCR, DNA binding dyes are used as fluorescent reporters to monitor the real-time PCR reaction. The fluorescence of the reporter dye increases as the product accumulates with each successive cycle of amplification. By recording the amount of fluorescence emission at each cycle, it is possible to monitor the PCR reaction during the exponential phase. If a graph is prepared between the log of the starting amount of template and the corresponding increase in fluorescence of the reporter dye during real-time PCR, a linear relationship is observed. SYBR Green is the most widely used double-stranded DNA-binding dye for real-time PCR. SYBR Green binds to the minor groove of the DNA double helix, while in the solution, the unbound dye exhibits very little fluorescence. This fluorescence is substantially enhanced when the dye is bound to double-stranded DNA. The dye emits at 520 nm and fluorescence emitted can be detected and related to the amount of target. SYBR Green remains stable under PCR conditions and the optical filter of the thermocycler can be affixed to harmonize the excitation and emission wave lengths (Bhattacharya et al., 2014). Ethidium bromide can also be used for detection but due to its carcinogenic nature, it is not extensively used in real-time PCR.

20.2.5 Microsatellite analysis

Molecular characterization of germplasm is done by several tools of which one of the important techniques is microsatellite fingerprinting. Microsatellites, sometimes called variable number of tandem repeats, are short segments of DNA that have a repeated sequence and tend to be present mostly in the noncoding region of the genome. Microsatellites are the repeat sequences, which are either tandemly or interspersely repeated in the genome. These repeat sequences are very much prone to mutation and have very high polymorphism. The length of consensus sequences varies from 2 to 60 bases and may be repeated a number of times. Presently, microsatellites are the marker of choice for genetic diversity analysis due to hypervariability, codominant inheritance, reliability, easiness to perform, etc. They can be easily typed and can be used for estimating diversity and construction of linkage maps (Chatterjee et al., 2010). The mutation in microsatellites locus occurs in two ways, either unequal crossing over in meiosis or strand-slippage replication. But the strand-slippage replication probably remains the most frequent mechanism of a mutation at a microsatellite locus. The most common way to detect microsatellites is to design PCR primers that are unique to one locus in the genome and the base pair on either side of the repeated portion. Therefore a single pair of PCR primers will work for every individual in each species and produce different sized products for each of the different length microsatellites. The advantages of PCR-based microsatellite analysis are that this method requires a lower amount of DNA; the small size of the microsatellite loci detected from the PCR-based method improves the chance of obtaining precise result; discrete sizes of microsatellite alleles make the interpretation easier and efficient; the technique is easy to standardize and reproducible; it is less time-consuming; PCR-based microsatellite analysis can be performed by automated DNA sequencer, and many microsatellites can be analyzed at the same time. The Food and Agriculture Organization (FAO) has recommended the microsatellite markers for domestic species for identification of specific species and these markers may also be used for the detection of meat of these animals. Species-wise microsatellite markers are shown in Tables 20.1–20.6.

Table 20.1: Cattle-specific microsatellite markers recommended by FAO.

Locus name	Primers	Annealing temperature (°C)	Allele size (bp)	Chromosome number
BM1824	GAGCAAGGTGTTTTTCCAATC CATTCTCCAACTGCTTCCTTG	55–60	176–197	1
BM2113	GCTGCCTTCTACCAAATACCC CTTCCTGAGAGAAGCAACACC	55–60	122–156	2
INRA023	GAGTAGAGCTACAAGATAAACTTC TAACTACAGGGTGTTAGATGAACTC	55	195–225	3
ETH152	TACTCGTAGGGCAGGCTGCCTG GAGACCTCAGGGTTGGTGATCAG	55–60	181–211	5
ETH10	GTTCAGGACTGGCCCTGCTAACA CCTCCAGCCCACTTTCTCTTCTC	55–65	207–231	5
ILSTS006	TGTCTGTATTTCTGCTGTGG ACACGGAAGCGATCTAAACG	55	277–309	7
HEL9	CCCATTCAGTCTTCAGAGGT CACATCCATGTTCTCACCAC	52–57	141–173	8
MM12	CAAGACAGGTGTTTCAATCT ATCGACTCTGGGGATGATGT	50–55	101–145	9
ETH225	GATCACCTTGCCACTATTTCCT ACATGACAGCCAGCTGCTACT	55–65	131–159	9
CSRM60	AAGATGTGATCCAAGAGAGAGGCA AGGACCAGATCGTGAAAGGCATAG	55–65	79–115	10
ILSTS005	GGAAGCAATGAAATCTATAGCC TGTTCTGTGAGTTTGTAAGC	54–58	176–194	10
INRA037	GATCCTGCTTATATTTAACCAC AAAATTCCATGGAGAGAGAAAC	57–58	112–148	10
INRA032	AAACTGTATTCTCTAATAGCTAC GCAAGACATATCTCCATTCCTTT	55–58	160–204	11
HEL13	TAAGGACTTGAGATAAGGAG CCATCTACCTCCATCTTAAC	52–57	178–200	11
INRA005	CAATCTGCATGAAGTATAAATAT CTTCAGGCATACCCTACACC	55	135–149	12
CSSM66	ACACAAATCCTTTCTGCCAGCTGA AATTTAATGCACTGAGGAGCTTGG	55–65	171–209	14
HEL1	CAACAGCTATTTAACAAGGA AGGCTACAGTCCATGGGATT	54–57	99–119	15
INRA035	TTGTGCTTTATGACACTATCCG ATCCTTTGCAGCCTCCACATTG	55–60	100–124	16
SPS115	AAAGTGACACAACAGCTTCTCCAG AACGAGTGTCCTAGTTTGGCTGTG	55–60	234–258	15
TGLA53	GCTTTCAGAAATAGTTTGCATTCA ATCTTCACATGATATTACAGCAGA	55	143–191	16
ETH185	TGCATGGACAGAGCAGCCTGGC GCACCCCAACGAAAGCTCCCAG	58–67	214–246	17
TGLA227	CGAATTCCAAATCTGTTAATTTGCT ACAGACAGAAACTCAATGAAAGCA	55–56	75–105	18

(Continued)

Table 20.1: (Continued)

Locus name	Primers	Annealing temperature (°C)	Allele size (bp)	Chromosome number
INRA063	ATTTGCACAAGCTAAATCTAACC AAACCACAGAAATGCTTGGAAG	55–58	167–189	18
ETH3	GAACCTGCCTCTCCTGCATTGG ACTCTGCCTGTGGCCAAGTAGG	55–65	103–133	19
TGLA126	CTAATTTAGAATGAGAGAGGCTTCT TTGGTCTCTATTCTCTGAATATTCC	55–58	115–131	20
TGLA122	CCCTCCTCCAGGTAAATCAGC AATCACATGGCAAATAAGTACATAC	55–58	136–184	21
HEL5	GCAGGATCACTTGTTAGGGA AGACGTTAGTGTACATTAAC	52–57	145–171	21
HAUT24	CTCTCTGCCTTTGTCCCTGT AATACACTTTAGGAGAAAAATA	52–55	104–158	22
BM1818	AGCTGGGAATATAACCAAAGG AGTGCTTTCAAGGTCCATGC	56–60	248–278	23
HAUT27	AACTGCTGAAATCTCCATCTTA TTTTATGTTCATTTTTTGACTGG	57	120–158	26

Table 20.2: Buffalo-specific microsatellite markers recommended by FAO.

Locus name	Primers	Annealing temperature (°C)	Allele size (bp)	Chromosome number
CSSM019	TTGTCAGCAACTTCTTGTATCTTT TGTTTTAAGCCACCCAATTATTTG	55	131–161	1
CSSM032	TTATTTTCAGTGTTTCTAGAAAAC TATAATATTGCTATCTGGAAATCC	55	208–224	1
CSSM036	GGATAACTCAACCACACGTCTCTG AAGAAGTACTGGTTGCCAATCGTG	55	162–176	1
CSSM043	AAAACTCTGGGAACTTGAAAACTA GTTACAAATTTAAGAGACAGAGTT	55	222–258	1
CSSM045	TAGAGGCACAAGCAAACCTAACAC TTGGAAAGATGCAGTAGAACTCAT	60	102–122	2
DRB3	GAGAGTTTCACTGTGCAG CGCGAATTCCCAGAGTGAGTGAAGTATCT	50–55	142–198	2
ETH121	CCAACTCCTTACAGGAAATGTC ATTTAGAGCTGGCTGGTAAGTG	59	182–198	2
ILSTS030	CTGCAGTTCTGCATATGTGG CTTAGACAACAGGGGTTTGG	55	146–158	2
CSSM047	TCTCTGTCTCTATCACTATATGGC CTGGGCACCTGAAACTATCATCAT	55	127–162	3

(Continued)

Table 20.2: (Continued)

Locus name	Primers	Annealing temperature (°C)	Allele size (bp)	Chromosome number
ETH003	GAACCTGCCTCTCCTGCATTGG ACTCTGCCTGTGGCAAGTAGG	65	96–192	3
BMC1013	AAAAATGATGCCAACCAAATT TAGGTAGTGTTCCTTATTTCTCTGG	54	217–239	3
CSSME070	TTCTAACAGCTGTCACTCAGGC ATACAGATTAAATACCCACCTG	50–55	119–139	3
RM099	CCAAAGAGTCTAACACAACTGAG ATCCGAACCAAAATCCCATCAAG	60	87–119	3
CSSM022	TCTCTCTAATGGAGTTGGTTTTTG ATATCCCACTGAGGATAAGAATTC	55–60	203–213	4
CSSM013	ATAAGAGATTACCCTTCCTGACTG AGGTAAATGTTCCTATTTGCTAAC	55	162–172	5
CSSM029	GCTCCATTATGCACATGCCATGCT CGTGAGAACCGAAAGTCACACATTC	55	174–196	9
CSSM057	GTCGCTGGATAAACAATTTAAAGT TGTGGTGTTTAACCCTTGTAATCT	60	102–130	9
CSSM038	TTCATATAAGCAGTTTATAAACGC ATAGGATCTGGTAACTTACAGATG	55	163–187	11
CSSM046	GGCTATTAACTGTTTTCTAGGAAT TGCACAATCGGAACCTAGAATATT	55	152–160	11
CSRM060	AAGATGTGATCCAAGAGAGAGGCA AGGACCAGATCGTGAAAGGCATAG	60	95–135	11
ILSTS005	GGAAGCAATGAAATCTATAGCC TGTTCTGTGAGTTTGTAAGC	55	173–186	11
BRN	CCTCCACACAGGCTTCTCTGACTT CCTAACTTGCTTGAGTTATTGCCC	60	121–147	11
ILSTS033	TATTAGAGTGGCTCAGTGCC ATGCAGACAGTTTTAGAGGG	55	126–138	13
ILSTS008	GAATCATGGATTTTCTGGGG TAGCAGTGAGTGAGGTTGGC	58	168–176	15
CSSM033	CACTGTGAATGCATGTGTGAGC CCCATGATAAGAGTGCAGATGACT	65	154–175	17
HMH1R	GGCTTCAACTCACTGTAACACATT TTCTTCAAGTATCACCTCTGTGGCC	60	169–187	21
CSSM041	AATTTCAAAGAACCGTTACACAGC AAGGGACTTGCAGGGACTAAAACA	55	129–147	21
CSSM061	AGGCCATATAGGAGGCAAGCTTAC TTCAGAAGAGGGCAGAGAATACAC	60	100–126	Unknown
CSSM008	CTTGGTGTTACTAGCCCTGGG GATATATTTGCCAGAGATTCTGCA	55	179–193	Unknown
CSSM062	GTTTAAACCCCAGATTCTCCCTTG AGATGTAACAGCATCATGACTGAA	55	124–136	Unknown

Table 20.3: Sheep-specific microsatellite markers recommended by FAO.

Locus name	Primers	Annealing temperature (°C)	Allele size (bp)	Chromosome number
BM1824	GAGCAAGGTGTTTTTCCAATC CATTCTCCAACTGCTTCCTTG	58		1
OarFCB20	AAATGTGTTTAAGATTCCATACAGTG GGAAAACCCCCATATATACCTATAC	56	95–120	2
OarFCB128	ATTAAAGCATCTTCTCTTTATTTCCTCGC CAGCTGAGCAACTAAGACATACATGCG	55	96–130	2
OarCB226	CTATATGTTGCCTTTCCCTTCCTGC GTGAGTCCCATAGAGCATAAGCTC	60	119–153	2
ILSTS28	TCCAGATTTTGTACCAGACC GTCATGTCATACCTTTGAGC	53	105–177	3
OarCP34	GCTGAACAATGTGATATGTTCAGG GGGACAATACTGTCTTAGATGCTGC	50	112–130	3
MAF70	CACGGAGTCACAAAGAGTCAGACC GCAGGACTCTACGGGGCCTTTGC	60	124–166	4
OarAE129	AATCCAGTGTGTGAAAGACTAATCCAG GTAGATCAAGATATAGAATATTTTTCAACACC	54	133–159	5
BM1329	TTGTTTAGGCAAGTCCAAAGTC AACACCGCAGCTTCATCC	50	160–182	6
MCM140	GTTCGTACTTCTGGGTACTGGTCTC GTCCATGGATTTGCAGAGTCAG	60	167–193	6
ILSTS5	GGAAGCAATGAAATCTATAGCC TGTTCTGTGAGTTTGTAAGC	55	174–221	7
ILSTS11	GCTTGCTACATGGAAAGTGC CTAAAATGCAGAGCCCTACC	55	256–294	9
MAF33	GATCTTTGTTTCAATCTATTCCAATTTC GATCATCTGAGTGTGAGTATATACAG	60	121–141	9
OarCP38	CAACTTTGGTGCATATTCAAGGTTGC GCAGTCGCAGCAGGCTGAAGAGG	52	117–129	10
OarFCB193	TTCATCTCAGACTGGGATTCAGAAAGGC GCTTGGAAATAACCCTCCTGCATCCC	54	96–136	11
SRCRSP9	AGAGGATCTGGAAATGGAATC GCACTCTTTTCAGCCCTAATG	55	99–135	12
HUJ616	TTCAAACTACACATTGACAGGG GGACCTTTGGCAATGGAAGG	54	114–160	13
SRCRSP1	TGCAAGAAGTTTTTCCAGAGC ACCCTGGTTTCACAAAAGG	54	116–148	13
INRA063	ATTTGCACAAGCTAAATCTAACC AAACCACAGAAATGCTTGGAAG	58		14
MAF65	AAAGGCCAGAGTATGCAATTAGGAG CCACTCCTCCTGAGAATATAACATG	60	123–127	15
MAF214	GGGTGATCTTAGGGAGGTTTTGGAGG AATGCAGGAGATCTGAGGCAGGGACG	58	174–282	16

(Continued)

Table 20.3: (Continued)

Locus name	Primers	Annealing temperature (°C)	Allele size (bp)	Chromosome number
BM8125	CTCTATCTGTGGAAAAGGTGGG GGGGGTTAGACTTCAACATACG	50	110–130	17
MAF209	GATCACAAAAAGTTGGATACAACCGTGG TCATGCACTTAAGTATGTAGGATGCTG	63		17
OarHH47	TTATTGACAAACTCTCTTCCTAACTCCACC GTAGTTATTTAAAAAAATATCATACCTCTTAAG	58	130–152	18
SRCRSP5	GGACTCTACCAACTGAGCTACAAG GTTTCTTTGAAATGAAGCTAAAGCAATGC	56	126–158	18
OarFCB304	CCCTAGGAGCTTTCAATAAAGAATCGG CGCTGCTGTCAACTGGGTCAGGG	56	150–188	19
DYMS1	AACAACATCAAACAGTAAGAG CATAGTAACAGATCTTCCTACA	59	159–211	20
OarJMP29	GTATACACGTGGACACCGCTTTGTAC GAAGTGGCAAGATTCAGAGGGGAAG	56	96–150	24
OarVH72	GGCCTCTCAAGGGGCAAGAGCAGG CTCTAGAGGATCTGGAATGCAAAGCTC	57	121–145	25
OarJMP58	GAAGTCATTGAGGGGTCGCTAACC CTTCATGTTCACAGGACTTTCTCTG	58	145–169	26

Table 20.4: Goat-specific microsatellite markers recommended by FAO.

Locus name	Primers	Annealing temperature (°C)	Allele size (bp)	Chromosome number
OarFCB20	GGAAAACCCCCATATATACCTATAC AAATGTGTTTAAGATTCCATACATGTG	58	93–112	2
BM6444	CTCTGGGTACAACACTGAGTCC TAGAGAGTTTCCCTGTCCATCC	65	118–200	2
INRA023	GAGTAGAGCTACAAGATAAACTTC TAACTACAGGGTGTTAGATGAACT	58	196–215	3
ILSTS029	TGTTTTGATGGAACACAG TGGATTTAGACCAGGGTTGG	55	148–170	3
MAF70	CACGGAGTCACAAAGAGTCAGACC GCAGGACTCTACGGGGCCTTTGC	65	134–168	4
MCM527	GTCCATTGCCTCAAATCAATTC AAACCACTTGACTACTCCCCAA	58	165–187	5
ETH10	GTTCAGGACTGGCCCTGCTAACA CCTCCAGCCCACTTTCTCTTCTC	55	200–210	5
ILSTS087	AGCAGACATGATGACTCAGC CTGCCTCTTTTCTTGAGAG	58	135–155	6
SRCRSP7	TCTCAGCACCTTAATTGCTCT GGTCAACACTCCAATGGTGAG	55	117–131	6

(Continued)

Table 20.4: (Continued)

Locus name	Primers	Annealing temperature (°C)	Allele size (bp)	Chromosome number
ILSTS005	GGAAGCAATTGAAATCTATAGCC TGTTCTGTGAGTTTGTAAGC	55	172–218	10
SPS113	CCTCCACACAGGCTTCTCTGACTT CCTAACTTGCTTGAGTTATTGCCC	58	134–158	10
TCRVB6	GAGTCCTCAGCAAGCAGGTC CCAGGAATTGGATCACACCT	55	217–255	10
SRCRSP3	CGGGGATCTGTTCTATGAAC TGATTAGCTGGCTGAATGTCC	55	98–122	10
SRCRSP9	AGAGGATCTGGAAATGGAATC GCACTCTTTTCAGCCCTAATG	58	99–135	12
ILSTS011	GCTTGCTACATGGAAAGTGC CTAAAATGCAGAGCCCTACC	58	250–300	14
CSRD247	GGACTTGCCAGAACTCTGCAAT CACTGTGGTTTGTATTAGTCAGG	58	220–247	14
MAF065	AAAGGCCAGAGTATGCAATTAGGAG CCACTCCTCCTGAGAATATAACATG	58	116–158	15
TGLA53	GCTTTCAGAAATAGTTTGCATTCA ATCTTCACATGATATTACAGCAGA	55	126–160	16
OarFCB48	GAGTTAGTACAAGGATGACAAGAGGCAC GACTCTAGAGGATCGCAAAGAACCAG	58	149–173	17
MAF209	GATCACAAAAAGTTGGATACAACCGTG TCATGCACTTAAGTATGTAGGATGCTG	55	100–104	17
INRA063	GACCACAAAGGGATTTGCACAAGC AAACCACAGAAATGCTTGGAAG	58	164–186	18
INRABERN185	CAATCTTGCTCCCACTATGC CTCCTAAAACACTCCCACACTA	55	261–289	18
SRCRSP5	GGACTCTACCAACTGAGCTACAAG TGAAATGAAGCTAAAGCAATGC	55	156–178	21
DRBP1	ATGGTGCAGCAGCAAGGTGAGCA GGGACTCAGTCTCTCTATCTCTTTG	58	195–229	23
OarAE54	TACTAAAGAAACATGAAGCTCCCA GGAAACATTTATTCTTATTCCTCAGTG	58	115–138	25
INABERN172	CCACTTCCCTGTATCCTCCT GGTGCTCCCATTGTGTAGAC	58	234–256	26
SRCRSP23	TGAACGGGTAAAGATGTG TGTTTTTAATGGCTGAGTAG	58	81–119	Unknown
SRCRSP8	TGCGGTCTGGTTCTGATTTCAC GTTTCTTCCTGCATGAGAAAGTCGATGCTTAG	55	215–255	Unknown
SRCRSP15	CTTTACTTCTGACATGGTATTTCC TGCCACTCAATTTAGCAAGC	55	172–198	Unknown
P19 (DYA)	AACACCATCAAACAGTAAGAG CATAGTAACAGATCTTCCTACA	55	160–196	Unknown

Table 20.5: Pig-specific microsatellite markers recommended by FAO.

Locus name	Primers	Annealing temperature (°C)	Allele size (bp)	Chromosome number
S0155	TGTTCTCTGTTTCTCCTCTGTTTG AAAGTGGAAAGAGTCAATGGCTAT	55	116–158	1
Sw1828	AATGCATTGTCTTCATTCAACC TTAACCGGGGCACTTGTG	55	100–104	1
S0226	GCACTTTTAACTTTCATGATACTCC GGTTAAACTTTTNCCCCAATACA	55	172–198	2
Sw240	AGAAATTAGTGCCTCAAATTGG AAACCATTAAGTCCCTAGCAAA	55	164–186	2
Sw72	ATCAGAACAGTGCGCCGT TTTGAAAATGGGGTGTTTCC	55	172–218	3
S0002	GAAGCCAAAGAGACAACTGC GTTCTTTACCCACTGAGCCA	60	195–229	3
S0097	GACCTATCTAATGTCATTATAGT TTCCTCCTAGAGTTGACAAACTT	55	135–155	4
IGF1	GCTTGGATGGACCATGTTG CATATTTTCTGCATAACTTGAACCT	55	256–294	5
S0005	TCCTTCCCTCCTGGTAACTA GCACTTCCTGATTCTGGGTA	55	134–168	5
S0228	GGCATAGGCTGGCAGCAACA AGCCCACCTCATCTTATCTACACT	55	93–112	6
Sw122	TTGTCTTTTTATTTTGCTTTTGG CAAAAAAGGCAAAAGATTGACA	55	220–247	6
Sw2406	AATGTCACCTTTAAGACGTGGG AATGCGAAACTCCTGAATTAGC	55	117–131	6
Sw632	TGGGTTGAAAGATTTCCCAA GGAGTCAGTACTTTGGCTTGA	55	115–138	7
S0101	GAATGCAAAGAGTTCAGTGTAGG GTCTCCCTCACACTTACCGCAG	55	200–210	7
S0178	TAGCCTGGGAACCTCCACACGCTG GGCACCAGGAATCTGCAATCCAGT	60	160–196	8
Sw2410	ATTTGCCCCCAAGGTATTTC CAGGGTGTGGAGGGTAGAAG	50	81–119	8
Sw911	CTCAGTTCTTTGGGACTGAACC CATCTGTGGAAAAAAAAGCC	60	217–255	9
Sw830	AAGTACCATGGAGAGGGAAATG ACATGGTTCCAAAGACCTGTG	50	149–173	10
Sw2008	CAGGCCAGAGTAGCGTGC CAGTCCTCCCAAAAATAACATG	55	148–170	11
S0090	CCAAGACTGCCTTGTAGGTGAATA GCTATCAAGTATTGTACCATTAGG	55	227–253	12
S0143	ACTCACAGCTTGTCCTGGGTGT CAGTCAGCAGGCTGACAAAAAC	55	261–289	12
S0068	AGTGGTCTCTCTCCCTCTTGCT CCTTCAACCTTTGAGCAAGAAC	55	118–200	13
Swr1941	AGAAAGCAATTTGATTTGCATAATC ACAAGGACCTACTGTATAGCACAGG	55	215–255	13

(Continued)

Table 20.5: (Continued)

Locus name	Primers	Annealing temperature (°C)	Allele size (bp)	Chromosome number
Sw857	TGAGAGGTCAGTTACAGAAGACC GATCCTCCTCCAAATCCCAT	55	165–187	14
S0355	TCTGGCTCCTACACTCCTTCTTGATG TTGGGTGGGTGCTGAAAAATAGGA	50	196–215	15
Sw936	TCTGGAGCTAGCATAAGTGCC GTGCAAGTACACATGCAGGG	55	134–158	15
S0026	AACCTTCCCTTCCCAATCAC CACAGACTGCTTTTTACTCC	55	156–178	16
Sw24	CTTTGGGTGGAGTGTGTGC ATCCAAATGCTGCAAGCG	55	99–135	17
S0218	GTGTAGGCTGGCGGTTGT CCCTGAAACCTAAAGCAAAG	55	234–256	X

Table 20.6: Chicken-specific microsatellite markers recommended by FAO.

Locus name	Primers	Annealing temperature (°C)	Allele size (bp)	Chromosome number
ADL0268	CTCCACCCCTCTCAGAACTA CAACTTCCCATCTACCTACT	60	102–116	1
MCW0020	TCTTCTTTGACATGAATTGGCA GCAAGGAAGATTTTGTACAAAATC	60	179–185	1
MCW0248	GTTGTTCAAAAGAAGATGCATG TTGCATTAACTGGGCACTTTC	60	205–225	1
MCW0111	GCTCCATGTGAAGTGGTTTA ATGTCCACTTGTCAATGATG	60	96–120	1
MCW0034	TGCACGCACTTACATACTTAGAGA TGTCCTTCCAATTACATTCATGGG	60	212–246	2
LEI0234	ATGCATCAGATTGGTATTCAA CGTGGCTGTGAACAAATATG	60	216–364	2
MCW0206	CTTGACAGTGATGCATTAAATG ACATCTAGAATTGACTGTTCAC	60	221–249	2
LEI0166	CTCCTGCCCTTAGCTACGCA TATCCCCTGGCTGGGAGTTT	60	354–370	3
MCW0103	AACTGCGTTGAGAGTGAATGC TTTCCTAACTGGATGCTTCTG	64	266–270	3
MCW0222	GCAGTTACATTGAAATGATTCC TTCTCAAAACACCTAGAAGAC	60	220–226	3
MCW0016	ATGGCGCAGAAGGCAAAGCGATAT TGGCTTCTGAAGCAGTTGCTATGG	60	162–206	3
MCW0037	ACCGGTGCCATCAATTACCTATTA GAAAGCTCACATGACACTGCGAAA	64	154–160	3
MCW0098	GGCTGCTTTGTGCTCTTCTCG CGATGGTCGTAATTCTCACGT	60	261–265	4

(Continued)

Table 20.6: (Continued)

Locus name	Primers	Annealing temperature (°C)	Allele size (bp)	Chromosome number
LEI0094	GATCTCACCAGTATGAGCTGC TCTCACACTGTAACACAGTGC	60	247–287	4
MCW0284	GCCTTAGGAAAAACTCCTAAGG CAGAGCTGGATTGGTGTCAAG	60	235–243	4
MCW0295	ATCACTACAGAACACCCTCTC TATGTATGCACGCAGATATCC	60	88–106	4
MCW0081	GTTGCTGAGAGCCTGGTGCAG CCTGTATGTGGAATTACTTCTC	60	112–135	5
MCW0078	CCACACGGAGAGGAGAAGGTCT TAGCATATGAGTGTACTGAGCTTC	60	135–147	5
MCW0014	TATTGGCTCTAGGAACTGTC GAAATGAAGGTAAGACTAGC	58	164–182	6
LEI0192	TGCCAGAGCTTCAGTCTGT GTCATTACTGTTATGTTTATTGC	60	244–370	6
MCW0183	ATCCCAGTGTCGAGTATCCGA TGAGATTTACTGGAGCCTGCC	58	296–326	7
ADL0278	CCAGCAGTCTACCTTCCTAT TGTCATCCAAGAACAGTGTG	60	114–126	8
MCW0067	GCACTACTGTGTGCTGCAGTTT GAGATGTAGTTGCCACATTCCGAC	60	176–186	10
ADL0112	GGCTTAAGCTGACCCATTAT ATCTCAAATGTAATGCGTGC	58	120–134	10
MCW0104	TAGCACAACTCAAGCTGTGAG AGACTTGCACAGCTGTGTACC	60	190–234	13
MCW0216	GGGTTTTACAGGATGGGACG AGTTTCACTCCCAGGGCTCG	60	139–149	13
MCW0123	CCACTAGAAAAGAACATCCTC GGCTGATGTAAGAAGGGATGA	60	76–100	14
MCW0330	TGGACCTCATCAGTCTGACAG AATGTTCTCATAGAGTTCCTGC	60	256–300	17
MCW0165	CAGACATGCATGCCCAGATGA GATCCAGTCCTGCAGGCTGC	60	114–118	23
MCW0069	GCACTCGAGAAAACTTCCTGCG ATTGCTTCAGCAAGCATGGGAGGA	60	158–176	E60C04W23

20.2.6 Agarose gel electrophoresis

DNA fragments can be separated based on their size and quality, for which one of the techniques is gel electrophoresis. Electrophoresis creates an electrical field in which negatively charged DNA moves towards the anode. For electrophoresis two types of carbohydrates are used and they are agarose or acrylamide and bis-acrylamide. AGE is a method of separating and analyzing DNA on an electric field created in an environment prepared by polysaccharides like agarose. This method is simple, quick, with the possibility of the recovery of DNA from gel and is capable of resolving fragments of DNA that cannot be separated by other techniques, such as density gradient centrifugation, etc. Agarose gel can separate a long range of DNA varying from 200 to 50,000 bp but has a relatively lower resolving power than polyacrylamide gel. Agarose is a polymer of 1,3-linked β D-galactose and 3,6-anhydro αL-galactose, which is extracted from a seaweed, namely red algae. Agarose which is commercially available is not completely pure but contains other polysaccharides, proteins, and salts, which can affect the migration of DNA as well as the quality of the DNA recovered from the gel. Nowadays, chemically modified agarose is commercially available and this is used for preparative electrophoresis of DNA and in situ digestion of DNA with restriction enzymes on gel. A special grade of low-melting agarose is also available to detect very small fragments of DNA up to 10 bp. Agarose gels are prepared by melting the agarose in an appropriate buffer. Agarose in solid state forms a matrix on which an electric field is applied for separating DNA. DNA, which is negatively charged at neutral pH, migrates towards the anode. The rate of migration of DNA is influenced by several factors including the size of the DNA fragment, the concentration of agarose in the gel, the temperature of the gel, the current flow, the presence of intercalating dyes, the composition of the buffer, the conformation of the DNA, etc.

20.2.7 Polyacrylamide gel electrophoresis

Polyacrylamide gel electrophoresis (PAGE) is a method of separating DNA fragments/proteins depending on size, structure, and molecular weight (MW). The gel is prepared by polymerizing acrylamide with the cross-linking agent N,N'-methylenebisacrylamide (bis-acrylamide). The polymerization process is accelerated by ammonium persulfate and is stabilized by N,N,N',N'-tetramethylethylenediamine. The gel porosity is determined by the length of the acrylamide chain and the degree of cross-linking, which are caused by the concentration of acrylamide and bis-acrylamide. The advantages of PAGE are the higher resolving power (sometimes 1 in 500 bp), the electrophoresis of larger quantities of DNA at a time (sometimes 10 μg), and the recovery of extremely pure DNA from the gel (sometimes used in microinjection of embryo). There are two types of PAGE commonly used for various purposes and they are nondenaturing or native PAGE and denaturing PAGE. In native PAGE, double-stranded DNA migrates according to its sizes and the voltage applied in the gel is low (1–8 V/cm). In the case of denaturing PAGE, denaturing agents like urea, formamide, etc., are used to denature DNA. The movement of DNA is based on the composition and sequence of the fragment. This sort of gel is used for DNA sequencing.

20.2.8 SDS-polyacrylamide gel electrophoresis

For the detection and analysis of proteins, SDS-polyacrylamide gel electrophoresis (SDS-PAGE) is performed. This is a kind of PAGE. SDS, a detergent is used in the gel to denature protein so that large-sized protein molecule without branching can pass through the gel easily. Protein lysate can be run directly in this gel and can be stained either by Coomassie blue or silver staining to visualize the protein band (Bhattacharya et al., 2014). Without staining, the protein bands can also be transferred onto a membrane so that a protein blot can be detected by chemiluminescence (Western blotting).

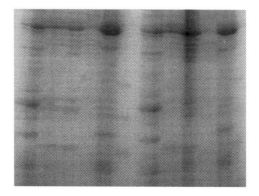

Muscle proteins separated in SDS-PAGE.

20.2.9 Western blot

After electrophoresis (PAGE), the gel is carefully removed from the glass plate and the protein is transferred to a PVDF membrane in the presence of tris-glysin–methanol buffer. Semidry transfer apparatus is used and transfer is carried out at 100 V for 1 hour. Following transfer, the membrane is washed in TBS buffer and then blocked with 3% BSA in TBS overnight at 4°C. The following day the blocked membrane is washed five times with TBS Tween. For assessment of the reactivity of the transferred protein, the blot is incubated with 1:500 dilutions (in TBS Tween 20) of anti-Rat antibodies under constant rocking at room temperature (Bhattacharya et al., 2016). Then the membrane is washed five times with TBS Tween buffer and treated with anti-Rat IGg HRP conjugate diluted to 1:1000 in TBS Tween 20 buffer (Satheesh et al., 2016). The conjugate is allowed to react for 1.5 hours under continuous rocking at room temperature. Then the membrane is washed and treated with DAB substrate. The substrate reaction is stopped using TBS buffer. The band appears if the antigen and antibody react, which shows the presence of protein in the lysate. Thus this technique can detect the presence of unwanted protein in the food materials.

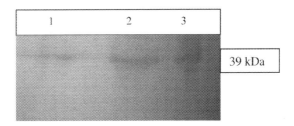

Western blot of a protein sample.

20.2.10 Two-dimensional gel electrophoresis

Two-dimensional (2D) gel electrophoresis is an electrophoresis technique to analyze proteins extracted from cells, tissues, or other biological samples by employing two properties in two-dimensions. This electrophoretic system was first introduced by O'Farrell and Klose in 1975 (O'Farrell, 1975). This technique separates proteins in two steps, that is, the first dimension is isoelectric focusing (IEF), which separates proteins according to their isoelectric points (pI); the second dimension is SDS-PAGE, which separates proteins according to their MWs. In this way, complex mixtures consisting of thousands of different proteins can be resolved and the relative amount of each protein can be determined. The procedure involves placing the sample in a gel with a pH gradient, and applying a potential difference across it. In the electrical field, the protein migrates along the pH gradient, until it carries no overall charge. There are two alternative methods to create the pH gradient—carrier ampholites and immobilized pH gradient gels. The IEF is the most critical step of the 2D electrophoresis process. The proteins must be soluble without charged detergents, usually in high concentrated urea solution, reducing agents, and chaotrophs. To obtain high-quality data it is essential to achieve low ionic strength conditions before the IEF itself. Since different types of samples differ in their ion content, it is necessary to adjust the IEF buffer and the electrical profile to each type of sample. In 1D electrophoresis proteins are separated in one dimension and they are engraved along a lane, while in 2D electrophoresis, molecules spread across the gel. Proteins are detected by two stains, that is, silver and Coomassie brilliant blue staining. Bioinformatic software tools analyze biomarkers by quantifying individual proteins, and showing the separation between one or more protein "spots" on a scanned image of a 2D gel. The important software packages are Bionumerics 2D, Delta 2D, Image master, Melanie, PDQuest, Progenesis, Redfin, etc. The main limitations in this method are incompletely separated spots or overlapping spots, weak spots, mismatched spots, etc. The protein spots may be sliced from the gel and proteins can be isolated from the gel slice and further used for mass spectrometry.

20.3 Some important requisites

20.3.1 Sterilization and disinfection of laboratory wares

The labwares like glasswares and plasticwares that we use in the laboratory should be dust-free, free from organic matter. They should not be contaminated with microorganisms. The following procedures are being adopted for the preparation of the labwares. The plasticwares like eppendorf tubes, microtips, PCR tubes, etc., are of single use and can be discarded after use. Different types of sterilization procedures are being practiced in the laboratory for obtaining accurate and precise results.

20.3.2 Sterilization of glasswares

Sterilization is the process of destroying all the living organisms, viruses and their spores, etc. Sterilization is of two types, viz., dry heat sterilization and moist heat sterilization.

20.3.3 Dry heat sterilization

It is carried out in hot air oven at 180°C for 2 hours. The metal items like forceps, spatulas, scissors, glasswares like beakers, glass measuring cylinders, glass pipettes, etc., can be autoclaved using dry heat sterilization.

20.3.4 Moist heat sterilization

Moist heat sterilization is achieved by subjecting the material to 121°C for 20 minutes at 15 lb pressure. The hot pressure of 15 lb raises the temperature to 121°C. Moist heat sterilization is more effective. The plasticware like eppendorf tube, PCR tubes, microtips, glasswares like reagent bottles, culture tubes, solution, reagents, media, etc., can be autoclaved using moist heat sterilization.

20.3.5 Disinfection and sterilization

In the molecular biology laboratory, we work with microorganisms. Cell cultures are grown for different types of scientific works. Many of the organisms may have artificially modified DNA. Such organisms are biohazards. DNA and proteins extracted from these organisms are also hazardous. We always work with certain chemicals which are mutagenic, carcinogenic, and may be neurotoxic also. Proper and safe handling of these cell cultures and contaminated labware is very important. The laboratory waste generated should be disposed of carefully. The contaminated waste generated can be decontaminated using chemical disinfectants (70% ethanol, 10% sodium hypochlorite, 3% phenol and its derivatives, etc.).

20.4 Conclusion

The genomic and proteomic tools are very efficient in detecting contamination, if present, in meat and other food products. These tools can be used as confirmatory diagnostic tools and need only very small quantities of materials to test. The only limitation is the requirement of sophisticated equipment and laboratory including expertise. The operational expenditure is a little higher than other subjective and biochemical assessment techniques. But continuous research on genomic and proteomic tools make these techniques a bit cheaper nowadays as compared to 20 years back. My suggestion is that this type of laboratory should be established at all corners of the country to check and examine the safety of all sorts of food materials before reaching the consumers. The food industry should encourage food safety testing through the molecular approach so that the results would be confirmatory and food becomes safer for consumption. Although the molecular tests by genomic and proteomic approaches are a bit expensive, their accuracy and efficiency make these tools more acceptable and adaptable not only in the developing countries but in developed nations also.

References

Bhattacharya, T.K., Kumar, P., Sharma, A., 2008. Animal Biotechnology: A Practical Guide for Molecular Biology, Cytogenetics and Immunogenetics. M/s Kalyani Publishers, Ludhiana, Punjab.

Bhattacharya, T.K., Chatterjee, R.N., Dushyanth, K., Shukla, R., 2014. Cloning, characterization and expression of myostatin (growth differentiating factor-8) gene in broiler and layer chicken (*Gallus gallus*). Mol. Biol. Rep. 42 (2), 319–327.

Bhattacharya, T.K., Shukla, R., Chatterjee, R.N., Dushyanth, K., 2016. Knock down of the myostatin gene by RNA interference increased body weight in chicken. J. Biotechnol. 241, 61–68.

Chatterjee, R.N., Bhattacharya, T.K., Dange, M., Rajkumar, U., 2010. Assessment of genetic relatedness of crossbred chicken populations using microsatellite markers. Biochem. Genet. Available from: https://doi.org/10.1007/s10528-010-9355-y.

Dhanasekaran, S., Bhattacharya, T.K., Chatterjee, R.N., Paswan, C., Dyushanth, K., 2014. Functional Genomics in Chicken (*Gallus gallus*) – status and implications in poultry. World Poult. Sci. J. 70, 45–56.

Guru Vishnu, P.B., Bhattacharya, T.K., Kumar, P., Chaterjee, R.N., et al., 2017. Expression profiling of activin type IIB receptor during ontogeny in broiler and indigenous chicken. Anim. Biotechnol. 28 (1), 26–36.

O'Farrell, P.H., 1975. High resolution two-dimensional electrophoresis of proteins. J. Biol. Chem. 250, 4007–4021.

Satheesh, P., Bhattacharya, T.K., Kumar, P., Chatterjee, R.N., et al., 2016. Gene expression and silencing of activin receptor type 2A (ACVR2A) in myoblast cells of chicken. Br. Poult. Sci. 57, 763–770.

SECTION 12

Sensory evaluation techniques

CHAPTER 21

Innovation in sensory assessment of meat and meat products

Sonia Ventanas[1], Alberto González-Mohino[1], Mario Estévez[1] and Leila Carvalho[2]

[1]IproCar Research Institute, University of Extremadura, Caceres, Spain [2]Postgraduate program in Food Science and Technology, Department of Food Engineering, Federal University of Paraiba, João Pessoa, Brazil

Chapter Outline

21.1 Introduction 393
21.2 Quick-fast descriptive sensory techniques 394
    21.2.1 Napping 396
    21.2.2 Flash profile 397
    21.2.3 Check-all-that-apply 400
    21.2.4 Rate-all-that-apply 401
21.3 Dynamic sensory techniques 402
    21.3.1 Time–intensity 402
    21.3.2 Temporal dominance of sensation 404
    21.3.3 Temporal check-all-that-apply 406
21.4 Emotions in sensory meat science 407
    21.4.1 Emotions 407
    21.4.2 Emotions elicited by food: methodological approaches 408
    21.4.3 Examining extrinsic factors that influence the emotions elicited by foods 409
    21.4.4 Emotional responses to meat and meat products 410
References 412
Further reading 418

21.1 Introduction

Sensory evaluation is often described using the definition from the Institute of Food Technologists—a scientific method used to evoke, measure, analyze, and interpret those responses to products as perceived through the senses of sight, smell, touch, taste, and hearing (Anonymous, 1975). Consumer preferences and satisfaction are principally dependent on the sensory properties of foods (Tuorila and Monteleone, 2009). Prior to any food production, marketing, or distribution, it seems unavoidable to estimate the extent to which its sensory quality would be accepted by consumers. In the field of meat science the understanding of consumer's responses to meat consumption is critical given the major role that organoleptic properties play in the purchasing decision and final acceptability of meat and meat products (Grunert et al., 2004). Satisfying sensory traits, as a reflection of what consumers understand by meat quality (freshness, safety, adequate production, storage conditions, etc.) are one of the driving forces for consumers to eat meat and meat products (Mcilveen and Buchanan, 2001; Verbeke et al., 2010). Yet this discipline has been underestimated and sensory assessment has been relegated to a minor role as complementary analyses to instrumental and/or biochemical studies on meat color (Faustman et al., 2010), flavor (Shahidi et al., 1986), and texture (Cavitt et al., 2005).

Sensory food science is a multidisciplinary and complex discipline in which both human sensory perceptions and emotional responses to foods are studied. While sensory evaluation is the main method of analysis, sensory food science also makes use of physicochemical, physiological, physiological, and consumer-based research methods. In recent decades this field has evolved to fulfill an increasing necessity for an in-depth comprehension of the sensory properties of foods and their impact on consumers. In this line, sensory-based studies have enabled a considerable progress in our understanding of consumer responses to foods and to emergent trends of food manufacture, handling, and consumption. Some advances rely on fast, accurate, informative, and intuitive descriptive methods such as napping, flash profile (FP) (Dehlholm et al., 2012), check-all-that-apply (CATA) (Jorge et al., 2015), and rate-all-that-apply (RATA) methods (Ares et al., 2018). Other innovative sensory methodologies aim to approach more realistic assessment models by analyzing how sensations vary over time. Among these temporal sensory techniques the time—intensity (TI) (Dijksterhuis and Piggott, 2001) and temporal dominance of sensations (TDS) (Ares et al., 2015a) are of particular scientific and technological interest. Finally, the assessment of the emotions evoked by foods in consumers and the influence of the scenario (external factors) is a true hot topic in the field in which food science finds support from other disciplines such as psychology, to build a new and exciting dimension of the sensory food science.

Yet meat scientists have generally not taken advantage of these innovative approaches, signifying sensory meat science as research field of great potential interest in these advanced methodologies. The present review aims to counteract the lack of comprehensive review articles on the recent advances of sensory evaluation of meat products. The most innovative sensory techniques and their applicability to muscle foods are reported.

21.2 Quick-fast descriptive sensory techniques

The sensory profiling of foods has been traditionally assessed by trained panels using descriptive sensory analyses. Although these sensory methodologies provide detailed, accurate, and consistent results, they are generally time-consuming and expensive. This makes it difficult for food companies to carry out product development and innovation based on sensory science (Ares and Jaeger, 2013). In this context, quick-fast descriptive techniques, including CATA, RATA, napping, and FP, were introduced from the necessity of developing reliable and rapid methods for sensory characterization of foods. Table 21.1 summarizes the application of these fast techniques to a variety of muscle foods.

Table 21.1: Application of fast sensory techniques to assorted muscle foods.

Food category	Product	Sensory technique	Reference
Meat	Chicken (images)	Napping	Daltoé et al. (2017)
Meat products	Bacon	CATA	Saldaña et al. (2019)
		Napping	Saldaña et al. (2018)
	Burgers	CATA	Heck et al. (2017), Neville et al. (2017), Heck et al. (2019)
		RATA	Schouteten et al. (2015)
	Chicken breasts marinades	CATA	Choi et al. (2015)
	Deli-style turkey	CATA	Grasso et al. (2017)
	Dry-cured loins	Flash profile	Lorido et al. (2018)
	Ham	CATA	Henrique et al. (2014), Oliveira et al. (2018)
	Kitoza	Flash profile	Pintado et al. (2016)
	Meatballs	CATA	Tan et al. (2017)
	Mortadella	CATA	Jorge et al. (2015), Saldaña et al. (2018)
		Flash profile	Santos et al. (2013)
	Salami	RATA	Ares et al. (2018)
	Sausages/fermented sausages	CATA	Santos et al. (2015), Meier-Dinkel et al. (2016), Yotsuyanagi et al. (2016), Alves et al. (2017), Neville et al. (2017), Massingue et al. (2018)
		Flash profile/ napping	Grossi et al. (2011, 2012)
	Sheep meat coppa	CATA	Andrade et al. (2018)
	White pudding	Flash profile	Fellendorf et al. (2015)
Fish	Fish (images)	Napping	Daltoé et al. (2017)
	Namely wreckfish, greater amberjack, gray mullet, meager, and pikeperch	CATA	Lazo et al. (2017)
Fish products	Nuggets	Flash profile/ napping	Albert et al. (2011)
Seafood	Mussels, lobster, squid, and abalone	CATA/RATA	Ares and Jaeger (2017)
Other	Liver pâté	Flash profile/ napping	Dehlholm et al. (2012)

21.2.1 Napping

Projective mapping was developed by Risvik et al. (1994) in order to avoid the disadvantages described for conventional descriptive techniques: time-consuming and intensive panel training. Pagès (2005) introduced the napping, a version of this rapid descriptive method. According to this method, untrained participants position the samples in the space enclosed by a sheet of paper considering their similarities and dissimilarities. Napping offers a quick discrimination between samples, while no description of the samples is obtained. Thus napping is commonly combined with other sensory tests such as FP, and in particularly ultra-flash profiling (UFP) (Perrin et al., 2007, 2008).

It is common to use a rectangular framework for samples positioning as panelists perceive the wider horizontal direction as being more important when they locate the samples on the map (Dehlholm, 2014). Assessors receive the whole sample set, allowing a straightforward comparison among samples. Subsequently, panelists have to place samples perceived as similar close to each other and samples perceived to be more different further apart. Hence each position for each sample represents two coordinates, X and Y axes. In the case of napping combined with UFP, the panelists should write the attributes that they consider that better described these samples, giving a qualitative component to the napping evaluation. Finally, all the data is collected in matrices. A sensory map is the outcome from the application of napping to food products and multifactorial analysis is commonly applied (Lê et al., 2016). Sorted napping is frequently supplemented by Euclidean distance with the sensory plot (Saldaña et al., 2018; Siegmund et al., 2018). This intuitive spatial discrimination of samples can be performed by consumers (Saldaña et al., 2018) and even children (Daltoé et al., 2017). However, napping shows certain drawbacks and limitations: (1) this method is inappropriate for samples that need processing prior to the presentation to the panelist and (2) the collection of the data takes longer than other rapid descriptive techniques, although recent works solved this problem (Lê et al., 2016). Like many other rapid sensory methods, napping has been subjected to modifications by subsequent scientific reports. For instance, napping can be used without any restriction (also known as "global napping"), or restricting/controlling some criteria ("partial napping"). While panelists have total freedom to locate the samples in global napping, in partial napping the assessors are instructed on the attributes or categories of attributes to be used to perform the positioning of the samples (Dehlholm et al., 2012).

Napping has been applied to several foods, including meat and meat products. Pagès (2005) originally applied napping to 10 different wines, showing the ability of this technique to discriminate among samples. However, results were interpreted with the help of conventional descriptive methods. Subsequently, Perrin et al. (2008) combined napping with UFP for discrimination between different types of wines and compared the results with those obtained from a conventional descriptive method. According to the authors, napping-UFP was a reliable method to describe a sample from a sensory point of view. Many recent studies reported the use of napping-UFP in assorted food items such as apples (Pickup et al., 2018), strawberries (Oliver et al., 2018), and honey (Siegmund et al., 2018). Grossi et al. (2011) analyzed the sensory properties of pork sausages with added carrot fiber and treated with high pressure using the napping-UFP methodology. Panelists were able to distinguish between groups of samples with napping positioning and furthermore, UFP identified homogeneity and firmness as discriminating attributes. In a subsequent study from the same authors, different levels of salt, carrot fiber, potato starch, and temperature treatments, were applied to pork sausages (Grossi et al., 2012). Napping-UFP proved again its discrimination ability, with texture attributes were the most cited by the panelists as relevant to discriminate between formulations and treatments. Dehlholm et al. (2012) used both variations of napping, global and partial napping, for characterizing liver pâtés with different formulations and brands. Napping-UFP had the ability to distinguish between the nine different types of liver pâté samples. Additionally, partial napping was found to exhibit a good correlation with conventional descriptive methods, highlighting the advantage of the fast technique versus the traditional time-consuming method. Santos et al. (2013) applied napping for describing sensory attributes of mortadella with different fat levels and adding different prebiotic ingredients. Each panelist generated his/her own vocabulary for the evaluation, with red color, hardness, and spicy taste being the most commonly used terms. While the results showed an efficient discrimination between samples, the authors stated that napping was limited compared to conventional descriptive methods. Finally, Saldaña et al. (2018) tested napping-UFP for the characterization of bacon samples smoked with different woods and correlated these results with volatile compounds. Napping successfully characterized the samples discriminating the woods used for smoking. In addition, sensory attributes were correlated with volatile compounds, showing that changes in the volatile profile of smoked bacon had an impact on the consumer perception of samples. Napping-UFP has been proved to be a quick and flexible sensory technique, that is intuitive and successful to describe and discriminate between different products with untrained panelists.

21.2.2 Flash profile

FP is a rapid descriptive technique developed by Sieffermann (2000). This sensory method, seemingly most related to conventional sensory techniques, is based on quantitative evaluation by means of sensory attributes freely chosen by the panelists (Delarue, 2015). The FP arises from free-choice profiling (FCP) (Williams and Langron, 1984), based, in turn, on the free choice of attributes by panelists. In addition to FCP, FP also enables a comparative evaluation of assorted types of product. This comparative feature allows the panelist to focus on the major differences between samples, showing an advantage compared to FCP. Therefore this technique consists of asking the panelists to describe a particular product using their own evocative terms in order to subsequently rank the products for each of those generated terms (Lorido et al., 2018). This procedure is generally performed in two steps, either in one or two sessions. In practice, samples are presented all together, coded with a letter or three-digits randomly. In the first step, assessors taste the products comparatively in order to generate defining and discriminating descriptors. Each participant generates his/her own set of attributes. No limitation of the number of attributes is indicated but panelists are instructed to avoid hedonic terms. In the second step panelists rank all samples from "low" to "high" (or "weak" to "strong," or "least" to "most," etc.) for each selected attribute, using a scale. Assessors could retaste the samples as often as required. The number of assessors depends on the objective of the study. Delarue and Sieffermann (2004) noted a minimum of four or five participants in FP evaluations. Individual matrices for each consumer (products × attributes) are built in order to enter product rankings from FP. A sensory map is the main result provided by FP, with the generalized procruster analysis (GPA) (Fig. 21.1) being the most used to compute the data provided by the panelists. Additionally, a hierarchical cluster analysis (HCA) is commonly performed to collect different groups of attributes, samples, etc. Finally, as Delarue (2015) reported, it is possible to run a multivariate analysis on variance or a discriminant analysis on the data after GPA.

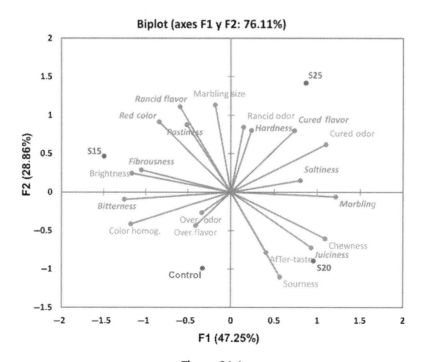

Figure 21.1
Example Biplot of General Procrustes Analysis (GPA) of data from ultra-flash profile of dry-cured loins (Lorido et al., 2018).

FP is one of the faster, more flexible, and novel sensory methodology to characterize food products. In fact, when the food products belong to the same or to similar product categories, FP can be more discriminating than conventional profiling (Delarue and Sieffermann, 2004). Furthermore, FP can be completed within a single session, while the duration of the session is usually longer, depending on the aim of the study. Another benefit of FP is the usage of untrained panelists who, yet, should be familiar with the product under evaluation. Besides that, it is not necessary an extensive training on the product properties, as other traditional descriptive methods. This fact and the freedom to generate attributes, allow assessors without any experience in sensory evaluation be part of the panel, facilitating the recruitment of panelists. Thus as other authors reported, this technique can be used for consumers (Bredie et al., 2018). However, FP have limitations too. FP lacks accuracy compared to conventional descriptive techniques, since the outcome of the former relies on the work of single assessors while the results from the latter relies on the consensus of the panel. Secondly, the results are frequently shown in sensory maps, which limit the comparison of FP results with those from other conventional descriptive methods. Lastly, having all samples at the same time, makes this technique inappropriate for certain products, for example, those that need to be freshly prepared before consumption. Assessors normally generate a large number of terms and inconsistencies and also redundant terms are commonly found. In order to ease the development of individual vocabularies and thus to improve the sample ranking and the interpretability of the results, a modification to the FP method was made, the UFP. Perrin et al. (2008) developed a simplified version of FP, in which the number of terms are reduced by grouping synonyms and terms with comparable meaning.

FP has been widely studied in food, and has become somewhat popular in food companies. Dairou and Sieffermann (2002) originally applied FP to characterize 14 different types of jams and compare the results from those obtained applying quantitative descriptive analysis (QDA). Both methods led to comparable results and FP appeared as a fast alternative method to characterize food products. To similar conclusions came Albert et al. (2011) comparing QDA and FP for assessing hot served food. This quick sensory method has also been proven as a suitable discriminative sensory technique in meat or meat products. Pintado et al. (2016) applied FP in Kitoza, traditional meat products from Madagascar made of pork or lean beef and bioprotective cultures. Beef and pork Kitoza were analyzed together with a traditional Portuguese sausage. FP method demonstrated an efficient discriminant ability, providing each sample with specific and defining attributes. Similar results were reported by Lorido et al. (2018) testing the ability of FP to discriminate between dry-cured loins with different levels of NaCl and partial replacement by KCl. For gaining additional insight into the sensory properties of the food products to be characterized, FP may also be used in combination with other sensory techniques, such as napping (Pintado et al., 2016), or dynamic sensory techniques (Lorido et al., 2018), for example.

21.2.3 Check-all-that-apply

The CATA is a questionnaire in a multiple choice question format (Ares and Jaerger, 2013). In this method, a list of terms previously generated by a group of consumers or by a trained sensory panel, is presented to consumers for selection of all the terms that they consider appropriate to define the sensory profile of a particular food product (Henrique et al., 2014). This method has the advantage of cost feasibility, effectiveness in consumer perception, compatible results with the descriptive analysis (Choi et al., 2015), requires minimum instructions (Henrique et al., 2014), is quick to design and execute, is simple to be performed (Saldaña et al., 2019), is less labor-intensive (Reinbach et al., 2014), and has versatility in terms of attributes since the descriptors are not limited to sensory attributes of the products, but may also be related to product use or their concept (Dooley et al., 2010). However, this method requires a larger group of consumers (Reinbach et al., 2014), and the CATA method may have low discrimination power in comparison with other techniques, because the attribute intensities are not measured (Dooley et al., 2010; Reinbach et al., 2014; Ares et al., 2014). In addition, there may be uncertainty over the meaning of some CATA terms (Jaeger et al., 2015), and the selection of the number and order of attributes used in the CATA list should be very carefully checked (Choi et al., 2015). The choice, order, and number of terms to be used in the CATA questions are extremely relevant issues. When consumers repeatedly assess terms in a particular order, they may fail to read and select relevant terms which were considered not applicable for describing the previous samples (Ares et al., 2015b). Furthermore, the order in which terms are included may affect the distribution of responses for some sensory attributes and the conclusions about product differences. The attributes located on the top of the question list tend to be more frequently selected (Ares and Jaerger, 2013). According to Jaeger et al. (2015), a long list of terms may provide a complete description of sensory attributes of the products and would help test administrators to check the validity of consumers' responses through the inclusion of antonym terms. These authors observed that the use of 10–28 CATA terms had little impact on the sensory characterizations of the products, but there was a dilution effect on citation frequency of the terms. However, a large number of terms can induce consumers to give satisfactory responses, allowing them to select the terms that most catch their attention (Jaeger and Ares, 2014).

Although CATA method has been applied for sensory characterization of a wide range of products it has not been extensively used for meat and meat products (Table 21.1). Some of these studies have compared CATA with QDA to show the effectiveness in consumer perception (Santos et al., 2015; Choi et al., 2015; Saldaña et al., 2018). Among others, CATA has been used to assess sensory profiling of fish (Lazo et al., 2017), seafood (Ares and Jaeger, 2017), and commercial meat products (Jorge et al., 2015); to analyze the acceptance of products with reduction or replacement of NaCl (Henrique et al., 2014; Santos et al., 2015; Yotsuyanagi et al., 2016; Alves et al., 2017); to evaluate products with partial protein replacement (Neville et al., 2017; Tan et al., 2017); to determine the impact of lipid replacement on the sensory profile (Heck et al., 2017, 2019; Oliveira et al., 2018; Saldaña et al., 2018); to study the development de new products (Meier-Dinkel et al., 2016; Andrade et al., 2018; Saldaña et al., 2019); to analyze the consumer perception and purchase intent of meat products with health claim (Grasso et al., 2017); to use ingredient of low cost (mechanically deboned meat) in meat products (Massingue et al., 2018); and to investigate the motivations behind everyday choices of different food groups (Phan and Chambers, 2016).

21.2.4 Rate-all-that-apply

The RATA questionnaire is a variant of CATA, with this new method consisting of asking consumers to rate the intensity of the sample attributes. Its purpose is to increase the ability to discriminate between samples with similar sensory profiling but different intensity of particular attributes (Ares et al., 2018). Including intensity scaling of attributes improves the accuracy of descriptive profiling and leads to a better product differentiation compared to the CATA questionnaire (Reinbach et al., 2014; Ares and Jaeger, 2017). According to Ares et al. (2014), the selected terms can be rated on intensity using a three-point scale ("low," "medium," or "high") and on applicability using a five-point scale (from "slightly applicable" to "very applicable"). RATA technique has the advantage of quick and easy data collection from a large number of consumers, and the reliability of the values for intensity assessment of sensory attributes (Schouteten et al., 2015). However, a comparison between descriptive analyses and RATA showed that the two methods generated different information in samples with complex sensorial properties, and trained assessors outperform consumers in the ability to discriminate samples for complex sensory attributes or attributes relating to specific flavors (Ares et al., 2018).

A limited number of studies on the application of the RATA method for sensory characterization of meat products by consumers is available in the scientific literature (Table 21.1). Yet these studies illustrate the feasibility of using a RATA questionnaire to obtain both emotional and sensory profiles of burgers (Schouteten et al., 2015), and to compare descriptive analysis by trained assessors with product characterizations from consumers (Ares et al., 2018). Ares and Jaeger (2017) compared RATA and CATA methods using emoji for seafood products.

21.3 Dynamic sensory techniques

The classic and most employed sensory technique in meat science, namely, QDA, assesses the perception of attributes as a "static" phenomenon. Therefore assessors integrate their responses to provide time-averaged single intensity values (Cliff and Heymann, 1993). However, sensory perception is a dynamic process, and hence the detection, intensity, and impact of particular sensory attributes on consumers vary over time during the experience of food consumption (Dijksterhuis and Piggott, 2001). In particular flavor and texture perceptions are dynamic, and hence an accurate and more realistic evaluation of consumer responses requires a temporal assessment of the variations of such attributes during food consumption. In this line, the first attempts to record dynamic sensory perceptions during food consumption were carried out by Sjöström (1954) and Jellinek (1964) who assessed temporal changes in the perception of bitterness in beers. The first TI curves were constructed by plotting the x (time) and $-y$ (bitterness perception) coordinates on graph paper. After 40 years of application in assorted food items, the first relevant review paper on the TI methodology was written by Cliff and Heymann (1993). The authors found, at that point, no application of the TI technique on muscle foods. Today, more than 25 years later, TI and other dynamic sensory techniques have been scarcely used to assess the temporal perception of flavor and texture in muscle foods. According to Scopus only 10 original research articles deal with TI ("article title" as search criteria) in meat and meat products ("title, abstract, keywords" as search criteria). In the following subsections various dynamic sensory techniques with straightforward applicability to meat products are described.

21.3.1 Time—intensity

The TI method is, among the dynamic sensory evaluation techniques, the first to be developed for assessing temporal perception of sensory attributes in food items (Sjöström, 1954). TI allows the temporal evaluation of the intensity of a given attribute by trained panelists (Cliff and Heymann, 1993). The outcome is a set of very intuitive graphical representations (TI curves; Fig. 21.2) of the increases and decreases of the intensity of the sensory perception over time (Dijksterhuis and Piggott, 2001). The record of the variations in perception is carried out by developed computerized TI systems and assorted hardware and software products are available. Yet the Fizz software is probably the most commonly used by food scientists. The assessor normally indicates the perceived response using a marker on a 10 cm nonstructured vertical scale and data is collected in real time over a period of time. The scales are commonly anchored with "less" and "more." The protocol followed during TI evaluation has to be fixed by panelists prior to the assessment of samples. Typically assessors would keep the sample in their mouths, chew, and start the evaluation. After swallowing, the panelists would continue the evaluation until perception of the attribute under study disappears. The panelists are required to move the cursor along the scale according to

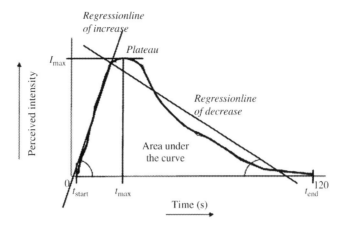

Figure 21.2
Graphical representation of a time—intensity curve (Dijksterhuis and Piggott, 2001).

the intensity of their perception. The intensity recordings start when assessors click on the scale and stop whenever the assessors return the marker to the lowest value in the scale, meaning that the attribute is not perceived any more.

A number of parameters, calculated from TI curves (Imax, maximum intensity; Tmax, time to achieve the maximum intensity; DurPl, duration of maximum intensity; Tend, total duration of perception; AreaTse, area under the curve, SIMInc, maximum slopes of the increasing portion of the curve; SIMDec, maximum slopes of the decreasing portion of the curve, etc.), enable an unbiased assessment of time-based changes as well as the comparison between TI curves obtained for different products, panelists, sessions, etc. However, there are individual differences between the curves generated by the panelists, requiring an exhaustive training to reduce this source of variation (van Buuren, 1992). According to Peyvieux and Dijksterhuis (2001) training assessors in the TI technique involves three stages. In the first stage panelists are introduced to the methodology and the computer system. At a second practical stage panelists are requested to assess solutions of basic tastes (sweet, salty, sour, bitter, and umami) at concentrations above the threshold of perception. Once panelists are familiar with the procedure and software interface, they are trained using the product of interest, which includes a basic sensory profiling using a static sensory method (i.e., QDA) and a subsequent specific training with TI technique.

The first applications of TI technique in a meat samples were carried out by Duizer et al. (1993) and Brown et al. (1996) to assess the temporal perception of tenderness in cooked pork and beef. Some of their early findings into the relationship between masticatory patterns and perceived tenderness were further confirmed by subsequent studies. Foster et al. (2011) reviewed the role of oral processing in the dynamic perception of flavor and texture in solid foods. Piggott (2000) collected information on the dynamism of flavor perception and comprehensively reviewed the interconnections between physiological patterns, flavor chemistry and release, and the outcome from the application of TI techniques. More recent applications of temporal approaches to the sensory evaluation of meat products have provided insight into assorted topics of technological interest such the role of salt and fat content on the temporal perception of flavor and texture in cooked bologna sausages (Ventanas et al., 2010) and the temporal perception of tenderness and juiciness in grilled beef strip loin steaks (Gomes et al., 2014). Other applications dealt with the interaction between oral burn, meat flavor, and texture in chili-spiced pork patties (Reinbach et al., 2007) and the effect of beverages on residual spiciness caused by spicy chicken meat (Samant et al., 2016). Fuentes et al. (2013) and Lorido et al. (2014) originally applied TI to assess the temporal sensory attributes to dry-cured meat products. In subsequent studies, the same authors characterized dry-cured hams (Lorido et al., 2015, 2016) and dry-cured loins (Lorido et al., 2018), comparing the outcome from the TI with those from other innovative sensory techniques such as FP and TDS.

21.3.2 Temporal dominance of sensation

A clear limitation of TI is its unidimensional nature as only one sensory attribute can be assessed at a time. TDS was introduced to achieve a temporal sensory profiling of a food item by producing a sequence of perceived attributes in which particular attributes are identified as "dominants" at particular points (or periods of time) during the dynamic assessment (Pineau et al., 2003). Schlich (2017) recently highlighted some of the benefits of applying TDS to food analyses and included (1) ease to apply within one intake (mouthful or sip) or during a longer intake process (i.e., food portion); (2) possibility of be paired with other techniques such as temporal liking, wanting, satiation, and emotions, and (3) feasibility to be applied with consumers with light training. It is worth noting that TDS does not ask panelists for intensity scores but dominant sensations, with this task being much simpler for untrained assessors. The complex concept of "dominance" was originally defined by the architects of TDS as the most intense attribute, and hence panelists were requested to score intensity of that dominant attribute (Labbe et al., 2009). In due course the authors redefined the dominant attribute as a "popping-up" sensation that "triggers the most of your attention" at a given moment and that was not hence quantified, simply identified among a total of 8–10 sensory attributes (Pineau et al., 2009, 2012). Naturally such a striking sensation may change during the consumption process. Assessors may select various dominant sensations over time, providing a temporal sensory profiling of the food under study. During the panel training, this concept can also be illustrated through sounds, with the attributes being the different instruments of an orchestra, as proposed by Lannuzel and Rogeaux (2007). The most triggering sound may not be necessarily the louder one, but the one conveying the most salient melody and catching the audience's attention.

The outcome of the assessment is the TDS curves which solely rely on the selection of an attribute as dominant and hence, no data concerning attribute intensity is shown. TDS curves show the dominance rates of attributes (y-axis) against time (x-axis) for each sample. The dominance rate is defined as the percentage of selections of an attribute as dominant at a particular time point and is calculated by dividing the number of citations of an attribute (all replications) by the number of runs (judge × replication) (Ng et al., 2012). The dominance rate reflects of consensus among judges and therefore can be regarded as a measurement of panel performance (Pineau et al., 2009). To assist the interpretation of TDS results, two lines representing the chance level and significance level are drawn on TDS curves. The chance level is the dominance rate that an attribute can obtain by chance (Pineau et al., 2009). It is calculated as $P_0 = 1/p$, where p is the number of attributes. The significance level indicates the value that must be reached (normally $\alpha < 0.05$) for the dominance rate to be considered significantly higher than the chance level (Pineau et al., 2009). It is calculated as $P_s = P_0 + 1.645 \, [P_0 \, (1 - P_0)/n]^{1/2}$, where n is the number of runs (assessors × replicates). Fig. 21.3 shows the data processing required to build TDS curves as developed by Pineau et al. (2009).

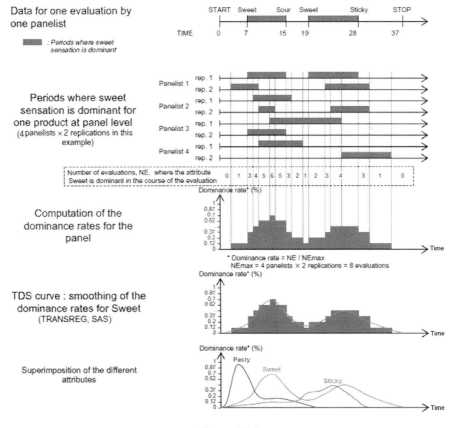

Figure 21.3
Methodology to compute TDS curves (Dijksterhuis and Piggott, 2001).

The application of this novel dynamic sensory technique to meat products, though scarce, has been found to offer sound and innovative results. A first approach was carried out by Paulsen et al. (2014) who studied the impact of NaCl substitution by assorted salt replacers on the temporal perception of flavor and texture in grilled sausages. Compared to the results from the classic QDA, TDS revealed unknown sensory descriptions of NaCl substitution in meat products. The dominant sensations during the mastication period changed in sausages with NaCl replacement with juiciness being one of the most affected textural attributes. The authors also found significant changes in the temporal aftertaste sensation in samples formulated with KCl. While the insight provided by the technique was emphasized, the authors claimed that a previous QDA may be required to have background sensory information on the samples. Other additional sensory analyses such as a hedonic test may also be performed to disclose the extent to which the changes on the temporal perception have an impact on the overall acceptance of the products. Lorido et al. (2016) simultaneously applied TI and TDS to dry-cured hams produced from pigs with different feeding background and varying in NaCl content. While both dynamic techniques provided consistent results, TDS enabled a more efficient discrimination between different types of ham. The authors highlighted the fact that TDS is a descriptive multiattribute methodology that addresses interactions among attributes during the process of selecting the dominant sensations. The study also indicated the feasibility of partially reducing NaCl content in dry-cured hams without affecting the temporal sensory properties of the final product. In a subsequent study, the same authors provided original insight into the dynamic sensory properties of dry-cured loins (Lorido et al., 2018). TI and TDS were found to provide complementary results which recommends using both temporal techniques when a thorough sensory examination of the samples is aimed.

21.3.3 Temporal check-all-that-apply

Temporal check-all-that-apply (TCATA), as a temporal extension of CATA, is the most recent dynamic sensory technique to be applied to food products. Originally described for cosmetic creams (Boinbaser et al., 2015), the method was subsequently applied to food products to assess multidimensional sensory properties as they evolve over time during consumption (Ares et al., 2015b; Castura et al., 2016). Trained panelists are requested to select (and deselect) sensory attributes freely and continuously over time, leading to a temporal characterization of the products. For instance, when assessing a salted and dried meat sample using TCATA, assessors may indicate a simultaneous perception of hardness and saltiness. As chewing progresses, assessors may indicate that the product is no longer hard, and the attribute is unchecked, while softness may become applicable and checked together with other new perceived sensations. Up to 10 attributes can be selected simultaneously and tracked over time, which allows the report of sensations that arise either sequentially or concurrently. According to Ares et al. (2015a), TCATA and TDS complement each other though the former provided a more detailed description of the dynamic sensory properties of assorted food products and enabled a more efficient discrimination between types of samples. Yet TCATA does not provide information on the dominant sensations and none of them predict consumers' hedonic perceptions of the products. The same authors subsequently proposed a variant of the TCATA in which some particular sensations are automatically unchecked as their presence fades over time (TCATA fading). While both techniques led to similar results and conclusions, the latter version seemed to offer more accurate temporal sensory profiling of the products.

The application of these dynamic methodologies is really scarce in meat products but has already been used to characterize various beverages such as sparkling wines (McMahon et al., 2017), fermented dairy products (Esmerino et al., 2017), and beer (Mitchell et al., 2019). Working with semisolid food samples, Nguyen et al. (2018) recently compared TCATA, TDS, and "TDS by modality" a TDS variant in which assessors evaluate temporal perception of flavor and texture in two consecutive and different steps. According to the authors, this modality facilitates evaluation as assessors concentrate on dominant sensations in each modality leading to a more detailed and precise sensory description. A recent study by Meyners and Castura (2018) delivers supportive information on how to analyze data obtained from the application of TCATA to food products. The valuable paper reports statistically valid analysis of data, defines chance and significance limits, and emphasizes that some of the outcomes can also be applied to other dynamic sensory techniques previously explained, such as TDS.

21.4 Emotions in sensory meat science

21.4.1 Emotions

Emotions and their relationship with food are complex and at least two different dimensions should be considered. One comprises how our mood and feelings influence the way we eat in terms of choices, motivation to eat, intake characteristics, etc., leading to an "emotional eating behavior" that can be associated with some eating disorders (Polivy and Herman, 1999). On the other hand, food has a great influence on people's mood and emotions (Köster and Mojet, 2015). Emotions elicited by food before, along, and after consumption has concentrated an increasing interest in sensory food science owing to the potential impact of this dimension on food acceptance or linking, on buying or choice predictions, and overall consumer's behavior (Lagast et al., 2017). Evidence confirms that the emotional profiles of food products add additional and valuable information to food acceptance (Gutjar et al., 2015) supporting that acceptability does not play an exclusive role in product success and sale predictions (Gutjar et al., 2015). In terms of the decision to purchase food or even consumption, two cognitive processes have traditionally been considered, intuitional and rational. Intuitive processes are fast, automatic, associative, and emotional in nature, while rational thinking or reasoning is a slow, effortful, and controlled process (Jiang et al., 2014). This intuitive dimension and the emotions elicited by food are directly related to food characteristics but also by other factors, such as the context or scenario, the ambiance and the previous consumption experiences with that particular food. King and Meiselman (2010) made a clear distinction between mood and emotion and defined emotion as an affective behavior characterized by being "brief, intense, and focused on a referent," whereas mood is "more enduring, builds up gradually, more diffuse, and not focused on a referent."

21.4.2 Emotions elicited by food: methodological approaches

From a methodological point of view, several techniques have been introduced to evaluate the emotional dimension of food exposure and consumption. Methods for emotions' elicitation can be classified as implicit or explicit based on how emotional responses are assessed (Lagast et al., 2017). The implicit methods are indirect and nonself-reported and register emotions while participants are consuming, smelling, or looking at food including physiological measures, expressive measures, or implicit behavioral tasks measures. In the explicit methods, participants report their own emotions or feelings through consumption using visual or verbal-self reported measurements—these methods are the most commonly used.

Explicit measurement of emotion was most frequently applied by using an emotional lexicon, meaning a questionnaire format with a list of emotional terms (or sentences) that can be checked [e.g., CATA or rated (RATA) or quantified (five-point rating scale)]. One of the most representative and applied methods among the explicit methods is the EsSense Profile questionnaire developed by King and Meiselman (2010) which incorporated both acceptability and emotions measurements. This method provides a list of 39 emotional terms selected by consumers to be appropriate to foods. In fact most existing available emotional questionnaires were developed for clinical purposes, which are focused on negative more than on positive terms. However, eating food is generally considered to be an enjoyable and pleasurable experience and thus mostly positive emotions (25) are included in the EsSense Profile questionnaire together with negative (3) and those classified as "unclear" (11). This list may be expanded or modified based on the specific product category or application. Data collection was made using CATA (frequency of citations) followed by rating the emotions in a five-point scale to obtain emotion scores. Most recently a short version of the EsSense Profile questionnaire has been produced by Nestrud et al. (2016) by reducing the emotional terms from 39 to 25 to create the EsSense25 with successful results in different product categories (den Uijl et al., 2016a,b; Borgogno et al., 2017; Mora et al., 2019).

Thomson and Crocker (2013) also developed a single lexicon of 59 terms for the description of feelings of any product category in four different languages (French, German, Italian, and English). Another tool for self-reported emotions consists of using a predefined list of terms, sentences, or descriptors to be checked or rated, such as the Emosemio multistep approach. This method involves the use of full sentences instead of nouns or adjectives in emotions questionnaires for a specific product category (Spinelli et al., 2014). Similarly the product emotion measurement instrument (PrEmo) (Desmet et al., 2000), and the 2.0 version (PrEmo2) (Laurans and Desmet, 2012) allow the reporting of product-evoked emotions in a more intuitive way by using a nonverbal web-based tool containing 12 animations of cartoon characters expressing specific emotions in approximately 1 second, with facial/bodily movement and vocal sound. Participants may report to which extent they experienced the emotions by rating on a five-point scale (anchoring 0 = "I do not feel this" to 4 = "I feel this strongly") (den Uijl et al., 2016a,b). On the other hand, another approach to emotion elicitation in a particular product category is using a lexicon defined by consumers based on preliminary or previous pretests. In general, several sessions are run with the participants including warm-up sessions in order to get familiar with the emotions' elicitation process using, for instance, pictures, short films, or similar strategies (Ng et al., 2013). After that focus discussion groups conducted by a moderator are carried out and consumers generate the emotional lexicon using their own words (Bhumiratana et al., 2014; Chaya et al., 2015). More recently Mora et al. (2019) proposed certain filters for improving the emotional lexicon defined by consumers in beers. Finally, the use of emojis has been recently introduced by Jaeger et al. (2017) as an interesting and innovative method for the measurement of product—emotion associations and the authors proposed the use of this tool as an alternative method but not as a direct substitution for word-based emotion surveys.

21.4.3 Examining extrinsic factors that influence the emotions elicited by foods

Over the past few years studies focused on understanding how some variables, besides the sensory characteristics, could influence the emotions elicited by food that has become popular. The effect of extrinsic food cues, such as the package (Ng et al., 2013; Gutjar et al., 2015), food name (Cardello et al., 2012), or brand (Spinelli et al., 2015), can effectively modify the emotional profiles with this outcome, which is valuable information for marketing purposes and success. Generally, three different conditions are considered for collecting the data related to the emotions elicited when extrinsic factors are evaluated: (1) blind (only tasting), (2) expected (only the package, name etc., without tasting the product), and (3) informed (taste and information) (Spinelli et al., 2015). Moreover, one of the most interesting factors to evaluate is the context, environment, or scenario in which the food is consumed, since the eating experience is certainly affected by the context and therefore emotional responses may be context-dependent (Piqueras-Fiszman and Jaeger, 2014). Different strategies have been described to elicit the different contexts or scenarios. Danner et al. (2016) used different real scenarios (sensory lab, restaurant, and at home) for testing the emotional dimension of four Australian commercial Shiraz wines in different quality categories. Piazza et al. (2015) constructed a written questionnaire with different sentences reflecting the different eating occasions by varying four contextual dimensions for kiwi fruit (product format, place of consumption, time available to eat, and health state) and ice cream (product format, place of consumption, social context, and hunger state) and asking for the appropriateness of each one. In a study carried out by Dorado et al. (2016), participants were requested to freely provide a written description of the particular occasion when they were having a beer in order to obtain information of the contextual dimensions and to generate the different scenarios in which beer is commonly consumed. Once it is accepted that context or scenario plays an important role in emotional responses, the great challenge is how to replicate the scenario under controlled laboratory circumstances as this is difficult to set up and control.

21.4.4 Emotional responses to meat and meat products

Despite the negative beliefs and attitudes towards them, meat and meat products continue to have a central role in the daily meals both in Western and non-Western countries (Font-i-Furnols and Guerrero, 2014). From a rational point of view, the four Ns theory would solve the meat attitude paradox since eating meat is considered *natural*, *normal*, *necessary*, and *nice* (Piazza et al., 2015). However, the emotional responses to meat and meat products have had limited study despite the crucial influence of the emotional dimension on food choice and product market success.

The first evidence of the emotions elicited by meat were reported by Narchi et al. (2008) using a large variety of food pictures as stimuli including meat pictures. In this study 19 emotional words (8 words depicting positive emotions and 11 with negative ones) were evaluated and elderly French men and woman participated. Lease et al. (2014) described the emotional responses to meatballs elaborated with recycled water. Consumers were asked to describe the emotions evoked depending on the original purpose of the water prior to being recycling and incorporated into the meatballs, that is, cleaning the equipment, cleaning the floors, or for all purposes. The emotional profiling of meat-based burger patties compared to insect-based and plant-based ones has been assessed using the EmoSensory Wheel questionnaire. Fourteen emotional terms were evaluated under blind, expected, and informed conditions (Schouteten et al., 2016). The EmoSensory Wheel was developed by Schouteten et al. (2015) and allows for obtaining both emotional and sensory profiles using a wheel-format questionnaire. The emotional terms were elicited for a particular product and the intensities were rated using a five-point scale ranging from 1 = "slightly" to 5 = "extremely." Veflen Olsen et al. (2014) carried out an online survey with Norwegian consumers in which the elicited emotions (fear, disgust, surprise, interest, pleasure, or none of these) were collected when consumers were confronted with pictures of hamburgers varying in the degrees of doneness. More recently, Cardoso Merlo et al. (in press) evaluated the influence of an extrinsic factor, the packaging color of hamburgers, on the emotional responses by using the temporal dominance of emotion methodology (TDE) from first sight of the packaging until eating of a portion of hamburger. This method, similarly to the dynamic descriptive technique TDS, allows the recording of the dominant emotions which are defined as the most striking emotions at a given moment. More recently, Lorido et al. (2019) described the emotional profiling of three different types of dry-cured hams using also the TDE method with Spanish consumers (Fig. 21.4).

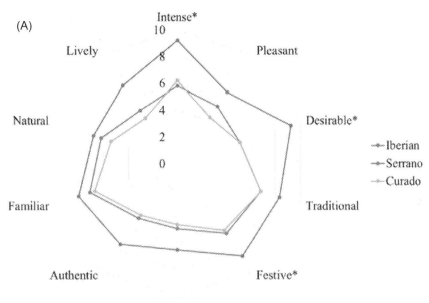

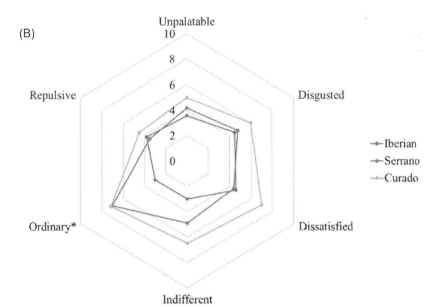

Figure 21.4
Positive (A) and negative-neutral emotions (B) evoked by Spanish consumers after tasting three types of dry-cured hams (Lorido et al., 2019).

A very interesting and original study was carried out by Kostyra et al. (2016) in smoked hams of different qualities. These authors measured the facial expression of participants by using a FaceReader software device. This equipment allows the assessment and analysis of facial reactions over time by recording physical reactions of eyes, mouth, eyebrows, and head. The facial expression obtained was interpreted basing on six basic types of emotions: happiness, sadness, anger, surprise, scared, and disgust. The lack of those emotions was recognized as neutrality (indifference). The emotional profiles of rib eye steaks from four different biological types of grass-fed cattle were assessed using the 39 terms of EsSense profile and CATA and rib eye steak pictures together with the description and potential health benefits comments were used as stimuli (Carabante et al., 2018). Similarly, Borgogno et al. (2017) used the validated and standardized questionnaire EsSense25 for the emotional characterization of beef varying in the breeding system information (conventional vs only Italian Simmental). Rousset et al. (2005) hypothesized that disgust with red meat is common in women and can lead to reduced meat intake and the negative emotions elicited during red meat consumption could contribute to a low meat intake. Based on this, 60 women were invited to assess the intensity of 26 emotions described by words and induced by 30 food pictures. The findings showed differences in emotions between the low and high meat-eating women. Authors concluded that low meat consumption was associated with specific negative emotions regarding meat and other foods.

References

Albert, A., Varela, P., Salvador, A., Hough, G., Fiszman, S., 2011. Overcoming the issues in the sensory description of hot served food with a complex texture. Application of QDA, flash profiling and projective mapping using panels with different degrees of training. Food Qual. Pref. 22, 463–473.

Alves, L.A.A.S., Lorenzo, J.M., Gonçalves, C.A.A., Santos, B.A., et al., 2017. Impact of lysine and liquid smoke as flavor enhancers on the quality of low-fat Bologna-type sausages with 50% replacement of NaCl by KCl. Meat Sci. 123, 50–56.

Andrade, J.C., Nalério, E.S., Giongo, C., Barcellos, M.D., Ares, G., Deliza, R., 2018. Consumer sensory and hedonic perception of sheep meat coppa under blind and informed conditions. Meat Sci. 137, 201–210.

Anonymous, 1975. Minutes of division business meeting. Institute of Food Technologists, Sensory Evaluation Division, IFT, Chicago, Illinois.

Ares, G., Jaeger, S.R., 2013. Check-all-that-apply questions: influence of attribute order on sensory product characterization. Food Qual. Pref. 28, 141–153.

Ares, G., Jaeger, S.R., 2017. A comparison of five methodological variants of emoji questionnaires for measuring product elicited emotional associations: an application with seafood among Chinese consumers. Food Res. Int. 99, 216–228.

Ares, G., Bruzzone, F., Vidal, L., et al., 2014. Evaluation of a rating-based variant of check-all-that-apply questions: rate-all-that-apply (RATA). Food Qual. Pref. 36, 87–95.

Ares, G., Jaeger, S.R., Antúnez, L., et al., 2015a. Comparison of TCATA and TDS for dynamic sensory characterization of food products. Food Res. Int. 78, 148–158.

Ares, G., Reis, F., Oliveira, D., Antúnez, L., et al., 2015b. Recommendations for use of balanced presentation order of terms in CATA questions. Food Qual. Pref. 46, 137–141.

Ares, G., Picallo, A., Coste, B., Antúnez, L., et al., 2018. A comparison of RATA questions with descriptive analysis: insights from three studies with complex/similar products. J. Sens. Stud. 33, e12458.

Bhumiratana, N., Adhikari, K., Chambers, E., 2014. The development of an emotion lexicon for the coffee drinking experience. Food Res. Int. 61, 83–92.

Boinbaser, L., Parente, M.E., Castura, J.C., Ares, G., 2015. Dynamic sensory characterization of cosmetic creams during application using Temporal Check-All-That-Apply (TCATA) questions. Food Qual. Pref. 45, 33–40.

Borgogno, M., Cardello, A.V., Favotto, S., Piasentier, E., 2017. An emotional approach to beef evaluation. Meat Sci. 127, 1–5.

Bredie, W.L.P., Liu, J., Dehlholm, C., et al., 2018. Flash profile method. Descriptive Analysis in Sensory Evaluation. John Wiley and Sons, Ltd, pp. 513–533.

Brown, W.E., Gérault, S., Wakeling, I., 1996. Diversity of perceptions of meat tenderness and juiciness by consumers: a Time-Intensity study. J. Text. Stud 27, 475–492.

Carabante, K.M., Ardoin, R., Scaglia, G., et al., 2018. Consumer acceptance, emotional response, and purchase intent of rib-eye steaks from grass-fed steers, and effects of health benefit information on consumer perception. J. Food Sci. 83, 2560–2570.

Cardello, A.V., Meiselman, H.L., Schutz, H.G., et al., 2012. Measuring emotional responses to foods and food names using questionnaires. Food Qual. Pref. 24, 243–250.

Cardoso Merlo, T., Soletti I., Saldaña E., et al., in press. Measuring dynamics of emotions evoked by the packaging colour of hamburgers using Temporal Dominance of Emotion (TDE). Food Res. Int. Available from: https://doi.org/10.1016/j.foodres.2018.08.007.

Castura, J.C., Antúnez, L., Giménez, A., Ares, G., 2016. Temporal Check-All-That-Apply (TCATA): a novel dynamic method for characterizing products. Food Qual. Pref. 47, 79–90.

Cavitt, L.C., Meullenet, J.-F.C., Xiong, R., Owens, C.M., 2005. The relationship of razor blade shear, allo-kramer shear, warner-bratzler shear and sensory tests to changes in tenderness of broiler breast fillets. J. Mus. Foods 16, 223–242.

Chaya, C., Eaton, C., Hewson, L., et al., 2015. Developing a reduced consumer-led lexicon to measure emotional response to beer. Food Qual. Pref. 45, 100–112.

Choi, J.H., Gwak, M.J., Chung, S.J., et al., 2015. Identifying the drivers of liking by investigating the reasons for (dis)liking using CATA in cross-cultural context: a case study on barbecue sauce. J. Sci. Food Agric. 95, 1613–1625.

Cliff, M., Heymann, H., 1993. Development and use of time-intensity methodology for sensory evaluation: a review. Food Res. Int. 26, 375–385.

Dairou, V., Sieffermann, J.M., 2002. A comparison of 14 jams characterized by conventional profile and a quick original method, the Flash Profile. J. Food Sci. 67, 826–834.

Daltoé, M.L., Breda, L.S., Belusso, A.C., et al., 2017. Projective mapping with food stickers: a good tool for better understanding perception of fish in children of different ages. Food Qual. Pref. 57, 87–96.

Danner, L., Ristic, R., Johnson, T.E., et al., 2016. Context and wine quality effects on consumers' mood, emotions, liking and willingness to pay for Australian shiraz wines. Food Res. Int. 89, 254–265.

Dijksterhuis, G.B., Piggott, J.R., 2001. Dynamic methods of sensory analysis. Trends Food Sci. Technol. 11 (8), 284–290.

Dehlholm, C., 2014. Projective mapping and napping. Novel Techniques in Sensory Characterization and Consumer Profiling. CRC Press, pp. 513–533.

Dehlholm, C., Brockhoff, P.B., Meinert, L., et al., 2012. Rapid descriptive sensory methods - comparison of free multiple sorting, partial napping, napping, flash profiling and conventional profiling. Food Qual. Pref. 26, 267–277.

Delarue, J., 2015. Flash Profile, its evolution and uses in sensory and consumer science. Rapid Sensory Profiling Techniques. Elsevier, pp. 121–151.

Delarue, J., Sieffermann, J.M., 2004. Sensory mapping using Flash profile. Comparison with a conventional descriptive method for the evaluation of the flavour of fruit dairy products. Food Qual. Pref. 15, 383–392.

Den Uijl, L.C., Jager, G., de Graaf, C., et al., 2016a. Emotion, olfaction, and age. A comparison of food-evoked emotion profiles of adults, older normosmic persons, and older hyposmic persons. Food Qual. Pref. 48, 199–209.

Den Uijl, L.C., Jager, G., Zandstra, E.H., et al., 2016b. Self-reported food-evoked emotions of younger adults, older normosmic adults, and older hyposmic adults as measured using the PrEmo2 tool and the Affect Grid. Food Qual. Pref. 51, 109–117.

Desmet, P.M., Hekkert, P., Jacobs, J.J., 2000. When a car makes you smile: development and application of an instrument to measure product emotions. Adv. Consumer Res. 27, 111–117.

Dooley, L., Lee, Y.S., Meullenet, J.F., 2010. The application of check-all-that-apply (CATA) consumer profiling to preference mapping of vanilla ice cream and its comparison to classical external preference mapping. Food Qual. Pref. 21, 394–401.

Dorado, R., Chaya, C., Tarrega, A., Hort, J., 2016. The impact of using a written scenario when measuring emotional response to beer. Food Qual. Pref. 50, 38–47.

Duizer, L.M., Gullett, E.A., Findlay, C.J., 1993. Time-Intensity methodology for beef tenderness perception. J Food Sci. 58, 943–947.

Esmerino, E.A., Castura, J.C., Ferraz, J.P., et al., 2017. Dynamic profiling of different ready-to-drink fermented dairy products: a comparative study using Temporal Check-All-That-Apply (TCATA), Temporal Dominance of Sensations (TDS) and Progressive Profile (PP). Food Res. Int. 101, 249–258.

Faustman, C., Sun, Q., Mancini, R., Suman, S.P., 2010. Myoglobin and lipid oxidation interactions: mechanistic bases and control. Meat Sci. 86, 86–94.

Fellendorf, S., O'Sullivan, M.G., Kerry, J.P., 2015. Impact of varying salt and fat levels on the physicochemical properties and sensory quality of white pudding. Meat Science 103, 75–82.

Font-i-Furnols, M., Guerrero, L., 2014. Consumer preference, behavior and perception about meat and meat products: an overview. Meat Sci. 98, 361–371.

Foster, K.D., Grigor, J.M., Cheong, J.N., et al., 2011. The role of oral processing in dynamic sensory perception. J. Food Sci. 76, R49–R61.

Fuentes, V., Ventanas, J., Morcuende, D., Ventanas, S., 2013. Effect of intramuscular fat content and serving temperature on temporal sensory perception of sliced and vacuum packaged dry-cured ham. Meat Sci. 93, 621–629.

Gomes, C.L., Pflanzer, S.B., de Felício, P.E., Bolini, H.M.A., 2014. Temporal changes of tenderness and juiciness of beef strip loin steaks. LWT - Food Sci. Technol. 59, 629–634.

Grasso, S., Monahan, F.J., Hutchings, S.C., Brunton, N.P., 2017. The effect of health claim information disclosure on the sensory characteristics of plant sterol-enriched turkey as assessed using the Check-All-That-Apply (CATA) methodology. Food Qual. Pref. 57, 69–78.

Grossi, A., Søltoft-Jensen, J., Knudsen, J.C., et al., 2011. Synergistic cooperation of high pressure and carrot dietary fibre on texture and colour of pork sausages. Meat Sci. 89, 195–201.

Grossi, A., Søltoft-Jensen, J., Knudsen, J.C., et al., 2012. Reduction of salt in pork sausages by the addition of carrot fibre or potato starch and high pressure treatment. Meat Sci. 92, 481–489.

Grunert, K.G., Bredahl, L., Brunsø, K., 2004. Consumer perception of meat quality and implications for product development in the meat sector - a review. Meat Sci. 66, 259–272.

Gutjar, S., Dalenberg, J.R., de Graaf, C., et al., 2015. What reported food-evoked emotions may add: a model to predict consumer food choice. Food Qual. Pref. 45, 140–148.

Heck, R.T., Vendruscolo, R.G., Etchepare, M.A., et al., 2017. Is it possible to produce a low-fat burger with a healthy n − 6/n − 3 PUFA ratio without affecting the technological and sensory properties?. Meat Sci. 130, 16–25.

Heck, R.T., Fagundes, M.B., Cichoski, A.J., et al., 2019. Volatile compounds and sensory profile of burgers with 50% fat replacement by microparticles of chia oil enriched with rosemary. Meat Sci. 158, 164–170.

Henrique, N.A., Deliza, R., Rosenthal, A., 2014. Consumer sensory characterization of cooked ham using the Check-All-That-Apply (CATA) methodology. Food Eng. Rev. 7, 265–273.

Jaeger, S.R., Ares, G., 2014. Lack of evidence that concurrent sensory product characterisation using CATA questions bias hedonic scores. Food Qual. Pref. 35, 1–5.

Jaeger, S.R., Beresford, M.K., Paisley, A.G., et al., 2015. Check-all-that-apply (CATA) questions for sensory product characterization by consumers: investigations into the number of terms used in CATA questions. Food Qual. Pref. 42, 154–164.

Jaeger, S.R., Min Lee, S., Kim, K.-O., et al., 2017. Measurement of product emotions using emoji surveys: case studies with tasted foods and beverages. Food Qual. Pref. 62, 46−59.

Jellinek, G., 1964. Introduction to and critical review of modern methods of sensory analysis (odor, taste and flavour evaluation) with special emphasis on descriptive analysis (flavour profile method). J. Nutr. Diet. 1, 219−260.

Jiang, Y., King, J.M., Prinyawiwatkul, W., 2014. A review of measurement and relationships between food, eating behavior and emotion. Trends Food Sci. Technol. 36, 15−28.

Jorge, E.C., Mendes, A.C.G., Auriema, B.E., et al., 2015. Application of a check-all-that-apply question for evaluating and characterizing meat products. Meat Sci. 100, 124−133.

King, S.C., Meiselman, H.L., 2010. Development of a method to measure consumer emotions associated with foods. Food Qual. Pref. 21, 168−177.

Köster, E.P., Mojet, J., 2015. From mood to food and from food to mood: a psychological perspective on the measurement of food-related emotions in consumer research. Food Res Int. 76, 180−191.

Kostyra, E., Rambuszek, M., Waszkiewicz-Robak, B., et al., 2016. Consumer facial expression in relation to smoked ham with the use of face reading technology. The methodological aspects and informative value of research results. Meat Sci. 119, 22−31.

Labbe, D., Schlich, P., Pineau, N., Gilbert, F., Martin, N., 2009. Temporal dominance of sensations and sensory profiling: a comparative study. Food Qual. Pref. 20, 216−221.

Lagast, S., Gellynck, X., Schouteten, J.J., De Herdt, V., De Steur, H., 2017. Consumers' emotions elicited by food: a systematic review of explicit and implicit methods. Trends Food Sci. Technol. 69, 172−189.

Lannuzel, C., Rogeaux M. 2007. How to speed up Temporal Dominance of Sensations training. In: Seventh Pangborn Symposium, p. 84.

Laurans, G.F.G., Desmet, P.M.A., 2012. Introducing PREMO2: new directions for the non-verbal measurement of emotion in design. In: Brassett, J., Hekkert, P., Ludden, G., Malpass, M., McDonnell, J. (Eds.), Proceedings of 8th International Design and Emotion Conference, pp. 11−14.

Lazo, O., Guerrero, L., Alexi, N., Grigorakis, K., et al., 2017. Sensory characterization, physico-chemical properties and somatic yields of five emerging fish species. Food Res. Int. 100, 396−406.

Lê, T.M., Husson, F., Lê, S., 2016. Digit-tracking: interpreting the evolution over time of sensory dimensions of an individual product space issued from Napping® and sorted Napping. Food Qual. Pref. 47, 73−78.

Lease, H.J., MacDonald, D.H., Cox, D.N., 2014. Consumers' acceptance of recycled water in meat products: the influence of tasting, attitudes and values on hedonic and emotional reactions. Food Qual. Pref. 37, 35−44.

Lorido, L., Estévez, M., Ventanas, S., 2014. A novel approach to assess temporal sensory perception of muscle foods: application of a time-intensity technique to diverse Iberian meat products. Meat Sci. 96, 385−393.

Lorido, L., Estévez, M., Ventanas, J., Ventanas, S., 2015. Salt and intramuscular fat modulate dynamic perception of flavour and texture in dry-cured hams. Meat Sci. 107, 39−48.

Lorido, L., Hort, J., Estévez, M., Ventanas, S., 2016. Reporting the sensory properties of dry-cured ham using a new language: time intensity (TI) and temporal dominance of sensations (TDS). Meat Sci. 121, 166−174.

Lorido, L., Estévez, M., Ventanas, S., 2018. Fast and dynamic descriptive techniques (Flash Profile, Time-intensity and Temporal Dominance of Sensations) for sensory characterization of dry-cured loins. Meat Sci. 145, 154−162.

Lorido, L., Pizarro, E., Estévez, M., Ventanas, S., 2019. Emotional responses to the consumption of dry-cured hams by Spanish consumers: a temporal approach. Meat Sci. 149, 126−133.

Massingue, A.A., Torres Filho, R.A., Fontes, P.R., et al., 2018. Effect of mechanically deboned poultry meat content on technological properties and sensory characteristics of lamb and mutton sausages. Asian-Australasian J. Anim. Sci. 4, 576−584.

Mcilveen, H., Buchanan, J., 2001. The impact of sensory factors on beef purchase and consumption. Nutr. Food Sci. 31, 286−292.

McMahon, K.M., Culver, C., Castura, J.C., Ross, C.F., 2017. Perception of carbonation in sparkling wines using descriptive analysis (DA) and temporal check-all-that-apply (TCATA). Food Qual. Pref. 59, 14−26.

Meier-Dinkel, L., Gertheiss, J., Schnäckel, W., Mörlein, D., 2016. Consumers' perception and acceptance of boiled and fermented sausages from strongly boar tainted meat. Meat Sci. 118, 34–42.

Meyners, M., Castura, J.C., 2018. The analysis of temporal check-all-that-apply (TCATA) data. Food Qual. Pref. 67, 67–76.

Mitchell, J., Castura, J.C., Thibodeau, M., Pickering, G., 2019. Application of TCATA to examine variation in beer perception due to thermal taste status. Food Qual. Pref. 73, 135–142. Available from: https://doi.org/10.1016/j.foodqual.2018.11.016.

Mora, M., Giussania, B., Pagliarini, E., Chaya, C., 2019. Improvement of an emotional lexicon for the evaluation of beers. Food Qual. Pref. 71, 158–162.

Narchi, I., Walrand, S., Boirie, Y., Rousset, S., 2008. Emotions generated by food in elderly French people. J. Nutr. Health Aging 12, 626–633.

Nestrud, M.A., Meiselman, H.L., King, S.C., et al., 2016. Development of EsSense25, a shorter version of the EsSense Profile®. Food Qual. Pref. 48, 107–117.

Neville, M., Tarrega, A., Hewson, L., Foster, T., 2017. Consumer-orientated development of hybrid beef burger and sausage analogues. Food Sci. Nutr. 5, 852–864.

Ng, M., Lawlor, J.B., Chandra, S., Chaya, C., Hewson, L., Hort, J., 2012. Using quantitative descriptive analysis and temporal dominance of sensations analysis as complementary methods for profiling commercial black-currant squashes. Food Qual. Pref. 25, 121–134.

Ng, M., Chaya, C., Hort, J., 2013. Beyond liking: comparing the measurement of emotional response using EsSense profile and consumer defined check-all-that-apply methodologies. Food Qual. Pref. 28, 193–205.

Nguyen, Q.C., Næs, T., Varela, P., 2018. When the choice of the temporal method does make a difference: TCATA, TDS and TDS by modality for characterizing semi-solid foods. Food Qual. Pref. 66, 95–106.

Oliveira, C.A., Massingue, A.A., Moura, A.P.R., et al., 2018. Restructured low-fat cooked ham containing liquid whey fortified with lactulose. J. Sci. Food Agric. 98, 807–816.

Oliver, P., Cicerale, S., Pang, E., Keast, R., 2018. Comparison of Quantitative Descriptive Analysis to the Napping methodology with and without product training. J. Sens. Stud. 33, e12331.

Pagès, J., 2005. Collection and analysis of perceived product inter-distances using multiple factor analysis: application to the study of 10 white wines from the Loire Valley. Food Qual. Pref. 16, 642–649.

Paulsen, M.T., Nys, A., Kvarberg, R., Hersleth, M., 2014. Effects of NaCl substitution on the sensory properties of sausages: temporal aspects. Meat Sci. 98, 164–170.

Perrin, L., Symoneaux, R., Maître, I., et al., 2007. Comparison of conventional profiling by a trained tasting panel and free profiling by wine professionals. Am. J. Enol. Vitic. 54, 508–517.

Perrin, L., Symoneaux, R., Maitre, I., et al., 2008. Comparison of three sensory Napping (R) procedure: case of methods for use with the ten wines from Loire valley. Food Qual. Pref. 19, 1–11.

Peyvieux, C., Dijksterhuis, G., 2001. Training a sensory panel for TI: a case study. Food Qual. Pref. 12, 19–28.

Phan, U.T.X., Chambers, E., 2016. Motivations for choosing various food groups based on individual foods. Appetite 105, 204–211.

Piazza, J., Ruby, M.B., Loughnan, S., et al., 2015. Rationalizing meat consumption. The 4Ns. Appetite 91, 114–128.

Pickup, W., Bremer, P., Peng, M., 2018. Comparing conventional Descriptive Analysis and Napping®-UFP against physiochemical measurements: a case study using apples. J. Sci. Food Agric. 98, 1476–1484.

Piggott, J.R., 2000. Dynamism in flavour science and sensory methodology. Food Res. Int. 33, 191–197.

Piqueras-Fiszman, B., Jaeger, S.R., 2014. The impact of evoked consumption contexts and appropriateness on emotion responses. Food Qual. Pref. 32, 277–288.

Pineau, N., Cordelle, S., Schlich, P., 2003. Temporal Dominance of Sensations: a new technique to record several sensory attributes simultaneously over time. In: 5th Pangborn symposium. July 20–24, p. 121.

Pineau, N., Schlich, P., Cordelle, S., et al., 2009. Temporal Dominance of Sensations: construction of the TDS curves and comparison with time-intensity. Food Qual. Pref. 20, 450–455.

Pineau, N., de Bouillé, A.G., Lepage, M., et al., 2012. Temporal Dominance of Sensations: what is a good attribute list? Food Qual. Pref. 26, 159–165.

Pintado, A.I.E., Monteiro, M.J.P., Talon, R., et al., 2016. Consumer acceptance and sensory profiling of reengineered kitoza products. Food Chem. 198, 75–84.

Polivy, J., Herman, C.P., 1999. Distress and eating: why do dieters overeat? Int. J Eat. Disord. 26, 153–164.

Reinbach, H.C., Meinert, L., Ballabio, D., et al., 2007. Interactions between oral burn, meat flavor and texture in chili spiced pork patties evaluated by time-intensity. Food Qual. Pref. 18, 909–919.

Reinbach, H.C., Giacalone, D., Ribeiro, L.M., et al., 2014. Comparison of three sensory profiling methods based on consumer perception: CATA, CATA with intensity and Napping. Food Qual. Pref. 32, 160–166.

Risvik, E., McEwan, J.A., Colwill, J.S., et al., 1994. Projective mapping: a tool for sensory analysis and consumer research. Food Qual. Pref. 5, 263–269.

Rousset, S., Deiss, V., Juillard, E., Schlich, P., Droit-Volet, S., 2005. Emotions generated by meat and other food products in women. Br. J. Nutr. 94, 609–619.

Saldaña, E., Garcia, A.O., Selani, M.M., et al., 2018. A sensometric approach to the development of mortadella with healthier fats. Meat Sci. 137, 176–190.

Saldaña, E., Saldarriaga, L., Cabrera, J., et al., 2019. Descriptive and hedonic sensory perception of Brazilian consumers for smoked bacon. Meat Sci. 147, 60–69.

Samant, S.S., Cho, S., Whitmore, A.D., et al., 2016. The influence of beverages on residual spiciness elicited by eating spicy chicken meat: time-intensity analysis. Int. J. Food Sci. Technol. 51, 2406–2415.

Santos, B.A., Pollonio, M.A.R., Cruz, A.G., et al., 2013. Ultra-flash profile and projective mapping for describing sensory attributes of prebiotic mortadellas. Food Res. Int. 54, 1705–1711.

Santos, B.A., Campagnol, P.C.B., Cruz, A.G., et al., 2015. Check all that apply and free listing to describe the sensory characteristics of low sodium dry fermented sausages: comparison with trained panel. Food Res. Int. 76, 725–734.

Schlich, P., 2017. Temporal Dominance of Sensations (TDS): a new deal for temporal sensory analysis. Curr. Opin. Food Sci. 15, 38–42.

Schouteten, J.J., Steur, H., Pelsmaeker, S., et al., 2015. An integrated method for the emotional conceptualization and sensory characterization of food products: the EmoSensory® Wheel. Food Res. Int. 78, 96–107.

Schouteten, J.J., De Steur, H., De Pelsmaeker, S., et al., 2016. Emotional and sensory profiling of insect-, plant- and meat-based burgers under blind, expected and informed conditions. Food Qual. Pref. 52, 27–31.

Shahidi, F., Rubin, L.J., D'Souza, L.A., 1986. Meat flavor volatiles: a review of the composition, techniques of analysis, and sensory evaluation. Crit. Rev. Food Sci. Nutr. 24, 141–243.

Sieffermann, J.M., 2000. Le profil flash: Un outil rapide et innovant d'évaluation sensorielle descriptive. In: AGORAL 2000 – XIIèmes Rencontres "L'innovation: De l'idée Au Succès".

Siegmund, B., Urdl, K., Jurek, A., Leitner, E., 2018. "More than Honey": investigation on volatiles from monovarietal honeys using new analytical and sensory approaches. J. Agric. Food Chem. 66, 2432–2442.

Sjöström, L.B., 1954. The descriptive analysis of flavor. In: Peryan, D., Pilgrim, F., Peterson, M. (Eds.), Food Acceptance Testing Methodology. Quartermaster Food and Container Institute, Chicago, pp. 25–61.

Spinelli, S., Masi, C., Dinnella, C., et al., 2014. How does it make you feel? A new approach to measuring emotions in food product experience. Food Qual. Pref. 37, 109–122.

Spinelli, S., Masi, C., Zoboli, G.P., et al., 2015. Emotional responses to branded and unbranded foods. Food Qual. Pref. 42, 1–11.

Tan, H.S.G., Verbaan, Y.T., Stieger, M., 2017. How will better products improve the sensory-liking and willingness to buy insect-based foods?. Food Res. Int. 92, 95–105.

Thomson, D.M.H., Crocker, C., 2013. A data-driven classification of feelings. Food Qual. Pref. 27, 137–152.

Tuorila, H., Monteleone, E., 2009. Sensory food science in the changing society: opportunities, needs, and challenges. Trends Food Sci. Technol. 20, 54–62.

van Buuren, S., 1992. Analyzing time intensity responses in sensory evaluation. Food Technol. 46, 101–104.

Veflen Olsen, N., Røssvoll, E., Langsrud, S., Scholderer, J., 2014. Hamburger hazards and emotions. Appetite 78, 95–101.

Ventanas, S., Puolanne, E., Tuorila, H., 2010. Temporal changes of flavour and texture in cooked bologna type sausages as affected by fat and salt content. Meat Sci. 85, 410–419.

Verbeke, W., Pérez-Cueto, F.J.A., Barcellos, M.D.D., et al., 2010. European citizen and consumer attitudes and preferences regarding beef and pork. Meat Sci. 84, 284–292.

Williams, A.A., Langron, S.P., 1984. The use of free-choice profiling for the evaluation of commercial ports. J. Sci. Food Agric. 35, 558–568.

Yotsuyanagi, S.E., Contreras-Castillo, C.J., Haguiwara, M.M.H., et al., 2016. Technological, sensory and microbiological impacts of sodium reduction in frankfurters. Meat Sci. 115, 50–59.

Further reading

Piqueras-Fiszman, B., Jaeger, S.R., 2015. The effect of product-context appropriateness on emotion associations in evoked eating occasions. Food Qual. Pref. 40, 49–60.

Saldaña, E., Saldarriaga, L., Cabrera, J., et al., 2019. Relationship between volatile compounds and consumer-based sensory characteristics of bacon smoked with different Brazilian woods. Food Res. Int. 119, 839–849. Available from: 10.1016/j.foodres.2018.10.067.

Index

Note: Page numbers followed by "*f*" and "*t*" refer to figures and tables, respectively.

A

Absolute quantification of proteins, 357
Absorption, 39
Accelerated solvent extraction (ASE), 248
Acceptable daily intake (ADI), 200
Acetic acid, 194–195
Acetoin, 316
Achilles tendon, 70
Acid
  acid-catalyzed esterification methods, 221–222
  extraction, 227
  hydrolysis, 30, 218–219
Acidic compounds, 270–271
Acidic-pH solutions, 121–122
Acidified sodium chloride (ASC), 192
Acidified sodium chlorite (ASC), 271
Acoustic
  methods, 46
  waves, 12, 248
Actin, 24
Additives in foods, 188–189
Adenosine triphosphate (ATP), 68
  bioluminescence, 291, 323, 338–339
ADI. *See* Acceptable daily intake (ADI)
Aerobic bacteria, 314–315
Aerobically packaged meat, spoilage microflora with, 310–311
Aeromonas hydrophila, 251–252
Aflatoxin B_1, 174

AFLP. *See* Amplified fragment length polymorphism (AFLP)
African buffalo. *See Syncerus caffer* (African buffalo)
Agarose gel electrophoresis (AGE), 374, 384–385
AGE. *See* Agarose gel electrophoresis (AGE)
Aging, 68, 72–74, 82–83
  dry, 72–73
  effects on
    buffalo meat quality, 77–78
    color, 76–77
    tenderness, 68–76
  wet, 74
Airflow, 73
Alcohols, 315
Aldehydes, 122, 315
Alkaline hydrolysis, 30
Alkylsulfite salt, 27
Allele specific polymerase PCR, 373
α-actinin, 106
American Meat Science Association (AMSA), 54
  Color Guidelines, 123
Amino acid(s), 68–69, 191–192, 314–315
  determination, 29–30
Ammonia (NH_3), 84–85, 337
Amplicon/PCR product, 137
Amplified fragment length polymorphism (AFLP), 143
AMR. *See* Antimicrobial resistance (AMR)
AMSA. *See* American Meat Science Association (AMSA)

AMV. *See* Avian myeloblastosis virus (AMV)
Analytical chemistry, 4
Anchored PCR, 373
Angle interrogation, 294
Animal
  animal-derived foods, 177
  and human health, 153–154
  identification systems, 154–155, 157–158
    bar-coded tags, 156
    methods for, 155–156
    quick response code-based tags, 156
    RFID, 156
    visual tagging, 156
Antemortem stress, 117–118
Anthraquinone, 249
Antibodies, 10–11
  antibody-based automatic techniques, 177
  antibody-based methods, 290
Antimicrobial resistance (AMR), 10–12
Antimicrobial(s), 176, 192
  activity, 195–196
  hurdles, 320
Antioxidants, 190, 242–243
  in animal products during postharvest stage
    flavor and shelf life, 251
    nutritional values, 252
    reducing microbiological contamination, 251–252
  effect, 178
  mechanisms of action of, 252–253
  from plant sources, 243–250

AOAC. *See* Association of Official Analytical Chemist (AOAC)
AOAC method 992.15, 28–29
AP-PCR. *See* Arbitrarily primed PCR (AP-PCR)
APCI. *See* Atmospheric pressure chemical ionization (APCI)
Apoptosis, 84
Arbitrarily primed PCR (AP-PCR), 138
Armor tenderometer, 12
Aroma of meat, 42
Arrays-based techniques, 12–13
ASC. *See* Acidified sodium chloride (ASC); Acidified sodium chlorite (ASC)
$AscH_2$, 254–255
ASE. *See* Accelerated solvent extraction (ASE)
Ash, 25
  content determination, 31–32
Aspergillus oryzae, 71–72
Association of Official Analytical Chemist (AOAC), 202, 218–219
Asymmetric PCR, 373
Atlantic cod. *See Gadus morhua* (Atlantic cod)
Atmospheric pressure chemical ionization (APCI), 222–223
ATP. *See* Adenosine triphosphate (ATP)
Australia, livestock traceability system in, 159
Authenticity of meat and meat products, 133–134
Autolytic domain, 102–103
Automated microbial identification systems, 276–277
Autooxidative process, 180
Avian myeloblastosis virus (AMV), 279

B

B vitamins, 32–33
Bacillus subtilis, 71–72
Bacterial bioluminescence, 291
Bacterial metabolite detection biosensors, 325
  electronic nose, 325
  gas chromatography and mass spectroscopy, 324–325
  spectroscopic methods, 324
Bacterial pathogens, 288
Bacteriocins, 196–197
Bacteriophage, biocontrol with, 275
Bactometer, 294
Banteng. *See Bos javanicus* (Banteng)
Bar-coded tags, 156
Basic Local Alignment Search Tool (BLAST), 138–139
Beef, 23
  labeling, 159
  loins, 83
  semimembranosus, 360
BeefCam, 74
Belonostomus indicus (Zebu), 142
Belt grill, 55
Benz(a)anthracene, 175
Benzo-γ-pyran ring, 256
Benzo(a)pyrene, 175
Benzo(b) fluoranthene, 175
Benzoic acids, 195, 257
β-carotenoids, 255–256
BHA. *See* Butylated hydroxyanisole (BHA)
BHT. *See* Butylated hydroxytoluene (BHT)
Binders, 191
Bioactive components, 248
Bioactive compounds, 256
Biochemical
  method, 108–109
  tests, 110
Biocontrol with bacteriophage, 275
Biodegradable polymer, 197
Biogenic amines, 173
Bioimpedance, 12
Bioinformatics, 134–135
  software tools, 387
Biological sensors for food-quality monitoring, 345
Bioluminescence
  assay, 109
  sensors, 291
Biomarkers
  identification, 110
  of postmortem aging, 91–93
  genotyping for marker gene, 93
  metabolomics, 93
  proteomic markers in conversion of muscle to meat, 91–92
Bionumerics 2D (software package), 387
Biopreservation, 194
Biosensors, 11–12, 175, 325
  bioluminescence sensors, 291
  biosensors-based detection techniques, 290–296, 290*f*
  cell-based sensor, 295
  electrical impedance biosensor, 294
  fiber optic biosensor, 291–292
  flow cytometry, 296
  FTIR, 295–296
  impedance-based biochip sensor, 294–295
  piezoelectric biosensors, 295
  SPC, 296
  SPR biosensor, 293–294
Biotechnology, 134–135
Biotin-labeled chromosomal DNA fragments, 136
Bison bonasus (Wisent), 142
Blade tenderization, 71
BLAST. *See* Basic Local Alignment Search Tool (BLAST)
BlastN. *See* Nucleotide blast (BlastN)
Bone marrow discoloration, 120
Bos frontalis (Mithun), 139–140
Bos indicus (cattle), 139–140
  cattle-specific microsatellite markers, 376*t*
  traceability systems in different countries, 161*t*
Bos javanicus (Banteng), 142
Bos taurus. *See* Taurine cattle (*Bos taurus*)
Bovine mitochondrial oxygen uptake, 127
Bovine spongiform encephalopathy (BSE), 157

Brocothrix, 311
  B. campestris, 339–340
  B. thermosphacta, 339–340
Bromelain, 71–72
Brown discoloration, 120
BSE. *See* Bovine spongiform encephalopathy (BSE)
Buffalo (*Bubalus bubalis*), 139–140, 142
  buffalo-specific microsatellite markers, 377t
  meat, 166
    aging effects on buffalo meat quality, 77–78
Butylated hydroxyanisole (BHA), 190
Butylated hydroxytoluene (BHT), 190

C

C S. *See* Continuous Soxhlet (C S)
C-protein, 69
CAD. *See* Charged aerosol detector (CAD)
Calcium ions (Ca^{2+}), 102–103
Calpain-Glo protease assay. *See* Bioluminescence—assay
Calpains (CAPNs), 69, 83–84, 102, 103f
  CAPN1, 93, 102–103
    purification, 106–107
  CAPN2, 102–103
    purification, 106–107
  CAPN3, 102
  detection and quantification, 108–110
  extraction, 90
  pathways of calpain activity in muscle tissues, 103–104
  postmortem proteolysis of skeletal muscle by, 104–106
  scope of future work, 111
  structure and functions, 102–103
  system, 7, 101–102
Calpastatin, 102
  purification, 106–107
  structure and functions, 102–103, 104f
Calpastatin, 102
  purification, 106–107
  structure and functions, 102–103, 104f
Camera, 38–39
Capillary electrophoresis methods (CE methods), 203
CAPNs. *See* Calpains (CAPNs)
Capra hircus (goat), 139–140
Carabeef, 23
Carbon monoxide, 119–120
Carboxymethyl cellulose, 191
Carboxymyoglobin, 122
Carotenoid, 255–256
Carrageenan, 191
Casein zymography, 90, 108
Casing for imaging medium, 43
Caspases, 83–84
CATA method. *See* Check-all-that-apply method (CATA method)
Catalase, 253
Cathepsins, 69, 83–84
  activity, 89–90
  cathepsin B and B-/L-like enzymes, 89–90
Cation-exchange HPLC, 29–30
Cattle. *See* Bos indicus (cattle)
CBAs. *See* Cell-based assays (CBAs)
CCC. *See* Countercurrent chromatography (CCC)
CDC. *See* Center for Disease Control and Prevention (CDC)
CDs. *See* Conjugated dienes (CDs)
CE methods. *See* Capillary electrophoresis methods (CE methods)
Cell death, 84
Cell-based assays (CBAs), 295
Cell-based sensor, 295
Center for Disease Control and Prevention (CDC), 288
Center for Science in Public Interest (CSPI), 200
Cetylpyridinium chloride (CPC), 192

CFR. *See* Code of Federal Regulations (CFR)
Charged aerosol detector (CAD), 223–224
CHD. *See* Coronary heart disease (CHD)
Check-all-that-apply method (CATA method), 394, 400–401
Chelators, 198–199
Chemical additives, 199
Chemical analysis, 26–27
Chemical contaminants, 10–12
Chemical interventions, 192
Chemical preservatives, 194–196
  nitrites, 195–196
  organic acids and derivatives, 194–195
  sulfites, 196
Chemical sensors for food-quality monitoring, 345
Chevon, 23
Chicken-specific microsatellite markers, 383t
Chiral chromatography, 222
Chitosan, 197
Chlorine
  dioxide, 271
  and related chemicals, 271
Chloroform, 218–219
  chloroform–methanol procedure, 30
Cholesterol, 179–180
Chromatography, 202
  linear retention index LRI and MS, 229–230
  techniques, 9, 228–229, 354–355
CIE. *See* Commission Internationale de l'Eclairage (CIE)
Cinnamic acids, 257
Clostridium botulinum, 123, 195–196, 321
Code of Federal Regulations (CFR), 199
Codex Alimentarius, 189
Cold plasma. *See* Nonthermal plasma
Cold shortening, 69

Collagen, 24
Color, 85
　additives detection, 204
　Atlases and charts, 40
　deviations and approaches to improve, 121–122
　　dark-cutting beef, 121
　　lactate-induced darkening, 121
　　myoglobin and lipid oxidation, 122
　　packaging techniques to improve appearance, 121–122
　effects of aging on, 76–77
　　biochemical basis of lower color stability in aged meat, 77
　　practical approaches to improve color stability of aged steaks, 77
Colorimeter, 39, 74–75
Colorimetric method, 9, 203
Coloring agents, 193
Commission Internationale de l'Eclairage (CIE), 74–75
L^*, a^*, and b^* values, 40, 74–75
Compressive techniques, 12
Compulsory beef labeling system, 159
Computed tomography (CT), 43–44, 225
Computer hardware and software, 38–39
Computer vision system, 38–39, 46–47
Conditioning, 82
Conducting organic polymer (CP), 340–341
Conjugated dienes (CDs), 226–227
Consumer
　evaluation, 54
　panels, 75–76
Continuous Soxhlet (C S), 219f
Conventional culture-based techniques, 276–277
　automated microbial identification systems, 276–277

Conventional descriptive techniques, 396
Conventional liquid–liquid extraction strategy, 250
Conventional methods, 247. See also Detection methods of meat tenderness
　HD, 247
　maceration extraction, 247
　Soxhlet extraction or hot continuous extraction, 247
Conventional microbiological approach, 338
Conventional PCR, 374
Conventional protein electrophoresis, 134
Cooked color, 122–123
　measurements, 128–129
Coomassie Blue, 355–356
Coomassie Blue R-250, 108
Coronary heart disease (CHD), 22
Countercurrent chromatography (CCC), 250
Counterimmunoelectrophoresis, 134
CP. See Conducting organic polymer (CP)
CPC. See Cetylpyridinium chloride (CPC)
Cross-hybridizing species, 136
Cross-links, 89
CSPI. See Center for Science in Public Interest (CSPI)
CT. See Computed tomography (CT)
Cure accelerators, 192–193
Curing agents, 192–193
Cyclohexenyl, 255–256
Cysteine, 180
Cystine, 180
Cytochrome b gene, 141
Cytochrome c gene, 117–118
Cytoskeletal proteins, 69

D

DAD. See Diode array detector (DAD)
DALP. See Direct amplified length polymorphisms (DALP)

Dark meat, 23
Dark-cutting beef, 121
　packaging techniques to improve appearance, 121–122
DEAE-sephacel, 107
Decorin, 89
Delta 2D (software package), 387
Denatured proteins, 84
Denaturing PAGE, 385
Dense phase carbon dioxide, 274
Deoxymyoglobin, 76, 119, 122
Deoxyribonucleic acid (DNA), 135, 162, 370–371. See also Ribonucleic acid (RNA)
　binding dyes, 374–375
　DNA-based
　　meat traceability, 162
　　molecular typing methods, 297
　　techniques, 134–135, 145
　　tools, 134–135
　hybridization, 135–137, 136f, 297–298
　　for animal species identification, 137t
　invader assays, 143
　microarrays, 142, 279–280, 298–299
　　hybridization, 142
　polymerase-mediated strand-displacement synthesis, 144
　quantification and purity measurement of, 372
Deproteinization process, 29–30
Descriptive analysis, 54
Desmin, 68–69, 104–105
Detection methods of meat tenderness, 54–61
　sensory analysis, 54
　shear force measurements, 55–57
DHA. See L-dihydroascorbic acid (DHA)
Diacetyl ($C_4H_6O_2$), 316, 337
Dibenzo-p-dioxins, 174
Dibenzofurans, 174
Dietary
　fibers, 252
　intake, 243

Differential pulse polarography, 203
Diffusion-weighted imaging, 5
Digital PCR, 8
Diglycerides, 191–192
Dimethylsulfide (C_2H_6S), 337
Diode array detector (DAD), 223–224
Dioxins, 174
"Dipstick" tests. *See* Lateral flow—tests
Direct amplified length polymorphisms (DALP), 143
Direct epifluorescent filter technique, 338
Discoloration, 120
Discriminant model, 60
Discrimination analysis, 54
Dispersive liquid–liquid microextraction (DLLME), 33
DNA. *See* Deoxyribonucleic acid (DNA)
Dot-blot, immunological protein quantification by, 90
Dry aging, 72–73, 82–83
　airflow, 73
　bag, 73, 83
　cost, 73
　relative humidity, 73
　temperature, 72
　tenderness, 72
　time, 72
Dry ashing, 31–32
Dry heat sterilization, 388
Dual X-ray energy imaging (DXA), 5
Dumas' method, 28–29
DXA. *See* Dual X-ray energy imaging (DXA)
Dynamic sensory techniques, 402–407
　TCATA, 406–407
　TDS, 404–406
　TI technique, 402–404
Dystrophin, 104–105

E

e-nose. *See* Electronic nose (e-nose)
EC-SOD. *See* Extracellular SOD (EC-SOD)
ECN. *See* Equivalent carbon number (ECN)
Edman degradation, 358
EDTA. *See* Ethylenediamine tetraacetate (EDTA)
EI. *See* Electrical impedance (EI)
EIS. *See* Electrical impedance spectroscopy (EIS)
Elastin, 24
Electric tongue, 12–13
Electrical
　stimulation, 69
　tenderization, 69
Electrical impedance (EI), 295
　biosensor, 294
Electrical impedance spectroscopy (EIS), 44–46, 343
Electrochemical biosensors, 325
Electrochemiluminescence-based qPCR, 326–327
Electrolyzed oxidizing water (EOW), 274
Electron spin resonance technology, 231
Electron transport chain, 117–118, 120, 242
Electron transport-mediated metmyoglobin reduction, 127–128
Electronic nose (e-nose), 12–13, 42–43, 340–341
Electronic spin resonance method, 9, 202
Electrophoresis, 9, 202, 384–385
Electrophoretic methods, 134, 387
Electrospray ionization source (ESI), 223–224
ELISA. *See* Enzyme-linked immunosorbent assay (ELISA)
Emosemio multistep approach, 408–409
EmoSensory Wheel, 410–411
Emotions, 407–408
　eliciting by food, 408–409
　emotional responses to meat and meat products, 410–412
　extrinsic factors influencing emotions elicited by foods, 409–410
　in sensory meat science, 407–412
Emulsifiers, 191–192
Emulsifying agents, 191
Endogenous antioxidants, 242–243
Endogenous proteinases, 69
Energy efficiency, 275
Enhanced solvent extraction. *See* Accelerated solvent extraction (ASE)
Enumeration methods, 323, 338–339
Environmental contaminants, 174–175
Enzymatic antioxidants, 243, 253–254
　catalase, 253
　glutathione peroxidase, 253–254
　SOD, 253
Enzymatic hydrolysis, 222
Enzymatic metmyoglobin
　reductase activity, 128
　reduction, 125
　systems, 125
Enzyme-linked fluorescent assay, 326
Enzyme-linked immunosorbent assay (ELISA), 10–12, 175, 277, 326
Enzymes, 117–118
EOW. *See* Electrolyzed oxidizing water (EOW)
Equivalent carbon number (ECN), 223
ESI. *See* Electrospray ionization source (ESI)
EsSense Profile questionnaire, 408
Essential trace metals, 175
Esters, 316
Ethidium bromide, 374–375
Ethylenediamine tetraacetate (EDTA), 28–29
Eukaryotic organisms, 242
European Union (EU), 157, 174
　livestock traceability system in, 157–159
　beef labeling, 159
　identification of animals, 157–158

European Union (EU) (*Continued*)
    inspection by authorities, 159
    passport, 158
    registers, 158
Even kit-based tests, 110
Extracellular matrix, 354
Extracellular SOD (EC-SOD), 253
Extraction, 29
    methodologies/strategies of antioxidants from plant sources, 243–250
    of plant polyphenolic compounds, 246–249
    purification and fractionation, 249–250

F

FA. *See* Fatty acid (FA)
FaceReader software device, 412
FAME. *See* Fatty acid methyl ester (FAME)
FAO. *See* Food and Agricultural Organization (FAO)
Farm-to-fork traceability, 154–155
Fat(s), 188
    discoloration, 120
    fat-soluble vitamins, 32
Fatty acid (FA), 217, 221–224
    composition of meat, 24–25
    determination, 31
Fatty acid methyl ester (FAME), 31, 221
FCP. *See* Free-choice profiling (FCP)
FD&C Act. *See* Federal Food, Drug, and Cosmetic Act (FD&C Act)
FDA. *See* Food and Drug Administration (FDA)
Federal Food, Drug, and Cosmetic Act (FD&C Act), 189–190
Fermented meat products, 195
Ferrous oxidation–xylenol orange, 226–227
FFFS. *See* Front-face fluorescence spectroscopy (FFFS)
Fiber optic biosensor, 291–292
Ficin, 71–72
Fingerprints, 138

FINS. *See* Forensically informative nucleotide sequencing (FINS)
Firefly. *See* Photinus pyralis (firefly)
Firmness indices, 46
Fizz software, 402–403
Flash profile (FP), 394, 397–400
Flavonoids, 190, 256–257
Flavor, 84–85, 251
Flavoring agents, 193, 197–198
Flavylium cation. *See* Benzo-γ-pyran ring
Flow cytometry, 296, 326
Flow injection methods, 9, 202–203
Fluorescence imaging, 12–13
Fluorescence resonance energy transfer (FRET), 143
Fluorescent dye-labeled nucleotides, 135
Fluorescent-labeled analytes, 291–292
Fluorometric method, 109
Fluoroscence polarization, 91
FMD. *See* Foot and mouth disease (FMD)
Folch method, 30
Food additives, 189–193
    antimicrobials, 192
    antioxidants, 190
    binders, 191
    coloring agents, 193
    curing agents and cure accelerators, 192–193
    emulsifiers, 191–192
    flavoring agents, 193
Food and Agricultural Organization (FAO), 308, 375
Food and Drug Administration (FDA), 188–189
Food Safety Inspection Service (FSIS), 71–72, 177, 199
Food(s), 308
    chemical and biological sensors for food-quality monitoring, 345
    food-grade hydrocolloid, 191
    food-producing animals, 176
    preservatives and additives, 9

Foodborne diseases, 275–276
Foodborne pathogens, 10, 295–296, 339
    elimination, 268, 280
    application of low temperature, 273
    high-pressure processing (HPP) for, 273
    meat irradiation for eliminating microbial hazards, 272–273
    physical methods of, 268–270
Foot and mouth disease (FMD), 160
Forensically informative nucleotide sequencing (FINS), 138–140
Formaldehyde, 175
Fourier transform infrared spectroscopy (FTIR spectroscopy), 295–296, 324, 342–343
FP. *See* Flash profile (FP)
Fractionation, 249–250
Free radicals, 242, 252–253, 257
Free-choice profiling (FCP), 397–398
Freeze-drying, 26
FRET. *See* Fluorescence resonance energy transfer (FRET)
Front-face fluorescence spectroscopy (FFFS), 231
FSIS. *See* Food Safety Inspection Service (FSIS)
FTIR spectroscopy. *See* Fourier transform infrared spectroscopy (FTIR spectroscopy)
Functional proteomics, 361

G

G2 tenderometer, 12
Gadus morhua (Atlantic cod), 46
γ-tocopherol, 254
GAPDH. *See* Glyceraldehyde-3-phosphate dehydrogenase (GAPDH)
Gas chromatography (GC), 31, 175, 226–227, 324–325

Gas chromatography with flame ionization detection (GC-FID), 221
Gaseous atmosphere in microbial meat spoilage, 322
GC. See Gas chromatography (GC)
GC-FID. See Gas chromatography with flame ionization detection (GC-FID)
Gel
　electrophoresis, 138
　filtration chromatography, 126
　gel-based techniques, 110, 355–357
　gel-free approaches, 357–358
Gene chip technology, 298–299
General Procrustes Analysis (GPA), 397–398, 398f
Generally Recognized As Safe (GRAS), 71–72, 189–190
Genetic tests, 110
Genomic(s), 134–135
　disinfection and sterilization, 388
　DNA isolation, 371
　dry heat sterilization, 388
　for meat quality assessment
　　AGE, 384–385
　　isolation of nucleic acid, 370–372
　　microsatellite analysis, 375–383
　　PAGE, 385
　　PCR, 372–374
　　quantification and purity measurement of DNA, 372
　　SDS-PAGE, 386
　　2D gel electrophoresis, 387
　　western blot, 386–387
　moist heat sterilization, 388
　sterilization
　　and disinfection of laboratory wares, 388
　　of glasswares, 388
　tools, 369–370
Genotyping for marker gene, 93
Global genetics–based analyses, 14
Global napping, 396–397
Glucose, 84–85, 314–315
Glutathione peroxidase, 253–254

Glyceraldehyde-3-phosphate dehydrogenase (GAPDH), 145
Glycolysis, 68
Glycolytic cycle, 124–125
Glycosaminoglycan quantification, 89
Goat. See Capra hircus (goat)
Goat-specific microsatellite markers, 380t
Gold-labeled polyclonal antibodies, 11
Golden tests, 13–14
GPA. See General Procrustes Analysis (GPA)
Gram-negative bacteria, 338
Gram-negative facultative anaerobes, 312
Gram-positive bacteria, 311
GRAS. See Generally Recognized As Safe (GRAS)
Gravimetric methods, 26
Green analytical techniques, 14

H

Handheld devices, 123–124
HCA. See Hierarchical cluster analysis (HCA)
HCL groups. See High cooking loss groups (HCL groups)
HD. See Hydrodistillation (HD)
Headspace (HS), 229
Heat Shock Protein-27 (HSP-27), 361–362
Heat-induced denaturation, 123
Hemoglobin redox state, 120
"Heterogeneous" sensor arrays, 42–43
Hierarchical cluster analysis (HCA), 397–398
High cooking loss groups (HCL groups), 363
High-performance liquid chromatography (HPLC), 29–30, 203
High-performance liquid chromatography coupled to evaporative light scattering detector (HPLC-ELSD), 220

High-pressure processing (HPP) for pathogens elimination, 273
High-speed countercurrent chromatography (HSCCC), 250
High-throughput
　analytical techniques, 14
　mass spectrometry, 122
HNE. See 4-Hydroxy-2-nonenal (HNE)
"Horse-gate" in global media, 133–134
Hot continuous extraction, 247
Hot start PCR, 373
HP. See Hydroxylysyl pyridinoline (HP)
HPLC. See High-performance liquid chromatography (HPLC)
HS. See Headspace (HS)
HSCCC. See High-speed countercurrent chromatography (HSCCC)
HSI. See Hyperspectral imaging (HSI)
HSP-27. See Heat Shock Protein-27 (HSP-27)
Hybridization rate, 135
Hydrazine-based reagents, 228–229
Hydrodistillation (HD), 247
Hydrophobic
　domain, 102–103
　interaction chromatography, 106–107
Hydrosulfite-mediated reduction, 124–125
4-Hydroxy-2-nonenal (HNE), 359
Hydroxylysyl pyridinoline (HP), 89
Hyperspectral imaging (HSI), 12–13, 44, 46–47, 53–54, 59–60, 225, 231, 324, 342
　and analysis, 280–281
Hypoxanthine, 345

I

ICA. See Immunochromatography (ICA)
ICATs. See Isotope-coded affinity tags (ICATs)

IEC. *See* Ion exclusion chromatography (IEC)
IEF. *See* Isoelectric focusing (IEF)
Illumination system, 38–39
Image master (software package), 387
Image/imaging
　capture board, 38–39
　medium, 43
　processing techniques, 12–13
Immunoassay techniques, 10–12, 134
Immunochromatography (ICA), 11, 177
Immunological protein quantification by dot-blot, 90
Immunological techniques, 277–278, 326, 339
　ELISA, 277
　LAT, 277–278
Immunomagnetic separation, 326
Impedance
　impedance-based biochip sensor, 294–295
　microbiology, 294
In situ zymography, 108
In vitro DNA synthesis, 137
Infrared (IR), 293
　heating, 275
Innovative data-processing algorithms, 13
Inorganic phosphate, 84–85
Inosinic acid, 84–85
Instrumental color analysis, 123–124
Intensive panel training, 396
International Organization of the Flavor Industry (IOFI), 229–230
International Society for Animal Genetics-Food and Agriculture Organization (ISAG-FAO), 164
Interretrotransposon amplified polymorphisms, 143
Inverse PCR, 373
IOFI. *See* International Organization of the Flavor Industry (IOFI)

Ion exclusion chromatography (IEC), 203
Ion-exchange chromatography, 107
Ionization technique, 223–224
IR. *See* Infrared (IR)
Iron, 118
ISAG-FAO. *See* International Society for Animal Genetics-Food and Agriculture Organization (ISAG-FAO)
Isobaric tag for relative and absolute quantitation (iTRAQ), 358, 362
Isoelectric focusing (IEF), 387
Isotope-coded affinity tags (ICATs), 357
IT-based tools, 7–8
iTRAQ. *See* Isobaric tag for relative and absolute quantitation (iTRAQ)

J

Japan, livestock traceability system in, 160
Joint FAO/WHO Expert Committee on Food Additives (JECFA), 188–189, 202

K

Karl Fischer titration, 26–27
Ketones, 84–85, 316
Kjeldahl method, 28

L

L-dihydroascorbic acid (DHA), 254–255
LAB. *See* Lactic acid bacteria (LAB)
Label-free
　impedimetric immunosensor, 7–8
　mass spectrometry, 359–360
Lactate. *See* Lactic acid
Lactic acid, 195, 314–315
　lactate-induced darkening, 121
　spray, 270–271

Lactic acid bacteria (LAB), 309, 311, 318
Lactobacillus arabinosus, 321
LAMP. *See* Loop-mediated isothermal amplifications (LAMP)
Lanthionine, 196–197
Laser(s)
　in detecting food quality, 344
　laser-induced fluorescence imaging system, 12–13
　speckle imaging, 344
LAT. *See* Latex agglutination test (LAT)
Lateral flow
　assay, 11
　tests, 8
Latex agglutination test (LAT), 277–278
LC. *See* Liquid chromatography (LC)
LCD array. *See* Low-cost and low-density array (LCD array)
LD. *See* Longissimus dorsi (LD)
Lean color development, 7
Lecithin, 191–192
Level of concern (LOC), 200
LF-NMR transverse relaxometry. *See* Low field nuclear magnetic resonance transverse relaxometry (LF-NMR transverse relaxometry)
Linoleic acid (C18:2), 24–25
Lipid oxidation, 122, 179–180, 226–230
　advanced methodologies for novel spectrometric methodologies, 230
　spectroscopic and nondestructive methodologies, 230–231
　assessment
　　of primary oxidation products, 226–227
　　of secondary oxidation products, 227–230
　products, 124–125, 218
　advanced methodologies, 230–231

Index

advances in detection, 226–231
optimization of classic methods, 226–230
Lipid(s), 173, 217, 226–230, 251–252
  classes, 220–221
  content of meat, 24–25
  determination, 30–31
  extraction, 31, 227
  hydroperoxides, 226–227
  lipid-derived compounds, 179–180
  lipid-derived volatiles, 218, 229–230
  lipid-soluble antioxidants, 254
  optimization of classic methods for lipid composition fatty acids, 221–224
"Lipidome", 232
Lipidomics, 232
Lipoproteins, 179–180
Liquid chromatography (LC), 11–12, 33, 226–227, 357
  LC-tandem mass spectrometry, 204
Liquid–liquid extraction (LLE), 33, 176
Listeria monocytogenes, 251–252, 288
Listeriosis, 288
Livestock Production Assurance program (LPA), 159
Livestock traceability system, 153–155
  in Australia, 159
  in European Union, 157–159
  in Japan, 160
  regulatory systems in countries, 160–162
LLE. *See* Liquid–liquid extraction (LLE)
LMAP. *See* Loop-mediated isothermal amplifications (LAMP)
LOC. *See* Level of concern (LOC)
Longissimus dorsi (LD), 58–59
Longissimus thoracis (LT), 85
Loop-mediated isothermal amplifications (LAMP), 4, 144, 279, 300

Low field nuclear magnetic resonance transverse relaxometry (LF-NMR transverse relaxometry), 90
Low-cost and low-density array (LCD array), 8
LP. *See* Lysyl pyridinoline (LP)
LPA. *See* Livestock Production Assurance program (LPA)
LPA National Vendor Declaration (LPA NVD), 159
LPA NVD. *See* LPA National Vendor Declaration (LPA NVD)
LT. *See* Longissimus thoracis (LT)
Luminex's xMAP technology, 11
Lyophilization, 26
Lysozyme, 198–199
Lysyl pyridinoline (LP), 89

M

m-calpain, 69, 83–84, 102–103
M-protein, 69
Maceration extraction, 247
Macroscopic imaging, 12–13
MAE. *See* Microwave-assisted extraction (MAE)
Magnesium pyrophosphate, 144
Magnetic resonance imaging (MRI), 5, 41–42, 218
Maillard reaction-related sugar fragments, 84–85
MALDI-TOF MS. *See* Matrix-assisted laser desorption ionization-time of flight mass spectrometry (MALDI-TOF MS)
Malodorous substances, 337
Malonaldehyde (MDA), 227–229
Mammalian cells, 295
MAP. *See* Modified atmosphere packaging (MAP)
Marker-based tests, 110
Mass spectrometry (MS), 9, 202, 226–227, 357, 361–362
Mass spectroscopy, 324–325
Matrix-assisted laser desorption ionization-time of flight mass spectrometry (MALDI-TOF MS), 280

Mb. *See* Myoglobin (Mb)
MCCC. *See* Multilayer coil countercurrent chromatography (MCCC)
MDA. *See* Malonaldehyde (MDA)
Meat, 21–22, 25, 188
  analytical techniques, 201–204
  categorization, 22–23
  chemistry, 118–122
    color deviations and approaches to improve color, 121–122
    discoloration, 120
    red color development, 119–120
  color, 67–68, 76, 117–118, 359–361
    analysis, 123–129
    cooked color measurements, 128–129
    instrumental color analysis, 123–124
    measurement methodology, 123
    mitochondrial functional analysis, 125–128
    spectrophotometric techniques to study meat color, 124–125
    visual color analysis, 128
  composition, 320
  federal oversight, 199
  food additives, 189–193
  health concerns and safety assessment, 200–201
  irradiation for eliminating microbial hazards, 272–273
  and meat-based products, 308
  muscle conversion to, 68
  preservatives, 194–199
  quality, 37–38
    analysis, 4–5
    attributes, 353
    challenges, 13–15
    chemical contaminants, 10–12
    food preservatives and additives, 9
    freshness and pathogen identification, 9–10

Meat (*Continued*)
  meat traceability and authentication, 7−8
  nutritional composition, 5−6
  physical and structural quality, 6−7
  proteomic approaches to, 358−364
  sensory quality, 12−13
  species
    identification using LAMP, 144
    origin by DNA hybridization, 135−137
  systems, 217
  tenderization process, 101−102
  traceability and certification in meat supply chain
    benefits of livestock traceability system, 154−155
    livestock traceability around world, 157−162
    methods for identification of animals, 155−156
    molecular meat traceability, 162−168
Meat and Livestock Australia (MLA), 159
Meat lipids, 218−225
  nondestructive methods, 225
  optimization of classic methods
    for lipid composition, 220−224
    for total lipid quantification, 218−219
Meat spoilage
  causes of, 308−309
  chemistry of, 314−317
    microbiological spoilage of meat, 314−317
    nonmicrobial/biochemical spoilage of meat, 314
  factors affecting microbial, 319−322
    extrinsic factors, 321−322
    implicit factors, 322
    intrinsic factors, 320−321
  modern trends in meat spoilage detection

  chemical and biological sensors for food-quality monitoring, 345
  CP, 341
  lasers in detecting food quality, 344
  MOS, 341−342
  odor sensors and electronic nose technology, 340−341
  smartphone-based food diagnostic technologies, 345
  spectroscopy in advanced forms, 342−344
  pattern, 9−10
Meat tenderness, 53−54
  detection methods of, 54−61
    sensory analysis, 54
    shear force measurements, 55−57
Meatborne pathogens
  biosensors-based detection techniques, 290−296
  nucleic acid-based assays, 297−300
  rapid detection method
    limitations of, 301
    requirements for, 288−290, 289*t*, 300−301
    trends in, 289−290
Media-based impedance methods, 294−295
Melanie (software package), 387
Metabolic processes, 242
Metabolites, 93, 93*t*
Metabolome, 93, 232
Metabolomics, 93. *See also* Proteomics
  to study meat quality, 78
Metal oxide semiconductor (MOS), 340−342
Metavinculin, 104−105
Methanol, 218−219
Methanolic hydrochloric acid, 221−222
Methionine, 180
Methyl-lanthionine, 196−197
Methylation
  reactions, 222
  specific PCR, 374

Metmyoglobin, 76, 190
  accumulation, 120
  formation, 120, 124
Metmyoglobin reducing activity (MRA), 118
Micro-RNA (miRNA), 370−371
Microbes, 268
Microbial etiologies, 317
Microbial hazards, meat irradiation for eliminating, 272−273
Microbial inhibition, principles of, 188
Microbial metabolites, 337−338
  detection, 340
Microbial pathogens
  chemical processes for elimination, 270−272
    acidic compounds, 270−271
    chemical agents, 271−272
    chlorine and related chemicals, 271
    ozone, 271
  detection, 275−281
    assays, 281
    conventional culture-based techniques, 276−277
    hyperspectral imaging and analysis, 280−281
    immunological techniques, 277−278
    MALDI-TOF MS, 280
    nanotechnology-based approaches, 281
    nucleic acid-based techniques, 278−280
  elimination by UV light, 272
  emerging approaches for elimination, 273−275
    biocontrol with bacteriophage, 275
    dense phase carbon dioxide, 274
    EOW, 274
    IR heating, 275
    microwave and radio frequency, 274−275
    nonthermal plasma, 274
Microbial spoilage, 309
  detection in meat, 323−328
    bacterial metabolites detection, 324−325

Index   429

enumeration, 323
molecular methods, 326–328
indicators, 322–323
of meat, 314–317, 336–337
  alcohols, 315
  aldehydes, 315
  esters, 316
  ketones, 316
  sulfur compounds, 316–317
  volatile fatty acids, 316
Microbiome of spoiled meat, 309–314
  microflora of fresh meat, 309–310
  spoilage microflora
    with aerobically packaged meat, 310–311
    in MAP meat, 313–314
    in vacuum-packaged meat, 312–313
Micrococcus, 311
Microflora, 336–337
  of fresh meat, 309–310
Micronose microarray, 9–10
Microsatellite(s), 162–164
  analysis, 375–383
  genotyping, 162
    molecular meat traceability using, 162–166, 165t
  markers, 162–164
Microwave (MW), 274–275
Microwave-assisted extraction (MAE), 249
Mid-infrared range, 28
Milk, 22
Mineral(s), 188. *See also* Vitamins
  determination, 31–32
Miniaturization techniques, 14
miRNA. *See* Micro-RNA (miRNA)
Mithun. *See Bos frontalis* (Mithun)
Mitochondria(l), 117–119
  effects on oxymyoglobin, 127
  enzymes, 120
  functional analysis, 125–128
    electron transport-mediated metmyoglobin reduction, 127–128
    enzymatic metmyoglobin reductase activity, 128
    oxygen consumption measurements, 127

oxymyoglobin preparation, 126–127
genes, 140
isolation, 126
mitochondria–myoglobin mixture, 125
OC, 117–118
respiration, 68, 119
12S rRNA gene sequence analysis, 139–141
16S rRNA gene, 141
MLA. *See* Meat and Livestock Australia (MLA)
Mobile
  diagnostics, 345
  sensing approaches, 10
Modified atmosphere packaging (MAP), 77, 310
  spoilage microflora in MAP meat, 313–314
Moist heat sterilization, 388
Moisture, 24
  determination, 25–28
    chemical analysis, 26–27
    gravimetric methods, 26
    spectroscopic analysis, 27–28
Mojonnier extraction, 30
Molecular basis of meat color
  analysis of meat color, 123–129
  cooked color, 122–123
  meat chemistry, 118–122
Molecular biology, 134–135
Molecular imprinting, 11
Molecular meat traceability, 162–168
  DNA-based meat traceability, 162
  using microsatellite genotyping, 162–166
  using single nucleotide polymorphism genotyping, 166–168
Molecular or array-based techniques, 13–14
Molecular techniques for speciation of meat
  meat species identification using LAMP, 144
  origin of meat species by DNA hybridization, 135–137

PCR-based techniques for species identification of meat, 137–143
quantitative meat speciation, 145–147
Molecular weight (MW), 385
Mono-glycerides, 191–192
Monochloramine, 271–272
Monounsaturated fatty acids (MUFAs), 24–25
MOS. *See* Metal oxide semiconductor (MOS)
MRA. *See* Metmyoglobin reducing activity (MRA)
MRI. *See* Magnetic resonance imaging (MRI)
mRNA, 370–371
MS. *See* Mass spectrometry (MS)
μ-calpain, 69, 83–84, 93, 102–103
MUFAs. *See* Monounsaturated fatty acids (MUFAs)
Multidimensional protein identification technology (MudPIT), 357
Multidrug resistant *Salmonella typhimurium* DT-104, 288
Multilayer coil countercurrent chromatography (MCCC), 250
Multiplex PCR, 141, 373
Multivariate data analysis, 40
Munsell system, 40
Muscle
  biology, 218
  conversion to meat, 68
  foods, 337
  myofibrillar protein, 104–105, 105f
Mutagenicity, 227
Mutton, 23
MW. *See* Microwave (MW); Molecular weight (MW)
Mycotoxins, 175
Myofibrillar proteins, 68–69, 231, 354
Myoglobin (Mb), 7, 22–23, 76, 117–119, 122, 353–354
  content of meat in different livestock species, 23t

Myoglobin (Mb) (*Continued*)
 denaturation studies, 128–129
 oxidation, 120
 oxygenation, 119
 variation in Mb content with age within species, 23t
Myosin, 24, 191–192, 354
Myosin light chain 2 (MyLC2), 362–363

N

N-acetyl muramoyl hydrolase enzyme activity, 198–199
n-aldehydes, 84–85
N-nitrosamines, 173, 178–179
N-nitrosodibenzylamine (NDBzA), 178–179
N-nitrosodiisobutylamine (NDiBA), 178–179
N-nitrosodimethylamine, 178
N-Succinyl-Leu-Tyr-7-amido-4-methylcourmarin, 109
NADH. *See* Nicotinamide adenine dinucleotide (NADH)
Nanofabrication techniques, 281
Nanotechnology-based approaches, 281
Napping, 396–397
 napping-UFP, 396–397
NARP. *See* Nonaqueous reversed-phase (NARP)
NASBA. *See* Nucleic acid sequence based amplification (NASBA)
National Center for Biotechnology Information (NCBI), 138–139
National Feedlot Accreditation Scheme (NFAS), 159
National Livestock Identification System (NLIS), 159
Native PAGE, 385
Natural antioxidant-rich meat products, 252
Natural nonenzymatic antioxidants, 254–257
 flavonoids, 256–257
 phenolic acids, 257
 vitamin A, 255–256
 vitamin C, 254–255

vitamin E, 254
Natural preservatives, 196–199
 chitosan, 197
 lysozyme, 198–199
 nisin, 196–197
 plant-derived compounds, 197–198
NCBI. *See* National Center for Biotechnology Information (NCBI)
NDBzA. *See N*-nitrosodibenzylamine (NDBzA)
NDiBA. *See N*-nitrosodiisobutylamine (NDiBA)
Near-infrared range (NIR range), 28
Near-infrared reflectance spectroscopy. *See* Near-infrared spectroscopy (NIRS)
Near-infrared spectroscopy (NIRS), 5–6, 28, 40, 40f, 41f, 53–54, 58–59, 74–75, 180, 218, 225, 343–344
Nebulin, 104–105
New generation DNA invader assays, 143
Next-generation sequencing (NGS), 281
NFAS. *See* National Feedlot Accreditation Scheme (NFAS)
NGS. *See* Next-generation sequencing (NGS)
Niacin, 32–33
Nick translation, 297
Nicotinamide adenine dinucleotide (NADH), 120, 202
NIR range. *See* Near-infrared range (NIR range)
NIRS. *See* Near-infrared spectroscopy (NIRS)
Nisin, 196–197
 detection, 204
Nitrates, 195–196, 203
Nitric acid, 32
Nitric oxide (NO), 242
 synthase, 178

Nitrite(s), 123, 178, 195–196
 detection, 203
 nitrite-induced metmyoglobin reduction, 124
Nitrogen
 combustion method, 28–29
 content, 28–29
 oxides, 28–29
 reactive species, 254–255
Nitrosamines, 175–176
Nitrosylmyoglobin, 178
NLIS. *See* National Livestock Identification System (NLIS)
NMR. *See* Nuclear magnetic resonance (NMR)
NMRI. *See* Nuclear magnetic resonance imaging (NMRI)
NO. *See* Nitric oxide (NO)
No-observed-adverse-effect level (NOAEL), 200
Non-polymerase chain reaction based molecular typing, 327–328
Nonaqueous reversed-phase (NARP), 223
Nonconventional species, 133–134
Nondestructive methods, 38–46, 225, 230–231
 acoustic methods, 46
 e-nose, 42–43
 electrical properties, 44–46
 future research, 46–47
 hyperspectral imaging, 44
 image processing techniques, 39t
 MRI, 41–42
 NIR spectroscopy, 40, 40f, 41f
 NMR, 41–42
 optical methods, 38–40
 X-ray and computed tomography, 43–44
Nondestructive testing, 38
Nonenzymatic
 antioxidants, 243, 254
 metmyoglobin reduction, 125
Nonmicrobial/biochemical spoilage of meat, 314
Nonpolar solvents, 249
Nonthermal plasma, 274

Normal phase (NP), 223–224
Novel bioinformatics-based proteomic approach, 110
Novel spectrometric methodologies, 230
NP. See Normal phase (NP)
Nuclear magnetic resonance (NMR), 41–42, 225
   spectroscopy methods, 9–10
Nuclear magnetic resonance imaging (NMRI), 38
Nuclear magnetic resonance transverse relaxation measurements, 90
Nucleic acid
   amplification techniques, 297
   detection method, 339
   hybridization, 297
   isolation, 370–372
      genomic DNA, 371
      RNA, 371–372
   molecules, 142
   nucleic acid-based assays, 297–300
      DNA hybridization, 297–298
      DNA microarrays, 298–299
      LMAP, 300
      PCR, 298
   nucleic acid-based techniques, 278–280, 326–328
      DNA microarray, 279–280
      loop-mediated isothermal amplification assay, 279
      molecular techniques, 134–135
      NASBA, 279
      non-polymerase chain reaction based molecular typing, 327–328
      PCR, 278
      PCR-based molecular typing, 326–327
      real-time PCR, 278
Nucleic acid sequence based amplification (NASBA), 279
Nucleotide
   collection, 138–139
   sequencing method, 167
Nucleotide blast (BlastN), 138–139

Nuitrosamines, 179
Nutrient(s), 317
   nutrient-rich matrix meat, 335–336
Nutritional quality analysis of meat, 22
   meat categorization, 22–23
   methodology for assessing, 25–33
      amino acid determination, 29–30
      ash content and mineral determination, 31–32
      fatty acid determination, 31
      lipid determination, 30–31
      moisture determination, 25–28
      protein determination, 28–29
      vitamin determination, 32–33
   nutritional composition of meat, 23–25, 24t

O

OC. See Oxygen consumption (OC)
Odor
   recognition, 43
   sensors, 340–341
Oleic acid (C18:1), 24–25
Olfaction, 42
Oligonucleotide probes, 297
"Omics" technologies, 4–5, 15
"One health program", 4
One-dimensional electrophoresis (1-DE). See Sodium dodecyl sulfate-polyacrylamide gel electrophoresis (SDS-PAGE)
Online monitoring systems, 289–290
Optical
   biosensors, 325
   methods, 38–40, 175
Organic acids, 194–195, 204, 270–271
   and derivatives, 194–195
Organochlorine compounds, 174
Organoleptic spoilage, 337

Organophosphorus compounds, 174
Oven drying, 26
Ovis aries (sheep), 139–140
   sheep-specific microsatellite markers, 379t
Oxidation, 118, 173
   of FAs, 217
   of lipid-derived compounds, 179–180
   of protein-derived compounds, 180
"Oxidome", 232
Oxidomics, 232
Oxidoreductase enzymes, 254–255
Oxygen consumption (OC), 117–118
   measurements, 127
Oxygen reactive species, 254–255
Oxygenation, 119
Oxymyoglobin, 76, 119. See also Myoglobin (Mb)
   mitochondria effects on, 127
   oxidation, 124–125
   preparation, 126–127
Ozone, 271

P

PAA. See Peroxyacetic acid (PAA)
Packaging
   in microbial meat spoilage, 322
   techniques to improve appearance of dark-cutting beef, 121–122
PAGE. See Polyacrylamide gel electrophoresis (PAGE)
PAHs. See Polycyclic aromatic hydrocarbons (PAHs)
Pale, soft, exudative broiler breast meat (PSE broiler breast meat), 364
Palmitic acid (C16:0), 24–25
Papain, 71–72
Partial least squares (PLS), 60–61
Partial least squares regression model (PLSR model), 58–59
Partial napping, 396–397
PBS. See Phosphate buffer saline (PBS)

Index

PCBs. *See* Polychlorinated biphenyls (PCBs)
PCR. *See* Polymerase chain reaction (PCR); Principal component regression (PCR)
PCR-DGGE. *See* Polymerase chain reaction-denaturing gradient gel electrophoresis (PCR-DGGE)
PDQuest (software package), 387
Peptide mass fingerprint (PMF), 280
Peptides, 68–69
Perchloric acid, 32
Peroxyacetic acid (PAA), 192
PGs. *See* Proteoglycans (PGs)
pH, 194
  of myoglobin solution, 127–128
Phenol–chloroform extraction method, 370–371
Phenolic
  acids, 257
  compounds, 190, 256
Phosphate buffer saline (PBS), 371
Phosphates, 191
Phosphocreatine, 68
Phospholipids (PLs), 179–180, 218, 224t
Phosphorylation, 362–363
Photinus pyralis (firefly), 291
Physical methods of foodborne pathogen elimination
  carcass washing, 270
  hair removal, 269
  preslaughter washing, 268–269
  spot trimming of carcasses, 269
  vacuum-steam/water application, 269
PIC. *See* Polymorphism information content (PIC); Property identification code (PIC)
Piezoelectric biosensors, 295
Pig-specific microsatellite markers, 382t
Piwi-interacting RNA (piRNA), 370–371
Plant
  antioxidants, 243, 244t

antioxidants from plant sources, 243–250
enzymes, 71–72
plant-based industrial by-products, 251
plant-derived compounds, 197–198
plant-derived products, 242–243
polyphenolic compounds, 246–249, 246t
  conventional methods, 247
  MAE, 249
  PLE, 248–249
  UAE, 248
PLE. *See* Pressurized liquid extraction (PLE)
PLS. *See* Partial least squares (PLS)
PLs. *See* Phospholipids (PLs)
PLSR model. *See* Partial least squares regression model (PLSR model)
PMF. *See* Peptide mass fingerprint (PMF)
Polyacrylamide gel electrophoresis (PAGE), 164, 385
Polychlorinated biphenyls (PCBs), 174
Polycyclic aromatic hydrocarbons (PAHs), 173, 175–176
Polymerase chain reaction (PCR), 136–138, 278, 297–298, 326–327, 339, 369–370, 372–374. *See also* Quantitative PCR (qPCR)
PCR-based molecular typing, 326–327
PCR-based techniques for species identification of meat, 137–143
  AFLP, 143
  DALP, 143
  DNA invader assays, 143
  DNA microarrays, 142
  FINS, 138–140
  interretrotransposon amplified polymorphism, 143
  multiplex PCR, 141
  PCR-restriction fragment length polymorphism, 141–142

RAPD fingerprinting, 138
  species-specific PCR, 140
  SSCP, 143
  PCR-restriction fragment length polymorphism, 141–142
Polymerase chain reaction-denaturing gradient gel electrophoresis (PCR-DGGE), 327
Polymorphism information content (PIC), 164
Polyphenolic portion, 250
Polyunsaturated fatty acids (PUFAs), 24–25
"Popping-up" sensation, 404
Pork, 23
Postmortem aging, 68–69
  advanced methods for evaluation of postmortem proteolysis, 89–91
  biomarkers, 91–93
  changes in different meats, 86t
  evaluation/assessment, 85–88
  of meat, 82
  and meat quality, 83–85
Postmortem proteolysis
  advanced methods for evaluation of, 89–91
  of skeletal muscle by calpain, 104–106
Posttranslational modifications (PTMs), 354
Premature browning, 122
PrEmo. *See* Product emotion measurement instrument (PrEmo)
Preservatives, 194–199
  chemical, 194–196
  natural, 196–199
Preslaughter washing, 268–269
Pressurized liquid extraction (PLE), 248–249
Pressurized solvent extraction. *See* Accelerated solvent extraction (ASE)
Primary oxidation products, 226–227
Principal component regression (PCR), 58–59

Product emotion measurement instrument (PrEmo), 408–409
Progenesis (software package), 387
Projective mapping, 396
Property identification code (PIC), 159
Proteasomes, 83
Protein(s), 173, 188, 190, 353–355
   biomarkers, 91, 92f
   determination, 28–29
   extraction and reflectance, 118
   lysate, 386
   oxidation, 251–252
   protein-derived compounds, 180
Proteoglycans (PGs), 89
   quantification, 89
Proteolysis, 68–69, 84
Proteolytic digestion, 126
Proteome, 354
Proteomics, 4–5, 354–358
   approaches, 14
   gel-based approaches, 355–357
   gel-free approaches, 357–358
   markers in conversion of muscle to meat, 91–92
   proteomic approaches to meat quality, 358–364
      meat color, 359–361
      tenderness, 361–363
      WHC, 363–364
   proteomic-based analysis, 110
   to study meat quality, 78
   tools, 7
Proteosomes, 101–102
Protoporphyrin, 118
PSE broiler breast meat. *See* Pale, soft, exudative broiler breast meat (PSE broiler breast meat)
Pseudomonads, 316–317
Pseudomonas, 339
PTMs. *See* Posttranslational modifications (PTMs)
PUFAs. *See* Polyunsaturated fatty acids (PUFAs)
Pulsed UV light, 272
Purification, 249–250
   of calpain 1, calpain 2, and calpastatin, 106–107

liquid–liquid extraction, 250
solid phase extraction, 250
Pyrosequencing, 327

Q

QC-PCR. *See* Quantitative-competitive PCR (QC-PCR)
QDA. *See* Quantitative descriptive analysis (QDA)
qPCR. *See* Quantitative PCR (qPCR)
QR. *See* Quick response (QR)
Quality of meat, 369–370
Quantitative descriptive analysis (QDA), 399–400
Quantitative meat speciation, 145–147
Quantitative PCR (qPCR), 145–147, 278, 326–327, 374. *See also* Polymerase chain reaction (PCR)
   QC-PCR, 145
   qPCR-ELISA, 326–327
   real-time PCR, 145–147
Quantitative-competitive PCR (QC-PCR), 145
Quick response (QR), 156
   code-based tags, 156
Quick-fast descriptive sensory techniques, 394–401
   application, 395t
   CATA, 400–401
   FP, 397–400
   napping, 396–397
   RATA, 401

R

Radiation-powered beam of light, 74–75
Radio frequency (RF), 274–275
Radio frequency identification devices (RFIDs), 154–156
Radioactive metals, 272–273
Radiography, 43
Raman sensor, 9–10
Raman spectroscopy (RS), 53–54, 60–61, 231, 324
Random amplified polymorphic DNA (RAPD), 138

fingerprinting, 138, 139t
fingerprints, 138
markers, 138
Random priming technique, 297
Randomly amplified polymorphic DNA-PCR (RAPD-PCR), 327
RAPD. *See* Random amplified polymorphic DNA (RAPD)
Rapid detection method, 288–290, 289t
   requirements, 288–290, 289t, 300–301
      limitations of, 301
      trends in, 289–290
Rate-all-that-apply method (RATA method), 394, 401
REA-PFGE. *See* Restriction enzyme analysis coupled with pulsed-field gel electrophoresis (REA-PFGE)
Reactive alkoxyl radical, 255–256
Reactive nitrogen species (RNS), 242
Reactive oxygen species (ROS), 242
Ready-to-eat (RTE), 193
   foods, 336
   meats, 288
Real-time polymerase chain reaction (Real-time PCR), 145–147, 146t, 278, 373–374
Red meat, 22–23
Redfin (software package), 387
Redox reactions, 218
Reflectance variables, 123–124
Reflection, 39
Reflux extraction methods, 33
Refractive index (RI), 27–28
Refractometry, 27–28
Relative humidity, 73
Relative quantification of proteins, 357
Reliable culture-based techniques, 13–14
Repetitive extragenic palindromic-PCR (Rep-PCR), 327
REs. *See* Restriction enzymes (REs)

Reserved zymography, 108
Residues of harmful chemicals and
    detection techniques
  environmental contaminants,
    174–175
  N-nitrosamines, 178–179
  oxidation
    of lipid-derived compounds,
      179–180
    of protein-derived
      compounds, 180
  PAHs, 175–176
  veterinary drug residues,
    176–177
Restriction enzyme analysis
    coupled with pulsed-field
    gel electrophoresis (REA-
    PFGE), 327–328
Restriction enzymes (REs), 141
Restriction fragment length
    polymorphism (RFLP),
    167–168
Retinoid, 255–256
Reverse dot-blotting technique, 11
Reverse phase (RP), 223–224
Reverse transcriptase PCR, 373
Reverse-phase HPLC, 29–30
RF. See Radio frequency (RF)
RFIDs. See Radio frequency
    identification devices
    (RFIDs)
RFLP. See Restriction fragment
    length polymorphism
    (RFLP)
RI. See Refractive index (RI)
Riboflavin, 32–33, 272
Ribonucleic acid (RNA), 370–371
  isolation, 371–372
RMSEP. See Root mean square
    error of prediction
    (RMSEP)
RNA. See Ribonucleic acid (RNA)
RNS. See Reactive nitrogen
    species (RNS)
Root mean square error of
    prediction (RMSEP),
    60–61
ROS. See Reactive oxygen species
    (ROS)
RP. See Reverse phase (RP)

RP-HPLC, 223–224
rpoB gene, 340
rRNA, 370–371
RS. See Raman spectroscopy (RS)
RTE. See Ready-to-eat (RTE)

S

Safety of meat and meat products,
    9–10
Salt distribution analysis, 5–6
Sarcoplasm, 69
Sarcoplasmic proteins, 363
SBSE. See Stir-bar sorptive
    extraction (SBSE)
Scorpion probes, 147
SDS-PAGE. See Sodium dodecyl
    sulfate-polyacrylamide gel
    electrophoresis (SDS-
    PAGE)
Secondary oxidation products,
    227–230
  lipid-derived volatiles, 229–230
  MDA, 227–229
Secondary volatile compounds,
    190
Selected ion flow tube mass
    spectrometry (SIFT-MS),
    230
Sensation, temporal dominance of,
    404–406
Sensor-based techniques, 9–10
Sensory
  analysis, 54
    assessment of meat and meat
      products, 394
    dynamic sensory techniques,
      402–407
    emotions in sensory meat
      science, 407–412
    quick-fast descriptive sensory
      techniques, 394–401
  evaluation, 393–394
  food science, 394
  map, 396
  panels, 75–76
  perception, 402
Sequential injection analysis, 203
SFE. See Supercritical fluid
    extraction (SFE)
Shear force measurements, 55–57

spectroscopic methods, 58–61
SSF, 56–57
star probe measurement, 57–58
TPA, 57
WBSF, 55–56
sheep. See Ovis aries (sheep)
Shelf life, 251
Short tandem repeats, 162–164
SIFT-MS. See Selected ion flow
    tube mass spectrometry
    (SIFT-MS)
SILAC. See Stable isotopic
    labeling with amino acids
    in cell culture (SILAC)
Silica solid-phase extraction
    cartridge, 176
Silver staining, 355–356
Single nucleotide polymorphism
    (SNP), 142, 153–154
  genotyping, 162, 166–168
  methods, 167–168
  selection of single nucleotide
    polymorphism markers, 167
  markers, 167
Single sequence
  length polymorphisms, 162–164
  repeats, 162–164
Single-strand conformation
    polymorphism (SSCP), 143
Single-stranded DNA (ssDNA),
    142
siRNA. See Small interfering RNA
    (siRNA)
Slaughterhouse microbiome,
    309–310
Slice(d) shear force (SSF), 53–54,
    56–57, 75
Small interfering RNA (siRNA),
    370–371
Small nuclear RNAs (snRNAs),
    370–371
Smartphone
  applications, 156
  smartphone-based food
    diagnostic technologies, 10,
    345
SNP. See Single nucleotide
    polymorphism (SNP)
snRNAs. See Small nuclear RNAs
    (snRNAs)

SOD. *See* Superoxide dismutase (SOD)
Sodium dodecyl sulfate-polyacrylamide gel electrophoresis (SDS-PAGE), 355, 361, 386
Solid phase cytometry (SPC), 296
Solid stationary phase, 220
Solid-phase extraction (SPE), 33, 204, 220, 249–250
Solid-phase microextraction (SPME), 229–230
Solid-state immunochromatography. *See* Lateral flow—assay
Sorbate, 195
Sorbic acid, 195
  detection, 203–204
Sorted napping, 396
Soxhlet extraction, 247
Soxhlet method, 30, 218–219, 247
Spatial decision support system, 7–8
SPC. *See* Solid phase cytometry (SPC)
SPE. *See* Solid-phase extraction (SPE)
Species-specific PCR, 140, 140t
Species-wise microsatellite markers, 375, 376t
Spectral reflectance, 74–75
Spectrometric method, 9
Spectrophotometer, 125
Spectrophotometric
  detector, 228–229
  methods, 226–227
  techniques to study meat color, 124–125
Spectroscopic/spectroscopy
  in advanced forms, 342–344
    EIS, 343
    FTIR, 342–343
    NIR spectroscopy, 343–344
  analysis, 27–28
  methods, 58–61, 230–231, 324
    hyperspectral imaging technique, 59–60
    near-infrared reflectance, 58–59
    Raman spectroscopy, 60–61

Spin, 41
SPME. *See* Solid-phase microextraction (SPME)
Spoilage, 308
  bacteria, 73
    detection, 338–340
    enumeration methods, 338–339
    methods, 339
    by molecular methods, 339–340
  of food, 335–336
  microflora, 309
    associated with aerobically packaged meat, 310–311
    in MAP meat, 313–314
    in vacuum-packaged meat, 312–313
Spoiled meat
  characteristics, 317–319, 319t
    discoloration, 317
    filaments and slime formation, 318–319
    gas production, 318
    off-odors and off-flavors, 318
  microbiome, 309–314
SPR biosensor. *See* Surface plasmon resonance biosensor (SPR biosensor)
SSCP. *See* Single-strand conformation polymorphism (SSCP)
ssDNA. *See* Single-stranded DNA (ssDNA)
SSF. *See* Slice(d) shear force (SSF)
Stable isotopic labeling with amino acids in cell culture (SILAC), 357–358
Staphylococcal food poisoning or intoxication, 288
Staphylococcus aureus, 321
Star probe measurement, 57–58
Steam Pasteurization, 270
Stearic acid (C18:0), 24–25
Stepwise dry/wet aging, 83
Sterilization of glasswares, 388
Stir-bar sorptive extraction (SBSE), 201–202
Stromal proteins, 354

Sublimation, 26
Substrate-based zymography, 108
Suc-LLVY aminoluciferin, 109
Succinate dehydrogenase, 361–362
Succinyl-Leu-Tyr-7-amino-4-methylcoumarin (Succ-Leu-Tyr-AMC), 109
Sulfite(s), 196
  detection, 202–203
Sulfur
  compounds, 316–317
  sulfur-containing amino acids, 180
Sulfur dioxide (SO_2), 27
Sulfuric acid—hydrogen peroxide, 32
Supercritical fluid extraction (SFE), 33
Superoxide dismutase (SOD), 253
Superoxide molecule, 242
Surface plasmon resonance biosensor (SPR biosensor), 293–294, 293f
Surfaced-enhanced Raman spectroscopy, 342
SYBR Green, 147, 373–375
Syncerus caffer (African buffalo), 142
Synchronous front-face fluorescence spectroscopy, 9–10
Synergistic antimicrobial activity, 195
Synthetic antioxidants, 190
Synthetic phenolic antioxidants, detection of, 204
Systems biology, 14

T

Tamper-proof system, 155
Tandem mass spectrometry (MS/MS), 357
Taurine cattle (*Bos taurus*), 142
TBA. *See* Thiobarbituric acid (TBA)
TBARS. *See* Thiobarbituric acid reactive substances (TBARS)

TCATA method. *See* Temporal check-all-that-apply method (TCATA method)
TDE. *See* Temporal dominance of emotion methodology (TDE)
TDS. *See* Temporal dominance of sensations (TDS)
TEFs. *See* Toxic equivalency factors (TEFs)
Temperature in microbial meat spoilage, 321
Temporal check-all-that-apply method (TCATA method), 406–407
Temporal dominance of emotion methodology (TDE), 410–411
Temporal dominance of sensations (TDS), 394, 404–406
Tendercut, 70–71
Tenderization, 82
  of meat, 83
Tenderness, 12, 53–54, 84, 361–363
  effects of aging on, 68–76
    postmortem aging, 68–69
    techniques to improve, 69–72
    techniques to measure, 75–76
    techniques to quantify, 74–75
  of meat, 101–102
Tenderstretch(ing), 70
Tendertec mechanical penetrometer, 12
Texture profile analysis (TPA), 53–54, 57
TG. *See* Triacylglycerols (TG)
Thamnidium, 72
Thermal image/imaging, 12–13
  processing techniques, 5–6
Thermocycler, 137
Thermometric biosensors, 325
Thiamin, 32–33
Thin-layer chromatography (TLC), 220
Thiobarbituric acid (TBA), 227
Thiobarbituric acid reactive substances (TBARS), 251
2-Thiobarbituric acid reactive substances, 180
TI technique. *See* Time−intensity technique (TI technique)

Time-consuming, 396
Time-of-flight (TOF), 230
Time−intensity technique (TI technique), 394, 402–404
Titanium dioxide, 193
Titin, 68–69, 104–105
TLC. *See* Thin-layer chromatography (TLC)
TMT labeling, 360–361
Tocopherols, 254
Tocotrienols, 254
TOF. *See* Time-of-flight (TOF)
Total lipid quantification, optimization of classic methods for, 218–219
Toxic
  compounds, 173
  metals, 175
Toxic equivalency factors (TEFs), 174
TPA. *See* Texture profile analysis (TPA)
Traceability, 153–154
  regulatory systems in countries, 160–162
  traceability-based quality assurance programs, 154–155
Trade Control and Expert System (TRACES), 157
Trained panels, 75–76
Trans-fatty acids, 24–25
Trans-resveratrol, 249
Transporter, 158
Triacylglycerols (TG), 179–180, 218–219
Tricarboxylic acid cycle, 120, 124–125
Triglyceride molecules, 257
Trisodium phosphate (TSP), 271–272
Tristimulus, 74–75
  values, 40
tRNA, 370–371
Tropomyosine, 104–105
Troponin I, 104–105
Troponin T, 104–105
TSP. *See* Trisodium phosphate (TSP)
Two-dimensional differential gel electrophoresis (2D-DIGE), 356

Two-dimensional electrophoresis (2-DE). *See* Two-dimensional PAGE
Two-dimensional gel electrophoresis (2D gel electrophoresis), 387
Two-dimensional PAGE, 355–356, 362
Two-step column separation techniques, 107

U

UAE. *See* Ultrasonic-assisted extraction (UAE)
Ultra-flash profiling (UFP), 396
Ultrahigh purity acetanilide, 28–29
Ultrasonic-assisted extraction (UAE), 33, 248
Ultrasonics, 46
Ultrasound
  imaging, 5–6
  waves, 248
Ultraviolet, visible, and NIR spectroscopy (UV-vis-NIR spectroscopy), 40
Ultraviolet (UV), 272
  light, 204
  microbial pathogens elimination by, 272
UV-Vis spectrophotometers, 124–125
United States Department of Agriculture (USDA), 71–72, 188–189
UV-vis-NIR spectroscopy. *See* Ultraviolet, visible, and NIR spectroscopy (UV-vis-NIR spectroscopy)

V

Vacuum-packaged meat, spoilage microflora in, 312–313
Vapor phase Fourier transform infrared spectrometry, 203
Variable wavelength detectors (VWDs), 223–224
Veterinary drug residues, 176–177
Video image analysis (VIA), 74

Vinculin, 104–105
Visible (VIS)
  light, 58
  VIS/NIR reflectance spectroscopy, 58–59
Visual color analysis, 128
Visual spectroscopy, 40
Visual tagging, 156
Vitamins, 188
  vitamin A, 252, 255–256
  vitamin B6, 32–33
  vitamin B12, 32–33
  vitamin C, 254–255
  vitamin E, 254
Volatile fatty acids, 316
Volatile organic compounds (VOCs), 314–315, 324–325
Volatile sulfur compounds, 316–317
VWDs. See Variable wavelength detectors (VWDs)

W

Warner–Bratzler shear force (WBSF), 53–56, 70–71, 75
Water
  activity, 321
  water-soluble vitamins, 32
Water-holding capacity (WHC), 82, 85, 363–364
Wavelength interrogation, 294
WBSF. See Warner–Bratzler shear force (WBSF)
Western blot, 386–387
Wet aging, 74, 82. See also Dry aging
Wet ashing, 31–32
WHC. See Water-holding capacity (WHC)
White meat, 22–23
WHO. See World Health Organization (WHO)

Wisent*Bison bonasus* (Wisent)
Woody breast, 364
World Health Organization (WHO), 268
World Trade Organization (WTO), 288

X

X-ray, 43–44
  converter, 43
  crystallography, 358
  imaging, 5, 12–13
Xanthan gum, 191
Xylenol method, 203

Z

Zebu. See *Belonostomus indicus* (Zebu)
Zymography, 108

Vinculin, 104–105
Visible (VIS)
   light, 58
   VIS/NIR reflectance spectroscopy, 58–59
Visual color analysis, 128
Visual spectroscopy, 40
Visual tagging, 156
Vitamins, 188
   vitamin A, 252, 255–256
   vitamin B6, 32–33
   vitamin B12, 32–33
   vitamin C, 254–255
   vitamin E, 254
Volatile fatty acids, 316
Volatile organic compounds (VOCs), 314–315, 324–325
Volatile sulfur compounds, 316–317
VWDs. *See* Variable wavelength detectors (VWDs)

W

Warner–Bratzler shear force (WBSF), 53–56, 70–71, 75
Water
   activity, 321
   water-soluble vitamins, 32
Water-holding capacity (WHC), 82, 85, 363–364
Wavelength interrogation, 294
WBSF. *See* Warner–Bratzler shear force (WBSF)
Western blot, 386–387
Wet aging, 74, 82. *See also* Dry aging
Wet ashing, 31–32
WHC. *See* Water-holding capacity (WHC)
White meat, 22–23
WHO. *See* World Health Organization (WHO)
Wisent*Bison bonasus* (Wisent)
Woody breast, 364
World Health Organization (WHO), 268
World Trade Organization (WTO), 288

X

X-ray, 43–44
   converter, 43
   crystallography, 358
   imaging, 5, 12–13
Xanthan gum, 191
Xylenol method, 203

Z

Zebu. *See Belonostomus indicus* (Zebu)
Zymography, 108

Printed in the United States
By Bookmasters